Ecology and Palaeoecology of Benthic Foraminifera

Ecology and Palaeoecology of Benthic Foraminifera

JOHN W MURRAY

Co-published in the United States with
John Wiley & Sons Inc., New York

Longman Scientific & Technical,
Longman Group UK Limited,
Longman House, Burnt Mill, Harlow,
Essex CM20 2JE, England
and Associated Companies throughout the world.

Copublished in the United States with
John Wiley & Sons, Inc., 605 Third Avenue, New York, NY 10158

Published in collaboration with The Palaeontological Association

First published 1991

British Library Cataloguing in Publication Data
Murray, John W. (John William), *1937–*
Ecology and palaeoecology of benthic foraminifera.
1. Fossil Foraminiferida
I.Title
563.12

ISBN 0-582-05122-3

Library of Congress Cataloging-in-Publication Data
Murray, John William.
Ecology and palaeoecology of benthic foraminifera / John W. Murray.
p. cm.
Includes bibliographical references and index.
ISBN 0-470-21686-7
1. Foraminifera – Ecology. 2. Foraminifera, Fossil – Ecology. 3. Benthos. 4. Palaeoecology.
I. Title.
QL368.F6M83 1991
593.1'2045 – dc20 90-13377
CIP

Transferred to digital print on demand 2002

Printed and bound by Antony Rowe Ltd, Eastbourne

'Paleoecologic interpretations based directly or by analysis upon modern distributional patterns can be no better than our knowledge of live species.'

Robert G Douglas, 1979

Contents

Preface

The expansion in knowledge on foraminiferal ecology can be gauged by comparing the syntheses of Phleger (1960a), Boltovskoy (1965), Murray (1973), Boltovskoy and Wright (1976) and the present volume. New developments since the 1970s include better samplers (sealed grabs, box corers, multicorers), the use of scuba divers in shallow water both for sampling and the *in situ* observation of the environment, stable isotope studies and biochemical techniques. Much greater use is now made of computers in the statistical analysis of data. Improved position-fixing techniques at sea have made it possible to resample oceanic sites far from land to build up time-series studies. There has been a greater involvement of biologists with significant advances in understanding physiological processes. Multidisciplinary studies, especially of fossil material, have been actively promoted through the Deep Sea Drilling Project, International Phase of Ocean Drilling and Ocean Drilling Programme.

The aim of this book is to provide a synthesis of data on ecology as a basis for the interpretation of palaeoecology. The dynamics of living populations and assemblages and the dead assemblages drawn from them are discussed in Chapters 1–5. The regional distribution of faunas and the causes of the observed patterns are described and evaluated in Chapters 6–17 while Chapter 18 summarises the principal conclusions. The application of these results to palaeoecology is outlined in Chapter 19 and illustrated by a series of case-studies. Methods, ecological data for selected genera, a faunal reference list and a glossary are given as appendices.

Such a review of the literature inevitably points to gaps in knowledge. The best-studied areas are those adjacent to Western nations and little is known of the distribution of foraminifera in large parts of the seas and oceans. In order to understand ecological processes experimental studies are needed on environmental controls of distributions, life history, rate of production and postmortem effects. Long-term field investigation of faunal change in individual environments is an important theme which has scarcely been touched upon. Perhaps most important of all, more foraminiferal studies should be carried out alongside other ecological investigations as part of interdisciplinary research.

Exeter, August 1989 J.W.M.

Acknowledgements

It is a pleasure to acknowledge help received over the 3-year period during which this book was written. Bob Ellison gave much encouragement and discussed numerous topics during his sabbatical visit. Heather Eva and her library colleagues tracked down several hundred papers through the inter-library loans scheme. Ann Williams undertook the marathon task of word-processing. The University of Exeter granted a term of study leave in the autumn of 1988 which enabled me to concentrate on writing instead of running a department and teaching students. Finally, I am grateful to my wife for continued support.

We are indebted to the following people for permission to reproduce copyright material:

American Association for the Advancement of Science and the author for Fig. 3.9 (Wefer and Berger 1980) copyright 1980 by the AAAS; American Society of Limnology and Oceanography for Fig. 4.1 (Muller 1974); Cushman Foundation for Foraminiferal Research Inc for Table 4.5 (Murray 1983); Elf Aquitaine for Table 5.7 (Murray 1984a); Geological Society for Fig. 5.5 reproduced by permission of the Geological Society from Murray 1987a in *Journal of the Geological Society, London* Volume **144**; Geological Society for Table 19.1 reproduced by permission of the Geological Society from Murray, Weston, Haddon and Powell 1986 in *North Atlantic Palaeoceanography*; Micropaleontology Press for Table 6.14 (Murray 1969); Palaeontological Association for Table A2 (Rogers 1976); Society of Economic Paleontologists and Mineralogists – Pacific Section for Fig. 17.3 (Lagoe 1980).

Whilst every effort has been made to trace the owners of copyright material, in a few cases this has proved impossible and we take this opportunity to offer our apologies to any copyright holders whose rights we may have unwittingly infringed.

CHAPTER 1

Introduction

Ecology is the study of the causes of patterns of distribution and abundance of organisms. It is concerned with interactions between individuals and their physical and chemical environment, interactions between species and also interactions between individuals of the same species. The approach to ecology may be through field studies, laboratory experiments or mathematical modelling. Ideally ecological studies involve all three approaches because each provides new data or a new outlook which give rise to further questions or new ideas. During the 1970s the thrust of marine ecology moved away from the exploration of new habitats towards understanding more of the interrelations between organisms within the benthic community and with the overlying waters (Holme and McIntyre 1984), e.g. the ecosystem approach.

Distribution studies of benthic foraminifera have been carried out for well over a century, but a truly ecological approach did not develop until 1952 when Walton introduced the rose Bengal staining method for differentiating between protoplasm-containing (i.e. living or recently living) and empty (i.e. dead) tests. Now field studies of benthic foraminiferal ecology consider not only the living individuals but also the postmortem taphonomic processes which influence the preserved dead assemblages. This is because the principal application of ecological data on modern foraminifera is to aid the palaeoecological interpretation of fossil assemblages.

In spite of the large quantity of literature on modern benthic foraminifera there is still much work to be done to establish the basic distribution patterns. Most studies are field based; only a few involve laboratory experiments or mathematical modelling. However, the introduction of new techniques, such as the stable isotope composition of the test wall, and ultrastructural and biochemical studies of the soft parts, is throwing new light on their physiology and mode of life.

Although some foraminifera live on hard substrates such as shells or rocks, or attached to plants or other organic structures, most live on or in soft sedimentary substrates. In theory, competition between species for space and food may be ecological controls, but in practice there are no data to suggest that either takes place among small benthic sediment inhabitants. Most infaunal communities can undergo large increases in density without evidence of competitive exclusion Dayton, in Strong *et al.* 1984).

The niche

Lee (1974: 209–10) suggests that:

> The concept of niche is not easily understood except in very abstract terms. Often micropaleontologists and protozoologists think of niche in a spatial or habitat sense and commonly only from data gathered at a single point in time. While these studies are valuable in themselves, they give only very little insight into the biotic constraints which are important in the dynamic processes which establish the realized niches of foraminifera.

Further (Lee 1974: 219)

> . . . we really know very little about the realized niches of any group of foraminifera from any locality in the world. Hopefully in the future we will have enough field and laboratory data to describe the realised niches of a few foraminiferal species. It staggers the imagination to consider just the field work involved. . . .

Factors influencing the occurrence of organisms may be abiotic (temperature, salinity, dissolved oxygen availability, pH, etc.) or biotic (food, intra- and inter-specific competition, predation, etc.). Muller (in Lee 1974) attempted to quantify some of these

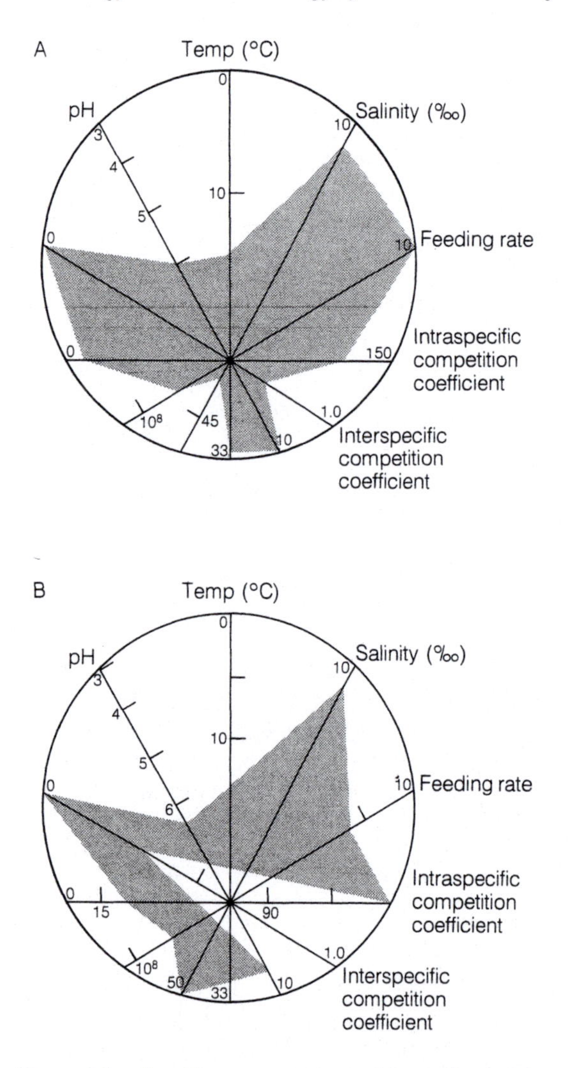

Figure 1.1 Graphic representations of the realised niches of (A) *Ammonia beccarii*; (B) *Spiroloculina hyalinea* (based on Lee 1974)

parameters for four species of foraminifera and he developed a polygonal graphical plot to represent the niche. The six most important factors are plotted along vectors, the centre being displaced from the centre of the circle in order to emphasize the importance of the three abiotic factors. He inferred that *Ammonia beccarii* occupies a broader niche (44%) than *Spiroloculina hyalina* (27%) (Fig. 1.1). The latter has a low interspecific competition coefficient and occurs in low numbers in the field. *Ammonia beccarri* is widely distributed and this is consistent with its niche breadth.

The niche concept is most applicable to microhabits. It still remains a theoretical ideal rather than a practical objective. Indeed, it may prove more useful to know broad distribution patterns rather than the minutiae of the niche of each species, for the interaction of environmental controls may cause the realised niche to vary from one region to another.

Environmental variability

It is likely that abiotic environmental parameters are of greater importance than biotic factors in shaping the communities which live in variable environments such as marginal marine settings. They also play a major role in delimiting biogeographic divisions.

Because many benthic foraminifera approach the size of detrital material in sedimentary substrates, they are very difficult to study experimentally under natural conditions. There is no evidence of strong interspecific competition for a potentially limiting resource at meiofaunal level (Warwick 1981; Dayton, in Strong *et al.* 1984). However, parasites (including viruses, bacteria and fungi) are considered to play a major role in shaping communities, even on a biogeographic scale (May, in Strong *et al.* 1984).

Environmental disturbance may be an important control on meiofaunal organisms. Apart from bottom disturbance from the action of waves or currents there may be localised disturbance by organisms. Fish, and especially rays, dig pits in the substrate to seek prey. In so doing, they displace the microfauna some of which may be eaten by 'picker' fishes. Apart from that, the disturbed patches may need to become recolonised by the microfauna so that the sea floor becomes a mosaic of successful patches (Dayton, in Strong *et al.* 1984). This process is not easy to quantify and it introduces variability in some environments which may otherwise seem to be stable. Similarly, the presence of tubes or burrows of macrofaunal species may provide niches for small species such as foraminifera. On ribbed bivalve shells, the grooves rather than the ribs are the parts colonised by clinging or fixed species (Plate 3c).

Diversity and abundance

Diversity refers to the number of species, but to be meaningful a diversity index must also take into account the number of individuals per species. One view is that greatest diversity is achieved in those assemblages in which all the species are equally abundant. In nature such a pattern is rarely achieved.

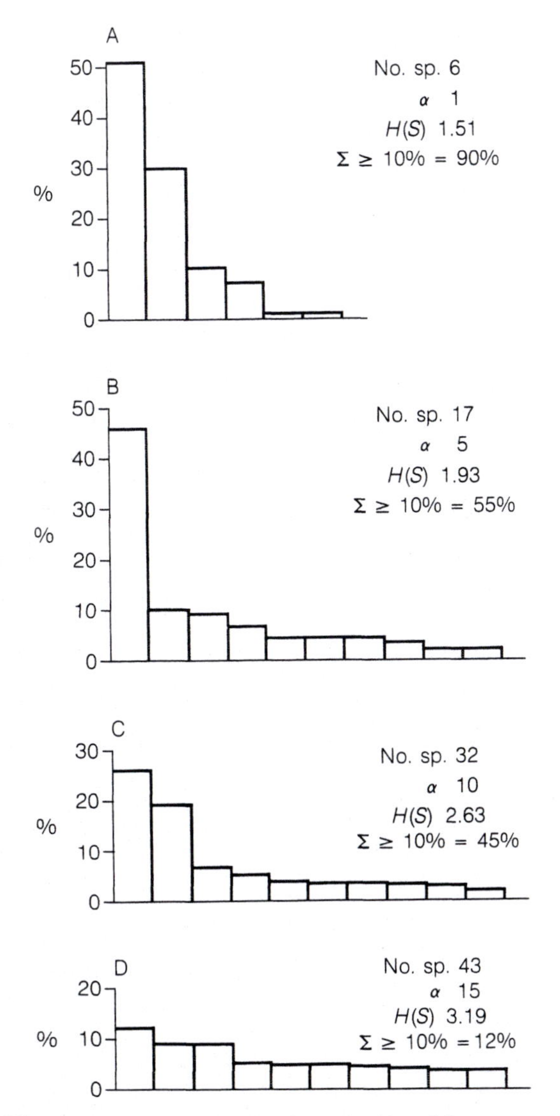

Figure 1.2 Histogram plots in rank order of the most abundant 10 species in assemblages of differing diversity. (A) Living, sta. 8, Long Island Sound (Buzas 1969); (B) living, sta. 1253; (C) dead, sta. 1245; (D) dead, sta. 1248, North Carolina Shelf (Murray 1969)

In assemblages with few species one is normally more abundant that the others and this can be readily shown by histogram plots of the rank order of species (Fig. 1.2A, B). As the number of species increases the percentage abundance of the dominant species decreases (Fig. 1.2C, D). In these examples diversity is expressed by two diversity indices: the Fisher α index and the information function $H(S)$ (see Appendix A for a description of them).

Synthesis of data

A principal objective of this book has been to synthesise the large volume of published data so that patterns may be defined which will assist the palaeoecological interpretation of fossil assemblage.

Most papers concern a relatively small area of sea floor and often the author has sought differences within the area rather than emphasising similarities. For example, Medioli *et al.* (1986) studied a small area of the Nova Scotia continental shelf and recognised seven assemblages based on factor analysis, but Williamson *et al.* (1984) who studied a larger area had a single (*Eggerella advena*) assemblage in place of the seven. In this book similarities are stressed.

The starting-point for ecological studies is the distribution of living assemblages for they can be related to environmental parameters. On death, or as a consequence of reproduction, empty tests are added to the sediment to give the dead assemblage. Consideration of the processes which influence the composition of the dead assemblage is important for understanding the preservation of ecological signals in the fossil record. Detailed discussion of this is given in Chapter 5.

Ecological studies are carried out at species level but much useful information can be gained by grouping data. Although fossil assemblages commonly consist of species which are no longer extant, genera are longer ranging through geological time. Some of the general attributes of modern assemblages are applicable to fossil examples. The methods of data analysis are discussed in Appendix A and include general attributes such as diversity, measured by the two commonly used methods of the information function $H(S)$ and the Fisher α index, the ratio of the three wall structure types (agglutinated, porcellaneous, hyaline), and generic predominance.

In considering the distribution of benthic foraminifera by geographic area it is most meaningful to recognise basic recurrent associations which, for convenience, are named according to the one or two key species. Such associations are most objectively defined by multivariate analysis and indeed this is possible within limited areas. However, it is not feasible to take the entire data set for a whole continental margin and to subject it to multivariate analysis.

The associations recognised here are based on the dominant species in terms of percentage abundance in either living, dead or total assemblages. The common additional species are those present in ≥10% abundance in some or all of the samples. It is a matter of observation that the common additional species differ from one geographic area to another and this reflects varying patterns of biogeographic distribution.

The data used are only from samples in which ≥100 individuals were counted as this is considered to be the minimum acceptable size. Because of this restriction the interpretations reached here sometimes differ from those of the original authors of the data. Not all papers give essential ecological data, but an attempt has been made to summarise basic environmental information for each of the associations.

The recognition of associations based on the dominant species might be taken to imply that each is discrete and has a sharp boundary with the next. This is clearly not the case for, as already noted, the common additional taxa vary from one region to another so the concept embraces variability. Also, on a local scale different associations may show seasonal replacement. Even with the often crude generalisation of environmental parameters, it is normally the case that the abiotic field differs from one association to another (e.g. see Fig. 6.4).

Objectives and structure of the book

Probably 99% of the papers on the ecology and distribution of modern benthic foraminifera are written by geologists. This is primarily because of the palaeoecological utility of foraminifera. In this book the objectives are to provide a state-of-the-art synthesis of the enormous quantity of available data, to provide it in a form which make it readily applicable to understanding the fossil record, and to demonstrate the major biogeographic patterns and their ecological causes.

Chapters 2–5 deal with the important aspects of life processes, stable isotope studies, population dynamics and the relationship between living and dead assemblages. These set the scene for a consideration of the geographical and ecological distribution of species and associations (Chs 6–7). These results are summarised in Chapter 18. As far as possible factual data have been concentrated into tables and figures and into the four appendices.

The section on palaeoecology is short because it is considered necessary only to define basic principles and to discuss some case-studies to illustrate the methods of interpretation. It is clearly beyond the scope of this book to attempt to synthesise global distribution patterns for each geological time interval.

CHAPTER 2

Life processes

Introduction

Consideration of the living animal and of life processes is an important starting-point for ecology. In this brief review emphasis is placed on those aspects believed to be most relevant to understanding the distribution of foraminiferal tests.

Cytology

The cytoplasm is differentiated into chamber cytoplasm or endoplasm, in which general metabolism takes place, and pseudopodial cytoplasm or ectoplasm, responsible for chamber formation, feeding, locomotion and attachment. It is beyond the scope of this work to consider the cytoplasm in detail. Useful reviews are those of Anderson and Bé (1978) and Alexander and Banner (1984).

The most conspicuous features are the pseudopodia which are noted for their ability to form a net (or reticulum) and which bear granules, hence the descriptive term granuloreticulose. The bidirectional movement of the granules along individual pseudopods has long been a source of puzzlement (Jahn and Rinaldi 1959) and ultrastructural studies are leading towards a better understanding of the mechanisms involved (Bowser 1985; Bowser *et al.* 1984; Bowser and Delaca 1985; Travis and Bowser 1988). Movement of the test is by means of the pseudopodia and rates of 9–500 μm Min^{-1} have been observed by Kitazato (1981, 1988a) and Severin and Lipps (1989) for epifaunal taxa. Pseudopodial development and ability to move differs in the same species according to whether it is epifaunal or infaunal (Kitazato 1988a). Observations by Frankel (1975) on *Miliammina fusca* showed that infaunal individuals developed an irregular mass of cytoplasm (podostyle) about the aperture and from this there arise short pseudopodia that gather food. Individuals kept in sediment-free containers lacked the podostyle and developed longer pseudopodia. Other infaunal species form burrows in cohesive sediments and displace grains in sands (Wetmore 1988).

Reproduction

Considering the large number of species of living benthic foraminifera, very little is yet known of their reproductive biology. Some species undergo an alternation of generations (see review by Grell 1979) while others reproduce only asexually. Test dimorphism is sometimes present: tests resulting from asexual multiple fission are megalospheric (large proloculus = A form) while tests resulting from the sexual fusion of two gametes are microspheric (small proloculus = B form). In fossil nummulitids, the A generation reaches a smaller test size than the B generation.

Röttger *et al.* (1986) have put forward the hypothesis that in *Heterostegina depressa* there is an alternation of generations, and that the megalospheric schizont, resulting from repeated asexual multiple fission, is not part of a trimorphic cycle (agamont–schizont–gamont–agamont) but a separate species. This has a separate depth distribution and some distinctive morphological features.

During reproduction the parent cytoplasm is divided up to form gametes (haploid nucleus) or agametes (diploid nucleus) and normally results in the empty parent test being added to the dead assemblage. However, this is not always the case. Mother tests have been reported to retain some cytoplasm after reproduction, to regenerate and reproduce again one or more times: a miliolid from California (Arnold 1955), *Bolivina doniezi* from California (Sliter 1970) and *Heterostegina depressa* from Hawaii (Röttger 1978).

In laboratory cultures, the temperature range over which reproduction will take place has been determined for a number of species but it is doubtful whether these values can be applied to the natural environment. This is because a plexus of interacting factors controls activities such as reproduction. If one factor is adverse, reproduction will be inhibited. The presence of large amounts of food is a stimulus which may override other factors. Thus, Myers (1936) determined the optimum reproductive range of *Spirillina vivipara* to be 18–26 °C yet this species lives on the inner shelf of the English Channel (Murray 1971a) where the temperature does not exceed 16 °C.

An attempt has been made to relate the ratio of microspheric to megalospheric tests of one species (*Brizalina skagerrakensis*) to climate inferred from oxygen isotope studies. A higher ratio of microspheric to megalospheric tests was correlated with warmer conditions (Nigam 1986a). However, some species are not dimorphic. Subsequent studies of *Rotalidium annectens* from the west coast of India have shown that mean proloculus size is inversely proportional to temperature and salinity (Nigam and Rao 1987).

On theoretical grounds, asexual reproduction should be the normal mode of reproduction in stable environments for it will provide individuals adapted to the environment. Sexual reproduction gives genetic variability which is advantageous under variable conditions in providing some individuals which will be able to adapt to the environment (Hallock 1985).

Behaviour

Phototaxis is the response of an organism to light. Those that move towards light are said to be positive while those that move away are negative. Experimental studies on *Amphistegina radiata* showed maximum response to blue-green light (505 nm) with the majority of those that responded being positive and only a few negative (Zmiri *et al.* 1974). *Amphistegina lobifera* is positively phototaxic at illumination of <10 klx and negatively phototaxic above this value. For *Amphisorus hemprichii* the values are less certain: positive at a point between 11 and 22 klx and between 1.7 and 6 klx (Lee *et al.* 1980). The phototactic responses of selected species are listed in Table 2.1.

Experimental studies on the effects of temperature on *Amphistegina radiata* and *A. madagascariensis*, kept in total darkness, showed that they moved towards higher temperatures from the starting-point of 18 °C. *Amphistegina radiata* moved to 29 °C while *A. madagascariensis* moved to 27 °C. The lowest temperatures at which either species moved were 13 °C and 17 °C respectively (Zmiri *et al.* 1974).

Function of the test

One of the attributes of foraminifera, which makes them so attractive to geologists, is the possession of a (generally) hard test capable of being preserved during fossilisation. To biologists, the test is an obstacle to examining the cytoplasm, hence their general neglect of the group. Five possible functions of the test are (after Marszalek *et al.* 1969):

1. To protect the foraminifera against predation;
2. To provide shelter against unfavourable physical or chemical conditions;
3. To serve as a receptacle for excreted matter;
4. To aid in the reproductive process;
5. To control the buoyancy of the organism.

In addition there is the possibility that the test assists growth of the cell.

Table 2.1 Phototactic response of selected species

Positive	Source	Negative	Source
Cibicides spp.	Kitazato 1981	*Ammonia* spp.	Kitazato 1981
Elphidium crispum	Kitazato 1981	*Astrononion* spp.	Kitazato 1981
Amphistegina radiata	Zmiri *et al.* 1974	*Bolivina* spp.	Kitazato 1981
Sorites orbiculus	Kloos 1980	*Buliminella elegantissima*	Kitazato 1981
		Fissurina spp.	Kitazato 1981
		Uvigerinella glabra	Kitazato 1981

Taking these points in turn, there is no direct evidence that the test protects foraminifera from predation. Indeed, certain species are selected by predators because of their form or colour. However, the test can certainly offer protection against unfavourable physical or chemical conditions. Marszalek *et al.* (1969) noted that many foraminifera can close off test openings for several hours by sealing them with debris. Also, planispirally coiled tests such as those of *Ammodiscus*, in which the cytoplasm occupies the early part of the test only, may offer protection as it would take a long while for chemicals to diffuse along the tube. However, this is a speculative comment rather than an observation. The organic lining of calcareous tests is undoubtedly a barrier. Bradshaw (1961) demonstrated that *Ammonia tepida* survived a pH of 2.0 for $1\frac{1}{4}$ hours even though dissolution of the test took place, and that it was able to repair the test afterwards. By contrast, *Spirillina*, which lacks an organic lining, did not survive. Marszalek *et al.* (1969) speculated that chambered tests would be a protection against sudden osmotic changes and quoted the example of *Quinqueloculina*. However, experimental work reveals this to be untrue for this genus (Murray 1968b). The protective value of the wall against certain wavelengths of light is a possibility in shallow-water foraminifera (Haynes 1965; Banner and Williams 1973). In low oxygen environments, the mitochondria and some symbionts are concentrated beneath the pore openings indicating that they serve as pathways for gas exchange (Leutenegger and Hansen 1979; Hottinger 1982). However, for the agglutinated genus *Clavulina*, it has been argued that the test pores function only during the construction of the chamber and have no other functional role (Coleman 1980).

The possibility that the organic wall-lining of rotaliine foraminifera may be the product of excretion has been put forward by Banner *et al.* (1973). This is based on the fact that the organic wall-lining is mainly mucopolysaccharide and that it may be pigmented with the same colour as the dominant food source. Brasier (1986) suggests that biomineralisation may be necessary to remove toxic Ca^{2+} from the cell although he also suggests that calcite biomineralisation may be more energy efficient than agglutination.

The value of the test in reproduction is debatable. If its use is confined to reproduction, why is the test present throughout the life of the individual?

The test may act as a buoyancy control. Marszalek *et al.* (1969) noted that the allogromiid *Iridia* stores low-density fats and lipids and the test may be necessary to compensate for this. It is a matter of observation that species living in environments of high physical energy either have heavy, robust tests or they are adapted to an epifaunal sessile or clinging mode of life. However, various forms gather up sedimentary detritus with their pseudopodia, if only as part of their feeding activity, and this may provide ballast even in the absence of a test. So it still remains unproven as to whether the test is essential to control buoyancy.

The possibility that possession of a hard test assists the growth of the cell is based on the supposition that there must be an upper limit to the size of a single cell if it is to maintain its integrity. Pseudopodia are easily broken by currents or turbulence when free in the water (Lipps 1982). They may be strengthened by secreted spines or elongate detritus along which they may extend. A test provides a container to enclose a cell and allows the organism to make more space than it needs. For instance, on average only 43% of the chamber space of *Alveolinella quoyi* is occupied by cytoplasm (Severin and Lipps 1989). It is noteworthy that foraminifera are the largest single-celled protistans. The test cannot be for support because the cell is constantly undergoing cytoplasmic flow and streaming and because the cytoplasm itself forms the template on to which new chambers are secreted.

The six possible functions of the test are not mutually exclusive. Some of the morphological features described by taxonomists as 'ornament' have obvious functional significance. Pustules on the ventral surface of rotaliid genera (e.g. *Asterigerina*) channel the food particles and aid in physical disaggregation (Haynes 1981: 249). Similarly, lines of pustules/tubercles around the apertural areas of planispiral rotaliids must filter out detritus and food particles that are too large (numerous illustrations in Hansen and Lykke-Andersen 1976). Ribs control the flow of extrathalamous cytoplasm. Spines support pseudopodia but may also reduce the sinking of tests into soft muddy substrates.

Many rotaliid foraminifera have canal systems within their chamber walls. In *Operculina* the canals allow communication between the chamber cavities and the lateral surfaces of the test. *Heterostegina* in addition has a three-dimensional network of canals within the peripheral keel or marginal cord, and this opens into the ambient seawater. The canal system replaces the true primary and secondary apertures (Röttger *et al.* 1984). Pseudopodia can be extruded from any point on the marginal cord by contrast with those forms having a single aperture from which the pseudopodia emerge. The canal system serves for the excretion of waste products, for the secretion of the transparent sheath which encloses the test, for the release of gametes during sexual reproduction and for the extrusion of cytoplasm and symbionts during asexual reproduction.

Test strength

Apart from the damage caused through predation, the principal cause of test breakage is the physical energy of the environment. Tests damaged during life may be repaired so differences in test strength may affect the difference between death or survival. Postmortem breakage is a separate matter which affects the likelihood of preservation in the fossil record.

Experimental analysis of test strength (Wetmore 1987) has shown that a complex relationship exists between crushing strength and morphology. Test architecture is more important than the thickness of the test wall. Test strength is greater in high-energy, coarse sedimentary substrates than on fine sediment or algal substrates. However, in the case of *Elphidiella hannai*, there was no difference in test weight between these environmental extremes, again pointing to an architectural control on strength.

Feeding

In order to live in such a diversity of environments benthic foraminifera have evolved many different types of feeding strategies. A useful review of current knowledge is that of Lipps (1983). To minimise competition within individual environments resource partitioning has evolved at least among herbivorous forms (Lee *et al.* 1977). Test form may reflect adaptation for specific modes of foraging (Lipps 1982; Price 1980).

Uptake of dissolved organic material

The agglutinated form, *Notodendrodes antarctikos*, has a branching test with one group of branches extending into the water and another group buried in the sediment. Although dissolved organic material is taken up by the entire cell, the majority is taken up by the root system (Delaca *et al.* 1981; Delaca 1982). Lipps (1983) suggests that many foraminifera may utilise dissolved organic matter, especially those species living where particulate organic matter is unobtainable and where productivity is low, e.g. coral reefs, deep sea. As long as the foraminiferid has a large cytoplasmic surface area, no distinctive morphology is required.

In rather artificial experiments, *Allogromia laticollaris* was deprived of its natural food (*Chlorella*), and placed in culture fluid containing D-glucose which it incorporated and metabolised. The cell membrane and enzyme activity of *Allogromia* showed the same characteristics as those of higher organisms (Schwab and Hofer 1979).

Lepidodeuterammina ochracea has a sessile infaunal habit living attached over depressions in sand grains, the mode of attachment being an organic membrane from the margin of the test to the substrate (so-called 'puffermasse'). Within the enclosed cavity beneath the test are living diatoms which are not eaten by the foraminiferid. Instead it is thought that their extracellular products serve as food for *Lepidodeuterammina* (= *Trochammina* Frankel 1974).

Herbivory

Active and passive herbivores can be distinguished. Active herbivores gather algae (and bacteria) as they move over the substrate. Movement is slow and they commonly extend the pseudopodia from the aperture in all directions when feeding. The test form is trochoid, lenticular or flattened, but not all foraminifera having these shapes are herbivores. Passive herbivores are sessile epifaunal forms that gather food from around the site of attachment. They have no distinctive test morphology. Herbivores are confined to the photic zone.

It has been observed in cultures, that when food (diatoms) is scarce, *Rosalina globularis* actively grazes but when diatoms are abundant it becomes passive (Sliter 1965). In nature, this species is epifaunal, sessile or clinging.

The circean phenomenon has been observed in laboratory cultures: the green alga *Dunaliella parva* stimulates rapid pseudopodial formation and is positively attracted to those of *Quinqueloculina lata* and *Ammonia beccarii*, immobilised and eaten (Lee *et al.* 1961).

Most littoral foraminifera feed on pennate diatoms, small chlorophytes and bacteria (Lee 1980). Experimental studies have shown that *Elphidium crispum* selects its food on the basis of size (Murray 1963). *Ammonia beccarii, Quinqueloculina* spp., *Elphidium poeyanum* and *Rosalina leei* are selective feeders on pennate diatoms and chlorophytes but do not eat yeasts, cyanobacteria, dinoflagellates, chrysophytes and most bacteria. Factors influencing feeding are the age of the food organism, the age of the foraminifera (juveniles vs. adults), and the concentration of food

organisms (Lee *et al.* 1966). There is variation in the nutritional value of different algae leading to faster growth when the appropriate food is available. For example, *Q. lata* grew three times as fast on diets of *Nitzschia acicularis* and *Chlorococcum* sp. than on other species (Muller and Lee 1969). Bacteria appear to be essential food sources to maintain reproductive activity. They must supply some nutritional element which is not otherwise available (Muller and Lee 1969; Muller 1975). Nevertheless, low numbers of bacteria (~20 000 ml^{-1}) seem optimal for littoral species (Lee *et al.* 1970).

Early studies by Myers (1943a) showed that growth and reproduction of *Elphidium crispum* were related to phytoplankton blooms. Many littoral foraminifera are bloom feeders. They feed and reproduce slowly when food is scarce, but when the food organisms are present as a vigorously reproducing bloom, the foraminifera feed, grow and reproduce (Lee *et al.* 1966). Because the blooms are patchy in their distribution, it is possible that this in turn causes patchiness in the abundance of foraminifera (Buzas 1965, 1969; Lee *et al.* 1969, 1977; Matera and Lee 1972).

Carnivory

The method of capturing food is by the spreading of the pseudopodial net as a trap into which active organisms move. Prey include small arthropods, small sea urchins and other foraminifera. In some cases, the prey are thought to die of exhaustion and are then digested extracellularly, e.g. *Astrorhiza limicola* (Buchanan and Hedley 1960) although this species and *A. arenaria* have also been reported to be suspension feeders (Tendal and Thomsen 1988, Cedhagen 1988). Normally the pseudopodia are adhesive during feeding (Buchanan and Hedley 1960; Banner 1978). Digestion of the prey takes about a day.

Passive suspension feeding

The mechanics of feeding on suspended particulate matter have been discussed by Conover (1981). To be efficient, the energy expended in search and capture of food must be small compared with the energy gained through feeding. For small organisms such as foraminifera, the viscosity of seawater is important (Reynolds number, $Re < 1$). The movement of a pseudopod may cause movement of food particles without actually achieving contact. A pseudopodial net is probably more efficient, but the pores must not be too small or the flow rate will be low. Foraminifera feeding in this way must rely on natural currents (passive suspension feeding) because they have no known means of drawing water through a pseudopodial net. During transport of the food particles towards the test, selection of trapped particles on the basis of size or density could take place.

Certain types of fragile agglutinated foraminifera (komaciaceans) living in the deep sea are sometimes found intergrown with paludicelline ctenostome bryozoans. Only 4% of the komaciaceans from the north-eastern Atlantic show this association. Gooday and Cook (1984) suggest that the particle-bearing feeding currents induced by the bryozoans may be of benefit to the foraminifera.

Passive suspension-feeding foraminifera are normally epifaunal and sessile on hard substrates or rooted in soft sediment with the test held erect and the aperture oriented away from the substrate (Lipps 1983). This is true even of discoid *Astrorhiza* (Tendal and Thomsen 1988, Cedhagen 1988).

Detritivores – detrital and bacterial scavenging

The majority of foraminifera living in fine-grained sediments deeper than the photic zone feed on detritus or bacteria. They are especially abundant in the top 1–2 cm of sediment and most are infaunal (Table 2.2). Deposit feeders may be selective or non-selective.

Table 2.2 Feeding strategies of agglutinated foraminifera (after Jones and Charnock 1985)

Herbivores	**Detritivores**
Passive	*Passive*
Hemisphaeramminінae	Saccamminids
Diffusilininae (part)	Hormosinids
Tolypammininae	Lituolids (uncoiling)
Placopsilininae	Textulariids
Coscinophragmatinae	Verneuilinids
Trochamminidae (part)	Eggerellids
Tritaxis	Ataxophragmiids
	Valvulinids
Active	*Active*
Lituolids (most)	Ammodiscids
Trochammina s. l.	? Rzehakinids
Trochamminidae (part)	Lituolids (most)
	Trochammina s. l.
Passive suspension feeders	
Komakiaceans	
Astrorhizidae (some)	

Omnivory

Lipps (1983) considers that many benthic foraminifera are probably opportunistic feeders eating algal, bacterial, detrital matter and possibly other animals. A varied diet seems essential for the well-being of foraminifera (Lee 1980).

Parasitism

There are few reliable reports of parasitism. *Fissurina marginata* (= *Entosolenia* of Le Calvez 1947) goes into the pseudopodial reticulum of *Discorbis villardeboanus* and takes out granules (ectoparasite). *Planorbulinopsis parasita* (Banner 1971) is an endoparasite of *Alveolinella quoyi*. *Lagena* has the aperture on a long neck which is considered to be a probe for a parasitic feeding strategy (Haynes 1981: 54). Similarly, *Cibicides refulgens* bores holes in scallop shells and feeds on the extrapallial fluid (Alexander and Delaca 1987).

Recyling waste products

Some agglutinated foraminifera retain waste products (stercomata) in their tests and may feed on bacteria which the stercomata attract (Tendal 1979).

Feeding and environment

In a review of the agglutinated foraminifera, Jones and Charnock (1985) noted the correlation between the three principal feeding strategies and environment (Fig. 2.1). There is a progression from marshes (50:50 herbivores:detritivores) to shelf and upper bathyal (mainly detritivores) to middle and lower bathyal and abyssal zone (detritivores and passive suspension feeders).

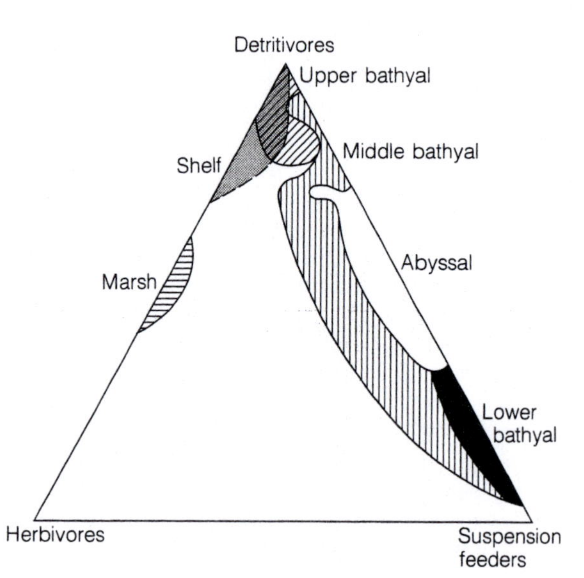

Figure 2.1 Feeding strategies of agglutinated foraminifera (after Jones and Charnock 1985)

Symbiosis

Symbiosis describes the association between different species of organism which is useful to both partners. The term 'endocytobiosis' refers to intracellular symbiosis in which the symbionts (smaller partner) live within host cells (larger partner). Two working hypotheses have been proposed:

1. Nutritional endocytobiosis in which the endocytobionts are chiefly responsible for the digestion of resistant food or poisonous metabolic wastes from the host; they convert these into vitamins, amino acids and enzymes for themselves and their hosts.
2. Organelle endocytobiosis in which it is assumed that there is integration of the symbiont's metabolism and developmental physiology into the host cell; such an integration requires genetic exchange between symbiont and host during their co-evolution (Schwemmler 1980; Cavelier-Smith and Lee 1985).

The colonisation of an organism by a symbiont can be divided into seven stages (Smith 1980):

1. Transmission – in asexual reproduction the agametes contain symbionts, but in sexual reproduction no mechansim has been evolved for transmitting symbionts in the host gametes. Therefore, if infection of new individuals by free-living symbionts is necessary, then the symbiont must be adapted not only to live in the host but also in the surrounding environment.
2. Entry – for each sexual generation the symbiont has to gain fresh entry to the host and this is invariably through the feeding process.
3. Countering host defences – most organisms have defensive mechanisms which the symbionts

must overcome. The symbionts may do this by possessing tough and indestructible outer coverings, by avoiding the digestive region soon after gaining access and/or repression of host reaction.

4. Positioning within the host – normally symbionts are not randomly distributed but confined to specific locations particularly to obtain optimum exposure to light
5. Providing net advantage to the host – the host must gain some advantage from the presence of the symbiont. The advantage is 'net' because the cost to the host is utilisation by symbionts of nutrients and space, and energy is expended in regulating the symbionts. Nutritive advantages are the flow of photosynthetically fixed carbon from the symbiont to the host and sometimes recycling nitrogen. Non-nutritive advantages include promoting calcification of the host.
6. Surviving in the host nutrient environment – symbionts have to survive on the nutrients available to them within the host.
7. Regulation of the symbiont by the host – in all associations the size of the symbiont population remains constant for a given set of environmental conditions; therefore, the hosts have evolved the means to control the symbionts. The three possibilities are digestion of surplus algae (thought to be rare), ejection and/or restriction of symbiont division.

Symbiosis in benthic foraminifera has been reviewed by Leutenegger (1984). Symbionts occur in large numbers in chamber endoplasm but are not normally present in ectoplasm. They lack cell envelopes, such as diatom frustules, dinophycean thecae or cysts. Non-symbiotic ingested food organisms possess a cell wall, are present in small numbers, in association with pseudopodial ectoplasm, usually near test openings. Each foraminiferal species has its own type of symbiont independent of locality, season or water depth. Therefore, the host is able to select specific algae from its environment. The algal symbiont type and host foraminifera are listed in Fig. 2.2.

During asexual reproduction, the symbionts are passed from the parent to the megalospheric juveniles in *Amphistegina lessonii, Heterostegina depressa* and *Peneroplis pertusus* as observed in laboratory cultures (Röttger 1974, 1981). In many larger foraminifera hosting symbionts, megalospheric forms predominate and sexual reproduction is greatly reduced (Röttger and Schmaljohann 1976). In modern and fossil sediments A forms are more abundant than B forms. Repeated asexual reproduction stabilises symbiosis as symbionts cannot be transmitted directly from parent to offspring during sexual reproduction.

Because the small peneroplids have compressed tests, the symbionts are abundant throughout the endoplasm. In discoid soritids, they lie below that side facing the light and in alveolinids they are most abundant in the chamberlets of the ultimate and penultimate whorls. In large biconvex Rotaliina, most symbionts are arranged beneath the lateral chamber walls of the last one or two whorls (Leutenegger 1984).

As the symbionts require light, symbiont-bearing benthic foraminifera are responsive to this. The different algal groups have different light requirements as can be seen from Fig. 2.2. However, light alone is not the only ecological control on the depth distribution and the shallow occurrence of the calcarinids is probably linked with high physical energy (Leutenegger 1984).

The idea that the test of foraminifera, especially those with a hyaline wall, acts as a glasshouse was first put forward by Haynes (1965). The observation that the symbionts are arranged side by side along the portion of the test facing the light supports the idea that compressed tests with large surface areas are well adapted to symbiosis. They provide the algae with favourable conditions for light (Leutenegger 1984). With increase in water depth, tests of *Amphistegina* show a decrease in sphericity (Hallock 1979) and thinning of the chamber walls (Hallock and Hansen 1979). Superimposed on this trend, water motion leads to the formation of thicker walls and more robust tests (Hallock *et al.* 1986). Even in porcellaneous forms, such as *Marginopora* and *Amphisorus*, the walls of the lateral chamberlets are effectively transparent when the cytoplasm-filled test is in water (Ross and Ross 1978).

Furthermore, the thin walls beneath the pits in *Amphisorus hemprichii* permit the passage of CO_2 for use by symbiotic algae (Hansen and Dalberg 1979). The thin walls of *Sorites* and *Marginopora*, the pitted walls of *Spirolina, Cyclorbiculina* and furrowed walls of *Peneroplis* may serve a similar function.

Symbiont-bearing larger foraminifera live in nutrient-poor, shallow, tropical waters. The most efficient use of the limited nutrients is to recycle them between the host animals and their algal symbionts. In such a system, the residence time of the recycled nutrients should be longer than that of nutrients taken up and excreted independently by host and symbiont. This can be tested using ^{14}C and experiments on

Suborder	Superfamily	Family	Subfamily	Species	Symbiont	Water depth range (m)
Miliolina	Soritacea	Peneroplidae		*Peneroplis planatus*	Rhodophyceans	0–60/70
				Peneroplis pertusus		
				Spirolina arietina		
				Laevipeneroplis proteus	Chlorophyceans	0–15
		Soritidae	Archaiasinae	*Cyclorbiculina compressa*		
				Archaias angulatus		
			Soritinae	*Sorites orbiculus*	Dinophyceans	0–60/70
				Sorites variabilis		
				Sorites marginalis		
				Amphisorus hemprichii		
				Marginopora vertebralis		
		Alveolinidae		*Borelis schlumbergeri*	Diatoms	2–70
				Alveolinella quoyi		
Rotaliina	Asterigerinacea	Amphisteginidae		*Amphistegina lobifera*		0–130
				Amphistegina lessonii		
				Amphistegina bicirculata		
				Amphistegina papillosa		
				Amphistegina radiata		
	Nummulitacea	Nummulitidae		*Operculina ammonoides*		0–130
				Heterostegina depressa		
				Heterocyclina tuberculata		
				Cycloclypeus carpenteri		
				Nummulites cumingii		
	Rotaliacea	Calcarinidae		*Calcarina calcar*		0–30
				Calcarina spengleri		
				Baculogypsina sphaerulata		

Figure 2.2 Foraminifera and their symbionts. Boxes indicate groups of related foraminifera with the same or similar symbionts (adapted from Leutenegger 1984)

Amphistegina lessonii showed that the algal cells were photosynthesising (Hallock Muller 1978). The foraminifera obtain organic carbon by release of algal assimilates rather than by digestion of the symbionts (Kremer *et al.* 1980). Not all symbiont-bearing foraminifera depend entirely on their symbionts for food; *Archaias angulatus* and *Sorites marginalis* both actively feed as well (Lee and Bock 1976).

Growth of the foraminifera is dependent on the symbiotic algae for zero or slow growth takes place in the dark (Hallock Muller 1978; Kuile and Erez 1984). Optimum growth conditions for *Heterostegina depressa* are 300 lx and too much as well as too little light reduces the growth rate (Röttger 1972b). For *Amphistegina lessonii* optimum growth occurs at 800 lx (Röttger *et al.* 1980). The three species *Amphisorus hemprichii*, *Amphistegina lobifera* and *H. depressa* are well adapted for life in shade or deeper water (Röttger 1976, Lee *et al.* 1980).

Resource partitioning or non-competitive feeding is a feature of herbivorous foraminifera. The choice of different symbionts permits different depth preferences for symbiont-bearing foraminifera (Fig. 2.2), but those forms with the same symbionts may still be in competition with one another. An example of niche separation has been described for two species of *Amphistegina* (Hallock 1981a). *Amphistegina lessonii* and *A. lobifera* have similar rates of carbon fixation by symbionts and of growth with respect to light intensity over a depth range of 3 m (80% surface light) and 15 m (30% surface light). *Amphistegina lessonii* has thinner test walls, is more biconvex, is inhibited by high light intensities and is therefore less competitive than *A. lobifera* at depths of <3 m. Also, *A. lobifera* requires higher light intensities for reproduction so favours shallower depths than *A. lessonii*. Experimental studies on the symbionts of *A. lessonii* and *Amphisorus hemprichii* have shown that the metabolites released (glucose and mannose) form less than 1% of the total carbon fixed by the algae. They therefore contribute little to the carbon budget of the host foraminifera (Saks 1981). Indeed *A. hemprichii* gains most of its carbon by feeding on unicellular algae from the substrate (Kuile *et al.* 1987).

Algal chloroplasts, which are intact but isolated from other algal cell organelles, occur in large numbers in the endoplasm of certain benthic foraminifera: *Haynesina germanica*, *Elphidium crispum*, *E. excavatum*, *E. macellum*, *E. williamsoni*, *E. craticulatus* and *Nonionella stella* (Leutenegger 1984). They are interpreted as symbionts. They are acquired during feeding on algae and continue to photosynthesise in their host (Lopez 1979). However, not all individuals of a species contain algal chloroplasts so they are not essential to the well-being of the foraminifera. Presumably the foraminifera gain photosynthetic products from their algal chloroplasts. Elphidiids and nonionids with symbiotic algal chloroplasts live in nutrient-rich environments in which the hosts depend on external food sources.

An agglutinated species, *Reophax moniliformis*, may also contain symbiotic algal chloroplasts as it has a high chlorophyll *a* content (Knight and Mantoura 1985).

In the past, foraminifera with coloured protoplasm were thought to contain symbionts (Haynes 1965) although experiments had shown that the colour of *Elphidium crispum* could be varied with the colour of its food (Murray 1963). Colour is not reliable for determining the presence of symbionts. Coloration can be caused by ingested food, storage products or pigmented shell components (Leutenegger 1984).

Commensalism

The positive interaction between two species in which one benefits and the other is unaffected is termed commensalism. Certain foraminifera are sessile on other animals (epizoic) such as sessile hydroids and bryozoans (Dobson and Haynes 1973), brachiopods (Zumwalt and Delaca 1980), clionid sponges (Voigt and Bromley 1974), tunicates (Nyholm 1961), and vagile decapods (Delaca and Lipps 1972), isopods, (Moore 1985) amphipods, pycnogonids, gastropods (Zumwalt and Delaca 1980) and bivalves (Bock and Moore 1968; Haward and Haynes 1976). The advantages to the foraminifera are that they are supported above the sediment and so protected from being overwhelmed by sedimentation. Furthermore, in some cases at least, the foraminifera select attachment sites near inhalent or exhalent currents of the host and may therefore benefit from an enhanced supply of food. Examples of commensalism in the fossil record include foraminifera and algae (Johnson 1950, Henbest 1963; Peryt and Peryt 1975), clionid sponges (Bromley 1970; Voigt 1970; Bromley and Nordmann 1971).

Foraminifera as food for other organisms

Benthic foraminifera are major contributors to meiofaunal biomass in many marine environments, yet little is known of this potential trophic resource. Foraminifera may be consumed by predators or by deposit feeders and other incidental consumers.

Predators include other foraminifera, nematodes, polychaetes, gastropods, scaphopods and crustaceans (see review by Lipps 1983) and fish.

Predation by other foraminifera is perhaps rare but an example is given by Christiansen (1964). Nematodes were thought to be responsible for boring holes in the test of *Rosalina globularis* and *Bolivina doniezi* and to feed on the cytoplasm (Sliter 1971). Fossil analogues occur in Cretaceous foraminifera. Gastropods are known to bore holes in tests and they are equipped with a radula to do this. However, most ingest the foraminifera and some are species-selective, e.g. *Olivella* feeds on *Elphidiella hannai* and agamonts of *Glabratella ornatissima* (Hickman and Lipps 1983). The tests suffer partial dissolution and scratching in the gastropod stomach and this led the authors to consider that they may be used as a $CaCO_3$ resource rather than a food (they called the process foraminiferivory to distinguish it from feeding).

The captacula of scaphopods are able to detect foraminiferal extrathalamous cytoplasm. Scaphopods can differentiate between species, between calcareous and agglutinated tests and between living and dead individuals. *Dentalium entale stimpsoni* preferentially selects *Quinqueloculina* spp. and *Islandiella islandica*. It consumes 2.4% of the standing crop annually, including 11.1% of *I. islandica* (Bilyard 1974).

The list of incidental consumers includes flatworms, polychaetes, chitons, gastropods, nudibranchs, bivalves, crustaceans, holothuroids, asteroids, ophiuroids, echinoids, crinoids, tunicates and fish (Lipps 1983). A study of the gut contents of bivalves, holothurians and echinoids showed that the tests, whether agglutinated or calcareous, are not destroyed (Boltovskoy and Zapata 1980). In a series of experiments, Buzas (1978, 1982) and Buzas and Carle (1979) demonstrated that a large number of deposit feeders consume foraminifera and that when these consumers are excluded, the foraminiferal densities increase. They therefore concluded that the predators limited the density.

Although most of the incidental consumers are deposit feeders, fish feeding on algae growing on reefs may consume attached foraminifera (Lipps 1983).

From the palaeoecological viewpoint, it is important to note that the majority of benthic foraminifera that are consumed are taken unselectively by deposit-feeders. Therefore no change is caused in the proportions of species between the original living standing crop and the defecated dead material. On the other hand, selective predation, as for instance by *Dentalium* on *Islandiella islandica*, might alter the species proportions especially if tests are destroyed in the feeding process.

Relationship with the substrate

Substrates range from firm surfaces, such as rocks, shells, and seaweeds or other plants, to soft unconsolidated sediment. Although some benthic foraminifera live on varied substrate types, others have specific requirements.

Ecologists recognise three levels:

1. Epifaunal: living on or above the sediment surface;
2. Semi-infaunal: living partly below and partly above the sediment surface;
3. Infaunal: living within the sediment.

Organisms which move around are said to be vagile or free living. Those which are attached are said to be sessile. The attachment may be permanent, e.g. cementation, or semi-permanent perhaps allowing for limited movement, e.g. clinging with the pseudopodia. Those living in association with plants are sometimes described as phytal.

A simple classification is as follows:

Epifaunal
- Sessile = attached immobile of Sturrock and Murray (1981)
- Clinging = attached mobile of Sturrock and Murray (1981); A and B of Kitazato (1981)
- Free = free (in part) of Sturrock and Murray (1981)

Semi-infaunal
- Sessile
- Clinging
- Free = free (in part) of Sturrock and Murray (1981)

Infaunal
- Sessile
- Clinging = ?C of Kitazato (1981)
- Free = free (in part) of Sturrock and Murray (1981); D of Kitazato (1981)

Table 2.3 Some examples of mode of life benthic foraminifera (based on Kitazato 1981, 1984, Christiansen 1971, Sturrock and Murray 1981)

	EPIFAUNAL		SEMI-INFAUNAL	INFAUNAL	
SESSILE	*Cibicides*	*Rosalina*		*Lepidodeuterammina**	
	Acervulina	*Haliphysema*			
	Placopsilina	*Vasiglobulina*			
	Nubecularia				
	Remaneica				
CLINGING	*Rosalina*	*Patellina*	*Saccammina*	*Trochammina*	
	Cribrostomoides	*Cancris*	*Marsipella*	*Elphidium*	
	Gavelinopsis	*Hanzawaia*	*Pelosina*		
	Planorbulina	*Pararotalia*	*Jaculella*		
	Glabratella	*Elphidium*	*Bathysiphon*		
FREE	*Elphidium*			*Elphidium*	*Fissurina*
	Peneroplis			*Ammonia*	*Reophax*
	Quinqueloculina			*Bolivina*	*Textularia*
	Triloculina			*Bulimina*	*Uvigerina*
				Cassidulina	

**Lepidodeuterammina ochracea* attached to sand grains within the sediment down to at least 7 cm (Frankel 1974).

Some examples are given in Table 2.3 but, for most genera and species, no direct observations have yet been made on their mode of life. Certain genera occupy more than one habitat, e.g. *Elphidium*. Others change particularly at the time of reproduction, e.g. *Rosalina globularis* is thought to be able to dissolve the organic membrane, by which it is attached to the substrate, during sexual reproduction (Delaca and Lipps 1972).

Certain benthic foraminifera are able to dissolve calcareous shells. *Rosalina carnivora* Todd (1965a) occurs attached to the valves of *Lima (Acesta) angolensis*. It sits in depressions which show evidence of pitting and corrosion and in some cases the bivalve shell is completely penetrated, which Todd believed may be due to the foraminifera utilising the calcium carbonate for test construction. *Rosalina globularis* attaches itself by means of an organic membrane to a great variety of firm substrates so that the test lies in a depression. Because *Rosalina* lives on varied substrates and because shallow seawater is normally saturated with $CaCO_3$ it seems unlikely that the foraminifera utilises dissolved biogenic carbonate in its own test formation (Delaca and Lipps 1972). Similarly, there seems no obvious function for the holes bored by *Cymbaloporella tabellaeformis* in gastropod shells except perhaps for protection (Smyth 1988).

A rather different example has been described by Poag (1969). The polymorphinid, *Vasiglobulina*, attaches itself to bivalve shell fragments by means of spines, with an hour-glass shape, that are wedged into dissolution cavities. The spines are thought to have been enveloped in pseudopodial cytoplasm that carried out the dissolution cavities.

The relationship with the substrate, the orientation of the test and to a certain extent test form are closely linked with feeding strategy and with the physical energy of the environment. One of the more stimulating approaches to understanding the diversity of growth plans in foraminifera is that of Brasier (1982). He recognised three main patterns: non-septate contained growth (i.e. a single near-spherical chamber), non-septate continous growth (i.e. a single tubular chamber) and septate growth (multi-chambered). He produced simple theoretical models to show the range of morphology, its evolution and its relationship to ecology. Table 2.4 summarises some of his conclusions. The main feature is that free-living forms exhibit a wider range of architecture than those constrained by a sessile mode of life.

Another innovative approach has been to subdivide the agglutinated foraminifera into 'morphogroups' (Jones and Charnock 1985, Jones 1986). Test form, life position and feeding strategy are considered to be interrelated (Table 2.5). The broad conclusions may be correct but there are, perhaps inevitably, some exceptions. Morphogroup A includes examples of detritus feeders (*Bathysiphon filiformis*), carnivores/

Table 2.4 Some examples of the relationship between mode of life and test architecture (adapted from Brasier 1982);

TEST ARCHITECTURE	EPIFAUNAL	
	Sessile	Clinging
1. Non-septate		
(a) Contained growth	One aperture, hemispherical (Hemisphaeramminae)	
(b) Continuous growth	Meandering, zigzag (Tolypammininae) Planispiral (*Ammodiscus, Cornuspira, Spirillina*) Branched (*Dendrophrya*)	
2. Septate		
(a) Uniserial		
(b) Biserial		
(c) Planispiral		
(d) Trochospiral	Low trochospiral, broad umbilical area (*Cibicides, Rosalina, Discorbis*)	Low trochospiral, biconvex (*Amphistegina*)
(e) Miliolid		x
(f) Annular, cyclic		Soritidae 'Larger' foraminifera
(g) Fusiform planispiral		

passive suspension feeders (*Astrorhiza limicola*) and dissolved organic material feeders (*Notodendrodes antarctikos*). Also, some of the feeding strategies and life positions are inferred rather than based on observation. Nevertheless, this classification provides a useful working hypothesis. The morphogroups show a distinctive distribution with respect to environment (see p. 21).

Microdistribution patterns

There are three main patterns of distribution: random, uniform and clumped. The clumped pattern is typical for most species of benthic foraminifera and the reasons for this are twofold: microenvironments and needs for reproduction. All environments have minor local differences which favour certain species against others and sexual reproduction requires aggregations of the same species.

Because benthic foraminifera occupy a range of habitats from epifaunal (including epizoic and epiphytic) to infaunal it follows that the microdistribution patterns are often three rather than two dimensional. The lateral distribution includes discussion of distributions within the top 1 cm of sediment and on plants, and depth of life in the sediment includes preferred depth of life beneath the top 1 cm.

Lateral distribution

Two theoretical models of species distributions are shown in Figs 2.3 and 2.4. In each, the distribution of species is taken as a circular area with greatest abundance near the centre. In model 1 (Fig. 2.3) four species have partially overlapping distributions. A sample taken at A would contain all four species, whereas at B only one is present. Sample C would be barren. Low-diversity, patchy distributions such as this are characteristic of marshes and other variable environments. There are big differences in the size of the standing crop and the dominant species varies from sample to sample. Model 2 (Fig. 2.4) has three or four dominant species present throughout and these are represented by the rectangular outline. Seven rarer species are indicated by circles. Sample A contain the dominant species plus two rarer forms and sample B the dominant species plus three rarer forms, but it is important to note that the rare species differ between

	INFAUNAL	
Free	Clinging	Free
One aperture, axis vertical (Gromiida)		One aperture (Gromiida, *Saccammina*)
Single and branched tubes (*Shepheardella, Nemogullmia, Astrorhiza*)		
Straight tubes and cones (*Hippocrepina*)		
x		x
x		x
x		x
High upright (*Bulimina*)		Low trochospiral, biconvex (*Ammonia, Asterigerina, Trochulina*)
x	x	x
'Larger' foraminifera		
Alveolinidae Free-wheeling rollers		

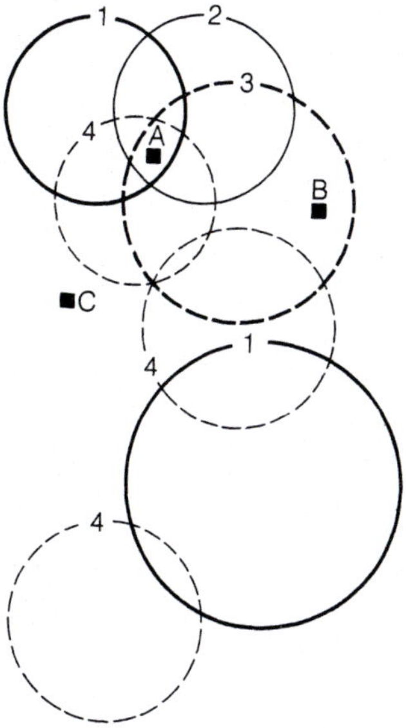

Figure 2.3 Theoretical model distribution of four species (Murray 1973)

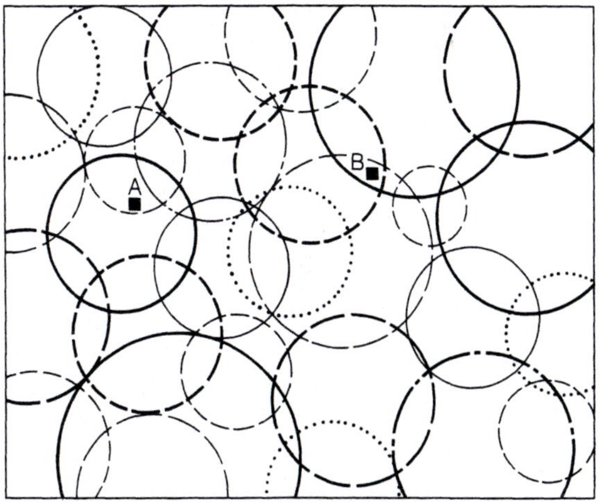

Figure 2.4 Theoretical model distribution of many species (Murray 1973)

the two samples. This model is thought to be typical of more stable environments such as the shelf, continental slope and deep sea. In reality, there is a spectrum of patterns between the extremes represented by the two models.

There is considerable field evidence which

Table 2.5 Morphogroups of agglutinated foraminifera, their postulated life positions and feeding strategies (adapted from Jones and Charnock 1985)

Morpho-group	Subgroup	Chambers 1	>1	Test form	Life position	Feeding strategy	Examples
A	A	x	x	Tubular or branching	Erect, sessile Epifaunal/ semi-infaunal	Primarily passive suspension feeders	Komakiaceans Astrorhizids
B	B1	x		Globular	Epifaunal/ infaunal	Passive detritivores	Saccamminids
	B2	x	?	Coiled (flattened)	Epifaunal	Active detritivores	Ammodiscids ? Rzehakinids
	B3		x	Planispiral/ trochospiral (lenticular)	Epifaunal	Active herbivores, detritivores, omnivores	Most lituolids *Trochammina s. l*
	B4	x	x	Irregular, limpet-like, conical	Epifaunal, sessile	Passive herbivores	Hemisphaerammininae Difusilininae (part) Tolypammininae Placopsilininae Coskinophragmatinae Trochamminidae (part)
C	C1		x	Elongate (mixed growth)	Infaunal	Detrital/bacterial detritivores	Hormosinids Uncoiling lituolids Textulariids Verneuilinids Eggerellids Ataxophragmiids Valvulinids
	C2		x	Elongate quinqueloculine	Infaunal	Detrital/bacterial detritivores	*Miliammina* ? Some rzehakinids
D	D		x	Trochospiral, conical	Epifaunal, clinging or free living	Herbivores	Most trochamminids

conforms with the models. Multiple samples from Poponesset Marsh, Massachusetts, showed that the distribution at each station was fairly uniform over a distance of 15 cm but there were big differences between stations (>200 m) (Parker and Athearn 1959). This conforms with model 1. The intertidal zone of Puerto Deseado, Argentina, was sampled on a grid pattern at 1 m intervals (Boltovskoy and Lena 1969a). Individual species showed very irregular distribution patterns with the centres of maximum abundance separated by several metres. One area sampled on a 10 cm grid had two centres of abundance 20 cm apart but these might represent two cusps on a single irregular patch of abundance. The number of species present in adjacent samples is sometimes the same and sometimes very different (11 and 30 in area 3). This again conforms to model 1.

The very shallow waters of Bottsand Lagoon, Germany (0.2 m depth) were sampled on a 1 m grid by Lutze (1968b). The two most abundant species, *Miliammina fusca* and *Elphidium williamsoni* (= *Cribrononion articulatum* of Lutze) showed clumped distribution with maximum abundance spread over several metres but also sharp gradients in standing crop over a distance of 1 m. In Eckernförder Bay, at a depth of 12 m, the patches were more widely spaced (3 m) but sharp gradients in standing crop on a 1 m scale were present. A more statistical approach was used by Lynts (1966) in this study of Buttonwood Sound, Florida (depth 0.9–2.7 m). The 19 sampling stations showed a range in standing crop of 4–1250 per 18 cm^2 and 9 samples were considered to be statistically different. The other 10 samples include individual species distributed over an area of at least 30 m^2.

Buzas (1968a) selected three 0.1 m^2 sites at a depth of 1 m in Rehoboth Bay, Delaware, and from these he took 28, 34 and 34 samples respectively. The more abundant species showed a clumped distribution which he attributed to asexual reproduction. The rarer species showed a random distribution and it is thought that they represent individuals that had settled out of suspension following transport to the area. In a further study of the same area, Buzas (1970) took samples on a grid of 10 m separation. Of the four species present, only one showed a clumped distribution with a patch size of 100 m^2. The four species together showed a homogeneous distribution over 1500 m^2. In general, the results from these shallow marginal marine environments conform with model 1.

Heald Bank, at a depth of 15–20 m in the Gulf of Mexico, was investigated by Shifflet (1961). Divers collected three samples within a radius of 30 m at four localities. She concluded that species were patchily distributed, but Lynts (1966) re-evaluated the data and inferred that the patches were of at least 2900 m^2 and that microenvironments remained fairly constant at two of the localities. Here the percentage of the living assemblage composed of species common to all three samples was 75–98% and 88–96%. At the other two localities, the values were 38–65% and 35–75%. Thus, this area comes closest to model 2.

The patch size of a single species of *Pelosina* from 32 m water depth in Oslo Fjord was from <50 to 400 cm^2 or 100 to >3200 cm^2 for standing crops of <17 and >21 per 10 cm^2 respectively (Gamito *et al.* 1988).

Buzas (1970) has emphasised how difficult it is to compare the different studies because of the varied sampling and statistical techniques used. However, a general conclusion is that intertidal and shallow marginal marine environments conform to model 1, whereas the deeper, and perhaps more stable, environments are closer to model 2. Furthermore, the size of individual species patches appears to be irregular and small (a few square metres) in shallow water and larger (>2000 m^2) in somewhat deeper water.

In regions of submarine vegetation, there is patchy distribution of benthic foraminifera because of the patchiness and discontinuous distribution of the plants. The lagoons of Curaçao support meadows of *Thalassia*. *Sorites orbiculus* lives in the protected areas, but populations normally cover <100 m^2 of a *Thalassia* field and outside these patches the species is rare or absent. The boundaries may be gradational or very sharp, with a decrease in specimens from 10 to 1 on a *Thalassia* leaf over a distance of 2 m (Kloos 1980).

The intertidal and subtidal salt marsh of Long Island supports large standing crops on various plants, including *Enteromorpha intestinalis*, *Zostera marina* and *Zanichellia palustris*, but they are rare on *Fucus* and *Codium*. The living foraminifera were patchy in distribution, and areas of high standing crop, termed 'blooms' by Lee *et al.* (1969), provided most of the individuals.

As pointed out by Murray (1973), the use of a diversity index eliminates the problem of patchy distribution of individual species. The distribution patterns for individual dominant species will be most reliable for stable environments but less so for variable ones. The distribution patterns of rarer species are not usually meaningful for the small number of samples on which they are normally based.

Depth of life in the sediment

It is well established that many benthic foraminifera are infaunal. There is considerable interest in knowing the depth to which living foraminifera can penetrate below the sediment–water interface. The factors which control this are the grain size of the sediment, the depth to which the oxic surface layer is developed and, presumably, the availability of food (bacteria or organic detritus).

It is generally observed that there is a marked downward decrease in abundance of living forms although certain species are most abundant below the surface few centimetres. In the Jade Bay, Germany, *Elphidium excavatum* forma *selseyense* lives at a depth of 0.5–6.0 cm in sandier sediments, but in muddy sediments where the surface oxidised layer is thin it is excluded (Richter 1964b). In the silty clays of Rhode River, Maryland, *Ammobaculites exiguus* is most abundant between 3 and 8m (Buzas 1974). The top 1–5 cm of the substrate in the Gulf of St Lawrence has living standing crops equal to or larger than that from 0 to 1 cm in 13% of the subsamples examined by Schafer (1971). Estimates of standing crop size based on a sample from the top 2 cm need to be multipled by a factor of 5. Collison (1980) reached a similar conclusion for *Fursenkoina fusiformis* living in the North Sea. But Gooday (1986b) found that 52–71% of the living fauna was in the top 1 cm in the deep sea.

In some cases a vertical distribution is apparent. In the deep sea, *Melonis barleeanum* peaked at 2–5 cm, *Chilostomella oolina* and *Globolulimina affinis* at 7–10 cm (Corliss 1985). However, in a statistical comparison of three cores from shallow water, Buzas (1977) concluded that no vertical distribution was present although it might appear to be so in individual cores. Other authors have observed uniform abundances in the top few centimetres and taken that as evidence of an infaunal mode of life (Brooks 1967; Richter 1964b).

Apart from downward changes in the number of living individuals, there is also clear evidence of a decrease in diversity from both shallow and deep water environments (Table 2.6).

The most remarkable occurrence of foraminifera living in sediment is that from the Kara-Kum Desert (Nikoljuk 1968). Small, thin-walled, protoplasm filled tests of '*Miliammina oblonga arenacea*', *Spiroloculina turcomanica* and *Fischerina* sp. were found living interstitially in waters of <30‰ salinity and temperatures of 17–20 °C. Previously a sea covered the area in the Pliocene.

Some species which normally live at the surface may become buried during storms. Because they are negatively geotropic, they try to crawl back to the surface. *Elphidium crispum* withstands burial to 1 cm (Myers 1943) but *Ammonia beccarii* can withstand burial to 8 cm (Lee *et al.* 1969) which is not perhaps surprising as the species is probably infaunal.

Experimental studies on *Quinqueloculina impressa* (Severin *et al.* 1982) showed that, following burial to a depth of 5 cm under silt or silty clay, individuals moved to the sediment surface over a period of hours. Two types of burrow were produced: straight vertical burrows formed during movement towards the sediment surface and a complex of near-horizontal burrows in the top 1 cm of sediment. It appears that this species is both epifaunal and infaunal to a depth of 1 cm. During periods of enhanced sedimentation it has the ability (negative geotropism and power of movement) to regain the near sediment surface. The mathematics of movement have been discussed by Severin and Erskian (1981). They concluded that the velocity of *Q. impressa* is inversely proportional to depth. In the surface layer, higher velocities enable individuals to escape from a rising anoxic layer and also allow increased foraging rates. These may be the reasons why benthic foraminifera do not live in large numbers far below the sediment surface. Further experimental

Table 2.6 Downward decrease in diversity within bottom sediment (from *Boltovskoy 1966, and †Corliss 1985)

Depth (cm) below sed. surface	Number of species	
	Intertidal*	Deep sea†
0–2	28	11
4–6	22	4
8–10	14	4
12–14	12	3
16	3	

Table 2.7 Relationship between environment and test architecture (partly after Brasier 1982)

Test Architecture	Salinity			Energy	
	Hypersaline	Normal marine	Brackish	High	Low
1. *Non-septate*					
(a) Contained growth		x	> 1 mm x		> 1 mm x
(b) Continuous growth		x			x
2. *Septate*					
(a) Uniserial		x			x
(b) Biserial		x	Rare	x	x
(c) Planispiral	x	x	x	x	x
(d) Trochospiral	x	x	x	x	x
(e) Miliolid	x	x	*Miliammina*	x	*Miliammina*
(f) Annular, cyclic	x	x		x	x
(g) Fusiform planispiral		x		x	

x = examples present

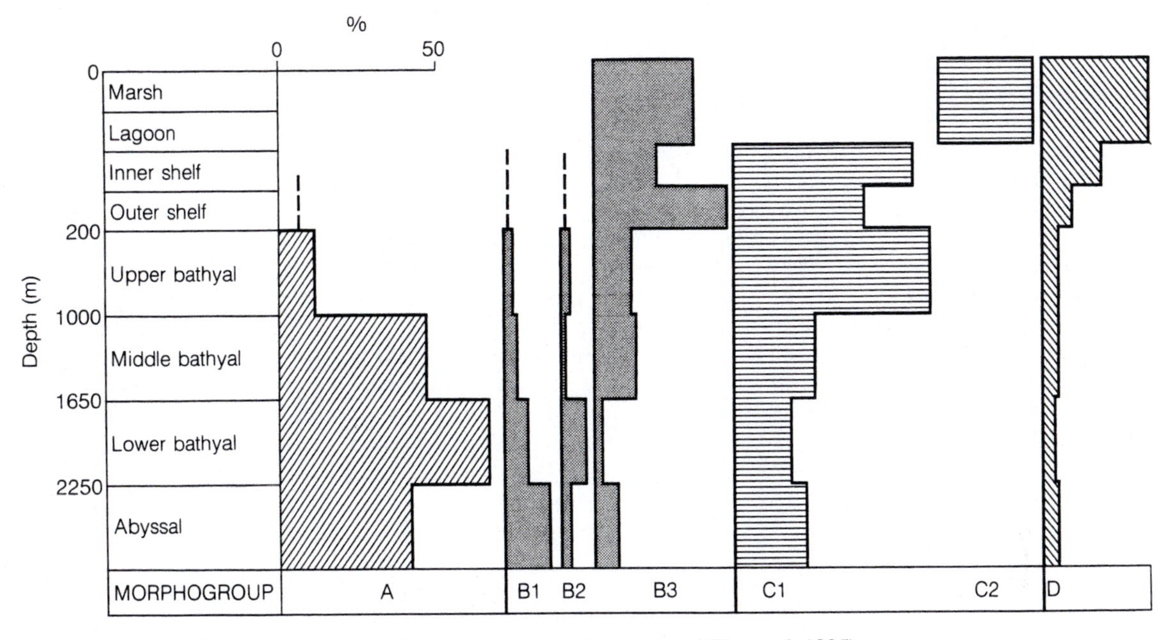

Figure 2.5 Distribution of agglutinated morphogroups (after Jones and Charnock 1985)

studies on the same species were somewhat inconclusive with respect to direction of movement, but they did show that when the foraminifera are of a size which requires them to displace sand grains during burrowing, then the rate of movement is constrained by the effective sediment overburden pressure (Severin 1987). Speeds of 0.5–7.7 mm^{-1} were recorded by Wetmore (1988) using time-lapse cinematography. Some individuals moved sand grains larger than themselves.

The occurrence of infaunal benthic foraminifera at depths below the top 1 cm of sediment is of considerable importance for sampling techniques and ecology.

Environment				
Lagoon	Nearshore	Shelf	Slope	Deep Sea
x	x	x		
	Spirillina		Straight and branched tubes (*Bathysiphon*)	
		x	x	x
	x	x	x	x
x	x	x	x	x
x	x	x	x	x
Miliammina	x	x	Rare	Rare
x	x			
x	x			

Test form and environment

Many organisms show a good correlation between their form and the environment in which they live. To a degree this is true of benthic foraminifera but the pattern is far from simple because, as with most ecological matters, shape is a compromise between a complex of requirements. An early attempt to relate form and environment is that of Bandy (1964a) but it is not reproduced here because it has been subsequently shown to be incorrect in several respects. A more recent attempt is that of Brasier (1982) summarised in Table 2.7. This shows that hypersaline and brackish environments exhibit a smaller ranger of morphological variety in their foraminiferal faunas than normal marine environments.

The classification of agglutinated foraminifera into morphogroups (already discussed, p. 18) shows that in modern environments they have a distinctive distribution. Group A (unilocular, tubular or branching) are characteristic of the deep sea. Group B3 (multilocular, planispiral/trochospiral, lenticular) are particularly common in shelf and marginal marine environments. Group C2 (elongate, quinqueloculine, e.g. *Miliammina*) is characteristic of marshes and lagoons (Fig. 2.5).

Ecophenotypes

Although it may be convenient to suppose that all populations of a given species will respond in the same way to the environmental controls, there is evidence to suggest that this is not always the case. Populations of some species are adapted to the environmental conditions of their local habitat – ecophenotypes. Furthermore, some ecophenotypes are morphologically distinct.

Experimental studies have shown that *Ammonia beccarii* can give rise to as many as seven morphotypes from a single parent stock and that one of these, *A. tepida*, has local populations adapted to local conditions (Schnitker 1974a) such as those described by Chang and Kaesler (1974). A complex of *Elphidium* 'species' from the Atlantic coasts of North America and Europe has provided a taxonomic minefield and has been interpreted by Feyling-Hanssen (1972), Poag *et al.* (1980) and Miller *et al.* (1982) as a group of ecophenotypes. The latter authors found that five ecophenotypes were both morphologically distinct and had different ecological requirements (Table 2.8). Poag *et al.* (1980) found living representatives of *E. excavatum* forma *clavatum* on the New Jersey shelf but another presumed ecophenotype, *E. excavatum* forma *album* (= forma *selseyensis*, Table 2.9), which was recorded only dead was thought possibly to live there during the summer and autumn.

Marginal marine environments, because of their very variable environmental conditions, are the sites of much ecophenotypic variation. Poag (1978) concluded that paired ecophenotypes exist in five species (Table 2.8). The calcareous species have one ecophenotype having a small test, thin walls and few chambers, considered to be living under optimum conditions, and another ecophenotype having a larger test, thick walls and more chambers, considered to live under less favourable conditions. Wang and Lutze (1986) drew attention to the presence of two ontogenetic stages in certain hyaline foraminifera; an early stage with compact, thick-walled chambers and a late stage with inflated, thin-walled chambers. Some ecophenotypes have only the early stage while others have variable developments of the late stage. They made observations on 11 species some of which, e.g. *Elphidium excavatum*, have already been discussed above. In the case of *Hyalinea balthica* the typical

Table 2.8 Ecophenotypes of *Elphidum excavatum* (from *Miller *et al.* 1982, and †Feyling-Hanssen 1972)

Forma *	Environment *	Distribution†
clavatum	Normal marine to slightly brackish, cold	Arctic
excavatum	Intertidal	
selseyensis	Estuarine, temperate to polar (1 to 16°C)	Boreal
lidoensis	Estuarine, warm to temperate	Lusitanian
magna	Nearshore, turbulent zone	—

Note: *Elphidium excavatum* forma *album* of Feyling-Hanssen (1972) is considered to be *E. excavatum* forma *selseyensis* by Miller *et al.* (1982).

Table 2.9 Paired ecophenotypes (formae) in Gulf of Mexico lagoons (from Poag 1978)

Species	Maximum salinity, maximum temperature	Minimum salinity, minimum temperature
Ammonia parkinsoniana	*tepida*	*typica*
Elphidium gunteri	*typicum*	*salsum*
Elphidium galvestonense	*typicum*	*mexicanum*
Palmerinella palmerae	*typica*	*gardenislandensis*
Ammotium salsum	*exilie*	*typicum*

form is of the compact variety, but an inflated type is widely distributed in shelf seas and has been given a separate name, *H. florenceae*.

A more objective approach using image analysis has enabled Gary and Healy-Williams (1988) to relate changes in individual features of shelf and slope *Bolivina* and *Uvigerina* species to particular environmental parameters. For instance, in *U. mediterranea* chamber lobateness is reduced in individuals from the lower boundary of the oxygen minimum zone.

It is probable that ecophenotypy is common in environments of all depths. Its recognition is important because it affects the assessment of diversity and because the formae have distinct ecological preferences which may be useful in palaeoecological interpretations, e.g. use of ecophenotypes of *Elphidium excavatum* in Quaternary studies. However, not all paired morphotypes are necessarily ecophenotypic. The two variants of *Nodulina dentaliniformis* (= *Reophax dentaliniformis*) which occur in the east and west Baltic have been interpreted by Werdelin and Hermelin (1983) as genetically determined due to the water circulation pattern restricting gene flow between the two groups.

CHAPTER 3

Stable isotope studies

Introduction

Studies of the stable isotopes of oxygen (^{18}O and ^{16}O) and carbon (^{13}C and ^{12}C) are proving to be of considerable interest in helping to interpret life processes of benthic foraminifera, their relationship with the ambient water and as a guide to palaeoceanography. This is because the tests of calcareous foraminifera preserve a record of the stable isotopic composition of the water in which they grew.

Analytical results are normally expressed using the δ notation:

$$\delta = \frac{\text{Isotopic ratio of sample} - 1}{\text{Isotopic ratio of standard}} \times 1000$$

A value of $\delta = +1.0$‰ indicates that the ratio is 1.0‰ higher than the standard. Seawater samples are normally standardised against Standard Mean Ocean Water (SMOW). Shell material is standardised with reference to the rostrum of a belemnite from the Cretaceous Pee Dee Formation of South Carolina (PDB). SMOW has $\delta^{18}O = -0.2$‰ PDB but the $\delta^{13}C$ of both is 0.0.

If the hydrographic and isotopic data of the bottom water are known, it is possible to calculate the $\delta^{18}O$ and $\delta^{13}C$ of equilibrium calcite. Any value from a calcite test that differs from the equilibrium value of its ambient water shows fractionation or disequilibrium, expressed as $\Delta\delta^{18}O$ and $\Delta\delta^{13}C$. Disequilibrium is calculated by subtracting the equilibrium δ value from the measured value (see Graham *et al.* 1981).

For a study of living material, it is essential that the samples are not stored in formalin (Ganssen 1981). Analyses are commonly carried out on a small number of individuals (<10) and in core samples mixing of tests through bioturbation may cause errors greater than the instrument error (Boyle 1984).

Biogenic carbonate

Biogenic carbonate shows relative variations in the proportions of oxygen isotopes dependent on:

1. The mineral phase secreted, i.e. calcite or aragonite;
2. The isotopic composition of the seawater;
3. The temperature of seawater during life;
4. Vital effects.

All modern calcareous benthic foraminifera are calcitic with the exception of the suborder Robertinina.

The isotopic composition of seawater is influenced by the ratio of evaporation to precipitation (particularly in the region of trade winds and the equator), and by dilution from surface runoff (which in high latitudes causes a strong lightening of oxygen isotopes as salinity decreases) (Turekian 1969).

The relationship between temperature and isotopic composition was first established by Epstein *et al.* (1953). A change in water temperature of 1 °C alters the $\delta^{18}O$ value of shell carbonate by 0.25‰.

Ideally, biogenic calcite would be secreted in isotopic equilibrium with the ambient seawater. However, commonly this is not so and the disequilibrium is caused by life processes collectively known as vital effects. These include:

1. Uptake of metabolic CO_2 during calcification;
2. Growth or calcification rate;
3. Physiological changes with ontogeny;
4. Kinetic isotope effects in the transport of carbonate ions to the site of calcification;
5. Photosynthetic activity of symbionts.

All these vital effects may lead to carbonates depleted

in ^{18}O relative to equilibrium conditions (Williams *et al.* 1981; Grossman 1987).

Variations in the proportions of ^{13}C : ^{12}C in biogenic carbonates are due to:

1. Global and regional changes in surface-water productivity;
2. Different water masses and circulation patterns;
3. Vital effects;
4. Microhabitat effects.

During photosynthesis, ^{12}C is used preferentially hence there is relative depletion of ^{13}C compared to CO_2 in seawater. In the oceans, photosynthesis takes place in the surface waters. Some of the organic matter sinks into the deeper water where it becomes oxidised. Thus, the surface waters are enriched in ^{13}C and the deeper waters are enriched in ^{12}C. The world average surface water has $\delta^{13}C$ = +2.0‰ PDB, but the deeper waters have lower values: $\delta^{13}C$ = +1.0 to +0.5‰ PDB for North Atlantic Deep Water (NADW), +0.5 to –0.5‰ for the deep Pacific and +0.5 to 0.0‰ for the Indian Ocean (Kroopnick 1985).

There is perhaps poor understanding of carbon isotopic equilibrium and this may partly explain why nearly all the benthic foraminifera studies show ^{13}C disequilibrium (Grossman 1987). Nevertheless, it appears that vital effect has a greater influence on $\delta^{13}C$ than on $\delta^{18}O$.

The term 'apparent oxygen utilisation' (AOU) is given to the quantity of oxygen used up in respiration and decay. This shows a straight-line relationship in modern ocean waters (Fig. 3.1). The difference between the Atlantic and Pacific oceans is that the deep waters of the Atlantic are young (well flushed) whereas those of the Pacific are older due to the slow circulation. It has been calculated that a shift of 1‰ $\delta^{13}C$ implies a loss of 3 ml l^{-1} oxygen (Berger and Vincent, 1986).

The AOU is inversely related to the $\delta^{13}C$ of the dissolved organic carbon (DIC). *Fontbotia wuellerstorfi* covaries with AOU (Belanger *et al.* 1981) and therefore with DIC. Grossman (1987) replotted the data with those of *Uvigerina* spp. and shower that those of *F. wuellerstorfi* approximate those of DIC whereas *Uvigerina* $\delta^{13}C$ values are ~0.6‰ less (Fig. 3.2).

It has been observed that off southern California the $\delta^{13}C$ values of living benthic foraminifera and ambient $\delta^{13}C$ DIC covary as a function of depth. Further, species of *Cassidulina* which are in ^{18}O equilibrium with the water, show a decrease in $\delta^{13}C$ with depth and are almost identical to those of DIC. Species not in ^{18}O equilibrium show no decrease in $\delta^{13}C$ with depth and therefore provide a less reliable record of past ^{13}C in the oceans (Grossman 1984ab, 1987).

Another complicating factor is the microhabitat effect (Grossman 1987). Oxidation of organic carbon within the sediment lowers the ^{13}C composition of

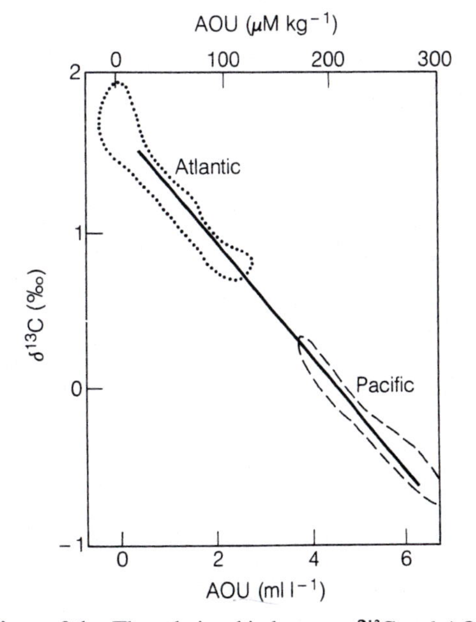

Figure 3.1 The relationship between $\delta^{13}C$ and AOU in ocean water (after Berger and Vincent 1986)

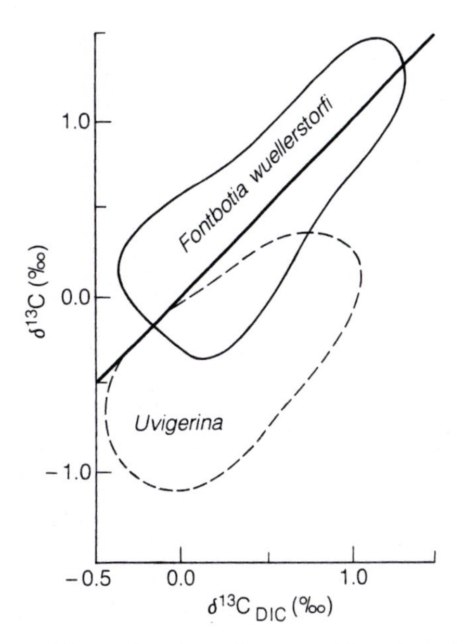

Figure 3.2 Carbon isotopic composition of two benthic taxa as a function of the $\delta^{13}C$ of the DIC. The straight line is a 1 : 1 correspondence (after Grossman 1987)

DIC. Off the southern California borderland where the sediment has a high organic content, the $\delta^{13}C$ of DIC in the upper 2 cm may be 2–3 ‰ less than that of the bottom water. Thus, for infaunal benthic species measurement of the DIC $\delta^{13}C$ of the pore water is more meaningful than that of the bottom water overlying the sediment. *Globobulimina pacifica* has an equilibrium ^{18}O value and a low $\delta^{13}C$ value due to the microhabitat effect (Grossman 1984a).

In summary, most biogenic carbonate shows ^{13}C disequilibrium due to vital effects. The maxium depletion is generally <2‰ with respect to DIC $\delta^{13}C$. Some species have compositions more or less equal to that of DIC, namely, *Pyrgo murrhina*, *Quinqueloculina auberiana*, *Fontbotia wuellerstorfi*, *Cibicidoides* spp. and *Cassidulina* spp. For ^{18}O some species are in equilibrium, but those showing disequilibrium are depleted by 0.5–1.0‰ and only rarely by >1.5‰. Taxa adapted to survival in low oxygen conditions, e.g. Buliminacea and Cassidulinacea, tend to precipitate tests in oxygen isotopic equilibrium with the ambient water (Grossman 1987).

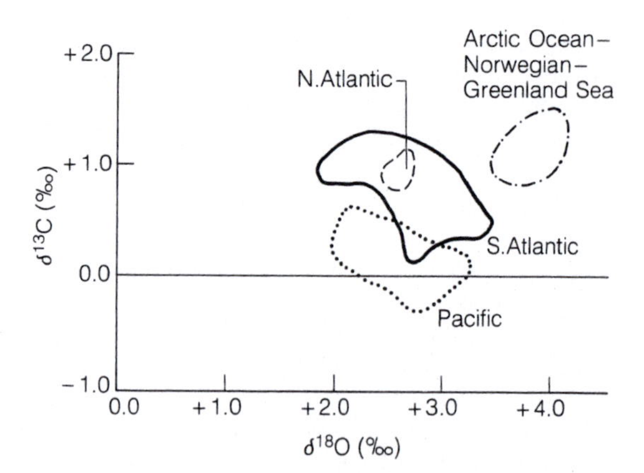

Figure 3.3 Isotopic data for *Fontbotia wuellerstorfi*. Arctic Ocean (Aksu and Vilks 1988), Norwegian–Greenland Sea (Belanger *et al.* 1981), Atlantic (Graham *et al.* 1981; Belanger *et al.* 1981; Ganssen 1983; samples deeper than 2000 m, Williams and Healy Williams 1984, NADW), Indian and Pacific oceans (Graham *et al.* 1981)

Deep-sea forms

Stable isotope studies have been restricted to relatively few species and genera. The $\delta^{18}C$ values range from +1.90 to +5.07 but most values are between +2.0 and +4.0‰. For $\delta^{13}C$ the range is from −2.70 to +2.32 with most values between −1.0 and 0.0‰. The most commonly studied species is *Fontbotia wuellerstorfi*. On a $\delta^{13}C/^{18}O$ plot this falls into three fields: the Arctic Ocean–Norwegian–Greenland Sea, the Atlantic and Pacific oceans (Fig. 3.3).

Deep-sea foraminifera live in an aphotic environment and are not therefore subject to the vital effect caused by symbiosis. Nevertheless, their calcitic tests still show disequilibrium. *Fontbotia wuellerstorfi* from the Atlantic and Pacific oceans have disequilibrium values ($\Delta\delta^{18}O$) with a range of −0.70 to −1.50, and mean −1.01‰ + 0.16 (1σ) (Graham *et al.* 1981). Other species with similar standard deviation values are *Oridorsalis tener*, *Uvigerina* spp., *Nuttallides umboniferus* and *Hoeglundina elegans* (aragonite test). *Pyrgo murrhina* has a $\delta^{18}O$ closest to ^{18}O equilibrium ($\Delta\delta^{18}O$ −0.39) but it has a large intraspecific standard deviation (0.28).

The $\Delta\,\delta^{18}O$ of *Fontbotia wuellerstorfi* (*Cibicides* of Aksu and Vilks 1988 and Belanger *et al.* 1981) in the Arctic Ocean–Norwegian–Greenland Sea is also around −1.0‰. Thus, it records the expected difference of 1.0‰ $\Delta\delta^{18}O$ calcite between the Norwegian–Greenland Sea and the Pacific. However, disequilibrium between *F. wuellerstorfi* and *Oridorsalis tener* and *Pyrgo* spp. is not constant in the Norwegian–Greenland Sea.

This has important consequences for the construction of down-core isotope curves (Belanger *et al.* 1981). It is clearly desirable to use a single species rather than a mix of different species. Woodruff *et al.* (1980) drew attention to the fact that the mean oxygen isotopic difference between *Uvigerina* spp. and *Pyrgo* spp. reported by different authors suggests that there may be a species-specific isotopic effect operating. On the evidence presented above, *Fontbotia wuellerstorfi* appears to be the species with the least intraspecific variation in ^{18}O. Palaeotemperature estimates based on this species would have a precision of ±0.7 °C (Graham *et al.* 1981).

Nevertheless, Shackleton (1974) concluded that at temperatures >7 °C *Uvigerina* spp. secrete their tests in oxygen isotopic equilibrium with the seawater in which they live. Woodruff *et al.* (1980) reached a similar conclusion. However, Ganssen (1983) found that off the northwest margin of Africa, *Uvigerina* $\delta^{18}O$ values differed from those of seawater by +1.0 to −0.5‰. Furthermore, for $\delta^{13}C$, *U. peregrina* is 0.38‰ lighter than *U. finisterrensis*. Results standardised to *U. finisterrensis* were compared with the bottom water and differed by +0.3 to +1.12‰. These differences can be correlated with water masses. In the

Gulf of Mexico, different morphotypes of *U. peregrina* from a single box core differ in their isotopic values and this casts doubt on their value in palaeoceanographic reconstructions (Williams *et al.* 1988). Furthermore, Dunbar and Wefer (1984) found that the isotopic values varied with specimen size.

In samples from the Atlantic, Pacific and Indian oceans, *Fontbotia wuellerstorfi* is nearly 1.0‰ less than the equilibrium ^{13}C value, but this is close to the fractionation between dissolved bicarbonate and solid $CaCO_3$ at the prevailing bottom-water temperature (Graham *et al.* 1981), i.e. close to $\delta^{13}C$ in bottom-water total CO_2 *Pyrgo murrhina* (Graham *et al.* 1981), *Pyrgo* spp., *Laticarinina pauperata* and *Nuttallides umboniferus* are all within –0.5‰ $\Delta\delta^{13}C$ (Woodruff *et al.* 1980).

In the Norwegian–Greenland Sea, *Fontbotia wuellerstorfi* has the smallest disequilibrium ($\Delta\delta^{13}C$ = –0.95‰). Belanger *et al.* (1981) point out that the morphologically similar genus *Cibicides* has an attached mode of life and that this may also apply to *Fontbotia*. This has been confirmed by Lutze and Thiel (1989).

Species showing higher $\Delta\delta^{13}C$ are thought to live in micro-environments depleted in ^{13}C due to oxidation of organic matter within the sediment (Woodruff *et al.* 1980; Belanger *et al.* 1981). This is shown graphically in Fig. 3.4 where the foraminifera are placed according to their $\delta^{13}C$ values. Species may, of course, occupy more than one microhabitat during the course of their life.

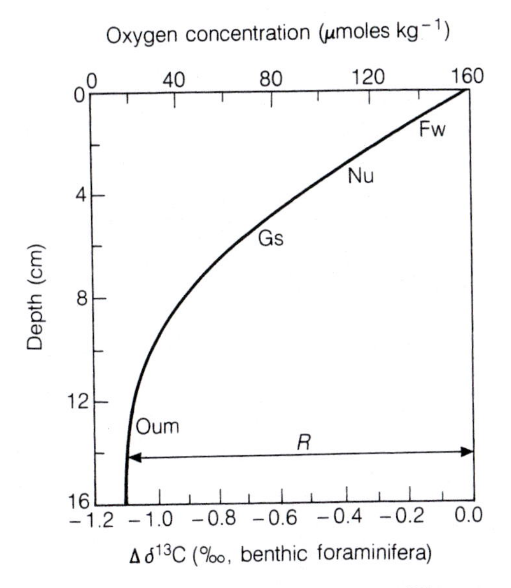

Figure 3.4 Explanation of differences in $\delta^{13}C$ values of tests based on hypothetical microhabitats of infaunal foraminifera: Fw = *Fontbotia wuellerstorfi*, Nu = *Nuttallides umboniferus*, Gs = *Globocassidulina subglobosa*, U = *Uvigerina* spp., Oum = *Oridorsalis umbonatus* (based on Vincent *et al.* 1981)

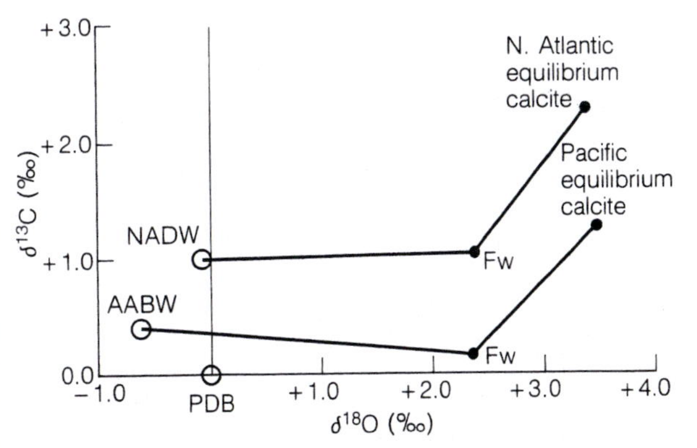

Figure 3.5 The relationship between the stable isotope signature of the bottom water, calcite precipitated in equilibrium with the water and biogenic calcite secreted in disequilibrium by *Fontbotia wuellerstorfi* (based on data from Graham *et al.* 1981)

The relationship between the stable isotope signature of bottom water, calcite precipitated in equilibrium with the water and biogenic calcite secreted in disequilibrium is shown in Fig. 3.5. The pattern is similar in both oceans, but the Pacific has a lower $\delta^{13}C$ value because the bottom water is further from source (i.e. older) than that of the Atlantic.

Continental slope

Data from the northwest African margin show differences between *Uvigerina finisterrensis* which lives from <200 to >2000 m and *U. peregrina* >470 m (Fig. 3.6). This is an area of complex hydrography with waters derived from north and south.

The southern California borderland is topographically complex with a series of basins separated by banks and shallow sills. At depths >100 m, the water shows little variation in $\delta^{18}O$ so temperature is the dominant abiotic factor affecting the oxygen isotopic composition of the benthic foraminifera. Grossman (1984a) recognised four groups:

I *Hoeglundina elegans*
II *Brizalina argentea*
Cassidulina braziliensis
III *Lenticulina cushmani*
Lenticulina sp.

Cassidulina limbata
Cassidulina tortuosa
Uvigerina curticosta
Uvigerina peregrina
Quinqueloculina vulgaris
Triloculina trigonula
Pyrgo sp.
IV *Globobulimina pacifica*

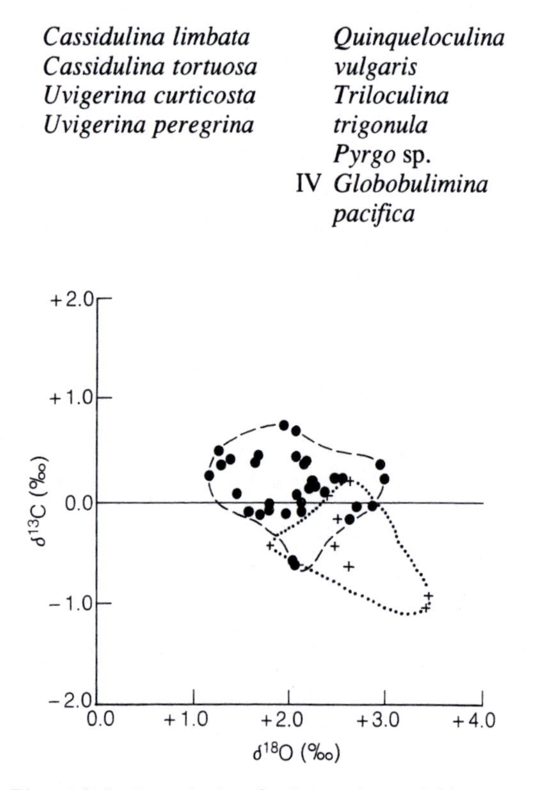

Figure 3.6 Isotopic data for the northwest African continental slope (data from Ganssen 1983)

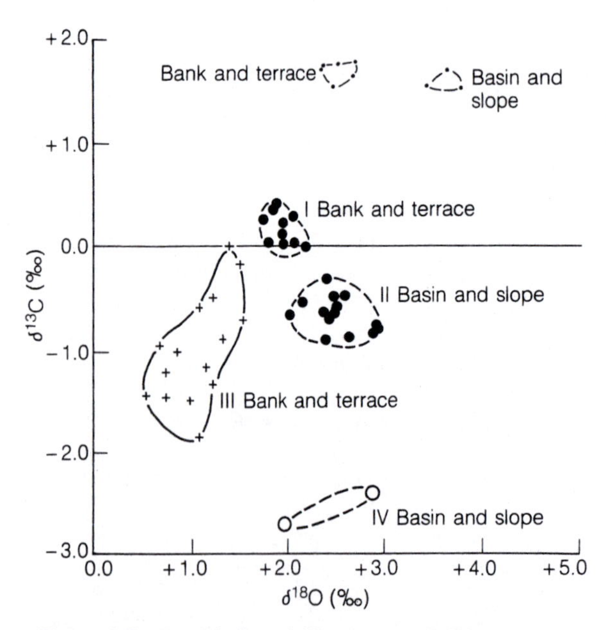

Figure 3.7 Benthic foraminiferal groups I–IV, southern California borderland (data from Grossman 1984a)

Group I, *Hoeglundia elegans*, is aragonitic and shows greatest enrichment in both ^{18}O and ^{13}C (Fig. 3.7). Group II are all calcitic and show enrichment in ^{18}O and ^{13}C in the bank and terrace samples, but enrichment in ^{18}O and depletion in ^{13}C in the basin and slope samples. Group III shows some enrichment in ^{18}O but small to moderate depletion in ^{13}C, while group IV shows considerable depletion in ^{13}C. Calculation of $\Delta^{18}O_{eq}$ and $\Delta^{13}C_{DIC}$ shows that group I is enriched in both ^{18}O (0.78‰) and ^{13}C (1.66‰). Group II has average $\Delta^{18}O_{eq}$ close to zero indicating no fractionation and $\Delta^{13}C_{DIC}$ is similar. Group III shows disequilibrium in ^{18}O and ^{13}C. Although group IV is in equilibrium with respect to ^{18}O it is in marked disequilibrium with respect to ^{13}C and this may be due to the microhabitat effect (infaunal).

Shelf seas – smaller foraminifera

Two genera, *Planorbulinella (*with symbionts) and *Bolivina* (without symbionts) from the Gulf of Aqaba show somewhat dissimilar results. Both have $\delta^{18}O$ close to 0.0, but the former has $\delta^{13}C = +1.92$‰ while the latter has $\delta^{13}C = +0.28$ to +1.28‰ (Buchardt and Hansen 1977). *Bolivina* was said to show a 0.6‰ depletion of ^{18}O attributed to vital effects. *Uvigerina finisterrensis* and *Trifarina fornasini* from the shelf (<200 m) of northwest Africa plot in separate fields (Fig. 3.8). Shallow-water foraminifera from the Mediterranean show a difference of $\delta^{18}O$ +2.0‰ even between species from the same sample (*Quinqueloculina viennensis* $\delta^{18}O$ −0.90‰, *Brizalina spathulata* $\delta^{18}O$ +0.79‰, *Valvulineria brady* $\delta^{18}O$ +0.49‰, *Am-*

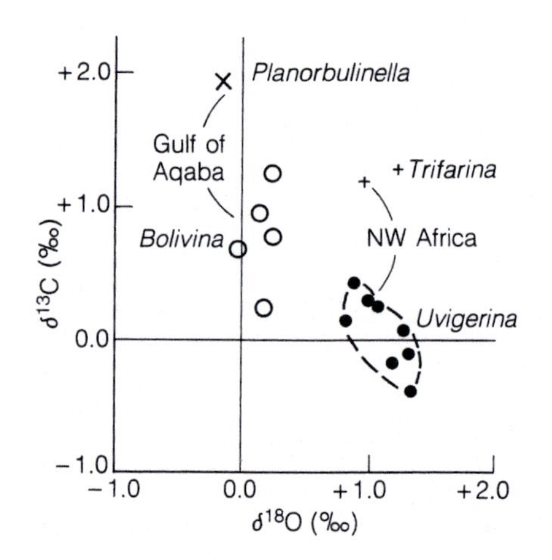

Figure 3.8 Plot of shelf sea stable isotope data

monia beccarii $\delta^{18}O$ +1.14‰; Vinot-Bertouille and Duplessy 1973). The $\delta^{18}O$ of equilibrium calcite in this case was ±1.7 +0.5‰ so all the values are depleted.

Larger foraminifera

Because of their contained symbionts, larger foraminifera have attracted particular interest. From an analysis of the stable isotopes of growth stages of various larger foraminifera, Wefer and Berger (1980) and Wefer *et al.* (1981) concluded that the oxygen isotope signal records seasonal temperature variation. Summer values are higher and winter values lower (Fig. 3.9).

Data from all oceans are given in Fig. 3.10. After correcting their data for seasonal temperature changes in the Gulf of Aqaba, Buchardt and Hansen (1977) found that the larger foraminifera had a constant ^{18}O depletion of 0.6‰ over the sampled depth range. They attributed this to a vital effect, the incorporation of metabolic CO_2 into the test carbonate. In addition, another 1.5‰ ^{18}O depletion was observed in symbiont-bearing species. They noted its positive correlation with light intensity and suggested that photosynthetic oxygen from the symbionts is incorporated into the test. The oxygen is derived from the metabolic uptake of CO_2. The same conclusion was reached by Williams *et al.* (1981) from their studies of field and laboratory clones of *Heterostegina depressa*. They demonstrated that the calcification rate was not an important vital effect. Thus, it may be concluded that the $\delta^{18}O$ values of test carbonate in the Gulf of Aqaba reflects productivity rather than temperature (Fermont *et al.* 1983).

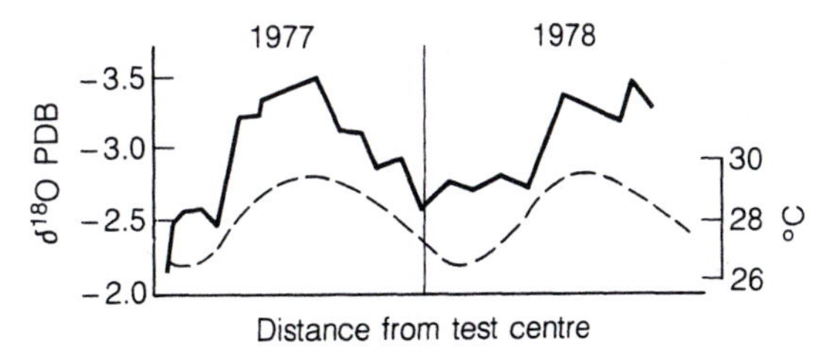

Figure 3.9 $\delta^{18}O$ of *Marginospora vertebralis* showing changes during growth and temperature (from Wefer and Berger 1980)

The alternative interpretations of the $\delta^{18}O$ signal as temperature controlled or production controlled

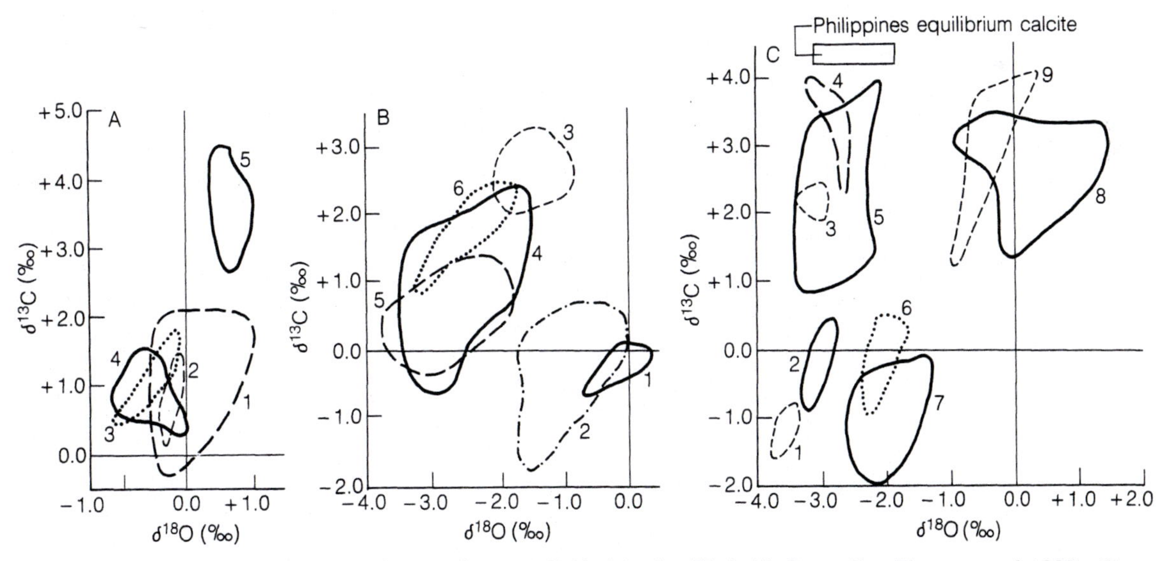

Figure 3.10 Isotopic plots for larger foraminifera. (A) Gulf of Aqaba (Elat). (1) *Operculina* (Fermont *et al.* 1983), (2) *Operculina*, (3) *Heterostegina*, (4) *Amphistegina* (Buchardt and Hansen 1977), (5) *Amphisorus hemprichii* (Luz *et al.* 1983). (B) (1,2) Persian Gulf, (3–6) Indian Ocean. (1) *Operculina*, (2) *Heterostegina depressa* (Wefer *et al.* 1981), (3) *Alveolinella quoyi*, (4) *Amphistegina radiata*, (5) *Baculogypsina sphaerulata*, (6) *Sorites marginalis* (Vinot-Bertouille and Duplessy 1973). (C) (1–5) Philippines, Pacific Ocean, (6–7) Hawaii, Pacific Ocean, (8–9) Bermuda, Atlantic Ocean. (1) *Calcarina spengleri*; (2) *Heterostegina depressa*; (3) *Parasorites orbitolitoides* cf. *monensis*; (4) *Laevipeneroplis proteus*; (5) *Marginopora vertebralis* (Wefer *et al.* 1981); (6) *Heterostegina depressa* (Zimmerman *et al.* 1983); (7) *H. depressa* (Williams *et al.* 1981); (8) *Cyclorbiculina compressa*; (9) *Archaias angulatus* (Wefer *et al.* 1981)

have not yet been reconciled. However, studies of both field specimens and laboratory cultures of *Heterostegina depressa* from Hawaii (Zimmerman *et al.* 1983) showed that tests formed at higher light intensities (1500 lx) resulted in further depletion in ^{18}C and ^{13}C due to higher rates of photosynthesis by the intracellular symbiotic algae.

There is some evidence that certain porcellaneous tests are close to equilibrium (Vinot-Bertouille and Duplessy 1973; Erez 1978; Wefer and Berger 1980; Wefer *et al.* 1981; Luz *et al.* 1983; Reiss and Hottinger 1984). This may be due to the fact that these forms have symbionts with very low photosynthetic rates. However, as pointed out by Luz *et al.* (1983), disequilibrium may depend on local conditions; tests from Elat and Bermuda are in equilibrium; those from the Philippines are depleted in ^{18}O. Productivity is higher in the Philippines than in Bermuda or the Gulf of Aqaba.

The stable isotopes of carbon appear to reflect metabolic activity but there are differences between porcellaneous and hyaline genera. The porcellaneous forms have $\delta^{13}C$ values of ~2.5‰ lighter than expected equilibrium values (+3.3 to +4.1‰), whereas the hyaline forms (*Heterostegina*, *Operculina*, *Calcarina*) are >2‰ lighter than expected equilibrium values. With age, the porcellaneous forms have higher $\delta^{13}C$ values while the hyaline forms (other than *Operculina*) have heavier $\delta^{13}C$ values.
These differences have been interpreted as an index of the amounts of metabolic CO_2 used for test construction (Wefer *et al.* 1981; Grossman 1987). More metabolic CO_2 is incorporated in the test with increasing photosynthesis of the symbionts, hence the carbon isotopic composition becomes lighter (Erez 1978). It thus appears that the photosynthetic rates of foraminifera with a hyaline wall structure is greater than that of those with porcellaneous walls. This is supported by experimental evidence.

The trend towards heavier carbon values in mature specimens of *Heterostegina* and *Calcarina* may represent a decrease in disequilibrium due to the slowing down of the growth rate. The trend towards lighter carbon values in porcellaneous forms is more difficult to explain. Wefer *et al.* (1981) postulated two causes: an increase in dependency of symbiotic algae during growth, and an increase in growth rate. They note that *Cyclorbiculina compressa* shows a late stage increase in growth with the formation of reproductive chambers.

Somewhat different results were obtained by Luz *et al.* (1983) from *Amphisorus hemprichii*. The $\delta^{13}C$ values were obtained on specimens collected from seagrass meadows in the Gulf of Aqaba. All fall within or above the equilibrium range ($\delta^{13}C$ + 3.2 to +4.1‰). They attribute this to the role of micro-environments. During spring and summer, ^{12}C is preferentially consumed during photosynthesis, so the water close to the seagrass leaves is enriched in ^{13}C. Specimens attached to the stems are depleted in ^{13}C by up to 1.0‰. However, an alternative to the photosynthesis explanation is that food sources on different parts of the plants may vary.

The role of food was discussed by Fermont *et al.* (1983). They found that *Operculina* from the Gulf of Aqaba had $\delta^{13}C$ values that did not correlate with any of their measured variables except that involute forms had more positive $\delta^{13}C$ than evolute forms. They attributed this to a difference of food requirements and ecological niches.

Summary of modern data

The fields of all environments for calcitic benthic foraminifera are shown on Fig. 3.11. Apart from any other controls, vital effects contribute to the very variable field of larger foraminifera, many of which are depleted in ^{18}O and enriched or depleted in ^{13}C. The limited data on shelf seas show some enrichment in ^{18}O and near equilibrium in ^{13}C. The continental slope of the Atlantic and the southern California borderland show enrichment in ^{18}O and enrichment or depletion in ^{13}C, depletion being especially marked in infaunal species. The deep-sea faunas of the Pacific, Atlantic and Norwegian–Greenland Sea are enriched in ^{18}O and generally show some enrichment in ^{13}C.

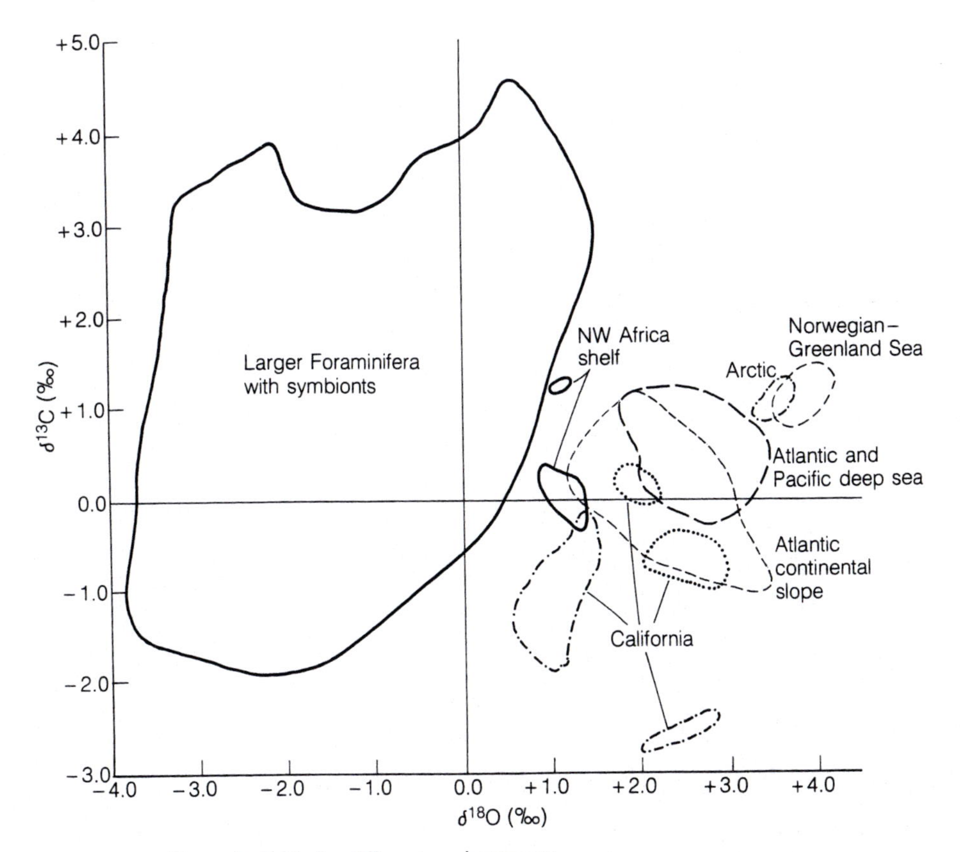

Figure 3.11 Summary of isotopic fields for different environments

CHAPTER 4

Population dynamics

Introduction

Communities are complex associations of species living together with some degree of interdependence. In this book, groups of species of benthic foraminifera living together and forming parts of communities are termed *assemblages*. Each species has its *population* of individuals. The term 'population dynamics' is used to describe the study of changes through time in the number of individuals of a species and the causes of those changes. It is important to the understanding of the development of the dead assemblages which have the potential to become fossilised.

Ecologists recognise four basic parameters in the study of population dynamics: birth (*B*), death (*D*), immigration (*I*) and emigration (*E*). A simple equation to describe changes over a period of time is

$$N_{t+1} = N_t + B - D + I - E,$$

where N_t is the number of individuals at time *t*, and N_{t+1} the number one time-period later at time t_{+1}, in the same unit area of habitat.

It is convenient to disregard gains and losses through immigration and emigration as benthic Foraminifera move very slowly and in random directions. If it is assumed that gains and losses are in-balance, the equation can be simplified to

$$N_{t+1} = N_t + B - D.$$

Three different results are possible:

1. Population increase (positive), where births exceed deaths;
2. A steady state, where births equal deaths;
3. A population decrease (negative), where births are exceeded by deaths.

Even where birth- and death-rates are not known, the relative importance of these two parameters can be gauged according to whether there has been an increase or a decrease of population size. In the short term, fluctuations in population size would be expected, especially in environments subject to seasonal variation. In the long term, the fluctuations may be smaller but there may still be changes from year to year and over periods of years.

Births

Foraminifera breed sexually and asexually but the details of the reproductive activity are known for very few species. However, asexual reproduction seems to be more common than sexual reproduction. The term 'fecundity' has been defined as 'the number of young produced by a parent individual during asexual multiple fission' (Hallock 1985). Values range from 12 to 2400 (Table 4.1) but in general smaller species

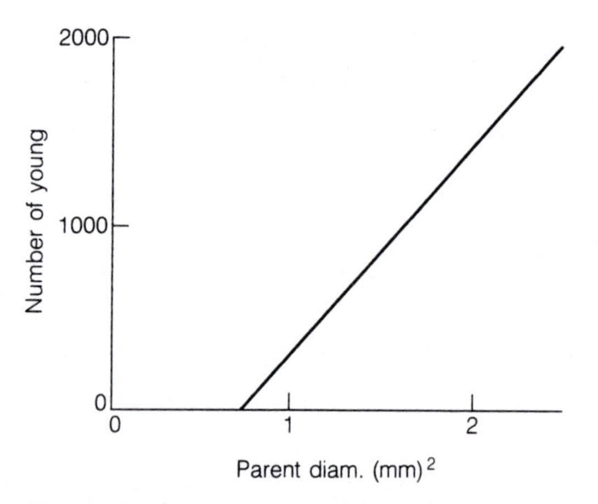

Figure 4.1 Size-specific fecundity in cultures of *Amphistegina madagascariensis* (from Muller 1974)

Table 4.1 Adult size, fecundity and longevity (modified from Hallock 1985)

Species	Adult size diam. (mm)	Fecundity	Longevity (months)	Source
Ammonia beccarii	0.25–0.35	17–44	2–3	Bradshaw 1957
Elphidium crispum	1–3	150	12	Jepps 1942
Glabratella patelliformis	0.3–0.5	30–120	1	Loeblich and Tappan 1964
Spirillina vivipara	0.1–0.2	12–16	0.3	Myers 1936
Amphistegina lessonii	1–2	200–900	4–12	Muller 1977 (Unpublished thesis)
Amphistegina lobifera	1.5–3	900–2400	6–12	Muller 1977 (Unpublished thesis)
Heterostegina depressa	3–5	60–400	4–12	Röttger 1974
Marginopora vertebralis	5–12	60–150	12–24	Ross 1972
Baculogypsina sphaerulata	1–2	568–979	18	Sakai and Nishihira 1981

produce fewer offspring than larger species. Within individual species, fecundity is proportional to parent size and is therefore dependent on growth rates and longevity (Fig. 4.1) (Röttger 1974; Muller 1974).

Growth of individuals

Most growth data have been derived not from single specimens but from populations sampled over a period of time. Growth curves generally show an initial period of faster growth (steeper part of curve) followed by a later period of slower growth (flatter part of curve). Growth measurements on a culture of *Amphistegina madagascariensis* show this relationship very clearly. The calculated maximum possible size for this species is given by the line Dα on Fig. 4.2 and is roughly 1800 μm diameter. Individuals start to become reproductively mature from ~700 μm and the rate of growth begins to decline from ~ 900 μm.

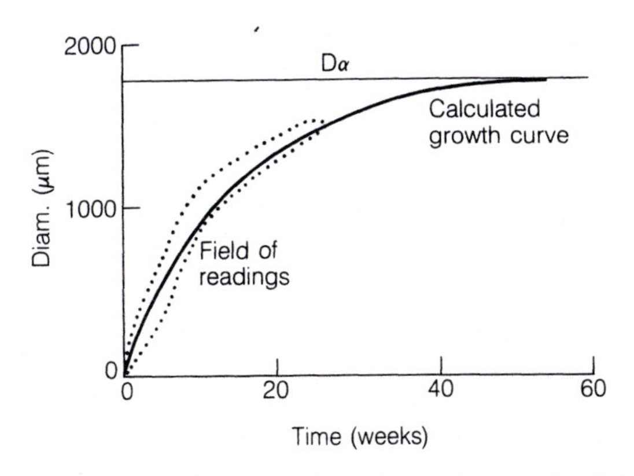

Figure 4.2 Growth of *Amphistegina madagascariensis* in culture (based on Muller 1974, Fig. 2)

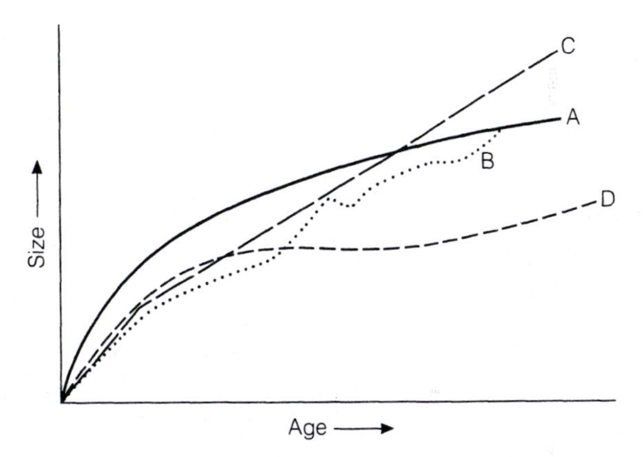

Figure 4.3 Growth curves. (A) *Marginopora vertebralis* (Ross 1972); (B) *Amphisorus hemprichii*, megalospheric (Zohary *et al.* 1980); (C) *Nonion depressulus* (Murray 1983); (D) *Elphidium crispum*, megalospheric (Myers 1942, sublittoral)

A selection of growth curves is shown in Fig. 4.3. In those which develop a pronounced flat part in the curve it can be concluded that conditions have been unfavourable for reproduction during this period (e.g. in megalospheric *Elphidium crispum*). Under these circumstances growth proceeds slowly until death whether or not due to reproduction.

Walford (1946) observed that a plot of the size of an individual at any age against its size one time unit later is approximately linear. This relationship is true of foraminifera and two examples are shown in Fig. 4.4. A summary of growth rates is given in Table 4.2.

Experimental studies have shown that the rate of growth is influenced by a number of factors which include the availability of food (Murray 1963), intensity and duration of illumination (for forms with symbionts, Röttger 1972b; Röttger *et al.* 1980; Hallock

Table 4.2 Summary of population dynamics data for selected species (+ indicates culture)

Taxon	Reproduces	No. of young per parent	Pre-productive deaths (%)	Size at maturity (μm)	Maturity (weeks)
Amphistegina madagascariensis	Feb–Aug	600–1800, av. 1100	>90	700	~9
Amphistegina lobifera	—	900	—	1400	40
Amphistegina lobifera	—	900	—	1400	18
Amphistegina lessonii	—	400	—	1200	14
Amphistegina lessonii	—	400	—	1400	14
Heterostegina depressa	—	1200–1800	—	7600–9200+	—
Baculogypsina sphaerulata	June–Aug	568–979	—	~1000	?56
Nonion depressulus	—	?	93–98	~220	13
Elphidium crispum	Mar–June	119	98	—	52
Archaias angulatus	? Summer	300+	99.5	1600	39–52
Marginopora vertebralis	Late spring	60–150+	—	15000	52–104
Haynesina germanica	—	?	—	270	4

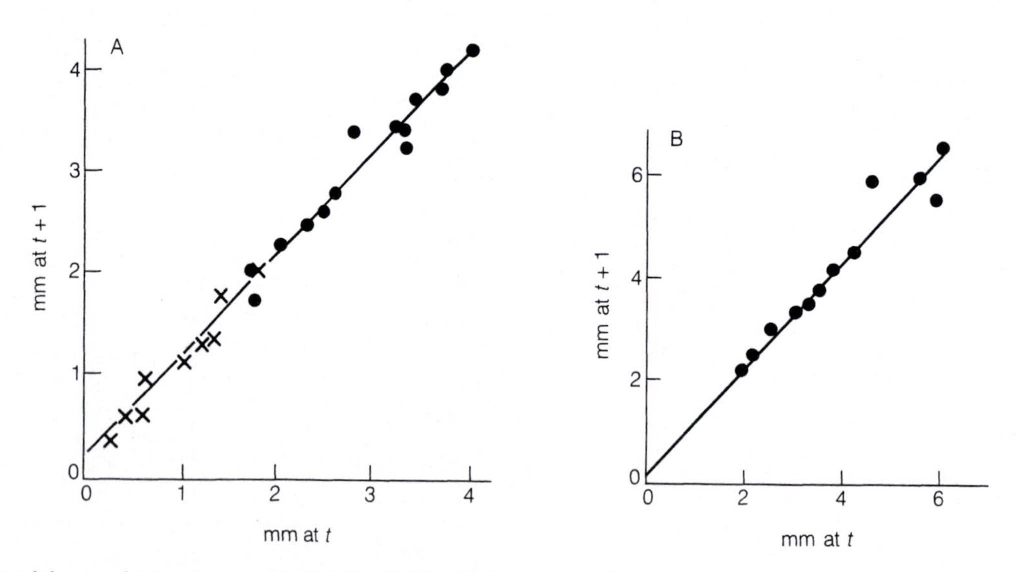

Figure 4.4 *Amphisorus hemprichii*, Red Sea. Walford plots of mean diameter (μm). (A) megalospheric (• generation 1, × generation 2); (B) microspheric (data from Zohary *et al.* 1980)

Growth rate (μm weeks)	Locality	Source
49	Hawaii	Muller 1974
46–81+	Hawaii	Hallock 1981a
52–81+	Palau	Hallock 1981a
58–116+	Hawaii	Hallock 1981a
40–70+	Palau	Hallock 1981a
—	Persian Gulf	Lutze *et al.* 1971
—	Japan	Sakai and Nisihira 1981
10–17	England	Murray 1983
—	England	Myers 1942
60–50	Florida	Hallock *et al.* 1986
400–70	Australia	Ross 1972
20–70	Spain	Cearreta 1988a

1981a), salinity (Murray 1963; Lutze *et al.* 1971), different types of seawater (Lutze *et al.* 1971) and addition of nutrients (Erdschrieber, Lutze *et al.* 1971).

Longevity, mortality and survivorship

In foraminifera, reproduction normally leads to the division of the parent cytoplasm either into gametes or into agamonts. In the latter case, the parent test is vacated and therefore 'dead'. This type of reproduction leading to the 'suicide' of the parent is sometimes termed 'semelparous'. From the point of view of population dynamics, it is important to distinguish between pre-productive and reproductive deaths. The former acts as an important counterbalance to reproduction in most populations.

Mortality is the rate of death and can be expressed as the percentage dying over a period of time, e.g. 10% per month. Survivorship is the opposite and a population with 10% mortality must have 90% survivors. Longevity is the life span of a generation (see Table 4.1 for examples).

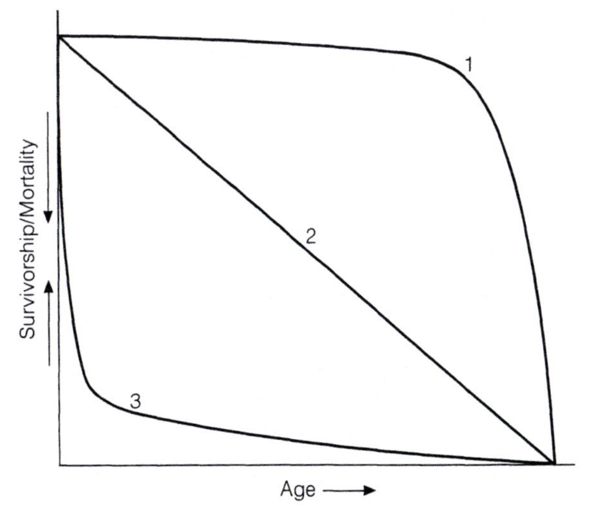

Figure 4.5 Patterns of survivorship and mortality

A spectrum of mortality/survivorship patterns is possible on theoretical grounds and three examples are shown in Fig. 4.5. Curve 1 shows mortality concentrated in old age, curve 2 a linear relationship and curve 3 high infant mortality. Benthic foraminifera generally follow a pattern close to curve 3 where high pre-reproductive mortality counterbalances a high birth-rate.

It is not feasible to observe death in foraminiferal populations so mortality rates have to be calculated. One method is to use size class data in conjunction with standing crop size to calculate a life table (Tables 4.3, 4.4). The standard notation is used where the age interval is x and all data relate to successive values of x. The cumulative values of M_x have been plotted as mortality/survivorship curves in Fig. 4.6. The longevity of *Nonion depressulus* is 3– 4 months and that of *Archaias angulatus* 12 months, although in the latter case some forms reproduce after 9 months and still others not until 16 months. The proportion entering the age class that die during that month is the age specific mortality rate q_x ($q_x = d_x/l_x$). Non-reproductive mortality take place in the months preceding reproductive maturity. In the case of *A. angulatus* age-specific mortality in months 1–8 is more or less constant at 0.24. During months 9–16 reproductive mortality (R_x) takes place and can be calculated from

$$R_x = q_x - 0.24.$$

The number of individuals reproducing (l_xR_x) peaks at 12–13 months, the longevity of the species. Hallock (1981b) considered that for *Amphistegina* half of the

Table 4.3 Life table, *Nonion depressulus*, England (after Murray 1983)

Age interval (months)	Diameter (μm)	Number entering age class (90 cm^{-2} month^{-1})	Number dying in age class (90cm^{-2} month^{-1})	Proportion dying (%)	Age specific mortality
x	D_x	l_x	d_x	M_x	q_x
0–1	—	536	0	0	0.00
1–2	140	350	186	35.2	0.53
2–3	160	99	251	47.5	0.39
3–4	200	23	76	14.4	0.30
4–5	220	8	15	2.8	0.53

Table 4.4 Life table, *Archaias angulatus*, Florida (after Hallock *et al.* 1986)

Age interval (months)	Diameter (μm)	Number entering age class (m^{-2} month^{-1})	Number dying in age class (m^{-2} months^{-1})	Proportion dying (%)	Age specific mortality	Proportion reproducing	Number reproducing (m^{-2} month^{-1})
x	D_x	l_x	d_x	M_x	q_x	R_x	l_xR_x
0–1	125–349	2.7 ×105	2.2 × 105	—	0.95		
1–2	350–599	12500	3050	24.4	0.24		
2–3	600–849	9450	2488	19.9	0.26		
3–4	850–1099	6962	1675	13.4	0.24		
4–5	1100–1349	5287	1287	10.3	0.24		
5–6	1350–1599	4000	863	6.9	0.22		
6–7	1600–1799	3137	625	5.0	0.20		
7–8	1800–1999	2512	575	4.6	0.23		
8–9	2000–2199	1937	487	3.9	0.25		
9–10	2200–2399	1450	425	3.4	0.29	0.05	72
10–11	2400–2599	1025	338	2.7	0.33	0.09	92
11–12	2600–2799	687	225	1.8	0.33	0.09	62
12–13	2800–2999	462	212	1.7	0.46	0.22	102
13–14	3000–3199	250	138	1.1	0.35	0.31	78
14–15	3200–3399	112	62	0.5	0.55	0.31	35
15–16	3400–3599	50	38	0.3	0.76	0.52	26
>16	≥3600	12	12	0.1	1.00	0.76	9
Observed total	(1–16)	5 × 104	1.25 × 104	100	—	—	—
Estimated total	(0–16)	2.8 × 105	2.3 × 105	—	—	—	—

adult mortality is a consequence of reproduction by multiple fission.

From the mortality data it is possible to plot a histogram of the size of the tests resulting from death each month. In the case of *Nonion depressulus* it can be seen that the pattern varies according to the time of year, but tests larger than 200 μm diameter make up only a small part of the total (Fig. 4.7). For larger foraminifera, Hallock (1981b) reported that 90% of the juveniles produced by multiple fission die before they are calcified. Data for *Archaias angulatus* in Hallock *et al.* (1986) have been recalculated to exclude the 96% uncalcified juvenile forms which died during the first month (Fig. 4.7A). The pattern shows a predominance of young tests and only a small proportion of forms larger than 2000 μm in diameter (15.5% of the calcified tests). It is probable that the youngest juveniles of all species are uncalcified or so lightly calcified that they are not preserved and this explains their general absence from dead assemblages.

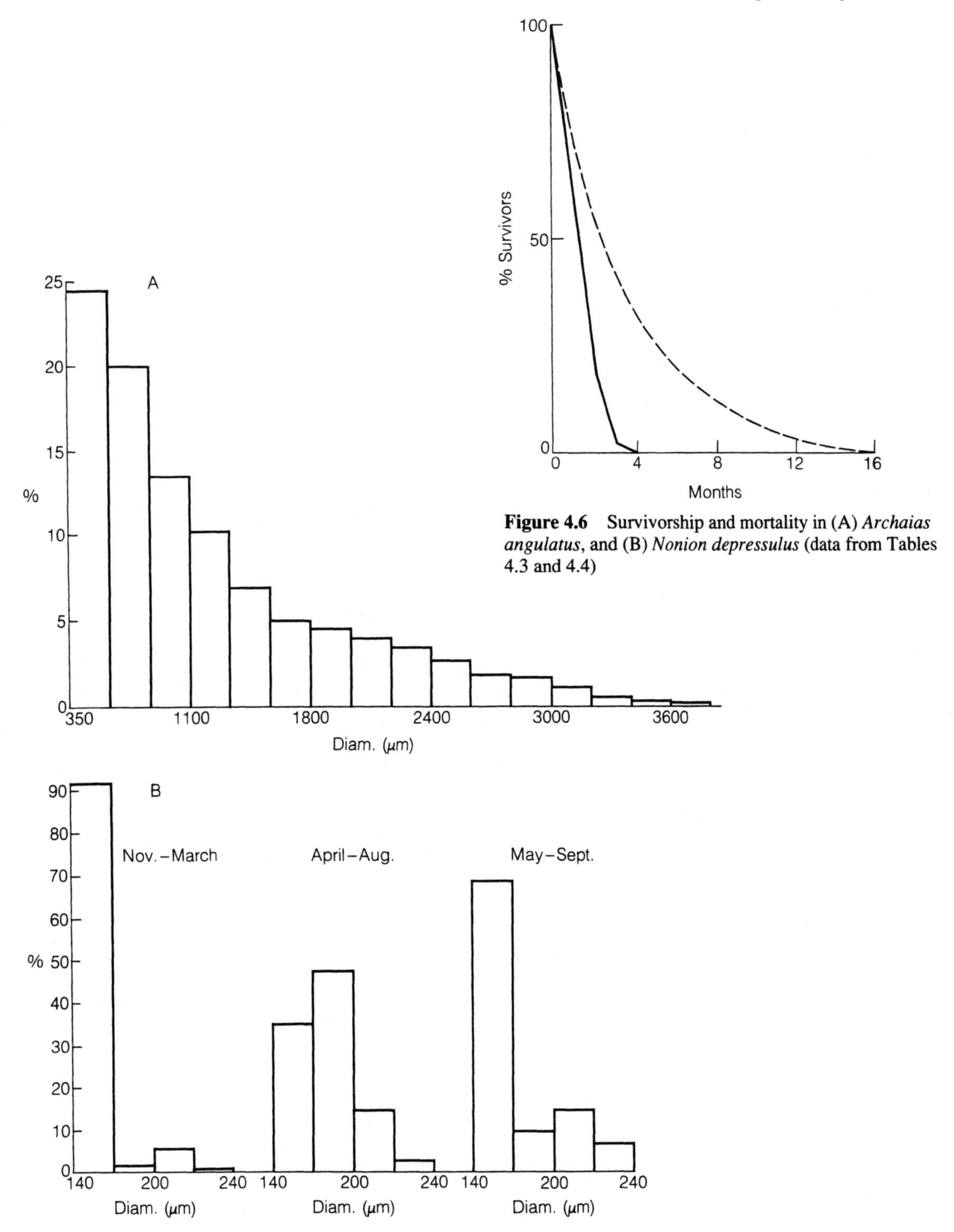

Figure 4.6 Survivorship and mortality in (A) *Archaias angulatus*, and (B) *Nonion depressulus* (data from Tables 4.3 and 4.4)

Figure 4.7 (A) Size distribution of dead *Archaias angulatus* from months 1 to 16 (data from Hallock *et al.* 1986; (B) size distribution of dead *Nonion depressulus* (data from Murray 1983)

Population growth

On theoretical grounds, if two parents in one generation give rise to four offspring then the net reproductive rate (R) will be two. Furthermore, R includes both birth and death, i.e. births minus deaths. The growth of a population through time can be calculated by multiplying the original population (N_0) by R. However, it is clearly impossible for a population to increase in size indefinitely. As the population size increases, there is progressively more competition between individuals and this reduces the net reproductive rate. A position of stability will be achieved when births and deaths are in equilibrium, i.e. each parent is replaced by one individual. This population size represents the carrying capacity (K). A simple model is the sigmoid logistic curve of population growth (Fig. 4.8).

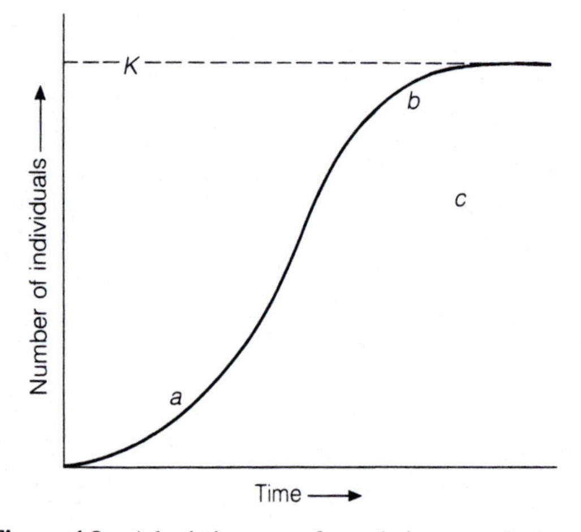

Figure 4.8 A logistic curve of population growth. At *a*, the population density is low and intraspecific competition is of minor importance in regulating the net reproductive rate. At *b*, the population density is high and intraspecific competition causes density-dependent regulation of the net reproductive rate. At *c*, the population has reached the carrying capacity (K)

It is a matter of observation that natural populations show fluctuations in size. The causes may be intrinsic or extrinsic.

Intrinsic: the carrying capacity (K) represents the maximum number of individuals that a given habitat can support and it is a state where births equal deaths. Population size increases when birth-rate is greater than death-rate below K and decreases when birth-rate is less than death-rate above K. This form of regulation is density dependent because K represents an

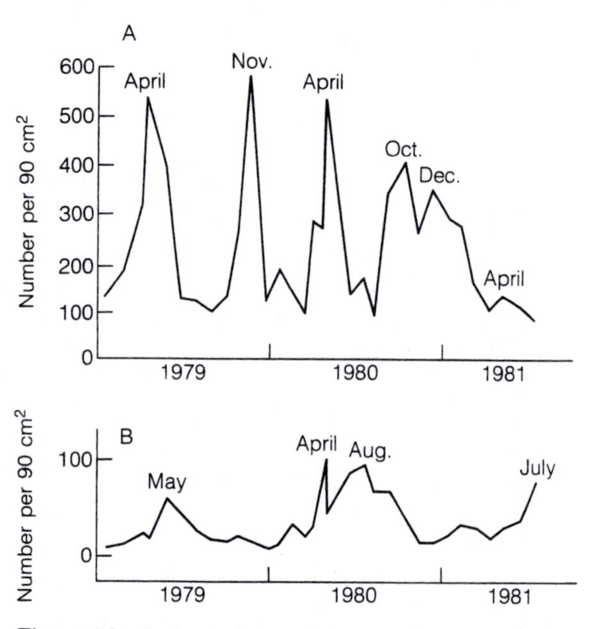

Figure 4.9 Periodic changes in the standing crop of (A) *Nonion depressulus*, and (B) *Haynesina germanica* (data from Murray 1983)

equilibrium condition. However, not all populations respond instantaneously to changes in their density. Some show a time-lag in response and this itself may cause greater fluctuations in population size.

Extrinsic: changes in environmental conditions and interaction with other species. Figure 4.9 shows two examples of the fluctuations in the number of individuals over a period of 31 months. The dominant form, *Nonion depressulus*, has peaks of abundance in April and November 1979, April and October/December 1980 but no peak in April 1981. The timing of the peaks in *Haynesina germanica* is much less regular although, with the exception of April 1980, the peaks are not coincident with those of *N. depressulus*.

In British coastal waters there are diatom blooms in the spring and autumn; the sediment surface is golden-brown with microflora during periods of low tide at these times. The peaks of abundance of *Nonion depressulus* in April and November may be in response to this increased food source. The rate of population growth can be measured by 'r', the intrinsic rate of increase (otherwise known as the Malthusian constant), in effect the excess of births over deaths:

$$\frac{(N_{t+1} - N_t)}{(N_t \times t + 1)}$$

where N_t is the original population size at time t, N_{t+1}

Table 4.5 Data for growth peaks of *Nonion depressulus* per 90 cm^2 in the Exe Estuary, England (from Murray 1983)

Days	Interval	Number	Log_{10}	*r*	*k*
14	0	129	2.11059		
				0.010	
56	42	185	2.26717		
				0.019	
94	38	318	2.50243		
				0.053	
107	13	536	2.72916		536
239	0	101	2.00432		
				0.009	
266	27	126	2.10037		
				0.055	
292	26	256	2.40824		
				0.043	
322	30	584	2.76641		584
440	0	95	1.97772		
				0.145	
454	14	288	2.45939		
				-0.002	
478	24	276	2.44091		
				0.082	
485	7	435	2.63849		435
583	0	92	1.96379		
				0.080	
677	34	341	2.53275		
				0.006	
651	34	410	2.61278		410

the new population at time t_{+1} and t+1 the number of days between t and t+1.

The results are given in Table 4.5. The values of r represent the realised intrinsic rate of increase and these values are less than the theoretical maximum rate of increase because of extrinsic influences. During these periods of rapid population increase, growth is approximately exponential. If it was truly exponential a plot of $\log_{10}$ numbers against days would be a straight line. This is almost the case for April and November 1979.

The carrying capacity of the environment (K) is reached when there is no further population increase (where $dN/dt = 0$) that is, births and deaths are equal (Fig. 4.10). In the case of *Nonion depressulus* the peak values of K vary from 410 to 584 per 90 cm^2 (Table 4.5).

Each phase of population increase is followed by

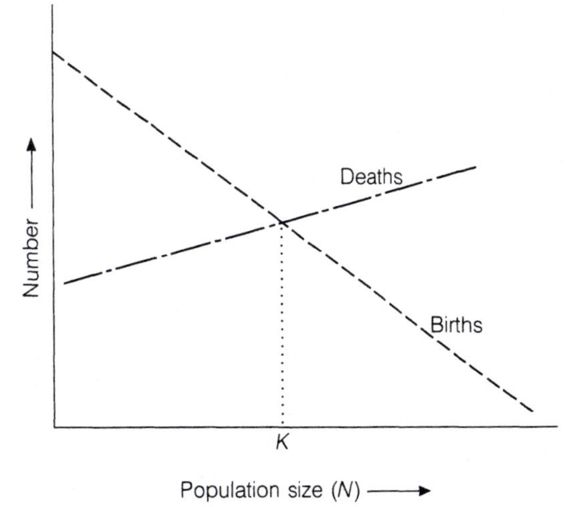

Figure 4.10 Relationship between births, deaths and carrying capacity (K)

a period of rapid decline in numbers (Fig. 4.9). The cause is likely to be density dependent due to intraspecific competition for a food source that is itself density dependent due to the limited availability of nutrients.

The fluctuations in the abundance of *Nonion depressulus* shows regularity from January 1979 until the summer of 1980, but thereafter it becomes irregular and in April 1981 there was scarcely any increase. The pattern for *Haynesina germanica* is irregular. These two species have different ecological requirements and therefore may not be directly competing with one another. *Haynesina germanica* is euryhaline with a broad tolerance to brackish water whereas *N. depressulus* is only slightly euryhaline. Both are opportunistic r-strategists, that is, they are able to increase their number rapidly in reponse to a short-term improvement in their environment.

In the only other studies extending over a period of more than 2 years, Lutze (1968a) found that there were major differences in the monthly standing crop sizes of *Elphidium williamsoni* (as *Cribrononion articulatum*) between 1965 (high), 1966 (low) and 1967 (high) although *Miliammina fusca* was less obviously affected. His study was carried out in a very shallow Baltic Sea lagoon and he concluded that because of this year-to-year variability, long-term studies over a period of 3–4 years would be necessary to determine production. Scott *et al.* (1980) took samples over the period 1976–78 (but at irregular intervals) and showed marked fluctuations in abundance of the living assemblages from marshes in Nova Scotia.

A 2-year study of *Glabratella ornatissima* from the shallow coastal zone of northern California (Erskian and Lipps 1987) revealed that this species grows and reproduces during the spring and summer and becomes dormant in winter. Wave activity is strong in the winter and there is a sparse population of plastogamous pairs and adult agamonts. Embryonic agamonts begin to develop within plastogamous pairs in early winter, and by January and February nearly all pairs contain embryos. The embryonic agamonts are released in February and March and they select an algal substrate to escape transport and damage to the test. Adult agamonts from the previous year reproduce by schizogamy and in March the first embryonic gamonts are produced. Upwelling leads to an increase in phytoplankton which in turns leads to an increase in abundance of *G. ornatissima* to reach a peak in the summer. Adult agamonts reach maturity in mid to late summer but those produced in the later summer do not reproduce; they overwinter until the next year. Populations of agamonts and gamonts occur on algae, embryonic agamonts in the winter and gamonts during the summer. But, by the end of the summer, adult gamonts and newly formed plastogamous pairs are mainly in the sediment. The timing and mode of reproduction of this species ensures that populations grow rapidly during periods of enhanced food supply. Also, by making use of the sediment as a refuge in winter it enables the species to overcome death and destruction by wave activity.

From the limited data to hand, it must be concluded that variations in standing crop size, whether cyclic or irregular, is normal for shallow-water species of benthic foraminifera.

Patterns of production

In the early theoretical analysis of production, Murray (1976b) suggested that there were four main factors: the initial size of the standing crop, the proportion of individuals which reproduce, the frequency of reproduction and the number of new individuals resulting from each reproductive phase. In 1967 there were data on standing crop but not on the other parameters. It was further suggested that there would be four basic patterns of relationship between standing crop and production under natural conditions:

Pattern 1. No seasonal change in the food supply hence uniform standing crop size throughout the year. Reproduction is annual and the number of offspring maturing to adulthood is the same as the number of parents. In this simple relationship, the standing crop would be approximately equivalent to the annual production.

Pattern 2. Similar to pattern 1, but reproduction occurs more than once per year in some or all species. The standing crop multiplied by the number of periods of reproduction would give the approximate annual production.

Pattern 3. Seasonal change in the food supply are matched by an increase in the size of the standing crop. Reproduction occurs annually, but not all species reproduce at the same time. The annual production differs significantly from the size of the standing crop.

Pattern 4. Similar to pattern 3, but having a higher frequency of reproduction. The standing crop is quite different from the annual production.

These four simple patterns were essentially for entire living assemblages but most of the data on production relate to single species. The range of environments studied extends from the intertidal zone to the continental slope and the majority follow pattern 4 with some pattern 3 (Table 4.6). As yet no examples have been reported of patterns 1 or 2.

It was further suggested that the annual production could be defined as the number of tests contributed to a unit area of sea floor during the course of 1 year. In practice, it is difficult to determine the figures accurately. Three possible methods are (after Murray 1983):

1. Production = initial standing crop + peak standing crop − final standing crop.
2. Production = sum of peak standing crop values for the year.
3. Production = sum of values from the simplified population dynamics equation $N_{i+1} = N_i + B - D$ (where N_{i+1} = the new population size, N_i = the original population size, B = births and D = deaths) over a period of 1 year.

Method 2 allows for the fact that the initial and final standing crop values are commonly nearly the same and can therefore be ignored. The production values in Table 4.6 are based on method 2. Method 3 is valid for observations taken over a period of several life cycles

Table 4.6 Production estimates prepared from tables or data or graphs in the named sources, using the sum of the peak standing crop values. (1) Walker expressed his results as number per gram of *Corallina*; (2) per 700 gm sample; (3) results expressed as number per gram of sediment; (4) value determined by taking the time period of 1 cm of sediment to be 2 years

Environment and author	Area	°Lat	Duration of study (months)	Average standing crop per 10 cm²	Annual production of test per 10 cm²	Frequency of reproduction per year	Pattern of production	Species or entire living assemblage
Intertidal–marine								
Muller 1974	Hawaii	20N	12	—	829	2	4	*Amphistegina madagascariensis*
Walker 1976	Nova Scotia	44N	12	141	844[1]	1–4	4	Entire assemblage
Intertidal–brackish								
Cearreta 1988a	Spain	43N	21	—	701	—	4	*Ammonia beccarii*
Cearreta 1988a	Spain	43N	21	—	376	—	4	*Haynesina germanica*
Scott and Medioli 1980b	Nova Scotia	44N	30	1976 : 382 1977 : 724	1976 : 1500 1977 : 2550	—	4	Entire assemblage
Boltovskoy 1964; Murray 1967	Argentina	48S	24	733[2]	1655	1	3	*Elphidium macellum*
Boltovskoy and Lena 1969b	Argentina	48S	25	92	1048	12	4	Entire assemblage
Jones and Ross 1979	Washington	49N	12	—	200[3]	—	—	Entire assemblage
Murray 1983	England	51N	31	26–41	144	8–9	4	*Nonion depressulus*
Haake 1967	North Sea	52N	12	2	8	—	—	*Elphidium excavatum*
Shallow marine lagoon								
Hallock *et al.* 1986	Florida	25N	13	5–170	125	1	3	*Archaias angulatus*
Daniels 1970	Adriatic	45N	14	98–263	400–1600	4–6	4	*Nonionella opima*
Shallow brackish lagoon								
Haake 1967	Baltic	54N	12	114–240	610–1510	?12	4	*Elphidium excavatum*
Lutze 1986a	Baltic	54N	32	1965 : 22 1966 : 6	1965 : 263 1966 : 67	1–2	4	*Elphidium williamsoni*
Wefer 1976a	Baltic	54N	23	233	467	2	4	*Elphidium excavatum clavatum*
Wefer 1976a	Baltic	54N	23	74	74	1	3	*Elphidium incertum*
Nyholm and Olsson 1973	Sweden	57N	13	<1–17	2–82	—	4	Rotaliina only
Inner shelf								
Walton 1955	Baja California	28N	8	—	650	—	4	Entire assemblage
Zohary *et al.* 1980	Gulf of Elat	29N	13	17 on *Halophila*	78	1	3	*Amphisorus hemprichii*
Myers 1942	England	50N	12	—	100	1	3	*Elphidium crispum*
Slope								
Walton 1955	Baja California	28N	8	—	100	—	4	Entire assemblage
Phleger and Soutar 1973	California	34N	1	—	111–3175[4]	—	4	Entire assemblage

Table 4.7 Production values obtained for *Nonion depressulus* using the three methods described in the text (all values = number of tests per unit area of 90 cm²)

Method 1
Feb. 1979–March 1980 = 1395
March 1980–April 1981 = 1264

Method 2
1979 Jan. 255, April 536, Nov. 584 = 1375
1980 Jan. 188, May 351, July 170, Oct. 410, Dec. 354 = 1373

Method 3
1979 N_{i+1} +ve (births > deaths), N_{i+1} –ve (births < deaths) for Jan–Dec.:
0, –87, +100, +268, –138, –275, –56, +18, +41, +230, +228, –482
= +885, –1038
1980 N_{i+1} +ve (births > deaths), N_{i+1} –ve (births < deaths) for Jan–Dec.:
+86, –52, –52, +224, +43, –218, +37, –78, +249, +69, –148, +92
= +800, –548

but for shorter periods it underestimates or overestimates either births or deaths.

Comparative results using the three methods on the same data for *Nonion depressulus* from Murray (1983) are given in Table 4.7. Methods 1 and 2 give similar values, whereas method 3 values for excess deaths over births (1038 and 548 for the 2 years) are somewhat lower.

The data in Table 4.6 are biased towards shallow-water environments simply because they are easily accessible and cheap to sample. The individual species studied are always the dominant species in the living assemblages so the differences between the results for species and entire assemblages may not be too great. The range of annual production values is surprisingly small. Most are between 100 and 1600 tests per 10 cm² substrate with a few below (North Sea, Haake 1967) or above (Nova Scotia, Scott and Medioli 1980b). The value for California (3195, Phleger and Soutar 1973) may or may not be reliable according to the assumption made about the sedimentation rate. In all cases, annual production greatly exceeds the standing crop observed at any one time.

Carbonate production

Estimates of the production of carbonate (calcite) by benthic foraminifera have been made by a number of authors. The results, summarised in Table 4.8, show that most observations relate to tropical or subtropical, shallow-water, larger foraminifera. These show a range of 40–2800 g m^{-2} year^{-1} and it is clear that in certain environments foraminifera are significant producers of carbonate. The data for smaller foraminifera are unrepresentative of normal marine environments.

Experimental studies of the effect of light on calcification in *Archaias angulatus* show the importance of the symbionts (Duguay and Taylor 1978). Both calcification and photosynthesis proceed at rates two to three times as fast in light as in dark.

Table 4.8 Carbonate production (based on Hallock *et al.* 1986) (+ = adults only)

Taxon	Standing crop (x 10⁴ m⁻²)	Carbonate production (g m⁻² yr⁻¹)	Location	Habitat	Depth (m)	Source
Cyclorbiculina compressa	0.5–0.8+	60	Bermuda	Bay	8–12	Lutze and Wefer 1980
Archaias angulatus	5	60 (100)*	Florida Keys	Bay	1	Hallock *et al.* 1986
Heterostegina depressa	4	150	Persian Gulf	Culture	0	Lutze *et al.* 1971
Amphisorus hemprichii	3	158	Gulf of Elat	*Halophila*	4	Zohary *et al.* 1980
Baculogypsina sphaerulata	40–60	500	Okinawa	Tidepool	0	Sakai and Nisihira 1981
Amphistegina spp.	15	300 (512)*	Hawaii	Rockpool	1	Muller 1974
Benthic foraminifera	17	156 (260)*	Hawaii	Nearshore	< 10	Muller 1976
Amphistegina spp. plus *Heterostegina depressa*	2–15	40–150	Hawaii	Reef flat and slope	< 20	Hallock 1981b, 1984
Large rotaliine foraminifera	5–106	100–2800	Palau	Reef flat	< 5	Hallock 1981b, 1984
Large rotaliine foraminifera	3–28	50–600	Palau	Reef slope	5–15	Hallock 1981b, 1984
Small rotaliine foraminifera	0.2–26	0.006–3.1	Baltic	Sediment	5–28	Wefer and Lutze 1978
Small rotaliine foraminifera	0.05–0.06	0.0005	Baltic	Epiphytic	6–13	Wefer and Lutze 1978
Small milioline foraminifera	0.9–1.6	0.002–0.003	Baltic	Epiphytic	6–13	Wefer and Lutze 1978

* Includes estimated production by juveniles.

CHAPTER 5

Relationship between living and dead assemblages

Introduction

Species are represented by *populations* of individuals. Groups of species form *assemblages*. Assemblages of benthic foraminifera form a part of benthic communities but these are normally defined on the macro-organisms present.

Living, dead and total assemblages

The published results on modern benthic foraminifera are reported in the literature in terms of living, dead or living plus dead (=total) assemblages.

Living assemblages. Living individuals are recognised by staining (using rose Bengal (Walton 1952) or Sudan black B (Walker *et al.* 1974)). The living assemblage sampled on one occasion may or may not be in equilibrium with its environment and may or may not be typical of living assemblages over a longer period of time. Nevertheless, it is customary to assume that it is both in equilibrium with its environment and that it is representative of a longer time period. These assumptions may be basically valid as syntheses of such results have shown (Murray 1973; Boltovskoy and Wright 1976)

Dead assemblages. The dead assemblage is built up over a long period of time. The relationships between living and dead assemblages are discussed on pp. 44–54. It suffices here to note that the composition of the dead assemblage may differ from the living assemblages from which it was drawn through life processes (such as production) and through postmortem changes. Although some dead assemblages may be in equilibrium with the depositional environment in which they are found others may not be so. For example, in the high-energy environment of the English Channel even coarse-grained sediments contain some small foraminiferal tests. They are present partly because they are trapped in the spaces between dead mollusc shells in the shell gravels and partly because they are in transit through the area but are caught by sampling.

Total assemblages. The use of the total assemblage is based on the belief that the variability of living assemblages over a period of years is sufficiently great to make meaningless a survey carried out at one time. It is therefore argued that to treat living and dead together is more realistic for the interpretation of the fossil record (Scott and Medioli 1980a; Smith 1987). The logic of this has been questioned by Murray (1982) because it disregards the changes which will affect the living element after its death. It is also sample dependent because the greater the vertical depth of a sample, the more important will be the dead contribution.

In spite of these objections, numerous authors have presented results in the form of total assemblages. It is therefore desirable to assess whether any useful conclusions can be drawn from such data. From simple mathematical models it can be shown that: (a) if the living and dead assemblages are closely similar they will both be similar to the total assemblage; (b) if the living and dead assemblages are dissimilar, their similarity with the total assemblage will depend on the proportions contributed by each; if the dead assemblage is say ten times more numerous than the living assemblage, the dead and total assemblages will be very similar whereas the live and total assemblages will be dissimilar.

In general, the first example characterises areas of rapid sediment accumulation and the second example characterises areas of slow sediment accumulation.

There are certain circumstances where it is feasible to study only the total assemblages. For example, in the deep sea the planktonic : benthic ratio is commonly 99 : 1, so to pick 200 benthic individuals means sorting through 20 000 specimens. Because the rate of sediment accumulation is slow and fertility is low, the proportion of benthic forms which are living is small. Under such conditions it would be very time-consuming to separate the living assemblages. Also, in some parts of the world the only material available is dry sediment. In this case it is better to have some data rather than none at all.

The relationship between living and dead assemblages

It is commonly observed that the living and dead assemblages from the same sediment sample are different to a greater or lesser extent. There is nothing surprising in this for the living assemblage represents only the time of sampling, whereas the dead assemblage represents many generations added over a long period of time (time averaging, Staff *et al.* 1986). Notwithstanding the importance of this topic for the understanding of fossil assemblages it has received very little attention. The principal contributions have been by Murray (1973, 1976a, 1983) and Douglas *et al.* (1980).

A living assemblage, through life processes such as production, gives rise to an ideal dead assemblage, and this, in turn, is invariably altered through postmortem changes to give the dead assemblage preserved in the sediments (Fig. 5.1).

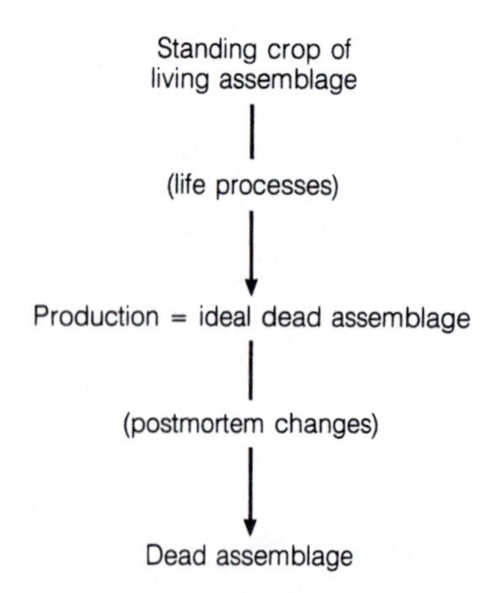

Figure 5.1 Simplified relationship between living and dead assemblages (based on Murray 1976a, 1983)

Role of life processes

Production

Production can be defined as the number of tests contributed to a unit area of sea floor during a specific period of time (see Ch. 4 for details). In foraminifera, the test is normally vacated by the animal during reproduction. Ultimately, all the empty tests, whether due to reproduction or death, are added to the ideal dead assemblage. If all the living species had the same production, the living and ideal dead assemblages would be identical. However, in reality, there are marked differences in production between species and this may alter the rank order of the most abundant species between the living and ideal dead assemblages. This is clearly seen in the example in Table 5.1 where the annual production is the ideal dead assemblage.

Table 5.1 Theoretical model of the effects of differential production on the abundance and rank order of species

	Standing crop			Annual production		
Species	Number	%	Rank	Number	%	Rank
A	75	37.5	1	150	35.0	1
B	50	25.0	2	50	11.6	4
C	30	15.0	3	90	20.9	3
D	30	15.0	3	120	27.9	2
E	10	5.0	5	10	2.3	5
F	5	2.5	6	10	2.3	5
Totals	200	100		430	100	

Time-averaging

In most studies, the living assemblage is studied at one season only. However, in those areas where successive samples have been taken over a period of months or years, the total number of living species greatly exceeds the number found at a single period of sampling. Generally, there are a few common species present at all times and an ever changing number of minor species. This is a consequence of the patchiness in distribution of species, the patches changing their distribution through time (see p. 16).

Because the ideal dead assemblage is made up of many successive contributions time-averaged from continually changing living assemblages, it invariably has a slightly higher diversity than the latter.

Determination of production rates

Where data on living and dead assemblages are available from the same sample point it is possible to estimate the different rates of production of the component species using the following method:

1. Tabulate the abundances of live and dead forms for those species occurring >4% (to eliminate rare occurrences).
2. For the species having the smallest percentage of dead compared with the percentage of living, let the production rate be taken as × 1 (i.e. the annual production = standing crop).
3. From this the number of individuals = 1% of the dead assemblage can be calculated because the percentage living = number of individuals equivalent to the percentage dead, e.g. living 50%, dead 10%, therefore 50 individuals = 10% dead, 1% dead = 5 individuals.
4. Multiply the percentage dead of all other species by this amount.
5. Finally divide the number of dead individuals by the living percentage for each species to give the rate of production.

Three examples are shown in Table 5.2. Sample 124 has almost identical living and dead assemblages and the two species >4% have very similar production rates. Samples 76 and 1212 both have dissimilar living and dead assemblages so postmortem influences rather than production differences are likely to be the cause of this. In sample 76 the production values range from 16 to 51 times that of *Eggerella advena.* In sample 1212, one living species is not represented in the dead assemblage so postmortem loss is evident. The other species show production rates from 2 to 42 times higher than that of *Portatrochammina lobata.* From production studies it seems unlikely that there are differential production rates greater than 10 or 12 times between species. Therefore, for most of the species in samples 76 and 1212 postmortem influences greatly exceed those brought about by different production rates.

Table 5.2 Calculation of production rates from data on living and dead assemblage. See text for discussion (data from Buzas 1965 and Murray 1969)

Sample 124 Long Island Sound

Species	Live (%)	Dead (%)		1% = 1.2 inds	Production
Buccella frigida	6	5	=	6	× 1
Elphidium clavatum	87	90	=	108	× 1.2

Sample 76 Long Island Sound

Species	Live (%)	Dead (%)		1% = 10.6 inds	Production
Buccella frigida	30	46	=	488	× 16.3
Eggerella advena	53	5	=	53	× 1
Elphidium clavatum	7	34	=	360	× 51.4
Elphidium varium	4	12	=	127	× 31.7

Sample 1212 Shelf off Long Island

Species	Live (%)	Dead (%)		1% = 1 inds	Production
Buccella frigida	4	1	=	8	× 2
Eggerella advena	18	33	=	264	× 14.9
Elphidium clavatum	10	3	=	24	× 2.4
Fursenkoina fusiformis	16	0		Lost by transport/dissolution	
Saccammina atlantica	23	53	=	424	× 24.4
Portatrochammina lobata	8	1	=	8	× 1

Colonisation

Following the return of favourable environmental conditions after an adverse period such as anoxia, or a heavy fall of volcanic ash, opportunistic benthic foraminifera will recolonise the area. The Baltic Sea undergoes periods of bottom-water anoxia and Lutze (1965) suggested that once oxygenation had been re-established foraminiferal recolonisation would be rapid. The mechanism of recolonisation was thought to be due to suspension and lateral transport of individuals from adjacent areas. Wefer and Richter (1976) tested this hypothesis by setting up three submerged artificial substrates at different water depths at varying levels above the sea floor. Of the four main species living in the western Baltic only one, *Elphidium excavatum clavatum*, colonised the substrates in any abundance. The first living forms to be observed were intermediate in size (~200 μm) and did not arrive until 7 months after the start of the experiment. This was attributed to the absence of strong water movements during this period. It was concluded that *E. excavatum clavatum* was successful because it is epifaunal and is easily suspended by water movement.

Similar experiments, carried out in St Georges Bay, Nova Scotia, involved placing sterile sand over the sea floor at a depth of 13 m and sampling it over an 8-day period. Living individuals were present on the sterile sand after 2 days and the species which recolonised the substrate included the *Elphidium clavatum* group, *Quinqueloculina semimula, Ammonia beccarii, Eggerella advena, Hemisphaerammina bradyi* and *Miliammina fusca*. The area is subject to periodic high turbidity and Schafer and Young (1977) speculated that these eipfaunal free-living forms were passively transported on to the substrate. Experiments in tanks using sterile sand adjacent to substrate from the bay showed that the most active forms were *Haynesina orbiculare* and the *Elphidium clavatum* group. The agglutinated species, *Eggerella advena, Saccammina atlantica* and *Trochammina* sp. were less active. The processes that lead to colonisation of new areas could also bring about changes in assemblages in already colonised areas.

Role of postmortem processes

In the majority of cases, the differences between the living assemblage and the ideal dead assemblages drawn from it are not very great. However, postmortem influences modify the ideal dead assemblage to give the dead assemblage observed in the sediment. The processes involved are transport, mixing and destruction of tests.

Transport

Transport mechanism include bed load and suspended load, floating plants, ice, turbidity currents and mass flow in the aqueous environments and by wind on land.

Suspended load. There are now numerous records of small, thin-walled tests of benthic foraminifera present in plankton sample (Table 5.3). These include the juveniles of larger species and the adults of small species. The total size range observed is from ~63 to 760 μm, but the majority have a diameter of <200 μm if rounded and a length of <300 μm if elongate. Such forms are extremely well preserved. They are thrown into suspension either in the intertidal zone (Loose 1970) or, more commonly, by the reworking of subtidal sediments during storms.

The most detailed study is that of the Western Approaches and English Channel (Murray *et al.* 1982). This is a continental shelf area influenced by powerful tidal currents throughout the year and by storms and waves during the winter months. Water samples with a volume of 7 l were taken at 40 m below the surface and 5 m above the bottom in October 1979 during a period of storms. The distribution pattern shows strong vertical mixing in the English Channel where the thermocline had been destroyed (Fig. 5.2), but in the Western Approaches the waters were still stratified so most of the suspended forms were in the lower part (Fig. 5.3). The size sorting of the foraminifera is clearly seen in Fig. 5.4. The average concentration of tests was 3800 per 1 m^3 water, thus enormous numbers are involved. Tests remain in suspension as long as there is sufficient turbulence to keep them there. They are moved back and forth by the tidal movement of water. Some tests, derived from the outer shelf, are ultimately deposited in the inner shelf and in estuaries. Other forms, derived from the English Channel, are transported in a near-bottom flow of water and are eventually deposited on the continental slope (Fig. 5.5). In this way, the apparent depth distribution may be greatly extended.

Many estuaries are the sites of deposition of suspended-load foraminifera derived from the shelf. This is because these marginal marine environments are of relatively low energy. Some examples are given

Table 5.3 Size range of tests transported in suspension

Size (μm)	Area	Source
150–200	English Channel *	Murray 1965d
	California	Lidz 1966
250–500	California †	Loose 1970
150–300	Argentina *	Watanabe 1982
100–150	Severn Estuary *	Murray and Hawkins 1976
	Bristol Channel *	Culver and Banner 1978
	Gulf of Mexico‡	Hueni *et al.* 1978
140–200	Exe Estuary *	Murray 1980
< 200 rounded	English Channel *	Murray *et al.* 1982
< 300 elongate	and Western Approaches	
75–375, av. 185	Elbe Estuary *	Wang and Murray 1983
< 270 μm	North Sea *	John 1987

* Dead only.
† Living and dead.
‡ Living.

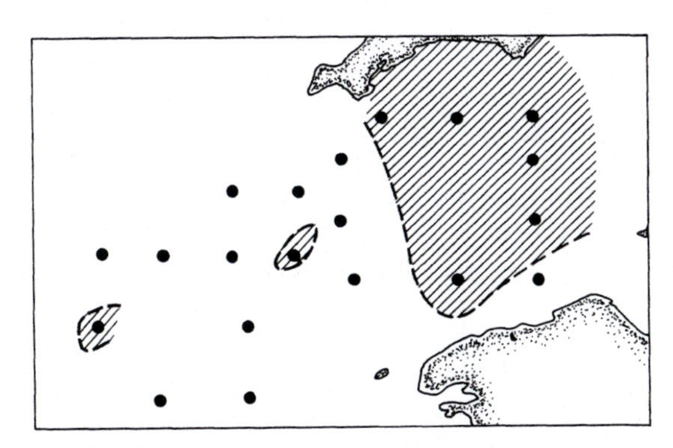

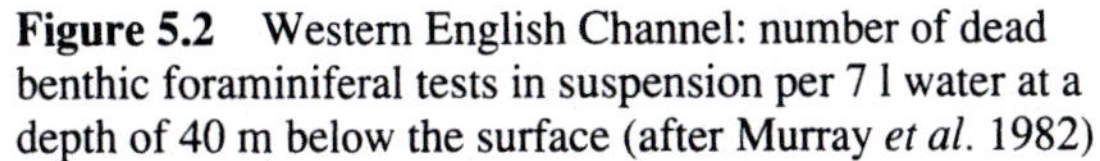

Figure 5.2 Western English Channel: number of dead benthic foraminiferal tests in suspension per 7 l water at a depth of 40 m below the surface (after Murray *et al.* 1982)

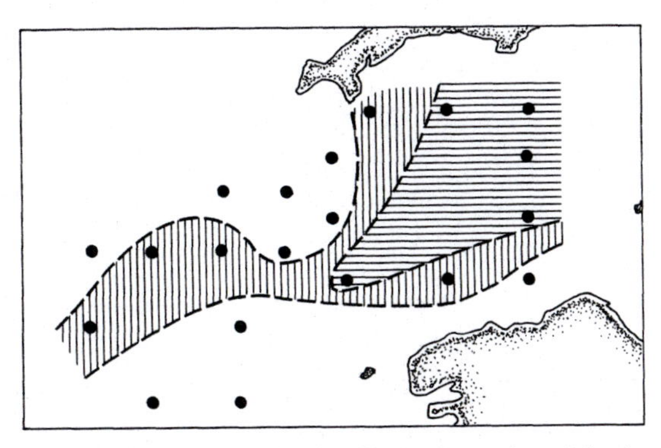

Figure 5.3 Western English Channel: number of dead benthic foraminiferal tests in suspension per 7 l at 5 m above the sea floor (after Murray *et al.* 1982)

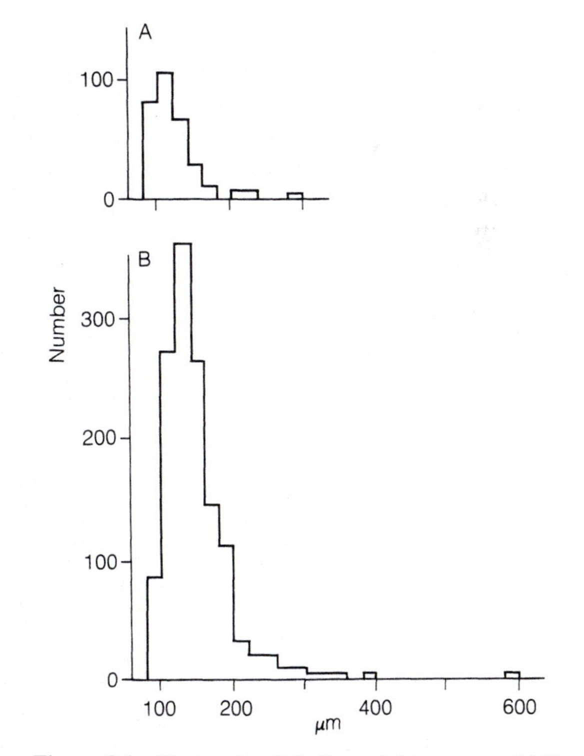

Figure 5.4 Western English Channel: histograms of (A) elongate tests, and (B) rounded tests of dead benthic foraminiferal tests in the water column (after Murray *et al.* 1982)

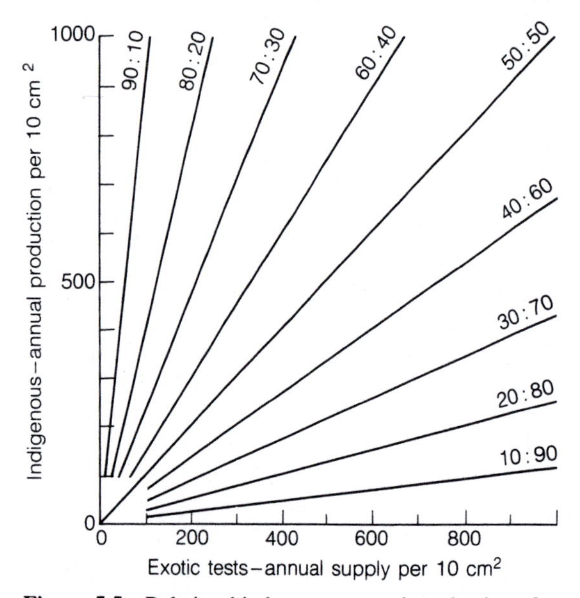

Figure 5.5 Relationship between annual production of indigenous foraminifera, annual supply of exotic foraminifera and indigenous: exotic ratio (oblique lines) (from Murray 1987a)

in Tables 5.4. A clear correlation can be seen between the tidal range and the proportion of exotic tests introduced as suspended load. Thus, microtidal estuaries have only a small component of transported tests, mesotidal examples have a moderate proportion and macrotidal estuaries commonly have >50% exotic tests in the muddy intertidal sediments (Wang and Murray 1983). In each case, the transported component of the dead assemblage is readily recognised because it is small in size (generally <200 μm) and because no living representatives of the species are encountered.

In the depositional area, a relationship exists between the annual production of the indigenous living assemblage, the annual supply of exotic tests as suspended load and the ratio indigenous : exotic tests in the dead assemblage (Fig. 5.5). Knowledge of any two of these variables allows the third to be estimated. For example, if the supply of exotic tests is 500 per 10 cm^2 year^{-1} and the indigenous : exotic ratio is 40 : 60, the annual production of the indigenous assemblage is estimated to be 334 per 10 cm^2.

If sufficient details of both the living and dead assemblages of the source area and depositional area are known, it is possible to attempt 'mass balance' calculations such as those in Table 5.5. The calculated supply of tests <200 μm in size from the English Channel is 1–6.5 per litre which brackets the observed concentration of 3.8 per litre. In the Exe Estuary 50 tests per 10 cm^2 are deposited annually and these would be provided by 7.7–50 l of seawater. In the Celtic Sea, the calculated supply of tests <200 μm is 0.4–1.1 per litre. Suspended sediment samples from the mouth of the Severn Estuary contain, on average, 4.4 tests per litre. Parker and Kirby (1982) have calculated that the turbid waters of the Severn Estuary hold 20 years' supply of river mud input. It is therefore probable that several years' supply of transported tests are also held in suspension. The calculated input of 100 exotic tests per 10 cm^2 would be supplied by

Table 5.4 Some examples of the occurrence of suspended load transport of exotic foraminiferal tests into estuaries. (a = subtidal, b = intertidal, x = muddy samples only). (1). Lankford 1959 (stas 797, 798, 128); (2). Murray 1968a; (3). Wang and Murray 1983; (4). Murray 1980; (5). Wang *et al.* 1980; (6). Brasier 1981; (7). Murray and Hawkins 1976

	River	Tidal range (m)	Exotic forms (%)	Test size (μm)	
				Range	Mean
	Mississippi, USA[1a] (fluvial marine)	0	0–9	—	—
Microtidal 0–2 m	Christchurch, England[2b]	0.6–1.5	0–22	80–300	166
	Pearl, China[3a]	1.59	0–9	75–375	130–200
	Elbe, West Germany[3a]	2	23–25	100–400	180–200
Mesotidal 2–4 m	Exe, England[4a]	1.5–3.7	0–36	63–200	140–180
	Yangtze, China[5a]	2.8	88–74	75–425	124–134
	Qiangtan, China[5a]	5.5	85–78	75–325	116–142
Macrotidal >4 m	Humber, England[6b]	6.1–7.6	2–54	63–325	100–150
	Severn, England[7b]	8–11	30–70	63–195	95–145

Table 5.5 'Mass balance' calculations for source and depositional areas (based on Murray 1987a)

Source area	English Channel	Celtic Sea
Production per 10 cm^2	200–250	220–550
% tests < 200 μm	75	20
Supply of tests < 200 μm	150–375	44–110
Average water depth	80	100
No. tests < 200 μm l^{-1}	1–6.5	0.4–1.1
Observed no. tests < 200 μm	Av. 3.8	No data
Depositional area	**Exe Estuary**	**Severn Estuary**
Production per 10 cm^2	200	? 100
Indigenous : exotic	80 : 20	50 : 50
Annual exotic input per 10 cm^2	50	? 100
	Mesotidal	Macrotidal

91–250 l of seawater containing 1.1–0.4 tests per litre or 23 l of seawater containing 4.4 tests per litre. These results seem of the right order of magnitude.

Apart from transport truly in suspension, it has also been suggested that tests may be transported in the foam of breaking waves (Loose 1970). Also, when tidal flats are dried out by the sun, the incoming tide may pick up air-filled foraminiferal tests and transport them shorewards. This has been reported from the Dee Estuary (Siddall 1878) and from the hypersaline Abu Dhabi Lagoon (Murray 1973).

Suspended load transport is not confined to recent foraminifera. Upper Cretaceous forms reworked from chalk are commonly transported in this way. They travel for great distances and are remarkably well preserved when deposited in recent sediment (Otvos and Bock 1976).

Bed load. Transport as bed load must be common in shallow-water areas. Test shape, size and excess density in seawater are factors which control transport and deposition. The settling velocities of tests ~300 μm in diameter have been determined experimentally as 0.009–0.018 m s^{-1}. These values are slightly artificial as the medium used was distilled water at 20 °C (Haake 1962; Grabert 1971). Nevertheless, differential settling velocities may lead to concentrations of species in separate areas. Nichols and Norton (1969) determined that *Elphidium clavatum* settles twice as fast as *Ammonia beccarii* and two to three times as fast as *Ammobaculites crassus*. Settling velocities can also be used to relate foraminiferal tests to the equivalent quartz sphere (Langus *et al.* 1972).

In reality, it is more useful to know the traction velocity, i.e. current speed at which individuals are moved. This has been determined experimentally for a number of species (Kontrovitz *et al.* 1978). The results were based on measurements of size, shape and weight for 25 individuals of each species. These were each then placed in a flume to determine the traction velocities and the mean value calculated. Distilled water at 21.0 ± 0.5 °C was the medium used. Although the results are of interest they can be criticised for several reasons. The most serious is that only a limited range in size was used, i.e. juveniles were ignored, and distilled water is less dense than seawater. The mean traction velocities ranged from 5.1 cm s^{-1} for *Quinqueloculina,* cf. *Q. vulgaris* to 14.3 cm s^{-1} for *Brizalina subaenariensis mexicana.*

The specific gravity of 12 smaller benthic foraminifera ranged from 0.27 for *Cibicides lobatulus* to 1.58 for *Hoeglundina elegans* for hyaline forms, 0.76 for *Quinqueloculina* cf. *Q. vulgaris* to 1.52 for *Laevipeneroplis proteus* and an extreme 5.70 for *Massilina decorata* for porcellaneous forms (Kontrovitz *et al.* 1978).

The specific gravity of larger foraminifera varies with genus and also according to the degree of damage suffered. Complete *Calcarina* has a mean value of 1.85 compared with a mean of 2.2 for eroded specimens. Complete *Baculogypsina* and *Marginopora* have mean values of 1.9 and 2.3 respectively. All other bioclasts fall in the range 2.4 and 3.0 (Jell *et al.* 1965), so foraminifera may be potentially more liable to transport in disturbed areas. The settling velocities of these large forms vary according to size but fall in the range 4–13 cm s^{-1} for sizes of 1–7mm intermediate diameter. Plates, much as *Marginopora*, settle more slowly than other shapes. Spherical forms (*Calcarina, Baculogypsina*) also settle slowly because of their low density. Rod-shaped *Alveolinella* settles fastest of all (Maiklem 1968).

Bed-load transport is prevalent in sediments of sand grade and coarser. It leads to abrasion of the tests which may cause breakage of chambers, a dull or polished texture. Bed-load transport is common in areas subject to powerful currents and to waves. Storms promote transport in shallow water. Vénec-Peyré and Le Calvez (1986) inferred that a hurricane which swept the Polynesian island of Moorea was responsible for the more widespread occurrence of dead large foraminiferal tests recorded in 1983 as compared with the 1977 survey.

Wind. Onshore winds may dry out sandy intertidal sediments and blow material inland to form dunes. The foraminifera transported in this way are derived from the adjacent coastal area and their presence in subaerial deposits is potentially misleading for palaeoecologists. Modern examples include Dog's Bay, Connemara, Ireland, where dunes composed of bioclastic material include abundant *Cibicides lobatulus*, and the United Arab Emirates where dunes up to 12 m high form adjacent to sublittoral oolith deltas (Murray 1970a). During the Pleistocene foraminifera were transported up to 800 km into the Thar Desert, India (Goudie and Sperling 1977).

Ice. Intertidal areas subject to freezing may undergo transport of material by ice. The surface layer of sediment and its included foraminiferids are frozen into the lower surface of the ice throughout the intertidal zone. As the tide rises, it floats the ice off and transports it shorewards. This process is known to occur in the Jade area of Germany (Richter 1965). Similarly, it has been reported that benthic foraminifera are incorporated in icebergs. This could be due either to grounding of the ice on the sea floor or to the incorporation of suspended sediment on to the base of an iceberg by freezing (Dieckmann *et al.* 1987).

Floating plants. Foraminifera living on submarine plants in shallow water are subject to transport when the plants are uprooted during storms. Transport is often towards the shore, but in some cases patches of uprooted vegetation are seen floating in the open sea. In the South Atlantic, the seaweed *Macrocystis pyrifera* lives along the Argentine coast from 42° S latitude. *Quinqueloculina intricata, Trifarina angulosa* (=*Angulogerian* of Boltovskoy and Lena 1969c) and *Cassidulina crassa* living in association with the weed are transported to the north by the Malvin Current.

Two species of benthic foraminifera, *Planorbulina acervalis* and *Rosalina globularis*, have been reported living attached to the pelagic seaweed *Sargassum natans* more than 5 km form Bermuda (Spindler 1980). The foraminifera have a patchy distribution on the *Sargassum* and all the *P. acervalis* were megalospheric resulting from asexual reproduction. Both generations of *R. globularis* were present. Both species are present throughout the year and *P. acervalis* averages 10 per gram *Sargassum* and reaches a size of up to 1.6 mm.

Turbidity currents and mass flow. Downslope movement by these means is known from ocean margins and shelf sea basins. It is commonly observed that the depth ranges exhibited by dead representatives is much greater than that of the living of the same species. Direct evidence of downslope movement is less readily obtained. However, sediment trap experiments in the Bedford Basin, Canada, revealed that the turbid bottom water contained 10 species of foraminiferids, 7 of which had living representatives. The size range was 140–660 μm (Schafer and Prakash 1968).

Transported shallow-water faunas in deep-sea cores are fairly easily recognised. Sometimes deep-water forms are mixed in with the shallow-water forms and this may be due to the recolonisation of the newly introduced sediment by the deep-water fauna (Phleger 1960a). The most likely transport pathway for shallow-water sediments into the ocean basins is via submarine valleys. From the presence of shallow-water microfaunas in short sediment cores taken from the axial channel in Wilmington Canyon, Stanley *et al.* (1986) concluded that the transport process involved several phases of transport and redeposition. On the Rhône deep-sea fan the sediments contain shallow-water benthic foraminifera at a depth of 1650 m (Vénec-Peyré and Le Calvez 1986).

Summary. It will be apparent form the above discussion that the most important process is transport by water, either in suspension or as bed load (Table 5.6). Tests transported in suspension are generally thin-walled and small in size. They include the adults of naturally small species and juveniles of larger species. Such tests are generaly <200 μm in diameter or length (Murray *et al.* 1982). Although some suspended tests may be redeposited close to source, in many cases they are transported elsewhere, sometimes over distances of >100 km. Both the source area and the depositional area will then have modified assemblages (Fig. 5.6). Bed-load transport normally affects sandy and coarser sediments. Tests may be size-sorted, abraded and, under severe conditions, destroyed. All medium to coarse sand substrates are likely to be influenced by bed-load transport and, if the sand is mobile, there will be no indigenous living foraminiferal fauna (e.g. a sandy beach).

Mixing

Mixing of assemblages may be brought about by transport (as already discussed), bioturbation and by coexisting substrates. Bioturbation can lead to upward or downward movement of particles. Upward movement occurs when material is excavated from depth

Table 5.6 Summary of transport effects on foraminifera

Transport mechanism	Size range transported (μm)	Preservation
Water		
Suspended load	Generally < 200 μm in diameter or < 300 μm in length	Good
Bed load	> 200 μm	Abraded, broken
Ice	Any size	Good
Floating plants	Any size	Good
Turbidity currents and mass flow	Any size	Variable
Wind	Generally < 200 μm	Often good

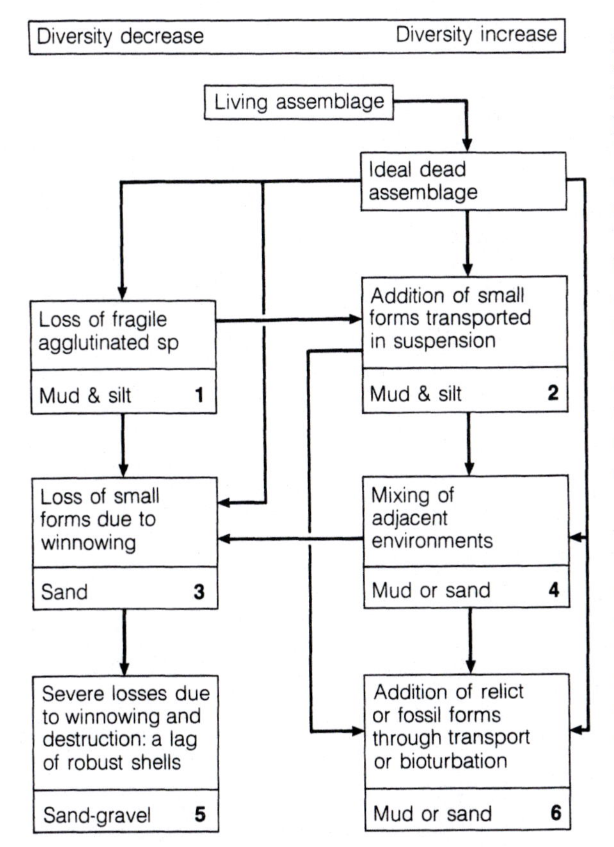

Figure 5.6 Summary of postmortem transport and mixing effects with examples, (from Murray 1983). (1) California borderland (Douglas *et al.* 1980); (2) Severn Estuary (Murray and Hawkins 1976); (3) Celtic Sea sand banks (Murray 1979a); (4) Abu Dhabi Lagoon (Murray 1970a), (5) English Channel (Sturrock and Murray 1981); 6. Bristol Channel (Culver and Banner 1978)

and dumped at the surface (e.g. crustacean burrows). Downward movement can be caused by surface material falling into open burrows or through biological pumping by organisms with a hydrostatic skeleton. Such organisms involuntarily move grains >2 mm in diameter upward through the surface sediment and finer material moves downwards to occupy the space. This process may operate down to a depth of 10 cm below the sediment surface (McCave 1988). Infaunal deposit feeders also cause mixing. The mathematics of mixing were discussed by Schiffelbein (1985). Mixing may not be important unless there are significant changes in the foraminiferal assemblages over a short time span and during a period of slow sedimentation. Otherwise mixing like with like will not cause a change in faunal composition.

In some environments there are coexisting substrates with their own separate assemblages, e.g. a sediment fauna and hard surface fauna on shells, tunicates, plants, etc. The firm substrate faunas include cemented and clinging (i.e. mobile attached) species. On death the clinging forms will be contributed to the sediment and this may also be true of some or all of the cemented forms. Attached faunas occur on shelves and also in bathyal and abyssal regions.

Destruction of tests

Destruction of tests through transport and abrasion has already been mentioned. Agglutinated tests which are weakly held together by organic material do not survive long after death. The organic matter is probably oxidised within the surface layer of sediment (Douglas *et al.* 1980). Losses of this kind are probably more common on fine-grained substrates for on coarser sediments the tests are generally more robust. Such losses are difficult to quantify but they are significant along the Californian borderland (Douglas *et al.* 1980; Smith 1987).

Dissolution

Dissolution of calcareous tests takes places in waters undersaturated with respect to calcium carbonate. Such waters may be in contact with the sea floor or interstitial within the sediment. It is commonly believed that dissolution always takes place in anoxic sediments but this is not so. If anoxia inhibits bioturbation, the interstitial waters may achieve saturation with minimum dissolution of (usually aragonitic) calcium carbonate. In this case, the faunas will be well preserved (Berger and Soutar, 1970). Also, some bacterial activity under anoxic conditions leads to alkaline interstitial waters in which preservation is excellent.

Benthic foraminifera appear to be more resistant to dissolution than are planktonic forms. Experimental studies show that some benthic species are more susceptible than others. For deep-sea species, Corliss and Honjo (1981) found the susceptibility to be high in *Pyrgo murrhina* and *Fontbotia wuellerstorfi*, moderate in *Cibicidoides kullenbergi, Nuttallides umboniferus, Hoeglundina elegans* and *Oridorsalis tener* and least in *Gyroidina orbicularis* and *Hansenisca soldanii*. Experiments on Californian borderland species by Heitman (reported by Douglas *et al.* 1980) showed that the rank order could be changed by dissolution. *Suggrunda eckisi* and *Cassidulina* spp. were more susceptible than *Brizalina argentea*. In general, the resistant species show an increase in the dead assemblages.

Hyaline tests undergo a progressive series of changes when subject to dissolution (Murray and Wright 1970; Corliss and Honjo 1981; Grobe and Futterer 1981). Initially, the surface acquires a dull texture and becomes opaque through etching; this then becomes pitted and the last chamber breaks; further breakage takes place and layers of wall are removed from unbroken chambers; this is followed by extensive chamber breakage and, finally, total destruction. Carbonate dissolution is most prevalent in the ocean basins but is also known to take place in some shelf seas (Alexandersson 1978; Lutze 1965). Similar dissolution effects on test appearance are caused by predators feeding on foraminifera (Mageau and Walker 1976).

The effects of dissolution can be calculated as follows:

$$\text{Residual } c = \text{Initial } c - \frac{(\text{Initial } c \times \%\ \text{Loss})}{100},$$

where c is the percentage calcareous component.

Thus, for an original assemblage having 20 agglutinated and 80 calcareous forms and subject to a 50% loss by dissolution the calculation is

$$\text{Residual } c = 80 - \frac{(80 \times 50)}{100} = 40.$$

The assemblage now totals 20 agglutinated and 40 calcareous forms, i.e. 60 individuals. Recalculating these values to percentages gives agglutinated 33% and calcareous 67%. Thus 50% dissolution of an assemblage having an original 80% calcareous component leads to a residual assemblage with an 67% calcareous component (Murray 1989).

The effects of varying degrees of dissolution on assemblages is shown in Fig. 5.7. Thus, a living assemblage having 20% agglutinated forms which gives rise to a dead assemblage having 45% agglutinated

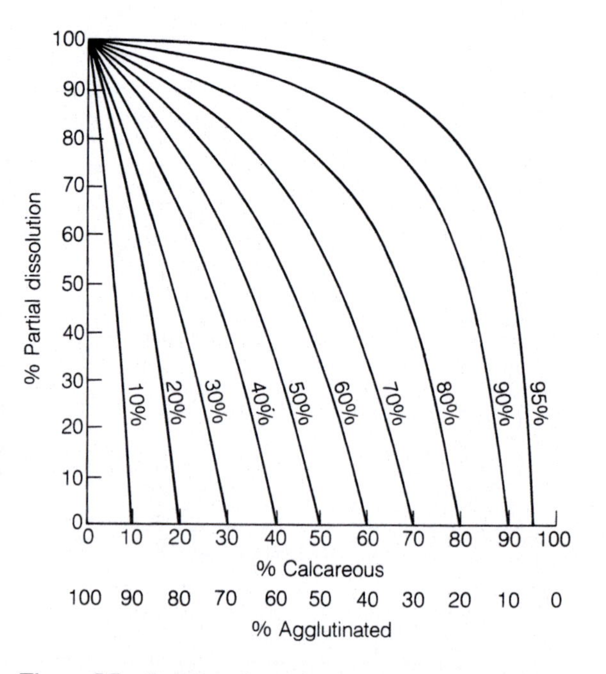

Figure 5.7 Graph to show changes in the proportions of calcareous and agglutinated foraminifera in assemblages subject to varying degrees of partial dissolution. For example, an assemblage with an initial 50% calcareous content, after 70% dissolution would then have only 23% calcareous tests

	Transport	Loss of Agglut. tests	Least altered except by transport	Loss of calcareous tests		Lithology
		Mainly poorly cemented forms		Moderate (30–90% loss)	Very severe (>90% loss)	
Gain ↑	Deposition of small individuals (<200 μm) from suspension	Abundant juveniles and small tests of calcareous forms II-A	Abundant juvenile and small tests, shiny to dull hyaline tests II	Few small calcareous tests. Opaque, broken hyaline tests II-C	Mainly agglutinated juveniles and adults II-CC	Mud, silt, muddy sand
	None or slight	Juveniles and adults of same calcareous species I-A	Juveniles and adults of same species I	Few small calcareous tests. Opaque, broken hyaline tests I-C	Mainly agglutinated juveniles and adults I-CC	
Loss ↓	Winnowed lag deposit – loss of small tests (<200 μm) – residue of medium – large tests	Medium to large calcareous tests and well-cemented agglutinated tests III-A	Medium to large calcareous and agglutinated tests III	Calcareous tests broken III-C	Medium to large agglutinated tests III-CC	Fine to medium sand
	Transported – abrasion and destruction in bed load – residue of large abraded tests	Large calcareous and well-cemented agglutinated tests IV-A	Large calcareous and agglutinated tests IV	Calcareous tests rare. Large agglutinated tests IV-C	Large agglutinated tests only IV-CC	Medium to coarse sand; gravel
			Decrease in calcareous component →			

Figure 5.8 Dead assemblages resulting from postmortem modifications of an original assemblage in which agglutinated tests found a minor part (= field I, outlined in black). The most severely altered assemblages have an open triangle at the bottom left corner

forms has suffered 70% postmortem dissolution of the calcareous component assuming no differences introduced by production or transport.

In tropical environments, tests of calcareous forms may be weakened by endolithic algae which bore into the wall. In the shallow lagoons of Curaçao, *Sorites orbiculus* tests, with a wall only 5–10 μm thick, show infestation of 40% of whole tests and 80% of the larger fragments. The tests become weakened and fragment down to silt-sized particles (Kloos 1982).

Recognition of postmortem changes

The most important causes of postmortem change are transport and destruction of tests (by dissolution of calcareous forms or oxidation of the organic cement of feebly cemented agglutinated forms). Figure 5.8 attempts to classify degrees of alteration. The best preserved assemblages are in field I, where all postmortem changes are at a minimum. Transport may cause gain (field II and variants), particularly of small tests whether naturally small adults or juveniles of larger species. It may also cause loss by winnowing out small tests (field III and variants) or by abrading and destroying all but the large, robust tests (field IV and variants). Loss of fragile agglutinated species alters the assemblages to fields I-A–IV-A. Loss of calcareous forms through dissolution may be moderate (fields I-C–IV-C) or very severe (fields I-CC–IV-CC).

Severe postmortem alteration is invariably obvious and therefore readily recognised. The intermediate stages of alteration are perhaps less obvious and may lead to misinterpretation if overlooked.

Degree of alteration

As shown above, life processes play a minor role in changing the composition of the dead assemblage drawn from living assemblages over a period of time. By contrast, postmortem changes may significantly alter dead assemblages. There is a correlation between the four major fields defined in Fig. 5.8 and environment. These are listed below together with some specific examples (Table 5.7).

Field I: microtidal estuaries and lagoons; shelf seas not subject to powerful tidal, current or wave disturbances; slope, rise, abyssal plain.

Field II: mesotidal and macrotidal estuaries and lagoons; protected areas within shelf seas affected by powerful tidal, current or wave disturbance; contourite deposits; some slopes.

Field III: areas under the influence of modest currents and waves; channels in lagoons and estuaries, some continental shelves, submarine canyons, areas beneath geostrophic currents.

Field IV: areas under the influence of powerful currents; channels in mesotidal and macrotidal lagoons and estuaries, beaches.

Table 5.7 Examples of some of the assemblage categories defined in Fig. 5.8 (from Murray 1984a)

Field	Examples	Source
I	Long Island Sound, USA	Buzas 1965
II	Foot of 36 m terrace, Arabian Gulf	Hughes-Clark and Keij 1973
I-C	Nazca Plate, Pacific, 3300–3700 m	Resig 1981
I-C	Baltic Sea	Jarke 1961a; Lutze 1965
I-CC	Intertidal marsh, USA	Parker and Athearn 1959
III	Nearshore shelf, Arabian Gulf	Hughes-Clark and Keij 1973
IV	Beach, Puerto Rico	Seiglie 1971a

CHAPTER 6

Atlantic seaboard of North America

The foraminifera of the eastern seaboard of North America have been intensively studied from the earliest papers by Ehrenberg (1843) and Bailey (1851) until the present day. A bibliography covering the period up to 1980 has been drawn up by Culver (1980). Many of the earlier papers were taxonomic with little ecological detail. Ecological studies became a primary objective in the 1940s with the works of Parker and Phleger and, especially after 1952 when Walton introduced the rose Bengal staining method, living and dead or total assemblages were distinguished. A synthesis of the data on living foraminifera was provided by Murray (1973). Culver and Buzas (1980) have plotted the geographical occurrence of 149 taxa. Although useful, this would have gained much if depth data had been included.

In the preparation of a synthesis problems of changing views of synonymies arise. The most difficult group is that of *Elphidium*, particularly the *clavatum–excavatum* complex. It now seems generally agreed that the correct name is *E. excavatum* (Levy *et al.* 1975; Buzas *et al.* 1985). Miller *et al.* (1982) recognise five formae which they regard as ecophenotypes (see Ch. 2, p. 22). Painter and Spencer (1984) thought they could distinguish six morphotypes but only two (*clavatum* and *selseyensis*) could be separated statistically. It is not always clear in other papers which is the correct form name and in this case the species is here referred to *E. excavatum*. However, as noted by Buzas *et al.* (1985) some references to this species are incorrect identifications of *E. williamsoni*. A simple key to some of the benthic Foraminifera is that of Todd and Low (1981).

Figure 6.1 shows the localities referred to in this chapter.

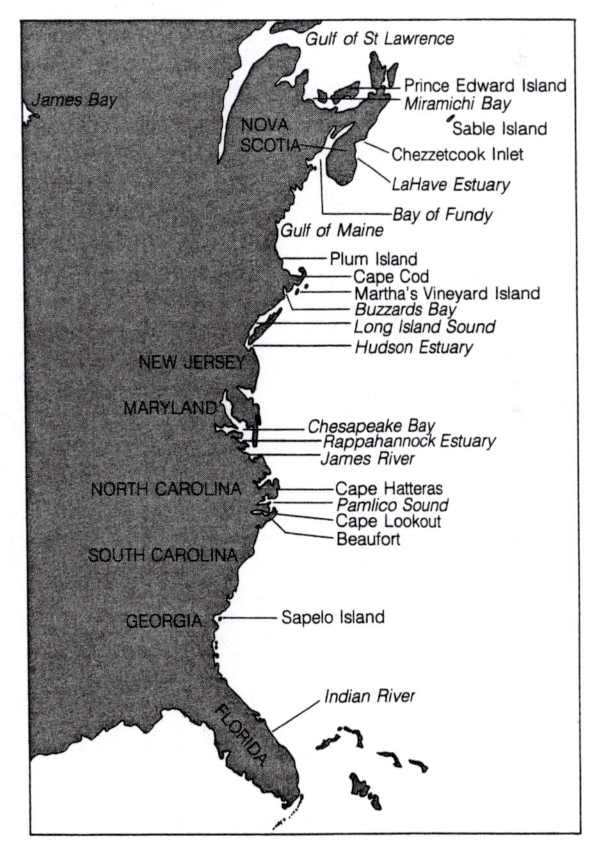

Figure 6.1 Localities referred to in this chapter

Marshes

Marshes are intertidal and sometimes supratidal areas colonised by halophytic (salt-tolerant) land plants. The sediment is normally of clay or silt grade because of the low-energy depositional environment provided by the wave-baffling effect of the vegetation. The marshes of the Atlantic seaboard of North America are dominantly brackish but pools may become hypersaline or evaporate entirely during hot summer weather.

Typical benthic species inhabiting marsh environments are listed in Table 6.1 and the details of assemblages are given in Tables 6.2 and 6.4. Most of the species are believed to be epifaunal and either herbivores or detritivores. Since organic detritus is abundant in marshes and algal blooms are present at least in the spring and autumn, food is not likely to be a major limiting parameter except perhaps for controlling the size of the standing crop. However, not all marsh species are epifaunal. Agglutinated species, notably *Ammobaculites exiguus*, *Trochammina inflata* and *Miliammina fusca*, were present at –5 and –10 cm below the surface on Hommocks Marsh, Long Island. The standing crop was 1–46 per 10 cm^2 at –5 cm and 1–33 at –10 cm, and diversity was low ($H(S)$ 0.14–1.93 at –5 cm, 0.30–1.59 at –10 cm; Steinbeck and Bergstein 1979). In another example, North Sea Harbor, Long Island, foraminifera were present to –10 cm although rare below –8 cm. *Ammotium salsum* was common down to –3 cm, *Elphidium excavatum* down to –6 cm and *T. inflata* down to –8 cm (Matera and Lee 1972). The occurrence of infaunal foraminifera in anoxic black sediments may be possible because burrows of fiddler crabs and the rhizomes of *Spartina* provide oxygen 'oases'.

Ability to withstand extreme environmental conditions (salinity, temperature, low dissolved oxygen levels and low pH) is probably the most important attribute for suvival. Marsh species should be opportunistic and reproduce rapidly during favourable environmental conditions.

Marsh species show patchy distribution patterns and blooms develop locally. On the North Sea Harbor Marsh, Long Island, the *Haynesina germanica* association was dominant in June while in late July and August it was replaced by the *Elphidium excavatum clavatum* association (Matera and Lee 1972). *Haynesina germanica* also peaked in June on Cape Cod (Parker and Athearn 1959).

All marsh assemblages, living and dead, have low diversity (α commonly <1, maximum 3; $H(S)$ <1.85). High marshes are characterised by agglutinated assemblages; hyaline forms are invariably totally absent. In terms of preservation high marsh dead assemblages fall in field I. Low marshes are commonly dominated by *Miliammina fusca* and have a

Table 6.1 Typical marsh species (feeding strategy form Ch. 2)

Species	Mode of life	Feeding strategy
Elphidium williamsoni	Epifaunal, free	Herbivore
Jadammina macrescens	Epifaunal, free	Herbivore or detritivore
Miliammina fusca	Infaunal, free	Detritivore
Tiphotrocha comprimata	Epifaunal, free	Herbivore or detritivore
Trochammina inflata	Epifaunal–infaunal, free	Herbivore or detritivore
Ammonia beccarii	Infaunal, free	? Herbivore
Haynesina germanica	Infaunal, free	? Herbivore

Table 6.2 Minor benthic associations on marshes along the Atlantic seaboard of North America (L=living, D=dead, T = total)

Association		Area	Source
Ammotium salsum	T	Poponesset Bay	Parker and Athearn 1959
	L	Long Island	Steineck and Bergstein 1979
Arenoparrella mexicana	T	Poponesset Bay	Parker and Athearn 1959
	T	Sapelo Island	Goldstein and Frey 1986
Ammobaculites exiguus	L	Long Island	Steineck and Bergstein 1979
Ammoastuta salsa	T	Rappahannock Estuary	Ellison and Nichols 1970
Triloculina oblonga	T	Sapelo Island, Georgia	Goldstein and Frey 1986
Haplophragmoides wilberti	T	Sapelo Island, Georgia	Goldstein and Frey 1986

Table 6.3 Principal features of marsh associations (L=living, D=dead)

Dominant species	Diversity α	% Agglutinated	% Hyaline
High marsh			
1. *Jadammina macrescens*	1–1 L	100 L	0 L
	1 D	100 D	0 D
2. *Jadammina macrescens–*	1 L	100 L	0 L
Tiphotrocha comprimata	1 D	100 D	0 D
Mid marsh			
3. *Trochammina inflata*	1.5–2 L	71–93 L	7–29 L
Low marsh			
4. *Elphidium williamsoni*	1–2 L	4–35 L	65–96 L
5. *Miliammina fusca*	1–1.5 L	68–100 L	0–32 L
	1–3 D	52–100 D	0–48 D
6. *Ammonia beccarii*	1–1.5 L	31–50 L	50–69 L
	1.5 D	57 D	43 D
7. *Haynesina germanica*	—	—	—

high agglutinated proportion in both living and dead assemblages. However, three hyaline living associations are present (Table 6.3), but due to postmortem dissolution of calcareous tests these are not normally preserved. Instead, they are represented by agglutinated assemblages often of the *M. fusca* association (field I-CC). Postmortem dissolution was noticed to be important by Parker and Athearn (1959) and Scott and Medioli (1980a).

The almost monospecific *Jadammina macrescens* association extends as far north as Hudson Bay (Fig. 6.2) where the marshes experience ice scour during the winter (Scott and Martini 1982). The *J. macrescens–Tiphotrocha comprimata* association is well developed on the high marshes of Prince Edward Island and Nova Scotia; there are few additional common species. The mid marsh level is sometimes colonised by the *Trochammina inflata* association. The additional common species are drawn from both high and low marsh and become more varied in the more southerly examples (Table 6.4). *Arenoparrella mexicana* makes its northernmost occurrence in this association at Percival River on Prince Edward Island (Scott *et al.* 1981).

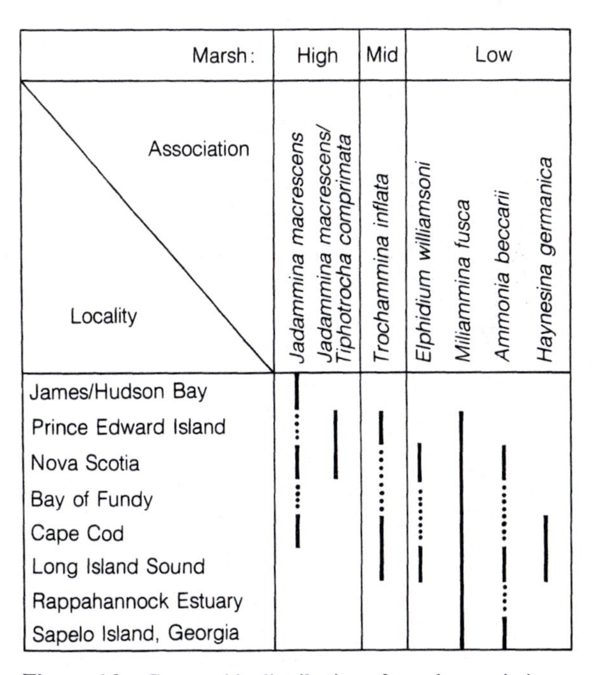

Figure 6.2 Geographic distribution of marsh associations

Unlike other marsh associations, the *Miliammina fusca* association is dominated by an infaunal species. On Prince Edward Island it is rarely present living but some of the dead assemblages are dominated by this species (Scott *et al.* 1981). In the macrotidal Bay of Fundy although the living assemblages are variable in composition, most of the dead assemblages are of the *M. fusca* association. The dissimilarity between the living and dead assemblages in this case may be the result of transport due to the extreme tidal conditions. From Table 6.4 it can be seen that the accessory species of the *M. fusca* association are different with passage from north to south, and there is a marked change between the Rappahannock Estuary and Sapelo Island.

Table 6.4 Marsh associations

Jadammina macrescens **association**

Salinity: variable

Temperature: extreme range

Substrate: silt, mud with a high organic (plant debris) content

Depth: intertidal–supratidal, high marsh

Distribution: (all as *Trochammina*)
1. James and Hudson bays (Scott and Martini 1982) living, total
2. Chezzetcook Inlet, Nova Scotia (Scott and Medioli 1980b) living, total
3. Plum Island, Massachusetts (Jones and Cameron 1987) total
4. Barnstable Harbor, Massachusetts (Phleger and Walton 1950) total

Additional common species:

Polysaccammina ipohalina	1
Trochammina inflata	3, 4
Tiphotrocha comprimata	3
Miliammina fusca	3

Trochammina inflata **association**

Salinity: very variable

Temperature: 1 to 20 °C

Substrate: silt, mud rich in organic detritus (plant)

Depth: intertidal, mid marsh

Distribution:
1. Wolfe Inlet and Percival River, Prince Edward Island (Scott *et al.* 1981) living, total
2. Barnstable Harbor, Massachusetts (Phleger and Walton 1950) total
3. Long Island (Steineck and Bergstein 1979) living

Additional common species:

Miliammina fusca	1, 2
Tiphotrocha comprimata	1, 3
Jadammina macrescens	1, 2, 3 (as *Trochammina*)
Ammotium salsum	3
Ammonia beccarii	3
Ammomarginulina fluvialis	3

Jadammina macrescens–Tiphotrocha comprimata **association**

Salinity: 0–20‰

Temperature: –1 to 20 °C

Substrate: silt, clay rich in organic detritus (plants)

Depth: high marsh

Distribution:
1. Prince Edward Island (Scott *et al.* 1981) living, total
2. Chezzetcook Inlet, Nova Scotia (Scott and Medioli 1980b) living, total
3. Nova Scotia (Deonarine 1979) total

Additional common species:

Pseudothurammina limnetis	1 (not preserved fossil)
Haplophragmoides bonplandi	1, 2
Trochammina inflata	2, 3
Miliammina fusca	3

Elphidium williamsoni **association**

Salinity: 20–30‰

Temperature: very variable

Substrate: silt, clay

Depth: low intertidal marsh and mudflat

Distribution:
1. Chezzetcook Inlet, Nova Scotia (Scott and Medioli 1980b) living only (dead = *Miliammina fusca* association)
2. Long Island (Steineck and Bergstein 1979) living

Additional common species:

Ammonia beccarii	1, 2
Miliammina fusca	2
Trochammina inflata	2

Table 6.4 *(continued)*

Miliammina fusca **association**

Salinity: 0–35‰

Temperature: –1 to 29 °C

Substrate: silt, clay with abundant organic detritus (plant)

Depth: intertidal, low marsh

Distribution:
1. Prince Edward Island (Scott *et al.* 1981) living, total
2. Chezzetcook Inlet, Nova Scotia (Scott and Medioli 1980b) living, total
3. Bay of Fundy (a) (Smith *et al.* 1984) living, total; (b) (Deonarine 1979) total
4. Plum Island, Massachusetts (Jones and Cameron 1987) total
5. Barnstable Harbor, Massachusetts (Phleger and Walton 1950) total
6. Poponesset Bay, Cape Cod (Parker and Athearn 1959) living, dead, total
7. Rappahannock Estuary (Ellison and Nichols 1970) total
8. Sapelo Island, Georgia (Goldstein and Frey 1986) total

Additional common species:

Tiphotrocha comprimata	2, 3a, 4, 5, 6
Trochammina inflata	3a, 4, 5, 6
Jadammina macrescens	3a, 4, 5, 6
Arenoparrella mexicana	4, 6, 7, 8
Ammotium salsum	4, 8
Ammobaculites crassus	7
Ammobaculites dilatatus	8
Ammotium pseudocassis	8
Lepidodeuterammina ochracea	8
Siphotrochammina lobata	8
Ammonia parkinsoniana	8
Triloculina oblonga	8

Haynesina germanica **association**

Salinity: 2–34‰

Temperature: 0 to 32 °C

Substrate: silt, clay rich in organic detritus (plant)

Depth: intertidal marsh

Distribution:
1. Poponesset Bay, Cape Cod (Parker and Athearn 1959) living only (as *Protelphidium tisburyense*)
2. North Sea Harbor, Long Island (Matera and Lee 1972) living (as *Protelphidium tisburyensis*)

Additional common species:

Miliammina fusca	1
Tiphotrocha comprimata	1
Ammotium salsum	1
Arenoparrella mexicana	1
Elphidium galvestonense	1
Elphidium excavatum clavatum	2

Ammonia beccarii **association**

Salinity: very variable

Temperature: very variable

Substrate: organic-rich clays

Depth: intertidal, low marsh

Distribution:
1. Chezzetcook Inlet, Nova Scotia (Scott and Medioli 1980b) living, total
2. Long Island (Steineck and Bergstein 1979) living
3. Sapelo Island, Georgia (Goldstein and Frey 1986) total (as *A. parkinsoniana*)

Additional common species:

Haynesina orbiculare	1
Miliammina fusca	1, 3
Ammotium salsum	2
Ammobaculites exiguus	2
Trochammina inflata	2
Triloculina oblonga	3
Elphidium incertum	3
Arenoparrella mexicana	3
Haynesina germanica	3 (as *Protelphidium tisburyense*)

Table 6.5 Standing crop per 10 cm^2 in the Chezzetcook Inlet marshes (data from Scott and Medioli 1980b)

	Upper estuary				Central estuary		Lower estuary
	Elevation				Elevation		
	+1.0 to +0.55 m		+0.55 to 0 m		+1.0 to +0.55 m	+0.55 to 0 m	
Transect	I	III	I	III	IV	IV	Stas 8, 9, 19
No. stations	62	56	6	4	20	20	26
Range	0–198	44–1194	1–128	99–295	0–706	96–1998	0–252
Average	22	485	18	153	167	386	48

Two of the hyaline low marsh associations have a very restricted geographic distribution. The *Elphidium williamsoni* association is present in Chezzetcook Inlet, Nova Scotia, and *E. williamsoni* makes up 50–80% of the living assemblages (with 17–37% *Miliammina fusca*). Because of 95% postmortem dissolution the dead assemblages are dominated by *M. fusca* (49–92%) and *E. williamsoni* is reduced to <20%. Similarly, the *Haynesina germanica* association is present only living. The *Ammonia beccarii* association is more widely distributed. Each area has a different set of additional common species (Table 6.4). Because of the high postmortem dissolution, low marsh living hyaline assemblages are generally preserved as agglutinated-rich assemblages of field I-CC.

The standing crop of the living assemblages is very variable as can be seen from the data for Chezzetcook Inlet (Table 6.5). In general the most extreme environments have the lowest values, e.g. James and Hudson bays range from 0 to 180 per 10 cm^2 but mostly <50 per 10 cm^2.

Estuaries and lagoons

The belt of major estuaries and lagoons of the Atlantic seaboard of North America extends from Canada to Cape Lookout. All the examples for which there are foraminiferal data are brackish to some degree and most have muddy substrates because of their sheltered low-energy setting.

Typical estuarine and lagoon species are infaunal, free-living detritivores (Table 6.6) feeding on the abundant organic detritus which is commonly present in the sediment. The depth of life extends to at least 10 cm (Rappahannock Estuary, Ellison 1972; Indian River, Buzas 1977; Rhode River, Buzas 1974). Seven major associations are recognised (Table 6.7 summary, Table 6.8 details). In addition, there area nine associations found only at a single locality (Table 6.9).

Estuaries and lagoons experience major seasonal variations in their physical conditions and in their salinity. They are less extreme than intertidal marshes but more variable than the adjacent continental shelf. The foraminifera which are successful are opportunistic species. Their lateral distribution patterns are patchy and replicate samples are commonly statistically different (see e.g. the publications by Buzas). The diversity is low, α <1 to 5, $H(S)$ <1.5. The associations show varying degrees of restriction in their geographic distributions (Fig. 6.3). The major controls appear to be abiotic as each association has its own temperature–salinity field although there is a considerable measure of overlap (Fig. 6.4).

Table 6.6 Typical estuarine and lagoon species (feeding strategy from Ch. 2)

Species	Mode of life	Feeding strategy
Ammobaculites crassus	Infaunal, free	Detritivore
Ammotium salsum	Infaunal, free	Detritivore
Ammonia beccarii	Infaunal, free	? Herbivore
Ammotium cassis	Infaunal, free	Detritivore
Eggerella advena	Infaunal, free	Detritivore
Elphidium excavatum clavatum	Infaunal, free	? Herbivore
Elphidium williamsoni	Infaunal, free	Herbivore
Miliammina fusca	Infaunal, free	Detritivore
Haynesina orbiculare	Infaunal, free	? Herbivore

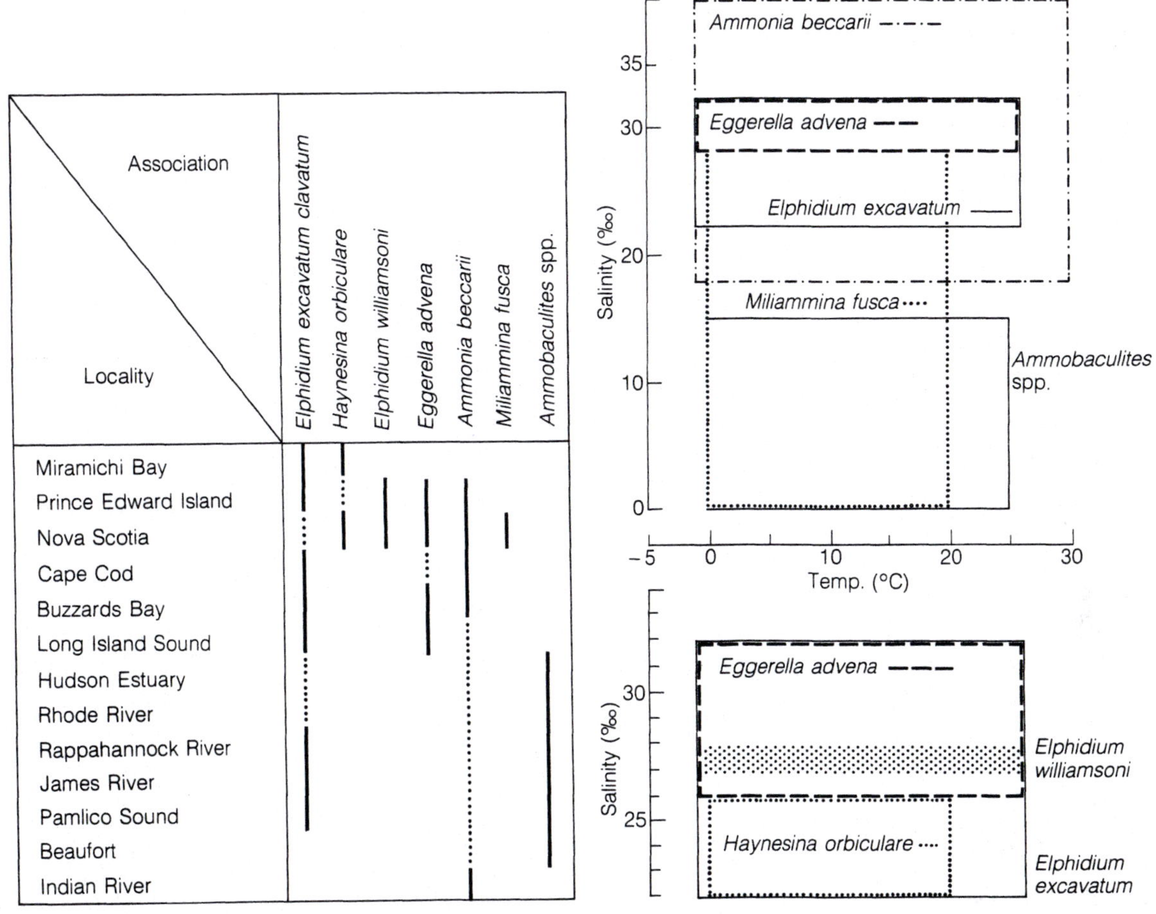

Figure 6.3 Geographic occurrence of estuary and lagoon associations

Figure 6.4 Temperature and salinity plots for estuary and lagoon associations

Table 6.7 Principal features of estuarine and lagoon associations from the Atlantic seaboard of North America (L = living, D = dead)

Dominant taxon	Diversity α	Agglutinated (%)	Porcellaneous (%)	Hyaline (%)
1. *Ammobaculites spp.*	1 L 1 D	100 L 40–100 D	0 L 0 D	0 L 0–60 L
2. *Ammonia beccarii*	1.5–2.5 L* 1.5–4 D	0–11 L 17–33 D	0–42 L 0 D	55–98 L 67–83 D
3. *Eggerella advena*	1.5 L 3–5 D	48–77 L 20–65 L	0 L 0–2 D	23–52 L 35–78 D
4. *Elphidium excavatum clavatum*	1–2 L 1–1.5 D	0–25 L 0–29 L	0–2 L 0 D	75–100 L 71–100 D
5. *Elphidium williamsoni*	1–1.5 L 2–4.5 D	12–24 L 20–42 D	0 L 0–2 D	76–98 L 58–78 D
6. *Miliammina fusca*	1–4 D	26–100 D	0 D	0–74 D
7. *Haynesina orbiculare*	1–2 L 1.5–4.5 D	0–38 L 3–38 L	0 L 0 D	62–100 L 62–97 D

*α 2.5–11 in the brackish to hypersaline Indian River, Florida

Table 6.8 Principal associations from estuaries and lagoons from the Atlantic seaboard of North America

***Ammobaculites* spp. association**

Salinity: 0–15‰

Temperature: 0 to 25 °C

Substrate: silty clay

Depth: 0–10 m

Distribution:
1. Hudson Estuary (Weiss 1976; Weiss *et al.* 1978) total
2. Rhode River, Maryland (Buzas 1974) living, dead (as *A. exiguus*)
3. Rappahannock Estuary, Virginia (Ellison and Nichols 1970) living, total (as *A. crassus*)
4. James River Estuary, Virginia (Nichols and Norton 1969) living, total (as *A. crassus*)
5. Pamlico Sound, N. Carolina (Grossman and Benson 1967) total (as *A. crassus*)
6. Beaufort area, N. Carolina (Akers 1971) living, dead (as *A. crassus*)

Additional common species:

Ammomarginulina fluvialis	1
Elphidium excavatum clavatum	1, 3
Ammonia beccarii	3
Ammobaculites exilis	5
Ammobaculites pamlicoensis	5
Ammoastuta inepta	5
Miliammina fusca	6
Ammotium salsum	6

***Ammonia beccarii* association**

Salinity: 18–40‰

Temperature: –1 to 30 °C

Substrate: silty sand

Depth: <20 m

Distribution:
1. Tracadie Bay, Prince Edward Island (Bartlett 1965) living, total
2. Chezzetcook Inlet, Nova Scotia (Scott *et al.* 1980) living, total
3. Hadley Harbor, Massachusetts (Buzas 1968b) living, total
4. Buzzards Bay, Massachusetts (Murray 1968b) living
5. Indian River, Florida (Buzas and Severin 1982) living

Additional common species:

Elphidium excavatum clavatum	1 (as *E. incertum*), 3, 4
Elphidium williamsoni	1 (as *E. margaritaceum*)
Miliammina fusca	1
Elphidium excavatum selseyensis	2 (as *Cribrononion*)
Eggerella advena	2, 3, 4
Haynesina orbiculare	2
Elphidium subarcticum	3
Fursenkoina fusiformis	4
Quinqueloculina seminula	5
Quinqueloculina impressa	5
Elphidium mexicanum	5
Brizalina striatula	5 (as *Bolivina*)

Table 6.8 *(continued)*

***Eggerella advena* association**

Salinity: 26–32‰

Temperature: –1 to 26 °C

Substrate: sand, clayey silt

Depth: <30 m

Distribution:

1. Restigouche Estuary, Gulf of St Lawrence (Schafer and Cole 1976) total
2. Tracadie Bay, Prince Edward Island, (Bartlett 1965) living
3. Chezzetcook Inlet, Nova Scotia (Scott *et al.* 1980) total
4. LaHave Estuary, Nova Scotia (Allen and Roda 1977) total
5. Buzzards Bay, Massachusetts (Murray 1968b) living
6. Long Island Sound (Buzas 1965) living, total

Additional common species:

Elphidium excavatum clavatum	1, 2 (as *E. incertum*)
Ammotium salsum	1 (as *Ammobaculites*)
Cuneata arctica	1 (as *Reophax*)
Reophax fusiformis	1
Ammotium cassis	1, 4
Spiroplectammina biformis	1
Ammonia beccarii	2, 3, 5
Elphidium williamsoni	3
Miliammina fusca	3
Haynesina orbiculare	3 (as *Protelphidium*)
Rosalina columbiensis	3
Buccella frigida	6
Elphidium pauciloculum	6

***Elphidium excavatum* forma *clavatum* association**

Salinity: 22–32‰

Temperature: –1 to 26 °C

Depth: <30 m

Substrate: clay, silt, sand

Distribution:

1. Miramichi Bay (Scott *et al.* 1977) living (as *Cribroelphidium*)
2. Tracadie Bay, Prince Edward Island (Bartlett 1965) living, total (as *E. incertum*).
3. Hadley Harbor, Massachusetts (Buzas 1968b) living, total
4. Buzzards Bay, Massachusetts (Murray 1968b) living
5. Long Island Sound: (a) (Parker 1952) total (as *E. incertum* and variants); (b) (Buzas 1965) living, total; (c) (Akpati 1975) living, total
6. Rappahannock Estuary, Virginia (Ellison and Nichols 1970) living, total
7. James River Estuary, Virginia (Nichols and Norton 1969) living, total
8. Pamlico Sound, N. Carolina (Grossman and Benson 1967) total

Additional common species:

Haynesina orbiculare	1, 2 (as *Protelphidium*)
Eggerella advena	1, 2, 4
Ammonia beccarii	2, 3, 4, 5c, 6
Elphidium subarcticum	3, 4
Buccella frigida	3, 5b, c
Elphidium pauciloculum	5b
Saccammina atlantica	5c (as *Proteonina*)
Quinqueloculina seminula	5c
Elphidium incertum	8

Table 6.8 *(continued)*

Elphidium williamsoni **association**

Salinity: 27–28‰

Temperature: –1 to 26 °C

Substrate: sand

Depth: <15 m

Distribution:
1. Tracadie Bay, Prince Edward Island (Bartlett 1965) dead (as *E. margaritaeum*)
2. Chezzetcook Inlet, Nova Scotia (Scott *et al.* 1980) living, total

Additional common species:

Elphidium excavatum clavatum	1 (as *E. incertum*)
Miliammina fusca	1
Ammonia beccarii	2
Haynesina orbiculare	2 (as *Protelphidium*)
Eggerella advena	2
Rosalina columbiensis	2
Elphidium excavatum selseyensis	2 (as *Cribrononion*)

Miliammina fusca **association**

Salinity: 0–28‰

Temperature: 0 to 20 °C

Substrate: silty clay

Depth: <10 m

Distribution:
1. Chezzetcook Inlet, Nova Scotia (Scott *et al.* 1980) total
2. LaHave Estuary, Nova Scotia (Allen and Roda 1977) total

Additional common species:

Ammotium salsum	1
Ammonia beccarii	1
Elphidium excavatum clavatum	1 (as *Cribrononion*)
Eggerella advena	1, 2
Haynesina orbiculare	1 (as *Protelphidium*)
Elphidium excavatum selseyensis	1 (as *Cribrononion*)
Saccammina atlantica	2
Ammotium cassis	2

Haynesina orbiculare **association**

Salinity: 22–26‰

Temperature: 0 to 20 °C

Substrate: sand, muddy sand

Depth: <10 m

Distribution:
1. Miramichi Bay (Scott *et al.* 1977) living
2. Chezzetcook Inlet, Nova Scotia (Scott *et al.* 1980) living, total
3. LaHave Estuary, Nova Scotia (Allen and Roda 1977) living, total

Additional common species:

Ammotium cassis	1
Elphidium williamsoni	2 (as *Cribrononion*)
Elphidium excavatum selseyensis	2, 3 (as *Cribrononion*)
Ammonia beccarii	2
Miliammina fusca	2
Buccella frigida	2
Hemisphaerammina bradyi	2

Table 6.9 Minor benthic associations from estuaries and lagoons from the Atlantic seaboard of North America; 1–8 are brackish, 9 hypersaline (L = living, D = dead, T = total)

Association		Area	Depth (m)	Source
1. *Ammotium cassis*	L	Miramichi Bay	?	Scott *et al.* 1977
2. *Hemisphaerammina bradyi*	L	Chezzetcook Inlet	10	Scott *et al.* 1980
3. *Elphidium excavatum selseyensis*	L, D, T	Chezzetcook Inlet	10	Scott *et al.* 1980
4. *Rosalina columbiensis*	L	Chezzetcook Inlet	45	Scott *et al.* 1980
5. *Ammotium salsum*	L	Restigouche Estuary	20	Schafer and Cole 1976
		Chezzetcook Inlet	10	Scott *et al.* 1980
6. *Buccella frigida*	L, T	Long Island Sound	15–28	Buzas 1965
7. *Elphidium galvestonense*	T	Pamlico Sound	6	Grossman and Benson 1967
8. *Haynesina germanica* (as *Protelphidium orbiculare*)	T	Pamlico Sound	5	Grossman and Benson 1967
9. *Quinqueloculina* spp.	L	Indian River	1–3	Buzas and Severin 1982

The *Elphidium excavatum clavatum* association is widely developed from the Gulf of St Lawrence to Cape Hatteras. It tolerates a broad seasonal temperature range but prefers a modest salinity range (22–32‰) and is present on various types of substrate. It occasionally extends into brackish marshes, e.g. Hommocks Marsh, Long Island Sound (Matera and Lee 1972) and is present on the inner continental shelf. The common additional species show a progressive change from north and south (Table 6.8).

The *Haynesina orbiculare* association is essentially a cool-water one although it has a broad temperature tolerance. It is present on sandy substrates in shallow waters with a salinity of 22–26‰. The *Elphidium williamsoni* association replaces it at salinities of 27–28‰. These two associations are restricted to the Gulf of St Lawrence and Nova Scotia.

The *Eggerella advena* association is characteristic of northern lagoons, from the Gulf of St Lawrence to Long Island Sound, and the adjacent inner shelf. It prefers moderate brackish salinities (26–32‰.). The *Ammonia beccarii* association embraces a number of ecophenotypes. The northern occurrences are of forms more tolerant of lower temperatures than those from the south. This species is moderately euryhaline. The additional common species differ from north to south. The southernmost occurrence, in Indian River, Florida, spans the brackish to hypersaline field (salinity 18–40‰) and because of this *Quinqueloculina* spp. are common additional species (porcellaneous 2–42‰). The diversity values are moderate (α 2.5–11) and much higher than those of living assemblages of this association elsewhere. Altogether 94 living species were present but only 15 make up 95% of the living assemblages (Buzas and Severin 1982).

The *Miliammina fusca* association so common on brackish marshes is present in estuaries and lagoons only in Nova Scotia (Fig. 6.4). The *Ammobaculites* spp. association is found only from the Hudson River to Cape Hatteras. The principal species is *A. crassus* which is so abundant in Chesapeake Bay that Ellison (1972) entitled a paper '*Ammobaculites*, foraminiferal proprietor of Chesapeake Bay estuaries'. This infaunal detritivore can tolerate wide ranges of temperature and salinity but its inferred optimum conditions are 5–15‰ salinity and 21–28 °C (Ellison and Murray 1987). It particularly favours silty mud with abundant organic detritus derived from adjacent marshes, but it also occurs on sediments rich in amphipod faecal pellets. Such conditions commonly occur beneath the turbidity maximum (= gradient zone of Nichols and Ellison 1967).

The standing crop values of the living assemblages are very variable. Experimental studies using open (to predators) and screened (predators excluded) cages placed on the bed of the Indian River showed, over a 4-year period, that standing crop values were higher in screened than unscreened cages. Deposit feeders consume benthic foraminifera and regulate their densities (Buzas 1978, 1982; Buzas and Carle 1979).

Postmortem changes

The dead assemblages of estuaries and lagoons may be closely similar or dissimilar to the living assemblages. In Long Island Sound, the living and dead assemblages are closely similar at depths of ≤13 m, but below this they show changes due either to dissolution of calcareous tests or to transport. In Chezzetcook Inlet calcareous living assemblages are commonly represented by agglutinated dead assemblages. Two species which are insignificant in the calcareous living assemblages, namely *Miliammina fusca* (0–16%) and *Eggerella advena* (0–6%), become important in the dead assemblages (0–82% and 0–40% respectively). Transport of tests from adjacent marshes and the inner shelf may lead to an increase in diversity. Transport may also blur boundaries between associations. Thus dead assemblages fall in fields I-C–I-CC.

Continental shelf

During the Plio-Pleistocene, glacial erosion affected the area from Newfoundland to the Gulf of Maine. Glacial and meltwater deposition extended as far south as the Cape Hatteras area. Marine or near-marine sediments are now accumulating along the coastal, shelf and slope areas. The shelf sediments are of terrigenous clastic type as far south as Florida where, on the narrow shelf, carbonate sediments predominate due to the lack of a source of detrital material.

Some attributes of the shelf seawater masses are shown in Fig. 6.5. To the south of Cape Cod the area is microtidal. The Gulf of Maine is mesotidal, and the Bay of Fundy macrotidal. The Scotian Shelf and the Gulf of St Lawrence are mainly microtidal. Bottom currents show gyres in the Gulf of St Lawrence and the Gulf of Maine. In mid shelf from Maine to Chesapeake Bay there is a line of divergence between currents on the inner and outer shelf.

Bottom-water temperatures are low in summer and winter throughout the Gulf of St Lawrence and the Gulf of Maine (≤ 5 °C) but part of the Scotian shelf reaches >7.5 °C. From Cape Cod to Cape Hatteras temperatures reach a maximum of 15 °C on the outer shelf, but the bottom waters along the coast are <7 °C in the winter because of the southward-moving Virginian Coastal Water (note the bottom currents in Fig. 6.5A). South of Cape Hatteras the bottom water is warm throughout the year and reaches values of 27.5 °C on the inner shelf during the summer. Surface salinity values show a general northward decrease due to the input of fresh water and the bottom salinities will be similar. Much additional oceanographic information is given in Emery and Uchupi (1972).

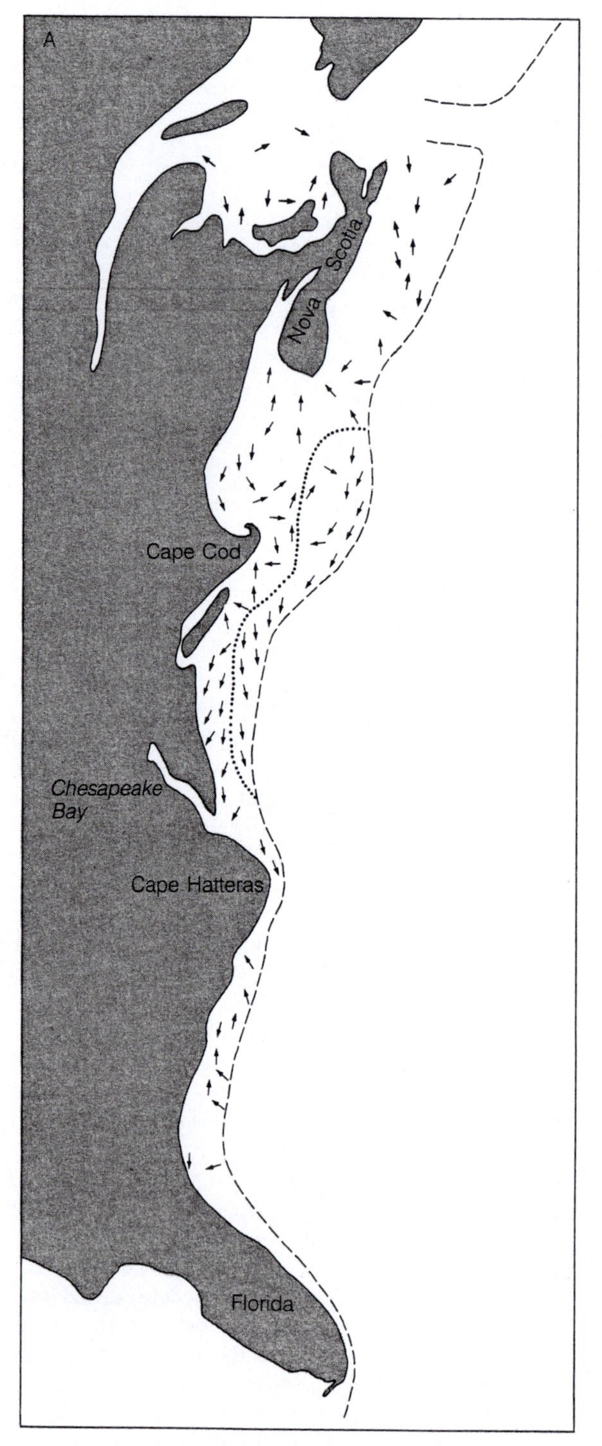

The basic biogeographic divisions are largely temperature based (Fig. 6.5C). For benthic foraminifera, Cape Hatteras marks a major faunal boundary

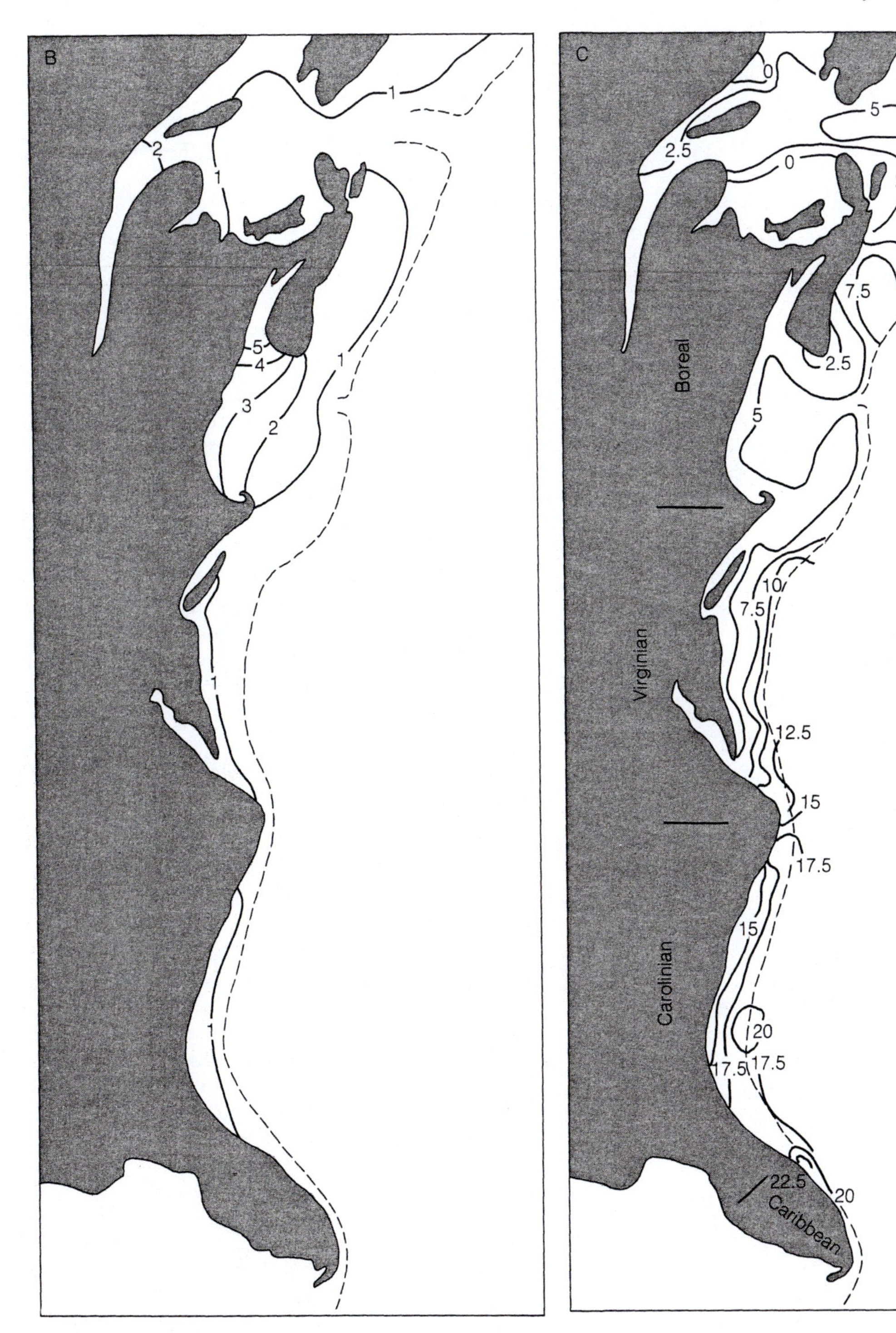

Figure 6.5 Oceanographic features of the Atlantic continental shelf of North America. (A) Bottom currents: (—) shelf edge, (...) zone of divergence; (B) tides (m); (C) bottom-water temperatures (°C) and biogeographic divisions; (D) surface salinity (‰), February (data from Emery and Uchupi 1972)

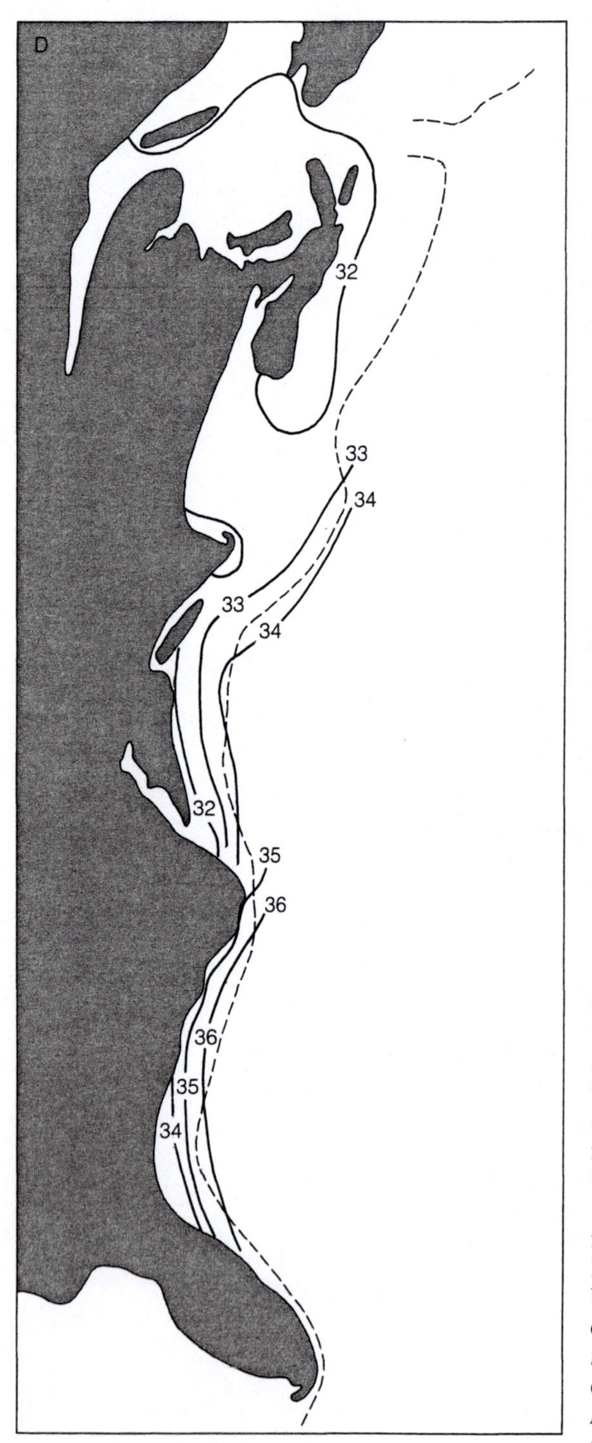

between northern and southern species (Murray 1973; Buzas and Culver 1980; Culver and Buzas 1981c, 1982a).

Labrador Sea to Cape Hatteras

The extensive literature on shelf assemblages is listed in Table 6.10. Sixteen minor associations present in only a single area (Table 6.11) will not be discussed further. Details of the 10 principal associations are given in Tables 6.12 and 6.13 and Figs 6.6 and 6.7. Of the abiotic environmental parameters for which there are details, salinity is not very variable from one association to another. The principal controls seem to be temperature and substrate. The upper temperature limits for the associations are as follows: *Islandiella helenae* 4 °C, *Adercotryma glomerata* 5.5 °C, *Trifarina angulosa* 8 °C and *I. islandica* 12 °C. The range for the *Bulimina marginata* association is 9–13 °C and all the other associations have a broad temperature tolerance. Coarse substrates are favoured by the *I. islandica* and *Cibicides lobatulus* associations. The latter species has an attached epifaunal mode of life. Muddy sand substrates are favoured by the *T. angulosa, B. marginata* and *Fursenkoina fusiformis* associations. The normal substrate for the *Elphidium excavatum clavatum* association is sand and the *Eggerella advena* association requires at least some silt or fine sand. Although some associations show a restricted depth range (Fig. 6.7: *I. islandica, C. lobatulus, Elphidium excavatum clavatum, Eggerella advena, F. fusiformis*) others are very variable (especially *Saccammina atlantica, I. helena, A. glomerata* and *B. marginata* associations).

Williamson *et al.* (1984) used regression analysis to attempt to explain the occurrence of their *Q*-mode varimax factor – defined assemblages on the Nova Scotia Shelf. Their conclusions are in general agreement with those reached here based on the broader distribution of the associations. Of particular interest is their observation that the *Saccammina atlantica* association shows no correlation with any of the measured environmental parameters. They noted that the association could represent the residue of an assemblage from which there had been selective removal of species through transport or dissolution. However, they were unable to test this hypothesis because they had studied total assemblages. The shelf off Long Island has dead assemblages of the *S. atlantica* association although the living assemblages are dominated in addition by *Eggerella advena* and *Fursenkoina fusiformis* (Murray 1969). At least part of the difference between the living and dead assemblages may be due to the inability of unilocular tests to grow in size. It is likely that when the old test is too small the animal leaves it and builds a new one

Tables 6.10 Sources of data on shelf assemblages, Labrador Sea to Cape Hatteras

Area	Source
Labrador Sea	Vilks *et al.* 1982
Grand Banks	Sen Gupta and McMullen 1969; Sen Gupta 1971; Sen Gupta and Hayes 1979
Gulf of St Lawrance	Rodrigues and Hooper 1982a, b; Schafer 1967, 1971; Buckley *et al.* 1974
Nova Scotia	Williamson *et al.* 1984; Medioli *et al.* 1986
Bay of Fundy	Thomas and Schafer 1982
Gulf of Maine	Parker 1948; Phleger 1952a
Martha's Vineyard–Long Island	Parker 1948, 1952; Murray 1969; Gervirtz *et al.* 1971
New Jersey	Poag *et al.* 1980; Miller and Ellison 1982
Maryland	Parker 1948
N. of Cape Hatteras	Murray 1969; Schnitker 1971

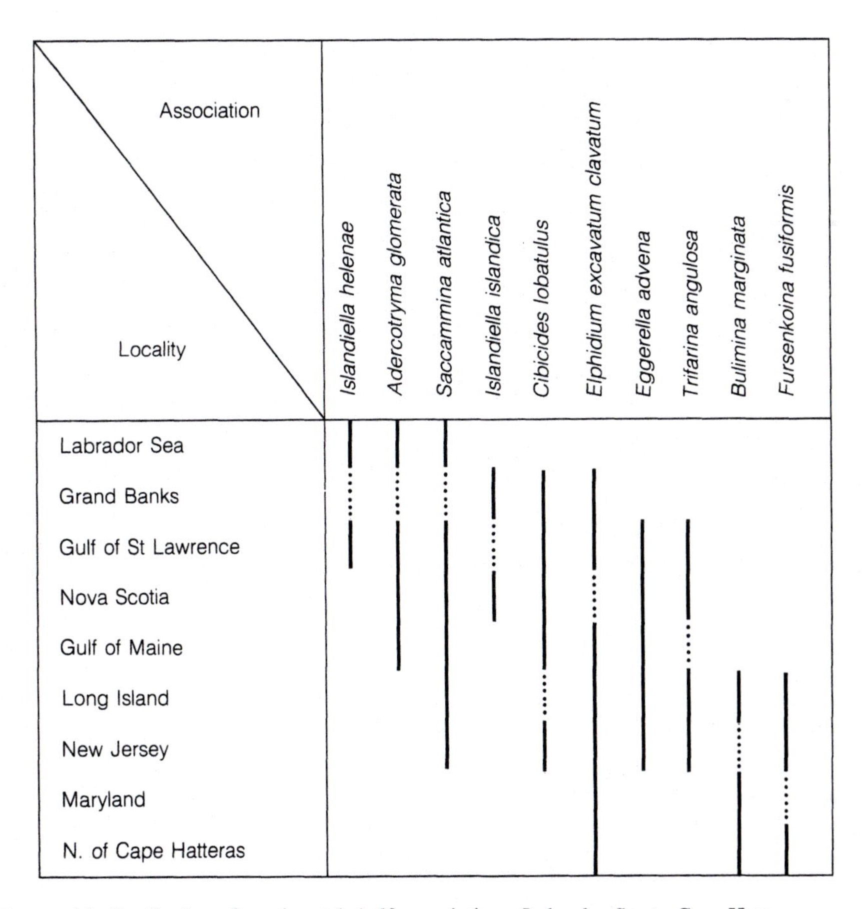

Figure 6.6 Geographic distribution of continental shelf associations, Labrador Sea to Cape Hatteras

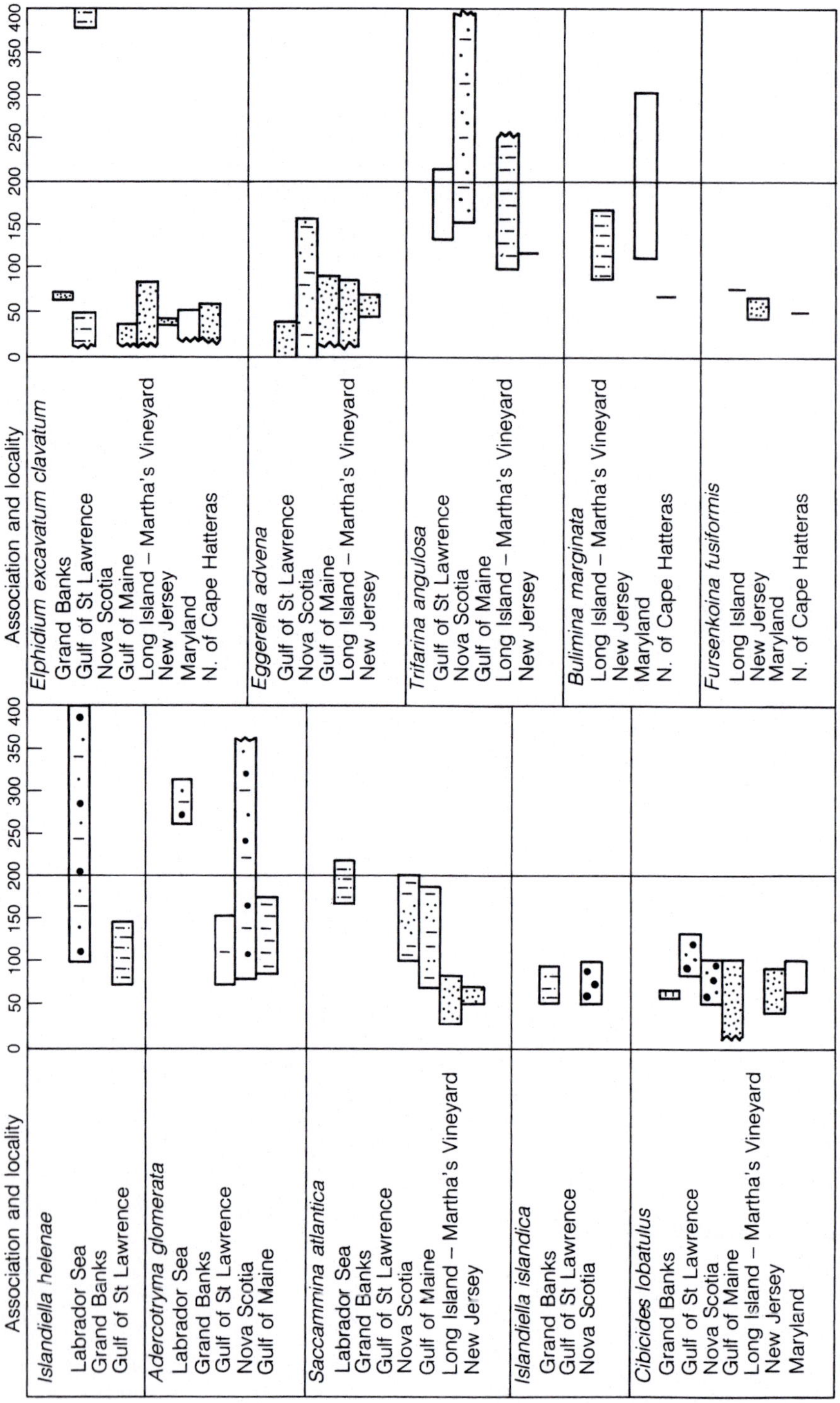

Association and locality	Association and locality
Islandiella helenae	*Elphidium excavatum clavatum*
Labrador Sea	Grand Banks
Grand Banks	Gulf of St Lawrence
Gulf of St Lawrence	Nova Scotia
Adercotryma glomerata	Gulf of Maine
Labrador Sea	Long Island – Martha's Vineyard
Grand Banks	New Jersey
Gulf of St Lawrence	Maryland
Nova Scotia	N. of Cape Hatteras
Gulf of Maine	*Eggerella advena*
Saccammina atlantica	Gulf of St Lawrence
Labrador Sea	Nova Scotia
Grand Banks	Gulf of Maine
Gulf of St Lawrence	Long Island – Martha's Vineyard
Nova Scotia	New Jersey
Gulf of Maine	*Trifarina angulosa*
Long Island – Martha's Vineyard	Gulf of St Lawrence
New Jersey	Nova Scotia
Islandiella islandica	Gulf of Maine
Grand Banks	Long Island – Martha's Vineyard
Gulf of St Lawrence	New Jersey
Nova Scotia	*Bulimina marginata*
Cibicides lobatulus	Long Island – Martha's Vineyard
Grand Banks	New Jersey
Gulf of St Lawrence	Maryland
Nova Scotia	N. of Cape Hatteras
Gulf of Maine	*Fursenkoina fusiformis*
Long Island – Martha's Vineyard	Long Island
New Jersey	New Jersey
Maryland	Maryland
	N. of Cape Hatteras

Depth (m): 0 50 100 150 200 250 300 350 400

Figure 6.7 Depth distribution of continental shelf associations

Tables 6.11 Minor benthic associations on the shelf from the Labrador Sea to Cape Hatteras (L = living, D = dead, T = total)

	Association		Area	Depth (m)	Source
1.	*Cuneata arctica*	T	Baffin Island	143	Schafer and Cole 1988
2.	*Textularia earlandi*	T	Baffin Island	302–486	Schafer and Cole 1988
3.	*Spiroplectammina biformis*	T	Baffin Island	294–440	Schafer and Cole 1988
4.	*Trochammina nana*	T	Baffin Island	253	Schafer and Cole 1988
5.	*Fursenkoina loeblichi*	L	Grand Banks	77–137	Sen Gupta 1971
6.	*Nonionellina labradorica*	L, D	Grand Banks	73–91	Sen Gupta 1971
7.	*Bulimina aculeata*	T	Gulf of St Lawrence	197–365	Rodrigues and Hooper 1982a
8.	*Bulimina exilis*	T	Gulf of St Lawrence	390–520	Rodrigues and Hooper 1982a
9.	*Globobulimina auriculata*	T	Nova Scotia	150–220	Williamson *et al.* 1984
10.	*Cassidulina algida*	T	Gulf of Maine	51–116	Phleger 1952a
11.	*Cribrostomoides crassimargo*	T	Gulf of Maine	98–137	Phleger 1952a
12.	*Haplophragmoides bradyi*	T	Gulf of Maine	104–183	Phleger 1952a
13.	*Reophax curtus*	T	Gulf of Maine	73–208	Phleger 1952a
14.	*Reophax scottii*	T	Gulf of Maine	66–82	Phleger 1952a
15.	*Textularia torquata*	T	Gulf of Maine	73–104	Phleger 1952a
16.	*Trochammina squamata*	T	Gulf of Maine	76–98	Phleger 1952a
17.	*Rosalina columbiensis*	T	Block Island	9–15	Parker 1952
18.	*Quinqueloculina seminula*	T	Long Island	30	Gevirtz *et al.* 1971
19.	*Elphidium subarcticum*	L	New Jersey	57–58	Poag *et al.* 1980, Poag 1982
20.	*Globocassidulina subglobosa*	T	N. of Cape Hatteras	80–140	Schnitker 1971

Table 6.12 Typical shelf species from the Atlantic seaboard of North America (feeding strategies from Ch. 2)

Species	Mode of life	Feeding strategy
Adercotryma glomerata	Epifaunal, free	Detritivore
Bulimina marginata	Epifaunal, infaunal, free	? Detritivore
Cibicides lobatulus	Epifaunal, attached	? Herbivore, detritivore
Eggerella advena	Infaunal, free	Detritivore
Elphidium excavatum clavatum	Infaunal, free	? Herbivore
Fursenkoina fusiformis	Infaunal, free	? Detritivore
Islandiella helenae	Infaunal, free	? Detritivore
Islandiella islandica	Infaunal, free	? Detritivore
Saccammina atlantica	Infaunal, free	Detritivore
Trifarina angulosa	? Infaunal	? Detritivore

Table 6.13 Details of the 10 major benthic foraminiferal associations on the continental shelf between the Labrador Sea and Cape Hatteras. The most important environmental controls are marked with an asterisk

***Islandiella helenae* association**

Salinity: 31–34‰

Temperature: –2 to 4 °C*

Substrate: sandy mud

Depth: 73–400 m

Distribution:
1. Labrador Sea (Vilks *et al.* 1982) total
2. Gulf of St Lawrence (Rodrigues and Hooper 1982a) total

Additional common species:
Elphidium excavatum clavatum

***Adercotryma glomerata* association**

Salinity: 31–34‰

Temperature: –2.0 to 5.5 °C*

Substrate: mud, sand, gravel

Depth: 72–512 m

Distribution:
1. Labrador Sea (Vilks *et al.* 1982) total
2. Newfoundland (Schafer and Cole 1982) living
3. Gulf of St Lawrence (Hooper 1975, Rodrigues and Hooper 1982a) total
4. Nova Scotia (Williamson *et al.* 1984) total
5. Gulf of Maine (Phleger 1952b) total

Additional common species:

Recurvoides turbinatus	1, 2, 3, 4, 5
Spiroplectammina biformis	2, 3, 4
Islandiella helenae	3
Trifarina angulosa	3 (as *T. hughesi*)
Cribrostomoides crassimargo	4
Saccammina atlantica	5
Reophax curtus	5

***Saccammina atlantica* association**

Salinity: 28–34, mainly 31–34‰

Temperature: –0.5 to 26 °C

Substrate: mud, sand

Depth: 30–210 m

Distribution:
1. Lake Melville, Labrador Sea (Vilks *et al.* 1982) total
2. Nova Scotia (Williamson *et al.* 1984) total
3. Gulf of Maine (Phleger 1952a) total
4. Cape Cod (Parker 1984) (as *Proteonina difflugiformis*) total
5. Long Island–Martha's Vineyard (Parker 1948, 1952) total (Murray 1969), dead
6. New Jersey (Poag *et al.* 1980) dead

Additional common species:

Reophax fusiformis	1
Reophax scorpiurus	2
Adercotryma glomerata	3
Cribrostomoides crassimargo	3, 5
Reophax curtus	3
Textularia torquata	3
Eggerella advena	5, 6
Elphidium excavatum clavatum	5 (as *E. inertum*), 6
Globobulimina auriculata	5
Cibicides lobatulus	6

Table 6.13 *(continued)*

***Islandiella islandica* association**

Salinity: 32–35‰

Temperature: 4 to 12 °C*

Depth: 50–485 m

Distribution:
1. Grand Banks (Sen Gupta 1971) living, dead, total
2. Nova Scotia (Williamson *et al.* 1984) total

Additional common species:

Buccella frigida	1
Cibicides lobatulus	1
Eggerella advena	1
Elphidium excavatum clavatum	1
Elphidium subarcticum	1
Islandiella teretis	1
Trochammina squamata	1

***Cibicides lobatulus* association**

Salinity: 32–34‰

Temperature: –2 to 26 °C

Substrate: sand, gravel, boulders*

Depth: 11–128 m

Distribution:
1. Grand Banks (Sen Gupta 1971) living and dead
2. Gulf of St Lawrence (Hooper 1975, Rodrigues and Hooper 1982a) total
3. Nova Scotia (Williamson *et al.* 1984) total
4. Gulf of Maine (Phleger 1952a) total
5. New Jersey (Poag *et al.* 1980) living and dead (Parker 1948) total (as *C. refulgens*).
6. Maryland (Parker 1948) total

Additional common species:

Islandiella islandica	1, 2, 3
Cassidulina algida	4
Elphidium excavatum clavatum	4, 5, 6 (as *E. incertum*)
Adercotryma glomerata	4
Eggerella advena	4, 5
Ammodiscus minutissimus	5
Saccammina atlantica	5 (as *S. difflugiformis typica + atlantica*)
Elphidium subarcticum	5, 6

***Elphidium excavatum* forma *clavatum* association**

Salinity: 30–34‰

Temperature: –2 to 26 °C

Substrate: muddy sand, sand

Depth: 0–82 m, 375–410 m (relict)

Distribution:
1. Grand Banks (Sen Gupta 1971) living and dead
2. Gulf of St Lawrence (Rodrigues and Hooper 1982a, b) total
3. Gulf of Maine (Parker 1948; Phleger 1952a) total
4. Long Island (Parker 1948, 1952) total (Gevirtz *et al.* 1971) living and total
5. New Jersey (Poag *et al.* 1980) living (Parker 1948) total (as *E. incertum*)
6. Maryland (Parker 1948) total (as *E. incertum*)
7. N. of Cape Hatteras (Murray 1969) dead (Schnitker 1971) total

Additional common species:

Islandiella islandica	1
Eggerella advena	1
Elphidium subarcticum	1, 3, 4
Eggerella advena	4, 5, 7
Saccammina atlantica	4 (as *Proteonina*), 5 (as *S. difflugiformis*)
Rosalina columbiensis	4 (as *Discorbis*)
Bulimina marginata	4 (only 80 m)
Fursenkoina fusiformis	5
Quinqueloculina seminula	5
Cibicides lobatulus	5 (as *C. refulgens*)
Brizalina sp.	7
Globocassidulina subglobosa	7
Hanzawaia concentrica	7
Melonis sp.	2 (in relict assemblage)
Bulimina exilis	2 (in relict assemblage)

Table 6.13 *(continued)*

***Eggerella advena* association**

Salinity: 31–34‰

Temperature: –2 to 26 °C

Substrate: muds, silts, well-sorted fine sand*

Depth: 0–160 m

Distribution:
1. Gulf of St Lawrence, Northumberland Strait (Schafer 1967) living
2. Canso Strait (Buckley *et al.* 1974) total
3. Nova Scotia (Williamson *et al.* 1984) total
4. Gulf of Maine (Parker 1984; Phleger 1952) total
5. Cape Cod (Parker 1948) total
6. Long Island (Parker 1952) total (Gevirtz *et al.* 1971) total and living
7. New Jersey (Poag *et al.* 1980) living

Additional common species:

Spiroplectammina biformis	3, 4
Cribrostomoides crassimargo	3
Cibicides lobatulus	4, 6
Elphidium subarcticum	4, 7
Reophax scottii	4
Textularia torquata	4
Saccammina atlantica	5 (as *Proteonina difflugiformis*), 6, 7 (as *S. difflugiformis typica* and *atlantica*)
Elphidium excavatum clavatum	6 (as *E. incertum*), 7
Fursenkoina fusiformis	6, 7

***Trifarina angulosa* association**

Salinity: 33–35‰

Temperature: 1 to 8 °C*

Substrate: mud, muddy sand*

Depth: 120–400 m

Distribution:
1. Gulf of St. Lawrence (Rodrigues and Hooper 1982a) total (as *T. hughesi*)
2. Nova Scotia (Williamson *et al.* 1984) total
3. Martha's Vineyard – Long Island (Parker 1984) total (as *Angulogerina*)
4. Long Island (Gevirtz *et al.* 1971) total
5. New Jersey (Parker 1948) total

Additional common species:

Islandiella helenae	1
Islandiella islandica	1
Buccella tenerrima	1
Brizalina spathulata	2
Buccella frigida	2
Elphidium excavatum clavatum	2
Brizalina sp.	3 (as *Bolivina*)
Bulimina marginata	3
Cibicidoides pseudoungerianus	3, 5 (as *Cibicides*)

Table 6.13 *(continued)*

***Bulimina marginata* association**

Salinity: 33–35‰

Temperature: 9 to 13 °C

Substrate: mud, muddy sand

Depth: 70–1000 m

Distribution:
1. Martha's Vineyard and Block Island (Parker 1948) total
2. Long Island (Gevirtz *et al.* 1971) total
3. Maryland (Parker 1948) total
4. N. of Cape Hatteras (Schnitker 1971) total
5. Nova Scotia – Cape Hatteras (Streeter and Lavery 1982) total

Additional common species:

Cibicidoides pseudoungerianus	1, 4
Trifarina angulosa	1 (as *Angulogerina*), 3
Elphidium excavatum clavatum	3 (as *E. incertum*), 4
Bolivina sp.	3
Globocassidulina subglobosa	3 (as *Cassidulina*), 4 (as *Islandiella*)
Globobulimina affinis	5

***Fursenkoina fusiformis* association**

Salinity: 33–34‰

Temperature: 3 to 26 °C

Substrate: muddy sand*

Depth: 42–76 m

Distribution:
1. Long Island (Murray 1969) living
2. New Jersey (Poag *et al.* 1980) living
3. N. of Cape Hatteras (Schnitker 1971) total

Additional common species:

Elphidium excavatum clavatum	1
Nonionella turgida	1
Saccammina atlantica	1, 2 (as *S. difflugiformis typica* and *atlantica*)
Cibicides lobatulus	2
Eggerella advena	2
Elphidium subarcticum	2

(Murray 1976a). Measurement of length and diameter of this somewhat variable form show at least four main peaks (Fig. 6.8) and a cumulative plot is very much like a survivorship curve. In this example, the 34 tests at growth stage 1 gives rise to a total of 100 tests (all growth stages added together). Thus, *S. atlantica* may be over-represented by three times, but this is not sufficient to account for the dominance of the *S. atlantica* association in the dead assemblages (see Table 6.14). The most likely cause is loss of calcareous tests through dissolution which may be as high as 95% (Table 6.14; Murray 1989).

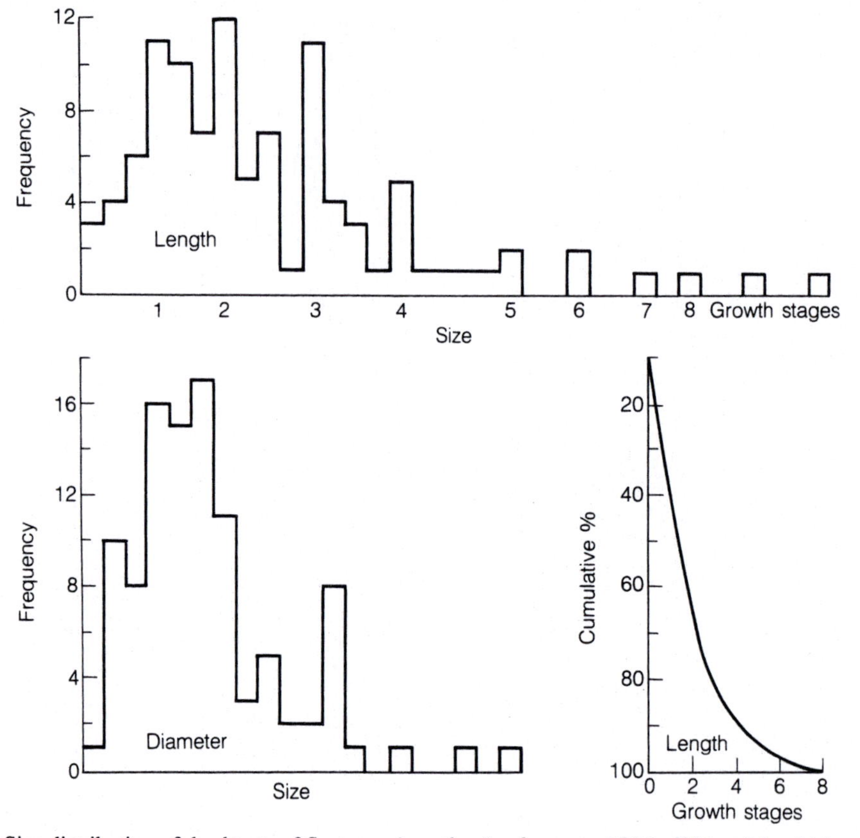

Figure 6.8 Size distribution of dead tests of *Saccammina atlantica* from sta. 1216 off Long Island (data from Murray 1969); $n = 103$

Table 6.14 Data on assemblages from the shelf off Long Island (from Murray 1969) and percentage dissolution of calcareous tests assuming no effects of transport or other postmortem modifying causes. In the lower part data on the abundance of *Saccammina atlantica* has been used to calculate the number of growth stages required to explain the increase in the dead assemblage

				Sample			
	1210	1212	1213	1214	1215	1216	1217
Living % agglutinated	83	60	67	49	21	15	26
Dead % agglutinated	71	95	93	91	84	71	51
(%) Dissolution	0	95	93	91	94	93	66
Living % *S. atlantica*	20	23	18	11	9	4	9
Dead % *S. atlantica*	55	53	40	60	49	41	32
Growth stages	×4	×4	×3	×12	×10	×16	×5

Miller and Ellison (1982) investigated possible controls on the abundance of *Saccammina atlantica* (as *Reophax*) on the New Jersey shelf. They found that no single factor accounted for variations in abundance between the ridges and troughs. However, topography was thought to play an important role for the troughs have higher organic carbon values and more variable bottom temperatures than the ridges. They considered that tests of *S. atlantica* are more readily broken than those of *Cibicides lobatulus* and that this may account for their lower abundance on the ridges. Further studies of living and dead assemblages on the New Jersey shelf by Poag *et al.* (1980) showed that the living fall into two main associations, *Eggerella advena* and *Fursenkoina fusiformis*, and three minor ones, (*C. lobatulus, Elphidium excavatum clavatum* and *E. subarcticum*). However, with one exception, the dead assemblages are represented by either the *C. lobatulus* or the *S. atlantica* associations. Furthermore, in every case the living and dead assemblages at the same station were completely different. In general, the agglutinated : calcareous ratio of both living and dead assembalges are similar although a few stations show losses of calcareous forms of 20–80%. The problem of *S. atlantica* is much the same as off Long Island. The abundance of *C. lobatulus* in the dead assemblages may be due to the addition of tests which during life were attached to firm substrates (such as bivalve shells, tunicates, hydroids, bryozoa) which were not sampled. However, Poag *et al.* (1980) favour the alternative view that *C. lobatulus* and *S. atlantica* have low standing crops in the spring (time of sampling) compared with other times of the year. This has been confirmed for *C. lobatulus* but not for *S. atlantica* (see Ellison and Peck 1983, Table 1).

The full extent of postmortem dissolution of tests cannot be assessed because so many studies have been based on total assemblages. However, the Gulf of Maine is characterised by agglutinated associations (Phleger 1952a) and it is likely that dissolution is prevalent here.

The *Elphidium excavatum clavatum* association is typically present on the inner shelf (Fig. 6.6). However, Rodrigues and Hooper (1982a) found that in the Gulf of St Lawrence it is present in total assemblages at 375–410 m; these may be relict. On the Nova Scotia shelf this association also occurs in total assemblages. The tests are worn and size-sorted and no living representatives were found. Williamson *et al.* (1984) therefore concluded that they were relict.

These continental shelf living assemblages commonly have low diversity (α 1–6.5) which probably reflects the influence of salinity values <35‰. Values of $H(S)$ for depths between 20 and 200 m based on total assemblages range from 0.75 to 3.0 (Buzas and Gibson 1969; Gibson and Buzas 1973).

The Minas Basin of the Bay of Fundy is macrotidal (~ 16 m) and the powerful bottom currents are responsible for much postmortem transport of tests. The total assemblages fall into three size groups (Thomas and Schafer 1982):

1. Relatively large and heavy species all preserved as abraded tests: *Ammonia beccarii, Pyrgo williamsoni, Quinqueloculina seminula;*
2. Species of average size which are abraded: *Cibicides lobatulus, Elphidium excavatum;* or undamaged, *E. margaritaceum, Miliammina fusca;*
3. Small species: *Eggerella advena, Elphidium frigidum.*

The first two groups (>340 μm in size) are thought to have been transported as bed load. The third group (<280 μm) is thought to have been transported in suspension. The deeper parts of the bay have coarse substrates and accumulations of large tests (field IV) while the finer grained sediments of the intertidal zone have tests derived from both bed and suspended load (fields II and III). Furthermore, the presence of planktonic tests in the sediments indicates that the source of the biogenic materials is from the shelf via the Bay of Fundy.

Cape Hatteras to Florida

This area has not yet been studied in detail. Off North Carolina, living and dead assemblages have been discussed by Murray (1969) and total assemblages by Wilcoxon (1964) and Schnitker (1971). For the area from South Carolina to Florida, Wilcoxon (1964) has provided data on total assemblages. The shelf off Georgia has been studied by Sen Gupta and Kilbourne (1974, 1976), Sen Gupta and Hayes (1979) and Arnold (1983). The occurrence of soritids and peneroplids on the Bahama Bank has been reviewed by Levy *et al.* (1988).

Nine associations of limited geographic occurrence are listed in Table 6.15 and will not be discussed further. Details of the seven principal associations are given in Tables 6.16 and 6.17. Most of the shelf sediments are sandy and the main environment variable appears to be bottom-water temperature (data from Wilcoxon 1964). All seven associations are present from North Carolina to Florida. Only the *Ammonia beccarii* association has been recorded as a living assemblage and then from only one station (Murray 1969, sta. 1253). The *Cibicides mollis, Hanzawaia concentrica* and *Textularia secasensis* associations are all present at depths <53 m. Their lower limit may be controlled by the base of the thermocline and the availability of microalgae for the herbivorous species.

The *Elphidium* spp. association off North Carolina is dominantly *E. excavatum clavatum* according to Schnitker (1971) or *E. rugulosum* and *E. subarcticum* (Wilcoxon 1964, sta. 78). Most of the occurrences are at depths of <65 m and the few deeper records may be relict. The *Quinqueloculina* spp. association is generally dominated by *Q. seminula* but includes 11 other species, of which *Q. ackneriana, Q. compacta, Q. dutemplei* and *Q. lamarckiana* are most common. The *Globocassidulina subglobosa* association is restricted to the outer shelf and includes *Trifarina angulosa* which is common in this setting to the north of Cape Hatteras also.

With the exception of the *Ammonia beccarii* association the other principal associations are diverse. For the depth range 0–100 m Gibson and Buzas (1973) reported a mean of 45 species per sample and a mean *H*(*S*) of 2.81 (observed range 1.51–3.50). Off Georgia *H*(*S*) of total assemblages ranges from 2.5 to 3.4 for the depth zone 10–100 m (Sen Gupta and Kilbourne 1974). Because of this high diversity, the majority of species are present in low abundance and the associated common species in each association frequently differ from one sample to another for any given association.

The terms of wall structure, all except the *Textularia secasensis* association are dominated by hyaline tests. Even the *Quinqueloculina* spp. association rarely exceeds 40% porcellaneous tests. The sediments are richer in biogenic carbonate than those to the north of Cape Hatteras (average 17.8% south of Cape Romain, average <3% to north, Emery and Uchupi 1972). In the absence of data on living and dead assemblages it is difficult to evaluate postmortem changes such as dissolution of calcareous tests and transport. However, dissolution does not appear to be a major process. As regards transport, the passage of a hurricane across the North Carolina Shelf left no measurable effect on bedform geometry (Mearns *et al.* 1988). Nevertheless, on this high-energy shelf winnowing and redistribution of small tests must take place.

In situ experiments at the shelf break off Florida showed that predation greatly reduces the standing crop size (Buzas *et al.* 1989). Furthermore, muddy substrates are recolonised more effectively than clean sand substrates probably due to the presence of organic material with mud and its absence from clean sand.

Taxa of southern origin which do not extend to the north of Cape Hatteras include *Asterigerina carinata, Elphidium poeyanum, Nonionoides grateloupi, Reussella spinulosa* and *Sagrina pulchella* (Murray 1973), *Articulina, Peneroplis, Vertebralina* plus *Amphistegina lessonii* (data from Wilcoxon 1964, Schnitker 1971). The greatest abundance of the latter taxon is at 65–70 m off North Carolina and Schnitker considers this to be relict from a period of lower sea-level. This may be true of the milioline genera too.

Table 6.15 Minor benthic associations on the shelf from Cape Hatteras to Florida (T = total, L = living, D = dead)

Association		Area	Depth (m)	Source
1. *Stetsonia minuta*	L	North Carolina	33–37	Murray 1969
2. *Nonionoides grateloupi*	L	North Carolina	19	Murray 1969
3. *Placopsilina confusa*	T	North Carolina	20–28	Schnitker 1971
4. *Laevipeneroplis proteus*	T	North Carolina	25	Schnitker 1971
5. *Triloculinella procera*	T	North Carolina	125	Schnitker 1971
6. *Planulina exorna*	T	Georgia	29	Wilcoxon 1964
7. *Nonionella atlantica*	T	Florida	34	Wilcoxon 1964
8. *Asterigerina carinata*	T	Georgia	18	Wilcoxon 1964
9. *Trochulina lomaensis*	T	South Carolina	12	Wilcoxon 1964

Table 6.16 Principal associations on the shelf from Cape Hatteras to Florida. The most important environmental controls are marked with an asterisk

***Ammonia beccarii* association**

Salinity: 34–36‰

Temperature: 15 to 27 °C

Substrate: sand

Depth: 0–20 m

Distribution:
1. North Carolina: (a) (Murray 1969) living; (b) (Wilcoxon 1964) total
2. South Carolina to Florida: (a) (Wilcoxon 1964) total (as A. tepida); (b) (Sen Gupta and Kilbourne 1974) total

Additional common species:

Quinqueloculina lata	1a
Cibicides mollis	2a
Quinqueloculina ackneriana	2a
Elphidium discoidale	2a
Bolivina paula	2a
Globocassidulina subglobosa	2a (as *Cassidulina*)
Textularia secasensis	2a

***Elphidium* spp. association**

Salinity: 35–35‰

Temperature: 20 to 27 °C

Substrate: sand

Depth: 12–159 m

Distribution:
1. North Carolina: (a) (Schnitker 1971) total; (b) (Wilcoxon 1964) total
2. South Carolina to Florida: (a) (Wilcoxon 1964) total; (b) (Sen Gupta and Kilbourne 1976) total

Additional common species:

Ammonia beccarii	1a, b, 2 (as *A. tepida*)
Hanzawaia concentrica	1a

***Quinqueloculina* spp. association**

Salinity: 35–36‰

Temperature: 20 to 27 °C

Substrate: sand

Depth: 14–140 m

Distribution:
1. North Carolina: (a) (Murray 1969) dead; (b) (Schnitker 1971) total; (c) (Wilcoxon 1964) total
2. South Carolina–Florida (Wilcoxon 1964) total

Additional common species:

Ammonia beccarii	1a, 2 (as *A. tepida*)
Elphidium poeyanum	1a
Placopsilina confusa	1a
Hanzawaia concentrica	1a, b, 2
Elphidium excavatum clavatum	1b
Laevipeneroplis proteus	1b (as *Peneroplis*)
Asterigerina carinata	1b
Reophax scorpiurus	1b
Planulina ornata	1b
Amphistegina lessonii	1b
Textularia secasensis	2
Cibicides mollis	2
Trochulina lomaensis	2 (as *Rotorbinella*)

***Cibicides mollis* association**

Salinity: 34–36‰

Temperature: 15 to 27 °C

Substrate: silty sand, sand

Depth: 14–32 m

Distribution:
1. North Carolina (Wilcoxon 1964) total
2. South Carolina–Florida (Wilcoxon 1964) total

Additional common species:

Textularia secasensis	1, 2
Quinqueloculina lamarckiana	1, 2
Hanzawaia concentrica	1, 2
Ammonia beccarii	1 (as *A. tepida*)
Bolivina paula	1
Asterigerina carinata	1

Table 6.16 *(continued)*

***Hanzawaia concentrica* association**

Salinity: 35–36‰

Temperature: 15 to 27 °C

Substrate: sand

Depth: 14–55 m

Distribution:
1. North Carolina: (a) (Murray 1969) dead (as *Cibicides*); (b) (Schnitker 1971) total; (c) (Wilcoxon 1964) total
2. South Carolina–Florida (Wilcoxon 1964) total

Additional common species:

Peneroplis carinatus	1a
Stetsonia minuta	1a
Elphidium excavatum clavatum	1b
Saccammina atlantica	1b (as *Reophax*)
Quinqueloculina seminula	1b
Reophax scorpiurus	1b
Laevipeneroplis proteus	1b (as *Peneroplis*)
Cibicidoides bradyi	1b (as *Cibicides*)
Planulina ornata	1b
Cibicides mollis	1c
Bolivina paula	1c
Ammonia beccarii	2 (as *A. tepida*)
Textularia secasensis	2
Bigenerina irregularis	2

***Textularia secasensis* association**

Salinity: 34–36‰

Temperature: 10 to 27 °C

Substrate: silty sand, sand, gravelly sand

Depth: 16–53 m

Distribution:
1. North Carolina (Wilcoxon 1964) total
2. South Carolina to Florida (Wilcoxon 1964) total

Additional common species:

Cibicides mollis	1, 2
Hanzawaia concentrica	1, 2
Planulina exorna	1, 2
Bolivina paula	2
Globocassidulina subglobosa	2 (as *Cassidulina*)

***Globocassidulina subglobosa* association**

Salinity: 35‰

Temperature: 10 to 20 °C*

Substrate: silty sand*

Depth: 145–173 m

Distribution:
1. North Carolina: (a) (Schnitker 1971) total (as *Islandiella*); (b) (Wilcoxon 1964) total
2. South Carolina–Florida (Wilcoxon 1964) total

Additional common species:

Trifarina angulosa	1b (as *Angulogerina*)
Bolivina paula	1b, 2
Cassidulina laevigata	2

Table 6.17 Typical species from the shelf from Cape Hatteras to Florida (feeding strategies from Ch. 2)

Species	Mode of life	Feeding strategy
Ammonia beccarii	Infaunal, free	? Herbivore
Cibicides mollis	Epifaunal	? Herbivore
Elphidium spp.	Infaunal, free	Herbivore, detritivore
Globocassidulina subglobosa	Infaunal, free	Detritivore
Hanzawaia concentrica	Epifaunal, clinging	? Herbivore
Quinqueloculina spp.	Epifaunal, free	Detritivore, herbivore
Textularia secasensis	? Infaunal	? Detritivore

Generic predominance

The available data (mainly total assemblages) are summarised in Fig. 6.9A–C. The pattern off Nova Scotia is complex partly because of the irregular topography and the varied bottom-water masses. The inner shelf (<100 m) in mainly dominated by *Cibicides*. At similar depths towards the shelf edge *Eggerella* is the main genus. At depths of ~100 to ~200 m are *Saccammina, Islandiella* and relict *Elphidium*. In the northern part of the area *Adercotryma*, a cold-water taxon, dominates. Basins deeper than 200 m have *Globobulimina*, but along the shelf edge *Trifarina* is the main form.

The Gulf of Maine has an inner shelf zone of *Eggerella* and a seaward zone mainly of *Reophax*. South of Cape Cod (Fig. 6.9B) *Elphidium* predominates to a depth of 20–40 m. Its distribution corresponds in general with lower salinity (~32‰) and vertically mixed waters which experience a large annual temperature range. Deeper than 20–40 m to ~100 m *Saccammina* dominates the area as far south as off Hudson River. This part of the shelf is beneath the summer thermocline. To the south *Saccammina* gives way to *Cibicides* but this dies out off Chesapeake Bay. The outer shelf is characterised by *Bulimina–Cassidulina* (plus some *Trifarina*). The sediment here is muddy, salinities normal and temperatures stable.

Cape Hatteras marks a major faunal change. *Elphidium* is dominant only off Georgia and its occurrence coincides with salinities of <34‰ (see Fig. 6.5D). Elsewhere the nearshore shelf is characterised by *Ammonia* or *Quinqueloculina*. Beyond this, at depths down to ~40 m, *Cibicides* and *Hanzawaia* are the key genera. *Textularia* is widely present in the central part of the area. The outer shelf is dominated by *Bolivina–Globocassidulina*.

Planktonic : benthic ratio

Off Nova Scotia and in the Gulf of Maine, planktonic tests are rare in the shelf sediments. South of Cape Cod to Cape Hatteras there is a progressive decrease in values from 50 to 60% at the shelf break to <2% close to the shore. The picture is similar to the south of Cape Hatteras except that shelf break values rarely exceed 25% (Fig. 6.9D).

Man's impact on the environment

If data on modern foraminiferal distributions are to be used as an aid to palaeoecological interpretation, then it is important to consider whether the influence of man is disturbing modern environments. The areas most likely to be affected are marshes, estuaries, lagoons and the inner shelf. In most cases we are aware of disturbance only if an investigation has been carried out especially with this in mind, i.e. there are other areas which have probably been disturbed but no one has written about it.

During the nineteenth century, the peak of activity in land clearance for agriculture on the piedmont and coastal plain caused extensive soil erosion. The sediment was transported to the sea where it led to build-up of nearshore areas and encouraged the growth of marshes. Since 1930 the process has been reversed: reafforestation, abandonment of agricultural land, construction of dams, etc. have reduced sediment yields especially in the southern USA. Because marsh sediments compact after accumulation there is a need for a continuing supply of sediment to maintain them even when sea-level is constant (Stevenson *et al.* 1988). Therefore southern marshes are undergoing erosion. Northern marshes are less affected because the higher tidal energy there continues to supply

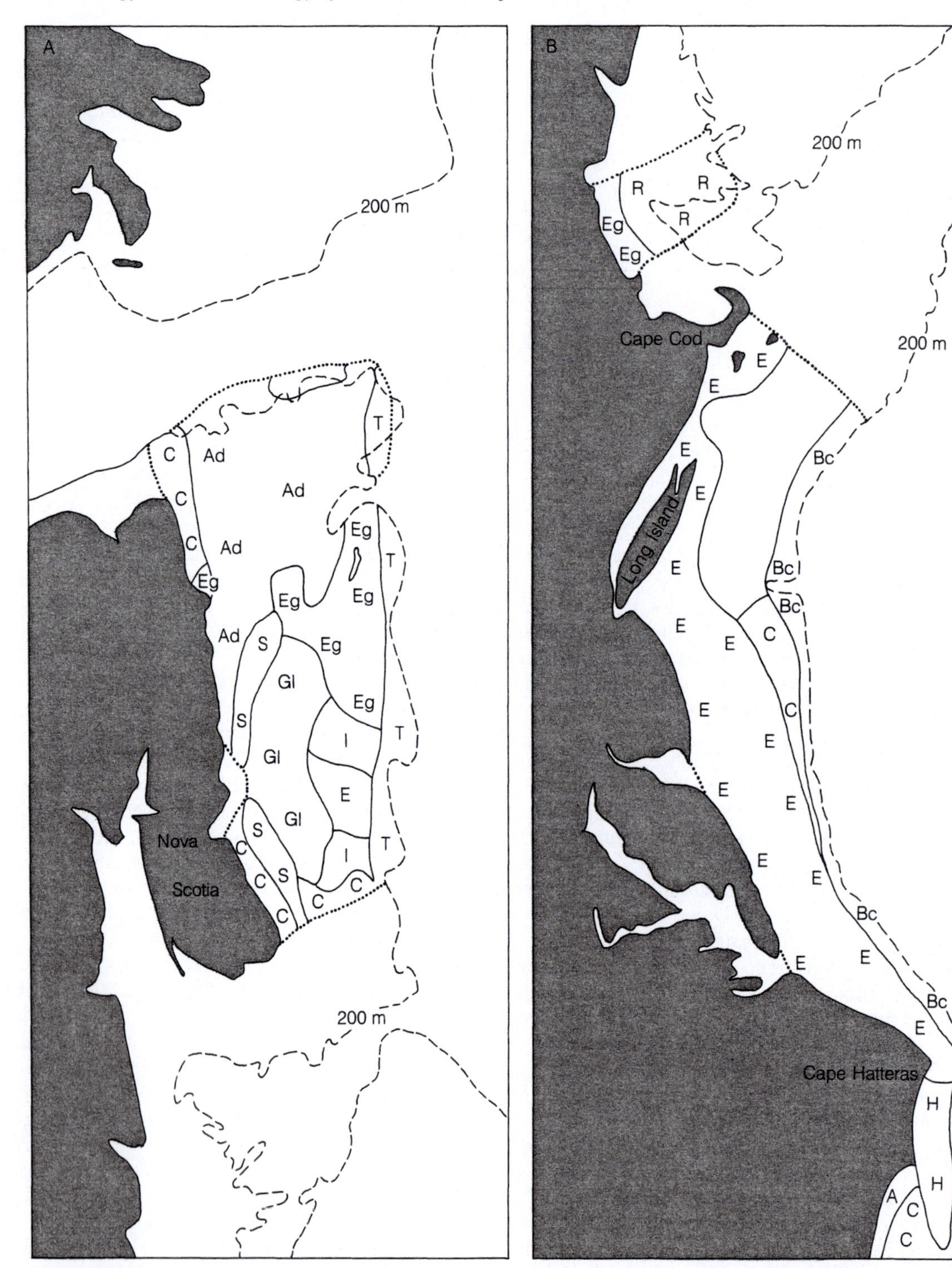

Figure 6.9 (A–C) Generic predominance on the continental shelf off North America. (A) Newfoundland to Gulf of Maine (data from Williamson *et al.* 1984). (B) Gulf of Maine to North Carolina (data from Phleger 1952a; Poag *et al.* 1980; Wilcoxon 1964). (C) North Carolina to Florida (data from Wilcoxon 1964). Key to letters A–C: A = *Ammonia*, Ad = *Adercotryma*, AE = *Ammonia–Elphidium*, BC = *Bulimina–Cassidulina*, BG = *Bolivina–Globocassidulina*, C = *Cibicides*, E = *Elphidium*, Eg = *Eggerella*, Gl = *Globobulimina*, H = *Hanzawaia*, I = *Islandiella*, Q = *Quinqueloculina*, R = *Reophax*, S = *Saccammina*, T = *Trifarina* and Te = *Textularia*. The dotted line shows the limit of data. (D) Planktonic : benthic ratio (as % planktonic of benthic plus planktonic) (data from Parker 1948; Wilcoxon 1964). R = rare planktonic tests

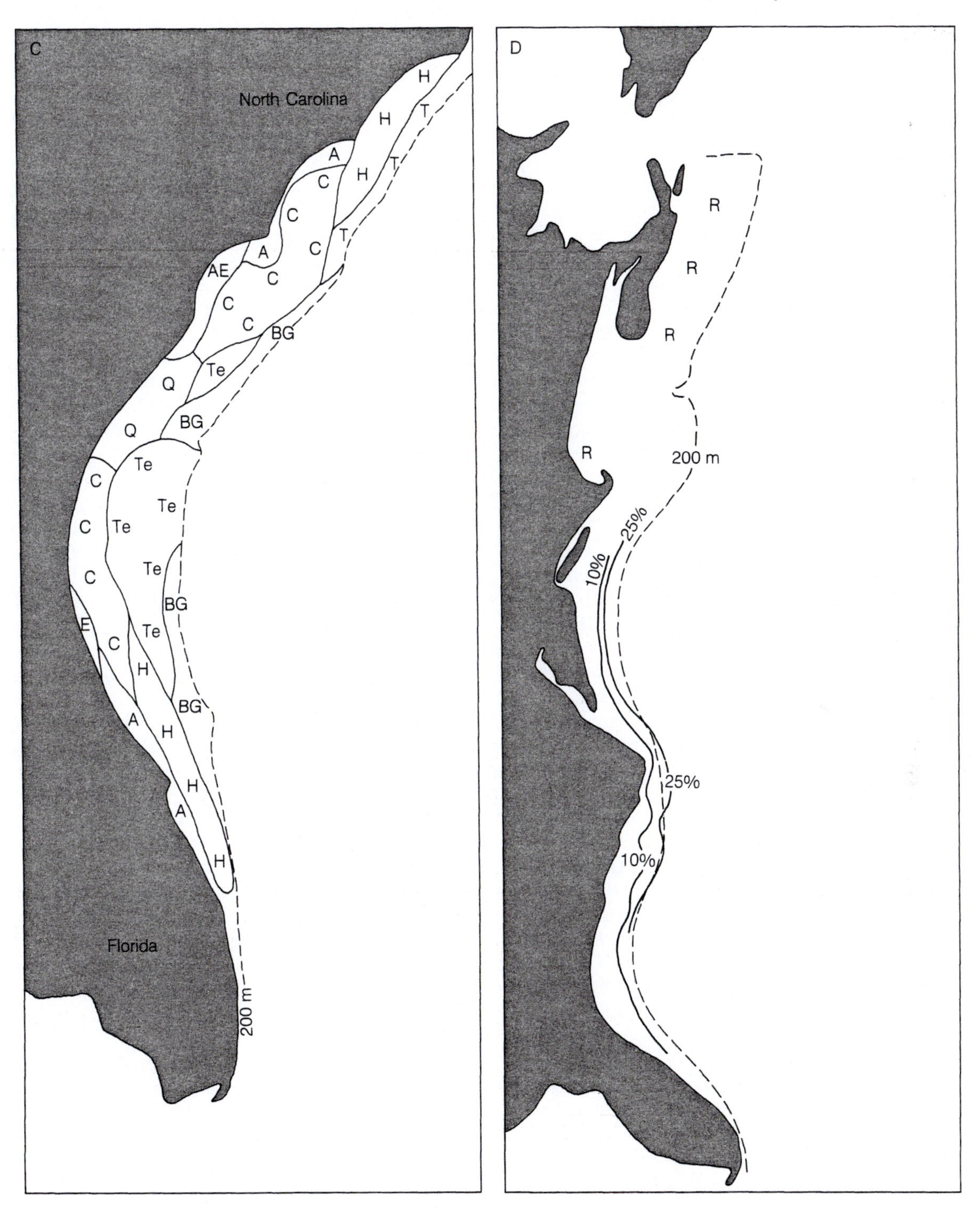
C
North Carolina
H
H
T
A
C
H
T
C
C
T
A
AE
C
C
C
BG
Q
Te
Q
BG
C
Te
Te
C
Te
C
Te
BG
E
C
Te
H
BG
A
H
H
A
H
Florida
200 m
D
R
R
R
R
200 m
25%
10%
25%
10%

sediment and the organic matter has a slower rate of decay because of the lower temperatures. Nevertheless, marshes are vulnerable to reclamation and for dumping waste.

Estuaries are commonly the sites of municipal and industrial pollution and several examples have been reported from Canada. Chaleur Bay receives filtered wood products and chemicals derived from processing pulpwood. Around the outfall the dominant living forms are *Eggerella advena* and *Miliammina fusca*, and other forms able to tolerate some degree of pollution are *Elphidium excavatum clavatum, Haynesina orbiculare* and *Buccella frigida* (Schafer 1970; Schafer and Cole 1974).

At another site where spoil rich in cadmium was deposited over a 6-year period, the foraminifera were sampled 1 month after the cessation of dumping and again 2 years later (Schafer 1982). The standing crop was 0–110, average 26 per 10 cm^2 in 1978 compared with an average of 250–270 per 10 cm^2 in 1980. There was also an increase in the number of species during this period although *Eggerella advena* remained the single most dominant species with *Ammotium cassis* being the second most abundant species in 1978 but less so in 1980. Schafer argued that *A. cassis* may favour environments rich in suspended particulate matter and that this declined between 1978 and 1980. *Eggerella advena* has high mobility and this may partly explain its ability to colonise dump areas.

The Miramichi Estuary has been studied since 1966 (Bartlett 1966; Tapley 1969). The area occupied by an open bay assemblage in the 1960s was occupied by a transitional assemblage in 1976–77 (Scott *et al.* 1977). The faunal change has been attributed to changes in the barrier islands and the consequences of this on the circulation pattern (Scott *et al.* 1977; Schafer and Smith 1983). However, Tapley (1969) drew attention to the presence of industrial pollutants and it is possible that they too have played a role. For instance, they consume dissolved oxygen and reduced circulation would enhance the magnitude of this effect.

Comparative studies of artificial ponds in a lagoon in North Carolina showed that *Elphidium excavatum clavatum* was the most commonly occurring species in both control and effluent ponds. All aspects of the foraminiferal assemblages were reduced in the effluent ponds, i.e. standing crop, production, diversity, etc. (Le Furgey and St Jean 1976). By a sewer outfall off Virginia, the changes to the foraminiferal assemblages were very localised and most were caused by seasonal effects (Bates and Spencer 1979).

As far as the author is aware, all the ecological data given in the main part of this chapter are from relatively unpolluted areas.

Continental slope–abyssal plain

Remarkably little is known of the slope assemblages. As can be seen from Table 6.13, the following associations extend from the shelf on to the slope: *Islandiella helenae* (73–400 m), *Adercotryma glomerata* (72–512 m), *I. islandica* (50–485 m), *Trifarina angulosa* (120–400 m) and *Bulimina marginata* (70–1000 m). Associations confined to the slope–abyssal plain are listed in Table 6.18.

The *Elphidium excavatum* (probably *clavatum*) association has been found living by Schafer and Cole (1982) thus removing any possibility that it is relict. The *Uvigerina peregrina* association is present on fine-grained sediments with >1% organic carbon, at depths of 1000–2660 m, beneath North East Atlantic Deep Water (NEADW). These waters are not poor in oxygen and oxygen availability does not appear to be an important environmental control (Streeter and Lavery 1982) as *U. peregrina* can tolerate lower oxygen values (Schnitker 1979).

Between ~2300 and 3000 m the slope and rise are swept by the Western Boundary Undercurrent which is a south-flowing geostrophic current. The *Epistominella exigua* and *Oridorsalis umbonatus* associations lie under its influence. The *Hoeglundina elegans* association (2500–3800 m) lies beneath NADW (Streeter and Lavery 1982). At depths >3800 m, the *Nuttallides umboniferus* association is present beneath Antarctic Bottom Water (AABW) with a temperature <1.9 °C.

The average standing crop on the Newfoundland Slope is 25 (at 512 m), 701 (at 1533 m) and 2356 per 10 cm^2 (at 2695 m) (Schafer and Cole 1982). Off Cape Lookout values of 1 (at 400 m) to 54 (at 1500 m) to 17 per 10 cm^2 (at 2500 m) and off Cape Fear of 2 (at 500 m), 62 (at 1000 m) and 19 per 10 cm^2 (at 2500 m) have been reported by Tietjen (1971). There is a good correlation between the size of the standing crop and the percentage organic carbon.

Sen Gupta and Strickert (1982) took seasonal samples from the Florida–South Carolina Slope. The dominant living species at all four seasons throughout the depth range samples, 285–520 m, was *Brizalina lowmani* (as *Bolivina*). It was particularly associated with muddy substrates. At one station off Daytona Beach, Florida, *B. lowmani* made up 40–63% of living assemblage. During winter and spring it was accompa-

Table 6.18 Associations from the slope to abyssal plain along the eastern margin of North America

***Elphidium excavatum* association**

Salinity: 34.91–34.95‰

Temperature: 3.16 to 3.70 °C

Substrate: silt, mud

Depth: 1000–3000 m

Distribution:
1. Newfoundland (Schafer *et al.* 1981) total (Schafer and Cole 1982) living
2. Nova Scotia–Maine (Streeter and Lavery 1982) total

Additional common species:

Epistominella exigua	1
'Virgulina'	1
Valvulineria	1
Nonionella atlantica	1

***Epistominella exigua* association**

Salinity: 34.9‰

Temperature: 2.4 to 3.7 °C

Substrate: biogenic sand and silt

Depth: 2200–3300 m

Distribution:
1. Newfoundland: (a) (Schafer and Cole 1982) living; (b) (Schafer *et al.* 1981) total

Additional common species:

Oridorsalis umbonatus	1
Epistominella vitrea	1
Eponides tumidulus	1

***Hoeglundina elegans* association**

Salinity: 34.5–35.8‰ NADW

Temperature: 2 to 3 °C

Substrate: silty clay

Depth: 2500–3800 m

Distribution:
1. Nova Scotia–Cape Hatteras (Streeter and Lavery 1982) total
2. Cape Cod (Miller and Lohman 1982) total
3. Delaware (Corliss 1985) living (top 2 cm)

Additional common species:

Fontbotia wuellerstorfi	1 (as *Cibicides*), 3 (as *Planulina*)
Oridorsalis umbonatus	1, 3 (as *O. tener*)
Melonis barleenanum	3

***Nuttallides umboniferus* association**

Salinity: AABW

Temperature: < 1.9 °C

Substrate: biogenic ooze

Depth: 3800–5128 m

Distribution:
1. Nova Scotia–Cape Hatteras (Streeter and Lavery 1982) total (as *Osangularia*)
2. W. North Atlantic (Phleger *et al.* 1953) total (as *Epistominella*)

Additional common species:

Fontbotia wuellerstorfi	2 (as *Planulina*)
Pullenia bulloides	2

***Oridorsalis umbonatus* association**

Salinity: 34.90‰

Temperature: 2.4 to 3 °C

Substrate: biogenic sand and silt

Depth: 2695 m

Distribution:
1. Newfoundland (Schafer and Cole 1982) living

Additional common species:

Epistominella exigua	1 (as *E. vitrea*)
Cibicides lobatulus	1

***Uvigerina peregrina* association**

Salinity: 34.95–34.97‰

Temperature: 3 to 3.5 °C

Substrate: silt with >1 % organic C

Depth: 1000–2660 m

Distribution:
1. Nova Scotia–Cape Hatteras (Streeter and Lavery 1982) total
2. Cape Cod (Miller and Lohman 1982) total

Additional common species:
none

nied by *B. subaenariensis* (as *Bolivina*) 11–26%, while during the summer and autumn by *Globocassidulina subglobosa* (as *Cassidulina*) 12–18%. Sen Gupta *et al.* (1981) attribute the unusual abundance of *B. subaenariensis* to its opportunistic response to increased food availability during seasonal upwelling.

Agglutinated foraminifera

The agglutinated component of the assemblages has received special attention in some areas.

Off Newfoundland between ~500 and 3000 m there are 43 genera of agglutinated and 63 of calcareous foraminifera. The latter dominate the assemblages and only one agglutinated genus, *Trochammina*, is common (Schafer *et al.* 1981). From 1000 to 3000 m the principal forms are *Trochammina, Textularia, Reophax, Rhabdammina, Saccorhiza* and *Hyperammina* with the addition of *Sigmoilopsis, Rhizammina* and *Ammobaculites* between 2200 and 3000 m. These distributions can be related in a general sense to NEADW down to ~2200 m and the Western Boundary Undercurrent from 2200 to 3000 m. The latter current causes the bottom sediments to be coarser (biogenic sand and silt) than those of the former.

A high-energy benthic boundary layer experiment (Hebble) site has been established on the continental rise off Nova Scotia. This area is intermittently affected by strong bottom currents which cause scour, formation of ripples, etc. Three total faunas were recognised on the basis of Varimax factor analysis. Fauna 1, *Glaphyrammina* cf. *G. americana* (as *Ammobaculites*) with *Saccammina tubulata, Reophax bilocularis* and *Nodulina dentaliniformis* (as *Reophax*); fauna 2, *S. tubulata, Psammosphaera* spp. and *Reophax sp.*; fauna 3, *Hormosinella distans.* Faunas 1 and 2 were from the Hebble area at a depth of 4815–4820 m, whereas fauna 3 was from a shallower site (4185 m) not affected by bottom currents. The latter had a higher diversity than the former (*H*(*S*) 3.13 and 2.16–3.04 respectively). Kaminski (1985) interpreted this as indicating stress through disturbance in the Hebble area. In a detailed study of burrows in the Hebble area, Aller and Aller (1986) noted that they are sufficiently long-lived to develop distinct biogeochemical fields around them. Bacterial abundances are almost double those of the normal sediment and foraminiferal standing crop may be 0–30 individuals per gram compared with normal values of 0–10. The influence of the burrows extends over a radius of ~3 m and affects 10–34% of the bottom.

Schröder *et al.* (1988) have listed those agglutinated foraminifera larger than 297 μm in Nares abyssal plain sediments. However, the most abundant species are smaller and include *Adercotryma glomerata, Nodellum membranaceum* and *Eratidus foliaceum* (as *Ammomarginulina*).

Diversity

In an early study of diversity, Buzas and Gibson (1969) noted that beyond 200 m there was a slight drop in both the number of species and *H*(*S*) down to ~1500–2000 m followed by an increase in both down to 5000 m (the limit of the data on a traverse from near New York to Bermuda). Additional data (Table 6.19) show that within the 100–1000 m range diversity increases from north and south. The position at greater depths is not clear because there are fewer data points. Carter *et al.* (1979) reported the numbers of species on the slope off Newfoundland (but gave no details of sample size): living assemblages 5–30 spp. down to 1000 m, 20–70 spp. 1000–3000 m, total assemblages 18–55 and 60–120 spp. respectively. These values are higher than those reported by Gibson and Buzas (1973) from the USA margin and cast some doubt on the apparent north–south trends.

Postmortem changes

Some comments have already been made in the discussion on individual environments. Few have

Table 6.19 Summary of diversity data, slope and abyssal plain of the North American seaboard. Total assemblages

Area	Depth (m)	*H (S)*	Depth (m)	*H (S)*	Sources
Georges Bank	100–1000	0.71–1.94			Gibson and Buzas 1973
Cape Cod–Maryland	100–1000	1.96–3.01	1000–10 000	1.18–3.55	Gibson and Buzas 1973
Cape Hatteras–Florida	100–1000	2.54–3.66	1000–10 000	3.17–3.26	Gibson and Buzas 1973
South Carolina	100–700	2.4–3.2			Arnold and Sen Gupta 1981

Table 6.20 Postmortem changes

Field I
High marsh, some low marshes
Some lagoons–Long Island Sound
Most slope areas
Abyssal plain
Field I–C to I–CC
Some low marshes
Some lagoons–Chezzetcook Inlet
Shelf off New York
Field II
Minas Basin–intertidal
Field III
Minas Basin–intertidal
Some parts of the slope, especially canyons
Field IV
Minas Basin–deep parts

similar living and dead assemblages (Table 6.20, field I). Most show some degree of alteration either through partial dissolution (fields I-C–I-CC) or the effects of transport. This is not confined to shallow-water areas for some downslope transport takes place on the continental slope. Schafer *et al.* (1981) consider that off Newfoundland the mid slope is the main area of mud and transported test deposition. Where submarine canyons are present they act as pathways down which transport takes place and this has been documented for Wilmington Canyon off New Jersey (Stanley *et al.* 1986). The transport process probably takes place in pulses with intervening phases of residence on the canyon floor. In this way shelf assemblages are transported into the canyon and progressively mixed with bathyal forms during their passage to the abyssal plain. Shallow-water asssemblages from the Bahama Platform are transported into deeper water in this way (Martin 1988).

At depths of 4100–5100 m in the Hebble area, intermittent currents with speeds of <1–73.5 cm s^{-1} have been recorded. At the higher velocities benthic and planktonic foraminifera are thrown into suspension. Forms >250 μm in size are present in large numbers in the water up to 10 m above the bottom, while forms <250 μm are resuspended to at least 1000 m (Gardner *et al.* 1984). In quieter periods this material is redeposited.

More details of postmortem processes awaits further data on living and dead assemblages.

CHAPTER 7

Gulf of Mexico and Caribbean Sea

The Gulf of Mexico is a small ocean substantially enclosed by land. It is microtidal, with most shelf areas having tides of <0.5 m although values of 1 m are attained along part of the Florida coast. As it lies between latitudes 18 and 30° N the climate is tropical to subtropical. Salinities are 35–36‰ except along those parts of the northern shelf under the influence of the Mississippi input of fresh water. Surface-water temperatures range from ~17 to 29 °C and a surface mixed layer extends down to ~100 m. Beneath this, from 100 to 300 m, is Subtropical Underwater with salinities of 36.2–36.6‰, temperature of 10–25 °C and oxygen of 3.8–3.4 ml l^{-1} (Nowlin 1971).

The Caribbean Sea has a general water flow from east to west in the upper 1500 m. The surface waters along the South American coast have salinities >36‰ during the winter due to upwelling of Subtropical Water from 50 to 200 m. At a depth of 400–600 m there is an oxygen minimum core (<3.0 ml l^{-1} O_2) separating the warm Subtropical Underwater above from the cool Subantarctic Intermediate Water below (Gordon, in Fairbridge 1966).

Table 7.1 Paired ecophenotypes from San Antonio Bay, Texas

Species	Ecophenotypes Temperature and salinity		Source
	Maximum	Minimum	
Ammonia parkinsoniana	*tepida*	*typica*	Poag 1978
Elphidium gunteri	*typicum*	*salsum*	Poag 1978
Elphidium galvestonense	*typicum*	*mexicanum*	Poag 1978
Palmerinella palmerae	*typica*	*gardenislandensis*	Poag 1978
Ammotium salsum	*exilie*	*typicum*	Poag 1978
Ammotium salsum	*tumidum*	*emaciatum*	Poag 1976

Table 7.2 Ecophenotypes from the Gulf of Mexico

Species	Ecophenotypes	Environment
Ammotium salsum	*dilatatum*	Coarse sand
Ammotium salsum	*fragile*	Near fresh water
Cibicidoides 'floridanus'	*bathyalis*	Slope
Cibicidoides 'floridanus'	*sublittoralis*	Shelf–slope
Elphidium discoidale	*translucens*	Shelf
Elphidium discoidale	*typicum*	Shelf
Hanzawaia concentrica	*strattoni*	Shelf
Hanzawaia concentrica	*typica*	Shelf
Spiroloculina soldanii	*typica*	Carbonate shelf
Spiroloculina soldanii	*dentata*	Carbonate shelf
Uvigerina peregrina	*hispidocostata*	Shelf edge to mid slope
Uvigerina peregrina	*parvula*	Shelf edge to upper slope
Uvigerina peregrina	*typica*	Outer shelf to upper slope

Fig. 7.1 Localities mentioned in the text

A useful taxonomic summary for the Gulf of Mexico has been provided by Poag (1981). He extended the paired ecophenotype principle from his 1978 study (Table 7.1) to include additional taxa (Table 7.2).

Compendia on the distribution of taxa in the Gulf of Mexico and the Caribbean Sea have been published by Culver and Buzas (1981b, 1982b).

The localities mentioned in this chapter are located on Fig. 7.1.

Marsh and mangrove swamp

Around the Gulf of Mexico most marshes have a flora of *Spartina* and related halophytes but in the Florida area mangroves are present. In general, the region is microtidal and although there are differences between berms, channel levees and marsh flats, there is not a zonation into high, mid and low marsh as in meso- and macrotidal regimes. Salinity and temperature data are sparse for the areas studied for their foraminiferal faunas.

In Table 7.3 minor associations found only in one area are listed. Two are confined to mangrove swamps and the remainder are from *Spartina* marshes. The mode of life and feeding strategies of the typical marsh species are summarised in Table 7.4 and details of the principal associations are given in Table 7.5.

The *Ammonia beccarii* association is the only one widely developed in the Florida mangrove swamps (Fig. 7.2). It is commonly accompanied by species of *Elphidium* (Table 7.5). The sediment is calcareous sand and silt and salinities are variable (5–36‰ Whitewater Bay, 28–47‰ Cape Sable). Although no data are available on dead assemblages, it seems likely that they have not undergone marked dissolution in this carbonate environment. The *A. beccarii* association is absent from marshes between Florida and Texas where it reappears in the San Antonio area. Here the sediments are terrigenous 'muds' with a flora of *Spartina* and *Salicornia*. Salinities range from 2 to 42‰ (Phleger 1966a). The associated species are somewhat different from those in Florida (Table 7.5).

Agglutinated taxa dominate the other three marsh associations. *Arenoparrella mexicana* is an epifaunal herbivore and this association is found in brackish to hypersaline marshes from Mississippi to Texas (Fig. 7.2). Diversity is low (α 1) and all the associated common species are agglutinated. Phleger (1965a) noted the preference of *A. mexicana* for well-drained areas of marsh.

The *Miliammina fusca* association is widely distributed in brackish marshes but is able to withstand periods of hypersalinity during dry, hot weather. Most of the additional species are agglutinated (Table 7.5) and diversity is low (α <1 to 2.5). Like *M. fusca*, *Ammotium salsum* is an infaunal detritivore, and this association is almost as widely distributed (Fig. 7.2). The additional species may be agglutinated, porcellaneous or hyaline (Table 7.5). This association is also present in the low-salinity lagoon, Lake Pontchartrain (Table 7.6).

A ternary plot of wall structure (Fig. 7.3) shows that Gulf of Mexico marshes, which span brackish to hypersaline environments, have assemblages in which all wall types are represented.

	Association	*Ammonia beccarii*	*Arenoparella mexicana*	*Miliammina fusca*	*Ammotium salsum*
Mangrove	Whitewater Bay, Florida	X			
	Ten Thousand Islands, Florida	X			
	Cape Sable, Florida	X			
Marsh	Mississippi Sound		X	X	
	Mississippi Delta			X	X
	Galveston Bay			X	X
	San Antonio area	X	X	X	X

Fig. 7.2 Distribution of mangrove and marsh associations in the Gulf of Mexico

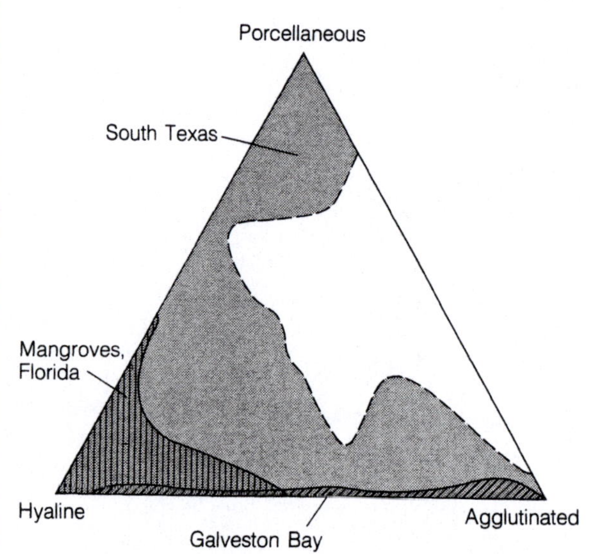

Fig. 7.3 Living assemblages. Marshes: Galveston Bay (Phleger 1965a), South Texas (Phleger 1966a); mangrove swamps: Florida (Phleger 1966b)

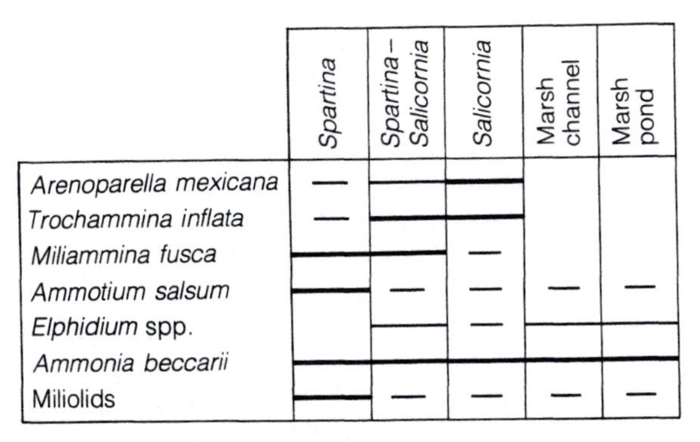

Fig. 7.4 South Texas marshes – generalised distribution of living taxa (based on Phleger 1966a)

The local distribution of taxa in south Texas marshes (Fig. 7.4) shows that agglutinated and hyaline forms are widely present but the porcellaneous miliolids are rare beyond the *Spartina* marsh.

Postmortem changes

Postmortem loss of calcareous tests does not appear to be a major factor in the development of the dead assemblages. Murray (1976a) concluded that for the Galveston Bay marshes most of the differences between the living and dead assemblages could be attributed to differing production rates of the component species.

Lagoons and estuaries

In this low-latitude region, the temperature of shallow enclosed waters may reach as high as 38 °C during the summer. Salinities may be consistently brackish, as in the Mississippi Sound–Delta area, or range from brackish during the wet summer months to hypersaline during the dry winter months. Such

Table 7.3 Minor benthic associations on marshes and mangrove swamps around the Gulf of Mexico and Caribbean Sea (L = living, D = dead, T = total)

Association		Area	Source
1. *Ammobaculites* spp.	T	Mississippi Sound	Phleger 1954
2. *Elphidium gunteri*	L	San Antonio area	Phleger 1966a
3. *Elphidium* cf. *koeboeense*	L	San Antonio area	Phleger 1966a
4. *Elphidium matagordanum*	L	Florida*	Phleger 1966b
5. *Massilina protea*	L	San Antonio area	Phleger 1966a
6. *Miliolinella microstoma*	L	San Antonio area	Phleger 1966a
7. *Palmerinella palmerae*	L	San Antonio area	Phleger 1966a
8. *Triloculina sidebottomi*	T	San Antonio area	Parker *et al.* 1952
		Mississippi Delta	Phleger 1955
9. *Discorinopsis aguayoi*	T	San Antonio area	Parker *et al.* 1953
10. *Triloculinella obliquinoda*	L	Florida*	Phleger 1966b
11. *Tiphotrocha comprimata*	T	Mississippi Sound	Phleger 1954
12. *Trochammina inflata*	L	San Antonio area	Parker *et al.* 1953
			Phleger 1966a
13. *Triloculina oblonga*	L	Barbuda, West Indies	Brasier 1975b
14. *Siphotrochammina lobata*	L	Tobago	Radford 1976a

* Mangrove swamps.

Table 7.4 Typical marsh and mangrove swamp species (feeding strategy from Ch. 2)

Species	Mode of life	Feeding strategy
Ammonia beccarii	Infaunal, free	? Herbivore
Ammotium salsum	Infaunal, free	Detritivore
Arenoparrella mexicana	Epifaunal, clinging or free	Herbivore
Miliammina fusca	Infaunal, free	Detritivore

Table 7.5 Principal associations from the Gulf of Mexico and Carribbean Sea marshes and mangrove swamps

***Ammonia beccarii* association**

Salinity: 2–47‰

Temperature: 5 to 32 °C

Substrate: calcareous sands and silts, 'muds'

Depth: intertidal mangrove swamp and marsh

Distribution:
1. Whitewater Bay, Florida (Phleger 1966b) living (as *Rotalia*)
2. Ten Thousand Islands, Florida (Phleger 1966b) living (as *Rotalia*)
3. Cape Sable, Florida (Gebelein 1977) total
4. San Antonio area: (a) (Phleger 1966a) living (as *Rotalia*); (b) (Parker *et al.* 1953) total (as *Rotalia*)

Additional common species:

'miliolids'	1
Elphidium galvestonense	1, 2, 3, 4a (as *E. incertum mexicanum*), 4b
Elphidium matagordanum	1, 2, 4a, b
Elphidium poeyanum	1, 3 (as *Cribroelphidium*)
Elphidium tumidum	1, 4a
Ammotium salsum	2, 4b (as *Ammobaculites*), 4a
Brizalina striatula	2 (as *Bolivina*)
Elphidium gunteri	4a, b
Miliammina fusca	4a
Elphidium cf. *koeboeense*	4a
Massilina protea	4a
Quinqueloculina poeyana	4a
Triloculina sidebottomi	4a
Palmerinella palmerae	4a

***Arenoparrella mexicana* association**

Salinity: 2–42‰

Temperature: variable

Substrate: sand

Depth: well-drained intertidal marsh

Distribution:
1. Mississippi Sound (Phleger 1954) total
2. Galveston Bay, Texas (Phleger 1965a) living, total
3. San Antonio area (Phleger 1966a) living

Additional common species:

Haplophragmoides subinvolutum	1
Ammoastuta inepta	1
Miliammina fusca	1, 2
Tiphotrocha comprimata	1 (as *Trochammina*)
Trochammina inflata	2, 3
Ammotium salsum	2

***Miliammina fusca* association**

Salinity: brackish, but 2–42‰ overall

Temperature: variable

Substrate: sand

Depth: intertidal marsh

Distribution:
1. Mississippi Sound (Phleger 1954) total
2. Mississippi Delta: (a) (Phleger 1955) total; (b) (Lankford 1959) living
3. Galveston Bay, Texas (Phleger 1965a) living, total
4. San Antonio area (Phleger 1966a) living

Additional common species:

Ammobaculites spp.	1
Arenoparrella mexicana	1, 3, 4
Ammomarginulina fluvialis	1 (as *Ammoscalaria*)
Ammoastuta inepta	2a, 3
Ammotium salsum	2b, 3, 4
Trochammina inflata	2b, 4
Ammonia beccarii	4

***Ammotium salsum* association**

Salinity: 2–42‰

Temperature: variable

Substrate: sand

Depth: intertidal marsh/mangrove

Distribution:
1. Mississippi Delta (Phleger 1955, as *Ammobaculites*) living
2. Galveston Bay, Texas (Phleger 1965a) living, total
3. San Antonio area (a) (Phleger 1966a) living; (b) (Parker *et al.* 1953) total (as *Ammobaculites*)
4. Cartagena, Colombia (Boltovskoy and Hincape de Martinez 1983) living

Additional common species:

Ammonia beccarii	1a (as *Rotalia*) 2, 3a
Miliammina fusca	1a, 2, 3a
Jadammina macrescens	1a (as *Trochammina*)
Elphidium matagordanum	2
miliolids	2, 3a
Arenoparrella mexicana	2, 3a, 4
Trochammina inflata	2
Elphidium gunteri	3a
Palmerinella palmerae	3a

environmental extremes are very stressful for the fauna and flora but, although this may limit the diversity, those organisms that are adapted to these conditions flourish.

Florida Bay, the Gulf of Batabano, Cuba, Barbuda and St Lucia are all areas of carbonate sediment accumulation. The water is shallow, clear, free of clastic detritus and the sediment is locally colonised by seagrass and calcareous algae (Bathurst 1971). The principal seagrass is *Thalassia testudinum* (turtle grass). It occurs throughout the Caribbean and around the Gulf of Mexico with the exception of the area between the Mississippi Delta and Galveston, Texas. *Thalassia* has broad flat blades which are commonly covered with organic detritus and growths of epiphytic diatoms (Grant *et al.* 1973). These serve as food and shelter for phytal foraminifera. Small species, such as *Discorbis candeianus*, occur mainly within patches of organic detritus. The large form *Marginopora vertebralis*, gathers the detritus into a peripheral ring (Steinker and Steinker 1976). Blades lacking epiphytic diatoms or organic detritus support few living foraminifera.

In a detailed study of two *Thalassia* beds in Jamaica, Buzas *et al.* (1977) found no statistical relationship between the standing crops of the 19 abundant species and the measured environmental variables. They concluded that '. . . the physical environment is sufficiently benign so that biotic variables are most important'. However, they found no evidence of either competition or spatial partitioning of the micro-habitat, contrary to that described below for calcareous algae.

The abundance of *Thalassia* varies seasonally from a peak in early spring and midsummer to a trough in the autumn (Zieman 1975). This has a profound effect on the bundance of living phytal foraminifera. During storms, *Thalassia* may be uprooted and transported over long distances. Phytal foraminifera commonly remain attached and may be dispersed into new areas by this means (Bock 1970). Some species, such as *Planorbulina acervalis* which is encrusting, remain firmly attached to *Thalassia* whether living or dead. *Archaias angulatus* attaches itself by its pseudopodia and can be detached during violent storms and by wave activity.

Other plants which are important as substrates for phytal foraminifera are calcareous green codiacean algae (Gancarz, in Bathurst 1971; Grant *et al.* 1973; Brasier 1975a; Steinker 1977; Steinker and Steinker 1976). *Penicillus capitatus* provides three habitats: the tuft traps fine-grained organic detritus and provides shelter for *Discorbis candeianus, Archaias angulatus, Quinqueloculina seminula* and *Flintinoides labiosa;* the stalk bears diatoms but not much detritus and supports *D. candeianus* and *A. angulatus*; and the rhizoids provide substrates for peneroplids. *Rhizocephalus phoenix* f. *breviolus* supports *D. candeianus, A. angulatus* and *F. labiosa. Udotea* and *Halimeda* support varying numbers of foraminifera.

The distribution of the flora is clumped and on each plant there are patches of phytal foraminifera. The composition of the fauna and the abundance of each species differs dramatically from one plant to another as shown by the data from Barbuda (Brasier 1975b). Some species are confined to plants, others to sediment and still others occur on both (7, 34 and 62 species respectively in Florida Bay, Martin and Wright 1988). The sediment-dwelling foraminifera also have a clumped distribution with patches of at least 30 m^2 in Florida Bay (Lynts 1966).

The dead assemblages preserved in the sediment are derived from two living assemblages: that of the sediment and that of the plants. The proportion of each may vary both temporally and spatially. In the case of total assemblages, these will consist of dead tests from plants and both living and dead tests from the sediment. They are therefore less than ideal records of the fauna but, unfortunately, in many cases only total assemblages have been studied.

Lagoons and estuaries which are the site of clastic sediment accumulation may support some seagrass, but no accounts have been given of phytal foraminiferal faunas in these areas. The foraminiferal assemblages are almost entirely free-living forms, herbivores or detritivores, epifaunal or infaunal (Table 7.7).

Foraminiferal associations of local occurrence only are listed in Table 7.6. Details of the principal associations are given in Table 7.8 and Fig. 7.5.

The *Archaias angulatus* association is confined to shallow carbonate environments and is commonly accompanied by species of *Quinqueloculina* and *Triloculina*. It is stenohaline and because *Archaias* has endosymbionts it is limited to depths <20 m (Lee *et al.* 1974; Martin 1986). The test is strong and resistant to abrasion and dissolution (Cottey and Hallock 1988). In the back reef lagoon of Molasses Reef, it is abundant on *Thalassia* in the nearshore region but even where there are small living populations *Archaias* can give rise to sediment assemblages dominated by this species because of its resistance to abrasion (Martin and Wright 1988). It also resists winnowing so it dominates lag assemblages (i.e. fields

Sediment	Area	*Quinqueloculina* spp.	*Archaias angulatus*	*Ammonia beccarii*	*Ammobaculites* spp.	*Elphidium matagordanum*	*Ammotium salsum*	*Elphidium gunteri*	*Miliammina fusca*	*Elphidium discoidale*	*Elphidium poeyanum*	*Triloculina oblonga*
Carbonate	Molasses Reef	x	x									
Carbonate	Florida Bay	x	x	x								
Carbonate	Cape Sable	x		x								
Carbonate	Cape Romano			x		x						
Clastic	Charlotte Harbor			x	x							
Clastic	Tampa Bay	x		x		x						
Clastic	Mississippi Sound			x	x		x	x	x			
Clastic	Mississippi Delta			x			x	x		x		
Clastic	Sabine Lake				x				x			
Clastic	Galveston Bay			x								
Clastic	San Antonio area	x		x			x	x			x	
Clastic	Matagorda Bay			x				x				
Clastic	Laguna Madre, Texas	x		x		x					x	x
Clastic	Laguna Madre, Mexico			x				x				
Clastic	Laguna de Tamiahua	x		x				x				
Clastic	Laguna de las Maritas			x								
Carbonate	St Lucia			x								
Carbonate	Barbuda	x										x
Carbonate	Cuba	x	x							x	x	

Fig. 7.5 Distribution of associations in Gulf of Mexico and Caribbean lagoons

III and IV). This association is widespread over a large area of the Gulf of Batabano, Cuba. It is also present in open shelf areas as described below.

The *Quinqueloculina* spp. association is widely developed in both carbonate and clastic environments. *Quinqueloculina* is a free-living epifaunal genus with both phytal and sediment-dwelling species. This association is absent from exclusively brackish areas (e.g. around the Mississippi Delta). It prefers normal marine or hypersaline environments but can tolerate periods of low salinity during the rainy season. The most common additional species are *Triloculina* spp. and *Ammonia beccarii* (Table 7.8). Living assemblages may have low diversity (e.g. α 1–2, Laguna Madre, Texas, Phleger 1960c) but total assemblages, which may include transported forms, are more diverse (α 15–20, Molasses Reef, Wright and Hay 1971; α 2.5–9, Florida Bay, Lynts 1962, 1971; $H(S)$ 4.3, Cape Sable, Gebelein 1977).

The most widely distributed association is that of *Ammonia beccarii*. It is present in brackish parts of the Mississippi Delta (salinity 20–34‰, Lankford 1959), in areas which range from brackish to hypersaline at different seasons of the year (4–41‰, San Antonio Bay, Phleger 1956) and in hypersaline lagoons (39–81‰, Laguna Madre, Texas, Phleger 1960c). The most common additional species are *Quinqueloculina* spp., *Elphidium poeyanum, E. gunteri* and *Ammotium salsum*. In the San Antonio area, where Phleger and Lankford (1957) recorded seasonal changes, this association was present at every period sampled.

The *Ammobaculites* spp. association is present only in the most brackish areas with salinities possibly down to 0‰ (Bandy 1956; Ellison and Murray 1987). It is a detritivore which prefers organic-rich sediments although Buzas (1969) has suggested that the correlation between densities of *A. exiguus* and chlorophyll *b* may mean that it feeds on green algae. The additional common species are principally agglutinated (Table 7.8).

There are four associations dominated by different species of *Elphidium*. The *E. gunteri* association occurs mainly in brackish areas (Mississippi Sound, Mississippi Delta) but is also present in slightly hypersaline environments (San Antonio area, Laguna de Tamiahua, Mexico). The *E. discoidale* association is present in normal marine salinities off the Baptiste Collette Subdelta of the Mississippi Delta, Puerto Rico and in the Gulf of Batabano, Cuba (Table 7.8). The *E. matagordanum* association and the *E. poeyanum* association range from brackish or normal marine to hypersaline environments, and both have other species of *Elphidium* plus *Ammonia beccarii* as additional common elements.

The *Ammotium salsum* association spans a broad range of salinities (4–41‰) but is mainly developed in brackish conditions. In upper San Antonio Bay this association was present at sta. 34 of Phleger and Lankford (1957) in August and November 1954, January, May and June 1955 and the standing crop varied from 320 (June) to 2608 (November) per 10 cm^2. Size measurements showed that juveniles and adults were present at all seasons so reproduction takes place throughout the year.

The *Miliammina fusca* association is restricted to brackish lagoons and is really more typical of marshes (see above). The *Triloculina oblonga* association is a low-diversity assemblage (*a* 2) which lives in normal marine to hypersaline environments of Barbuda and Laguna Madre, Texas. The additional common species are all porcellaneous. In Tobago, it occurs near a stream mouth but probably in waters of normal salinity.

In those lagoons subject to a marked seasonal change in salinity there may be different living assemblages at different seasons. For example, in Aransas Bay, Texas, the *Quinqueloculina* spp. association was present in January and March 1955 but the *Ammonia beccarii* association in May and June 1955 (sta. 1 of Phleger and Lankford 1957). Diversity for the San Antonio area ranged from α 1–4 except during March and May when values as high as α 8 were attained.

Table 7.6 Minor benthic associations in lagoons and estuaries in the Gulf of Mexico and Caribbean (L = living, D = dead, T = total)

Association		Area	Source
1. *Ammotium salsum*	T	Lake Pontchartrain	Otvos 1978
	L, D	Alvarado Lagoon, Mexico	Phleger and Lankford 1978
2. *Ammoscalaria pseudospiralis*	T	Mississippi Delta	Phleger 1955
3. *Amphistegina gibbosa*	T	St Lucia, West Indies	Sen Gupta and Schafer 1973
4. *Elphidium delicatulum*	L, T	Mississippi Delta	Lankford 1959
		Cape Romano, Florida	Benda and Puri 1962
5. *Elphidium galvestonense*	T	Upper Florida Bay	Lynts 1962
6. *Elphidium kugleri*	T	Laguna de las Maritas	Ruiz and Sellier de Civrieux 1972
7. *Palmerinella gardenislandensis*	L, T	Mississippi Delta	Lankford 1959
8. *Discorbis candeinus*	L	Molasses Reef, Florida	Wright and Hay 1971
9. *Triloculina* cf. *brevidentata*	L	Tampa Bay	Walton 1964b
10. *Triloculina* spp.	L, T	Galveston Bay	Phleger 1965
11. *Triloculinella obliquinoda*	L	Laguna Madre, Texas	Phleger 1960c
12. *Trochulina rosea*	L	Tobago	Radford 1976a, b
13. *Cymbaloporetta squamosa*	L	Tobago	Radford 1976a, b
14. *Elphidium poeyanum*	L	Tobago	Radford 1976a, b

Tables 7.7 Typical lagoonal and estuarine species from the Gulf of Mexico and Caribbean (feeding strategy from Ch. 2)

Species	Mode of life	Feeding strategy
Ammobaculites spp.	Infaunal, free	Detritivore
Ammonia beccarii	Infaunal, free	? Herbivore
Ammotium salsum	Infaunal, free	Detritivore
Archaias angulatus	Epifaunal, clinging, phytal	Symbiosis, herbivore
Elphidium discoidale	? Infaunal, free	
Elphidium gunteri	Infaunal, free	
Elphidium matagordanum	? Infaunal, free	
Elphidium poeyanum	? Infaunal, free	
Miliammina fusca	Infaunal, free	Detritivore
Quinqueloculina spp.	Epifaunal, free	Detritivore/herbivore
Triloculina oblonga	Epifaunal, free	Detritivore/herbivore

Table 7.8 Principal recurrent associations from the Gulf of Mexico and Caribbean Sea lagoons and estuaries

Archaias angulatus **association**

Salinity: 34–37‰

Temperature: 9 to 38 °C

Substrate: turtle grass, carbonate sand, mud

Depth: <12 m

Distribution:

1. Molasses Reef, Florida (Wright and Hay 1971) total
2. Card Sound, Florida Bay (Lynts 1962) total
3. Gulf of Batabano, Cuba (Bandy 1964b) total

Additional common species:

Species	Distribution
Quinqueloculina spp.	1, 2, 3
Triloculina spp.	1
Elphidium spp.	3

Quinqueloculina **spp. association**

Salinity: 10–47‰

Temperature: 15 to 38 °C

Substrate: turtle grass, carbonate sand, mud; fine quartz sand

Depth: <12 m

Distribution:

1. Molasses Reef, Florida (Wright and Hay 1971) total
2. Florida Bay (Lynts 1962, 1971) total
3. Cape Sable, Florida (Gebelein 1977) total
4. Tampa Bay, Florida: (a) (Bandy 1956) total; (b) (Walton 1964b) dead
5. San Antonio area (Phleger and Lankford 1957) living
6. Laguna Madre, Texas (Phleger 1960c) living, total
7. Laguna de Tamiahua, Mexico (Ayala-Castanares and Segura 1981) total
8. Barbuda, West Indies (Brasier 1975b) living
9. Gulf of Batabano, Cuba (Bandy 1964b) total
10. Puerto Rico (Seiglie 1971a) total

Additional common species:

Species	Distribution
Triloculina spp.	1, 2, 3, 4b, 5, 6, 8
Peneroplis carinatus	1
Archaias angulatus	1
Discorbis candeina	1 (as *Rosalina*)
Sigmoihauerina bradyi	2 (as *Hauerina*)
Elphidium poeyanum	2
Elphidium galvestonense	2
Ammonia beccarii	2, 5, 6 (as *Streblus*), 3, 4a (as *Streblus tepidus*), 7
Miliolinella circularis	2
Valvulina oviedoana	2
Neoeponides auberi	2 (as *'Rotalia' translucens*), 9
Elphidium gunteri	7
Bigenerina irregularis	10 (as *B. textularoides*)

Table 7.8 *(continued)*

***Ammonia beccarii* association**

Salinity: (2) 28–47, (3) 13–43, (4) 0–21, (7b) 20–34, (8) 7–19, (10b) 4–41, (11) 39–81‰

Temperature: 5 to 35 °C

Substrate: calcareous mud, fine quartz sand, sandy and silty muds

Depth: intertidal to 10 m

Distribution:

1. Blackwater Sound, Florida Bay (Lynts 1962) total
2. Cape Sable, Florida (Gebelein 1977) total
3. Cape Romano, Florida (Benda and Puri 1962) total
4. Charlotte Harbor area and Tampa Bay, Florida (Bandy 1956) total (as *Streblus tepidus* and *sobrinus*)
5. Tampa Bay, Florida (Walton 1964b) living, dead
6. Mississippi Sound (Phleger 1954) total
7. Mississippi Delta: (a) (Phleger 1955) total; (b) (Lankford 1959) living, total; (c) (Otvos 1978) total
8. Sabine Lake (Kane 1967) total
9. Galveston Bay, Texas (Phleger 1965a) living, total
10. San Antonio area: (a) (Parker *et al.* 1953) total; (b) (Phleger 1956) living, total; (c) (Phleger and Lankford 1957) living; (d) (Poag 1976) total
11. Matagorda Bay (Shenton 1957) total
12. Laguna Madre, Texas (Phleger 1960c) living, total
13. Laguna Madre, Mexico (Ayala-Castanares and Segura 1986) living, total
14. Laguna de Tamiahua, Mexico (Ayala-Castanares and Segura 1981) living, total
15. Alvarado Lagoon, Mexico (Phleger and Lankford 1978) living, dead
16. Laguna de Terminos, Mexico (Ayala-Castanares 1963) living, total
17. Laguna de las Maritas, Venezuela (Ruiz and Sellier de Civrieux 1972) total
18. St Lucia, West Indies (Sen Gupta and Schafer 1973) total
19. Puerto Rico (Seiglie 1971a) total
20. Guayanilla Bay, Puerto Rico (Seiglie 1975) living

Additional common species:

Quinqueloculina spp.	1, 3, 4, 5, 10c, 14, 15, 20
Triloculina spp.	1, 10c, 12
Elphidium poeyanum	1, 3 (as *Cribroelphidium*) 8, 10b, 11c, 12, 18, 19
Elphidium crispum	1
Elphidium matagordanum	2 (as *Nonion depressulum*), 5, 9, 10c
Elphidium galvestonense	2, 3, 5 (as *E. incertum mexicanum*), 9
Elphidium gunteri	3, 4, 5, 7a, 7b, 8b, 10a, 10c, 10d, 11, 12, 13, 14, 16
Elphidium rugulosum	4
Ammobaculites exiguus	4, 5
Ammotium salsum	4, 5, 6, 7a, 7c, 10a, 10b, 10c, 11 (as *Ammobaculites*), 7a, 7b, 9, 10d, 14
Trochammina cf. *multiloculata*	5
Trochammina cf. *rotaliformis*	5
Palmerinella gardenislandensis	7b, 16 (as *Eponidella*)
Ammobaculites spp.	8
miliolids	9, 13
Ammoscalaria pseudospiralis	10b, 10c
Triloculinella obliquinoda	12
Nonionella atlantica	18
Elphidium discoidale	19
Amphistegina gibbosa	19

Table 7.8 *(continued)*

***Ammobaculites* spp. association**

Salinity: 7–19‰

Temperature: 10 to 30 °C

Substrate: sandy and silty muds

Depth: <3 m

Distribution:
1. Charlotte Harbor area, Florida (Bandy 1956) total
2. Mississippi Sound (Phleger 1954) total
3. Sabine Lake (Kane 1967) total

Additional common species:

Ammonia beccarii	1 (as *Streblus tepidus*), 2 (as *Rotalia*)
Ammotium salsum	2 (as *Ammobaculites*)
Ammomarginulina fluvialis	2 (as *Ammoscalaria*)
Miliammina fusca	2, 3
Haplophragmoides wilberti	3

***Elphidium gunteri* association**

Salinity: (1) 1–35, (2) 20–34, (3) 4–41, (5) 25–31, (6) 33–37‰

Temperature: 5 to 35 °C

Substrate: sandy mud

Depth: <6 m

Distribution:
1. Mississippi Sound (Phleger 1954) total
2. Mississippi Delta: (a) (Phleger 1955) living, total; (b) (Lankford 1959) living
3. San Antonio area: (a) (Parker *et al.* 1953) total; (b) (Phleger 1956) living, total; (c) (Phleger and Lankford 1957) living; (d) (Poag 1976) total (as formae *gunteri* and *delicatulum*)
4. Matagorda Bay (Shenton 1957)
5. Laguna Madre, Mexico (Ayala-Castanares and Segura 1968) total
6. Laguna de Tamiahua, Mexico (Ayala-Castanares and Segura 1981) total

Additional common species:

Hanzawaia concentrica	1 (as *Cibicidina strattoni*)
Elphidium incertum	1
Ammonia beccarii	1, 2a, 3a, 3b, 4, (as *Rotalia*), 2b, 3c (as *Streblus*), 3d, 5, 6
Ammotium salsum	2a, 3c (as *Ammobaculites*), 2b, 3a, b, 3d
Palmerinella gardenislandensis	2b, 3b (as *P. palmerae*)
Elphidium matagordanum	2b, 3c
Triloculina sidebottomi	2b
Elphidium discoidale	3b
Ammonia pauciloculata	4 (as *Rotalia*)
Quinqueloculina seminula	6

***Elphidium discoidale* association**

Salinity: 35–37‰

Temperature: –

Substrate: muddy carbonate sand; muddy sand.

Depth: <12 m

Distribution:
1. Mississippi Delta (Phleger 1955) total
2. Gulf of Batabano, Cuba (Bandy 1946) total
3. Puerto Rico (Seiglie 1971a) total

Additional common species:

Ammonia beccarii	1 (as *Rotalia*), 3
Elphidium gunteri	1
Elphidium poeyanum	2
Quainqueloculina spp.	2, 3
Elphidium spp.	3

***Elphidium matogordanum* association**

Salinity: 22–81‰

Temperature: 10 to 35 °C

Substrate: fine quartz sand, organic-rich mud

Depth: <10 m

Distribution:
1. Cape Romano, Florida (Benda and Puri 1962) total
2. Tampa Bay, Florida (Walton 1964b) dead
3. Laguna Madre, Texas (Phleger 1960c) living

Additional common species:

Elphidium tumidum	1
Elphidium galvestonense	1, 2 (as *E. incertum*)
Elphidium poeyanum	2, 3
Ammonia beccarii	1, 3 (as *Streblus*), 2 (as *Rotalia*), 3
Triloculina cf. *brevidentata*	2

Table 7.8 *(continued)*

***Elphidium poeyanum* association**

Salinity: 35–81‰

Temperature: 10 to 35 °C

Substrate: muddy sand

Depth: <12 m

Distribution:
1. San Antonio area (Phleger and Lankford 1957) living
2. Laguna Madre, Texas (Phleger 1960c) total
3. Gulf of Batabano, Cuba (Bandy 1964b) total

Additional common species:

Ammonia beccarii	1 (as *Streblus*), 2
Elphidium gunteri	1
Ammotium salsum	1 (as *Ammobaculites*)
Elphidium discoidale	3
Quinqueloculina spp.	3

***Ammotium salsum* association**

Salinity: 4–41‰

Temperature: 16 to 41 °C

Substrate: muddy sediments

Depth: <6 m

Distribution:
1. Mississippi Sound (Phleger 1954) total (as *Ammobaculites*)
2. Mississippi Delta: (a) (Phleger 1955) total (as *Ammobaculites*); (b) (Lankford 1959) living, total
3. San Antonio area: (a) (Parker *et al.* 1953) total; (b) (Phleger 1956) living, total; (c) (Phleger and Lankford 1957) living; (d) (Poag 1976) total
4. Mayagüez Bay, Puerto Rico (Seiglie 1971c) total

Additional common species:

Ammobaculites spp.	1
Miliammina fusca	1, 2b
Nonionella atlantica	1
Gaudryina exilis	1
Elphidium discoidale	1
Ammonia beccarii	1, 2a, 3a, 3b, (as *Rotalia*), 2b, 3c (as *Streblus*), 3d
Nonionella opima	2a
Triloculina sidebottomi	2a
Brizalina lowmani	2b (as *Bolivina*)
Palmerinella gardenislandensis	2b, 3c (as *Eponidella*)
Elphidium gunteri	2b (as *E. delicatulum*), 3a, 3b, 3d
Elphidium matagordanum	3c

***Miliammina fusca* association**

Salinity: 7–19‰

Temperature: 10 to 30 °C

Substrate: sandy and silty muds

Depth: <6 m

Distribution:
1. Mississippi Sound (Phleger 1954) total
2. Sabine Lake (Kane 1967) total

Additional common species:

Ammobaculites spp.	1, 2
Ammomarginulina fluvialis	1 (as *Ammoscalaria*)
Ammoastuta inepta	1

***Triloculina oblonga* association**

Salinity: 35–81‰

Temperature: 10 to 35 °C

Substrate: fine sand

Depth: <5 m

Distribution:
1. Laguna Madre, Texas (Phleger 1960c) living, total
2. Barbuda, West Indies (Brasier 1975b) living, total
3. Tobago (Radford 1976a, b) living

Additional common species:

Triloculinella obliquinoda	1
Quinqueloculina spp.	2
Archaias angulatus	2

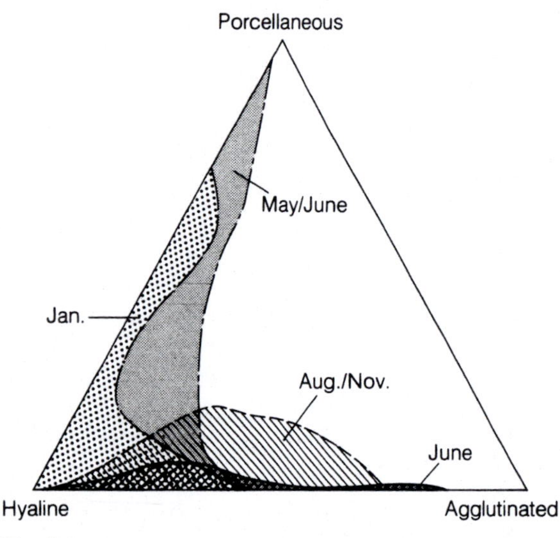

Fig. 7.6 Seasonal changes in the living assemblages from the San Antonio area (data from Phleger and Lankford 1957)

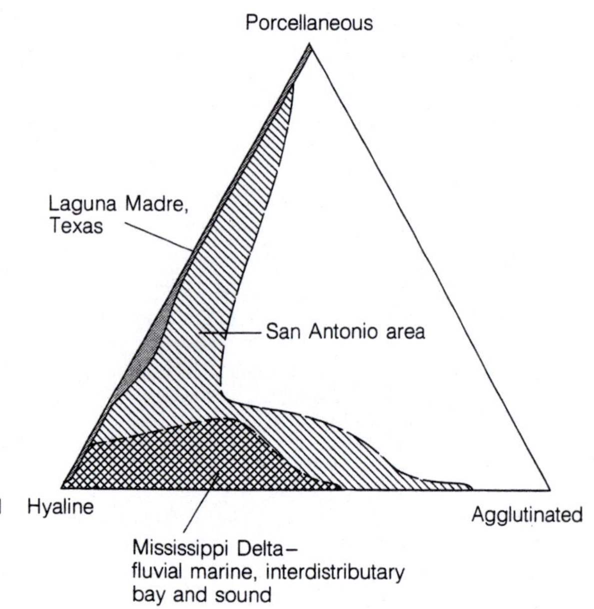

Fig. 7.7 Living assemblages from Gulf of Mexico lagoons

Seasonal changes in wall structure type and abundance are shown in Fig. 7.6. It is possible that with such constantly changing conditions the living assemblage is never in equilibrium with its environment (Murray 1968b).

A ternary plot of wall structure for living assemblages show the contrast between brackish to normal marine parts of the Mississippi Delta (fluvial marine, interdistributary bay, sound), dominated by hyaline forms, brackish to hypersaline lagoons from the San Antonio area, dominated by agglutinated, porcellaneous of hyaline forms, and the hypersaline Laguna Madre, Texas, dominated by porcellaneous and hyaline forms (Fig. 7.7). Greiner (1969, 1974) summarised the data on wall types of total assemblages from Gulf of Mexico lagoons and concluded that porcellaneous taxa are best able to secrete their tests where the waters are saturated or supersaturated with $CaCO_3$. He suggested that hyaline forms would not be able to secrete ordered crystallites in supersaturated environments. This does not appear to be true because hyaline and procellaneous taxa commonly occur together in these environments. However, the hypothesis could be tested experimentally.

Hiltermann and Tüxen (1978), Hiltermann (1982) and Hiltermann and Haman (1985) have used the Braun–Blanquet method of defining associations based on total assemblages from the Mississippi Delta and Mississippi Sound. There are similarities between these associations and those based on species dominance described here. Haman (1983) made a study of the Textulariina only from the Balize Delta, Louisiana, and concluded that the most favoured biotopes are interdistributary bays and levees. Foraminiferal tests containing some stained cytoplasm together with pyrite from Guayanilla Bay, Puerto Rico, were interpreted by Seiglie (1975) as living, but it seems more likely that they were recently dead when he collected them.

Generic predominance

A plot of generic predominance, based mainly on total assemblages (Fig. 7.8) demonstrates that few genera are important in the lagoons of this area. Figure 7.9 shows generic predominance in relation to salinity.

Postmortem modification of assemblages varies according to the lagoonal setting. The back-reef lagoon of Molasses Reef is subject to strong wave and current activity. Small tests are winnowed away and a lag of large robust *Archaias angulatus* is left (fields III, IV). In addition, other robust species from the fore-reef (*Amphistegina gibbosa. Asterigerina carinata, Homotrema rubra, Textularia agglutinans*) are transported into the back-reef lagoon (Martin and Wright 1988). Florida Bay in parts, at least, must undergo winnowing of tests from source areas with re-

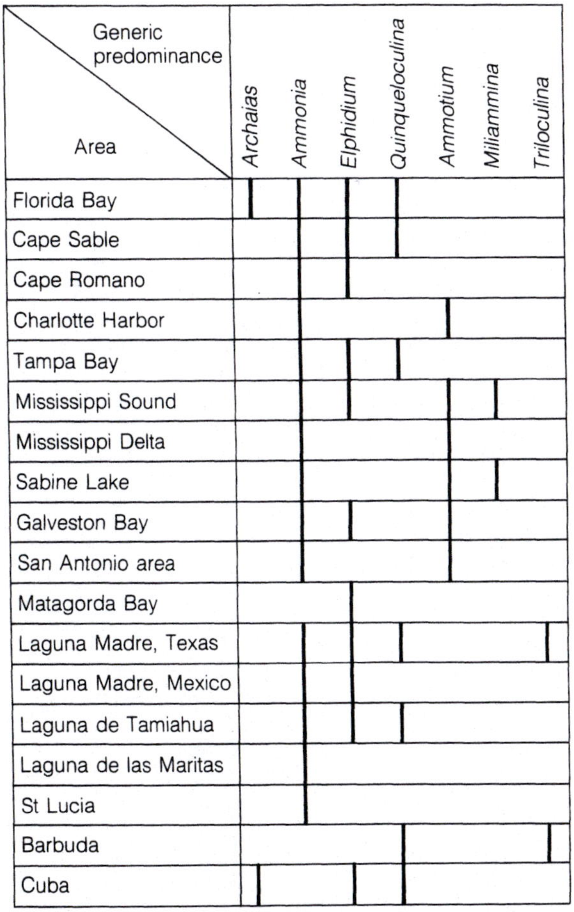

Fig. 7.8 Generic predominance in Gulf of Mexico and Caribbean lagoons

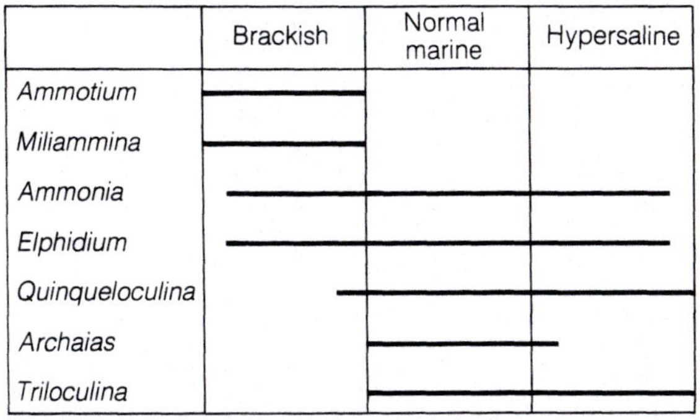

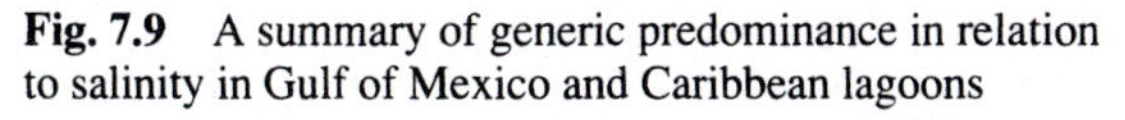

Fig. 7.9 A summary of generic predominance in relation to salinity in Gulf of Mexico and Caribbean lagoons

deposition in shallow and intertidal areas elsewhere (e.g. the subenvironments described from Cape Sable by Gebelein 1977). In those lagoons protected from wave attack by barrier islands, postmortem transport is negligible except near the passes which afford connection with the open seas. In many cases, the living and dead assemblages are closely similar and this has been demonstrated to be the case for Laguna Madre, Texas (Murray 1976a) (field I).

Shelf, slope and deep sea

Gulf of Mexico

Major regional studies giving data on total assemblages have been carried out by Phleger (1951), Parker (1954) and Pflum and Frerichs (1976). Walton (1964a) and Poag (1981) have presented their results mainly as maps. The distribution of individual species have been plotted by Culver and Buzas (1981b). Many new species were described by Phleger and Parker (1951) and a useful taxonomic revision has been provided by Poag (1981). A faunal reference list is given by Finger (1981).

Very few details of living assemblages are available for the Gulf of Mexico. Where samples were taken for this purpose, either they were too small to give 100 or more individuals or else the standing crop over much of the area was very low. The absence of these data makes it difficult to evaluate postmortem modification of assemblages.

Thirty-eight minor benthic associations of limited geographic distribution are listed in Table 7.9. Ten of these are confined to the Mississippi Delta area and they span a depth range of 9–1262 m. The principal associations are listed in Table 7.10 and their geographic distribution is summarised in Fig. 7.10. Although most authors give some information on temperature, salinity and depth (Table 7.11) Phleger, Parker and Bandy often gave no details of substrate type.

Off Florida, there is a nearshore zone of detrital quartz sand extending to a depth of ~20 m and beyond this the shelf sediments are dominantly carbonate, much being relict from lower sea-level. In northern Florida and off Alabama, terrigenous medium to fine quartz sand blankets most of the shelf (Gould and Stewart 1955). In the northwest gulf, active deposition of mud and silt takes place to the west of the Mississippi Delta following dispersion paths running along the shelf (Fig. 7.11), but off Texas much of the shelf is an area of slow or non-deposition with relict quartz sands (Van Andel 1960). On the upper slope there is a zone of fine-grained sediments with organic carbon

Table 7.9 Minor benthic associations from the shelf, slope and deep sea, Gulf of Mexico (L = living, T = total)

	Association		Depth (m)	Area	Source
1.	*Rosalina floridana*	T	12–14	Fort Myers	Bandy 1956
2.	*Archaias angulatus*	T	3–20	Tampa	Bandy 1956
3.	*Uvigerina hispidocostata*	T	79–134	Tampa	Bandy 1956
4.	*Brizalina daggarius*	T	177	Tarpon Springs	Bandy 1956
5.	*Nouria* sp.	L	58	Florida	Parker 1954
6.	*Cassidulina curvata*	L	75	Florida	Bock 1982
		T	232	Mississippi	Pflum and Frerichs 1976
7.	*Cibicides protruberans*	T	139–146	Florida	Parker 1954
8.	*Bulimina spicata*	T	950	Florida	Parker 1954
		T	227	Galveston	Phleger 1951
9.	*Textularia earlandi*	T	62	Chocktawatchee	Parker 1954
		T	22–88	Mississippi Delta	Parker 1954
10.	*Hoeglundina elegans*	T	2150–3280	Florida	Parker 1954
		T	2051–2443	Mississippi	Pflum and Frerichs 1976
11.	*Saccammina atlantica*	T	40	Mississippi	Parker 1954
		T	26	Matagorda	Phleger 1956
12.	*Goësella mississippiensis*	T	79–201	Mississippi Delta	Parker 1954
13.	*Bulimina marginata*	T	168	Mississippi Delta	Parker 1954
14.	*Trochamminopsis quadriloba*	T	208	Mississippi Delta	Parker 1954
15.	*Trochammina tasmanica*	T	314	Mississippi Delta	Parker 1954
16.	*Epistominella vitrea*	T	53	Mississippi Delta	Parker 1954, Phleger 1955
17.	*Eponides regularis*	T	205	Mississippi Delta	Parker 1954
18.	*Brizalina ordinaria*	T	373	Mississippi Delta	Parker 1954
19.	*Cassidulina carinata*	T	400	Mississippi Delta	Parker 1954
20.	*Bolivinella tessellata*	T	1262	Mississippi Delta	Parker 1954
21.	*Brizalina lowmani*	L, T	9–21	Mississippi Delta	Lankford 1959
22.	*Melonis pompilioides*	T	2126	Mississippi	Pflum and Frerichs 1976
		T	3294	Louisiana	Pflum and Frerichs 1976
23.	*Cibicidoides bradyi*	T	2315	Mississippi	Pflum and Frerichs 1976
24.	*Lagenammina difflugiformis*	T	26–42	Louisiana–Texas	Phleger 1951
25.	*Uvigerina laevis*	T	113–205	Louisiana	Phleger 1951
26.	*Cassidulina laevigata*	T	146	Louisiana	Phleger 1951
27.	*Sigmoilopsis schlumbergeri*	T	3107	Louisiana	Phleger 1951
28.	*Nodulina dentaliniformis*	T	3431	Louisiana	Phleger 1951
29.	*Haplophragmoides bradyi*	T	731	Galveston	Phleger 1951
30.	*Repmanina charoides*	T	1207–3475	Galveston	Phleger 1951
31.	*Eponides tumidulus*	T	3381–3548	Galveston	Phleger 1951
32.	*Brizalina minima*	T	128	Matagorda	Phleger 1951
33.	*Nonionella atlantica*	L	56	Matagorda	Phleger 1956
34.	*Brizalina striatula spinata*	L	59–75	Matagorda	Phleger 1956
35.	*Cancris oblonga*	L	66–73	Matagorda	Phleger 1956
36.	*Brizalina fragilis*	T	101	Brownsville	Phleger 1951
37.	*Adercotryma glomerata*	T	1208	Brownsville	Phleger 1951
38.	*Trifarina bella*	T	163–275	Texas	Pflum and Frerichs 1976

Table 7.10 Principal associations on the shelf, slope and deep sea, Gulf of Mexico and Caribbean

***Ammonia beccarii* association**

Salinity: (1) 34–36, (2) 34–37, (4) 34, (6) 34, (9b) 22–31, (9c) 21–34, (11b) 31–35‰

Temperature: (1) 21 to 30, (2) 26 to 32, (4) 20 to 26, (6) 20 to 26, (9a) 17 to 19, (9b) 10 to 30, (10a) 12 to 18, (11a) 14 to 18 °C

Substrate: sand, muddy sand

Depth: (1) 6–7, (2) 3–7, (3) 3, (4) 22, (5) 12, (6) 20, (9a) 26, (9c) 8–24, (10a) 20–31, (11a) 15–48, (11b) 7–28, (12) 2–5 m

Distribution:

1. Fort Myers, Florida (Bandy 1956) total (as *Streblus tepidus*)
2. Tampa, Florida (Bandy 1956) total (as *Streblus tepidus*)
3. Tarpon Springs, Florida (Bandy 1956) total (as *Streblus tepidus* and *sobrinus*)
4. Cape San Blas, Florida (Parker 1954, trav. VII) total (as '*Rotalia*')
5. Panama City, Florida (Bandy 1956) total (as *Streblus tepidus*)
6. Mobile, Alabama (Parker 1954, trav. VI) total (as '*Rotalia*')
7. Mississippi (Phleger 1954) total (as '*Rotalia*')
8. Mississippi Delta (Phleger 1955) total (as *Rotalia*)
9. Louisiana: (a) (Phleger 1951, trav. X) total (as *Rotalia beccarii*) vars. *parkinsonia* and *tepida*); (b) (Kane 1967) total (as *Streblus*); (c) (Bandy 1954) total (as *Streblus*)
10. Galveston, Texas, (Phleger 1951, trav. VI–VIII) total (as *Rotalia*)
11. Matagorda, Texas (a) (Phleger 1951, trav. III–V) total (as '*Rotalia*'), (b) (Phleger 1956, trav. III, V, VII) living, total (as '*Rotalia*')
12. Venezuela (Seiglie 1966) total (as *A. tepida* and *A. sarmiemtoi*)
13. Puerto Rico (Seiglie 1970, 1974) total (as *A. tepida*)

Additional common species:

Species	Distribution
Elphidium poeyanum	2, 5, 9c, 12
Quinqueloculina spp.	2, 4, 7, 9c
Elphidium gunteri	3, 7, 9b, 11b
Tretomphaloides concinnus	4, 6 (as *Rosalina*)
Rosalina floridana	6, 9c (as *Discorbis*)
Hanzawaia concentrica	7 (as *Cibicidina strattoni*)
Rosalina columbiensis	7 (as *Discorbis*)
Nonionella opima	8, 11b
Elphidium galvestonense	9a, c, 10, 11a
Elphidium discoidale	9a, c, 10, 11a, b
Planulina ornata	9c
Bigenerina irregularis	9c, 10, 11b
Ammonia rolshauseni	9c (as *Streblus*)
Brizalina subspinescens	10 (as *Bolivina*)
Buliminella morgani	10, 11a (as *Buliminella* cf. *bassendorfensis*)
Lagenammina difflugiformis	10 (as *Proteonina*)
Fursenkoina pontoni	10, 11a, b (as *Virgulina*), 13
Eratidus cf. *foliaceum*	10 (as *Ammobaculites*)
Brizalina lowmani	11a
Saccammina atlantica	11a (as *Proteonina comprima*), 11b (as *Proteonina*)
Ammoscalaria pseudospiralis	11b
Ammobaculites exiguus	11b
Nouria polymorphinoides	11b
Trochulina rosea	13 (as *Rotorbinella*)
Nonionoides grateloupi	13 (as *Florilus*)

Table 7.10 *(continued)*

***Asterigerina carinata* association**

Salinity: 35–37‰

Temperature: 18 to 31 °C

Substrate: sand

Depth: (1) 14–27, (2) 13–30, (3) 12–24, (4) 20–62, (5) 49, (6) 27, (7) 23 m

Distribution:
1. Fort Myers, Florida (Bandy 1956) total
2. Tampa, Florida (Bandy 1956) total
3. Tarpon Springs, Florida (Bandy 1956) total
4. Florida (Parker 1954, trav. VIII) total
5. Cape San Blas, Florida (Parker 1954, trav. VII) total
6. Chocktawatchee, Florida (Parker 1954, trav. VI) total
7. Lousiana (Bandy 1954) total
8. Puerto Rico (Seiglie 1971a) total

Additional common species:

Archaias angulatus	1, 2, 3
Tretomphaloides concinnus	1 (as *Discorbis*), 4 (as *Rosalina*)
Rosalina floridana	1, 2, 3, (as *Discorbis*)
Hanzawaia concentrica	1, 2, 3
Bigenerina irregularis	1
Elphidium poeyanum	3
Peneroplidae	4, 5
Miliolidae	4, 5, 8
Planulina exorna	4
Amphistegina spp.	6, 8
Trochulina rosea	8 (as *Rotorbinella*)
Textularia spp.	8

***Bigenerina irregularis* association**

Salinity: 36‰

Temperature: 16 to 23 °C

Substrate: sand

Depth: (1) 46, (2) 55–82, (3) 55–70, (4) 22 m

Distribution:
1. Fort Myers, Florida (Bandy 1956) total
2. Louisiana (Phleger 1951, trav. IX) total
3. Galveston, Texas (Phleger 1951, trav. VIII) total
4. Matagorda, Texas (Phleger 1956, trav. VII) living
5. Arayo-Los Testigos, Venezuela (Seiglie 1966) total (with *Textularia* spp.)

Additional common species:

Planulina ornata	1
Elphidium discoidale	2, 3
Hanzawaia concentrica	2, 3 (as *Cibicides*)
Ammonia beccarii	2, 3, 4 (as *Rotalia*)
Quinqueloculina spp.	5

***Brizalina lowmani* association**

Salinity: 31–35, mainly 35‰

Temperature: 5 to 15 °C

Substrate: muddy sand, mud

Depth: (1) 68–102, (2) 97–1426, (3a) 48–978, (3b) 27, (4) 53–1116 m

Distribution:
1. Louisiana (Phleger 1951, trav. X–XII), total (as *Bolivina*)
2. Galveston, Texas (Phleger 1951, trav. VII–IX) total (as *Bolivina*)
3. Matagorda, Texas: (a) (Phleger 1951, trav. III–IV) total (as *Bolivina*); (b) (Phleger 1956, trav. II) living (as *Bolivina*)
4. Brownsville, Texas (Phleger 1951, trav. I–II) total (as *Bolivina*)
5. Texas (Hueni *et al.* 1978) living (in plankton 0–25 m)

Additional common species:

Brizalina subspinescens	1 (as *Bolivina*)
Brizalina ordinaria	2, 3a, 4 (as *Bolivina simplex*)
Brizalina albatrossi	2 (as *Bolivina*)
Globocassidulina subglobosa	2 (as *Cassidulina*)
Bulimina aculeata	2
Brizalina minima	2, 3a (as *Bolivina*)
Cassidulina carinata	2
Islandiella norcrossi australis	2 (as *Cassidulina*)
Uvigerina laevis	2, 3a, 4
Osangularia culter	2 (as *Parrella*)
Fursenkoina pontoni	3a, b (as *Virgulina*)
Uvigerina peregrina	3a
Brizalina fragilis	3a (as *Bolivina*)
Brizalina subaenariensis mexicana	3a, 4 (as *Bolivina*)
Uvigerina peregrina parvula	3a, 4
Cibicidoides aff. *floridanus*	3a (as *Cibicides*)
Ammoscalaria pseudospiralis	3b
Bigenerina irregularis	3b
Hanzawaia concentrica	3b, 4 (as *H. strattoni*)

Table 7.10 *(continued)*

***Brizalina subaenariensis mexicana* association**

Salinity: 35–36‰

Temperature: 5 to 18 °C

Substrate: mud

Depth: (1) 186–274, (2) 165–223, (3a) 105–155, (3b) 152, (4) 187–265, (5) 1828, (6) 192 m

Distribution:
1. Cape San Blas, Florida (Parker 1954, trav. VII) total
2. Chocktawatchee, Florida (Parker 1954, trav. VI) total
3. Mississippi Delta (a) (Parker 1954, trav. III) total, (b) (Pflum and Frerichs 1976, trav. I) total
4. Louisiana (Phleger 1951, trav. X, XI) total
5. Galveston, Texas (Phleger 1951, trav. VII, IX) total
6. Matagorda, Texas (Phleger 1951, trav. III) total

Additional common species:

Cassidulina neocarinata	1
Trochulina translucens	1 (as '*Rotalia*')
Brizalina barbata	3a (as *Bolivina*)
Bulimina marginata	3a, b
Epistominella vitrea	3a
Bulimina aculeata	5
Uvigerina peregrina	5, 6
Brizalina lowmani	6 (as *Bolivina*)

***Buliminella morgani* association**

Salinity: 35–36‰

Temperature: 16 to 22 °C

Substrate: muddy sand, mud

Depth: (1a) 73–77, (1b) 33, (2) 33–46 m

Distribution:
1. Mississippi Delta: (a) (Parker 1954, trav. I–III) living, total (as *Buliminella* cf. *bassendorfensis*); (b) (Lankford 1959) living (as *B. bassendorfensis*)
2. Louisiana (Phleger 1951, trav. XI, XII) total (as *Buliminella* cf. *bassendorfensis*)

Additional common species:

Goësella mississippiensis	1a
Epistominella vitrea	1a
Nonionella opima	1a, b
Brizalina barbata	1a (as *Bolivina*)
Fursenkoina pontoni	1a (as *Virgulina*)
Brizalina lowmani	1b (as *Bolivina*)
Ammonia beccarii	2 (as *Rotalia*)

***Cibicidoides floridanus* association**

Salinity: 36‰

Temperature: 16 to 22 °C

Substrate: sand

Depth: 67–122 m

Distribution:
1. Fort Myers, Florida (Bandy 1956) total (as *Cibicides pseudoungerianus*)
2. Tampa, Florida (Bandy 1956) total (as *Cibicides pseudoungerianus*)
3. Mobile, Alabama (Bandy 1956) total (as *Cibicides pseudoungerianus*)
4. Louisiana (Phleger 1951, trav. X, XI) total (as *Cibicides*)
5. Galveston, Texas (Phleger 1951, trav. VII) total (as *Cibicides*)

Additional common species:

Planulina exorna	1
Bolivina goësii	1
Brizalina daggarius	1, 3 (as *Bolivina*)
Ehrenbergina spinea	1
Uvigerina hispido-costata	1, 3
Cassidulina spp.	2
Hanzawaia concentrica	3, 5 (as *Cibicides*)
Uvigerina peregrina parvula	4
Brizalina lowmani	4 (as *Bolivina*)
Lenticulina spp.	4 (as *Robulus*)
Brizalina fragilis	4 (as *Bolivina*)

Table 7.10 *(continued)*

***Elphidium discoidale* association**

Salinity: 35–36‰

Temperature: 14 to 20 °C

Substrate: sand, muddy sand

Depth: (1) 24–48, (2) 18–68, (3a) 49, (3b) 4–39, (4) 27–84, (5) 18 m

Distribution:
1. Louisiana (Phleger 1951, trav. X) total
2. Galveston, Texas (Phleger 1951, trav. VI–IX) total
3. Matagorda, Texas: (a) (Phleger 1951, trav. V) total; (b) (Phleger 1956, trav. V) total
4. Brownsville, Texas (Phleger 1951, trav. I–II) total
5. Heald Bank (Shifflett 1961) living, dead, total

Additional common species:

Hanzawaia concentrica	1, 2, 3a, 4 (as *Cibicides*)
Planulina exorna	1, 2
Ammonia beccarii	1, 2, 3a, 4 (as *Rotalia*)
Elphidium galvestonense	1, 2, 4
Fursenkoina pontoni	2, 3a, b, (as *Virgulina*)
Reussella atlantica	2
Bigenerina irregularis	2
Nonionella atlantica	3b
Quinqueloculina lamarckiana	4
Quinqueloculina ackneriana	5
Elphidium gunteri	5

***Fursenkoina pontoni* association**

Salinity: 35‰

Temperature: 14 to 18 °C

Substrate: sandy mud

Depth: (1–4) 22–73, (5) 10–46 m

Distribution:
1. Louisiana (Phleger 1951, trav. XI) total (as *Virgulina*)
2. Galveston, Texas (Phleger 1951, trav. VI) total (as *Virgulina*)
3. Matagorda, Texas: (a) (Phleger 1951, trav. IV–V) total (as *Virgulina*); (b) (Phleger 1956, trav. V, VII) living, total (as *Virgulina*)
4. Brownsville, Texas (Phleger 1951, trav. II) total (as *Virgulina*)
5. Belize (Wantland 1975) living
6. Puerto Rico (Seiglie 1974) living, total

Additional common species:

Cancris oblonga	1, 3b
Hanzawaia concentrica	1, 2 (as *Cibicides*), 3b (as *H. strattoni*)
Elphidium discoidale	1, 2, 3a, b
Ammonia beccarii	2, 3b (as *Rotalia*), 6 (as *A. tepida*)
Brizalina lowmani	3a, b, 4 (as *Bolivina*)
Buliminella elegantissima	3a, 6
Nonionella opima	3b
Nonionella atlantica	3b
Brizalina fragilis	3b (as *Bolivina*)
Brizalina striatula spinata	3b (as *Bolivina*), 6
Epistominella vitrea	4 (as *Pseudoparrella exigua*)
Quinqueloculina spp.	5, 6
Triloculina spp.	5
Nonionoides grateloupi	6 (as *Florilus*)
Cancris sagra	6
Reophax caribensis	6

Table 7.10 *(continued)*

***Globocassidulina subglobosa* association**

Salinity: 35–36‰

Temperature: 5 to 18 °C

Substrate: mud

Depth: (1) 421–2560, (2) 732, (3) 53–118, (4) 106–860, (5) 146–530, (6) 1793, (7) 128, (8) 1005, (9) 55–64 m

Distribution:

1. Florida (Parker 1954, trav. VIII, IX, X, XI), total (as *Cassidulina*)
2. Cape San Blas, Florida (Parker 1954, trav. VII) total (as *Cassidulina*)
3. Panama City, Florida (Bock 1982) living (as *Cassidulina*)
4. Chocktawatchee, Florida (Parker 1954, trav. VI) total
5. Mobile, Alabama (Parker 1954, trav. V) total
6. Mississippi Delta (Pflum and Frerichs 1976, trav. I) total
7. Louisiana (Phleger 1951, trav. IX) total (as *Cassidulina*)
8. Galveston, Texas (Phleger 1951, trav. VII) total (as *Cassidulina*)
9. Belize (Wantland 1975) total

Additional common species:

Nuttallides rugosus	1 (as *Epistominella*)
Nuttallides decoratus	1 (as *Epistominella*)
Trochulina translucens	2 (as *'Rotalia'*)
Cibicidoides floridanus	3 (as *Cibicides*)
Cassidulina curvata	3
Tretomphaloides concinnus	4 (as *Rosalina*)
Uvigerina peregrina	4
Bulimina aculeata	4
Brizalina subaenariensis mexicana	5 (as *Bolivina*)
Laticarinina pauperata	6
Fontbotia wuellerstorfi	6 (as *Cibicides*)
Cassidulina laevigata	7
Lenticulina sp.	7 (as *Robulus*)
Elphidium poeyanum	9 (as *Cribroelphidium*)
Fursenkoina pontoni	9
Reophax spp.	9
Reussella atlantica	9
Trifarina bella	9

***Hanzawai concentrica* association**

Salinity: 34–36‰

Temperature: 17 to 31 °C

Substrate: sand, muddy sand

Depth: (1) 30–37, (2) 25, (4) 31, (5a) 20–30, (5b) 28–53, (6) 21–77, (6b) 20–60, (7) 20, (8a) 24–48, (8b) 8–39, (9) 29–51), (10a) 48–59, (10b) 48–56, (11) 49, (12) 50–105, (13) 40–90 m

Distribution:

1. Fort Myers, Florida (Bandy 1956) total (as *H. concentrica/strattoni*)
2. Tarpon Springs, Florida (Bandy 1956) total (as *H. concentrica/strattoni*)
3. Florida (Bock 1976) total (as *H. strattoni*)
4. Cape San Blas, Florida (Parker 1954) total (as *Cibicidina strattoni*)
5. Panama City, Florida: (a) (Bandy 1956) total; (b) (Bock 1982) living
6. Mobile, Alabama: (a) (Parker 1954, trav. V) total; (b) (Bandy 1956) total (as *H. strattoni*)
7. Mississippi: (a) (Parker 1954, trav. IV) total (as *Cibicidina strattoni*); (b) (Phleger 1954) total
8. Louisiana: (a) (Phleger 1951, trav. X) total (as *Cibicides*); (b) (Bandy 1954) total (as *H. strattoni*)
9. Galveston, Texas (Phleger 1951, trav. VI-IX) total (as *Cibicides*)
10. Matagorda, Texas: (a) (Phleger 1951, trav. V) total (as *Cibicides*); (b) (Phleger 1956, trav. V) total (as *H. strattoni*)
11. Brownsville, Texas (Phleger 1951, trav II) total (as *Cibicides*)
12. Golfo de Santa Fe, Venezuela (Sellier de Civrieux and Bermudez 1973) total
13. Araya–Los Testigos, Venezuela (Seiglie 1966) total

Additional common species:

Rosalina floridana	1, 2, 6b, 8b (as *Discorbis*)
Planulina ornata	1, 5a, 6b, 8b
Cibicidoides floridanus	3, 9 (as *Cibicides*)
Rosalina columbiensis	3, 7b (as *Discorbis*)
Parasorites orbitolinoides	5a (as *Sorites*)
Quinqueloculina lamarckiana	5a, 6b, 8b
Tremomphaloides concinnus	5b, 6a, 7a, (as *Rosalina*), 6b (as *Discorbis*)
Globocassidulina subglobosa	5b (as *Cassidulina*)
Planulina exorna	5b, 6a, 9
Nonionella atlantica	6a, 9, 13 (as *Florilus*)
Reussella atlantica	6b
Bigenerina irregularis	6b, 7b, 8b, 9, 10a

Table 7.10 *(continued)*

Textularia mayori	6b
Ammonia beccarii	7a, b (as *'Rotalia'*), 8b (as *Streblus*)
Elphidium discoidale	8a, b, 9, 10a
Elphidium galvestonense	8a, b, 9
Asterigerina carinata	8b
Elphidium poeyanum	8b
Saccammina atlantica	8b (as *Proteonina*)
Fursenkoina pontoni	9, 10b, 11 (as *Virgulina*)
Globocassidulina subglobosa	13 (as *Cassidulina*)
Islandiella norcrossi australis	13 (as *Cassidulina*)

***Nonionella opima* association**

Salinity: 31–35‰

Temperature: 16 to 26 °C

Substrate: silty clay

Depth: 4–49 m

Distribution:
1. Mississippi (Parker 1954, trav. IV) living, total
2. Mississippi Delta: (a) (Phleger 1955) living; (b) (Lankford 1959) living, total
3. Matagorda (Phleger 1956, trav. II-III) living

Additional common species:

Nonionella atlantica	1
Hanzawaia concentrica	1 (as *Cibicidina strattoni*)
Nouria polymorphinoides	1, 3
Brizalina striatula spinata	1 (as *Bolivina*)
Saccammina atlantica	1 (as *Proteonina*)
Ammonia beccarii	2b (as *Streblus*), 3 (as *Rotalia*)
Brizalina lowmani	2b (as *Bolivina*)
Buliminella morgani	2b, 3 (as *Buliminella* cf. *bassendorfensis*)
Epistominella vitrea	2b, 3
Fursenkoina pontoni	3 (as *Virgulina*)
Bigenerina irregularis	3

***Nonionoides grateloupi* association**

Salinity: 34–35‰

Temperature: –

Substrate: fine sand

Depth: 9–21 m

Distribution:
1. Tobago (Radford 1976a, b) living (as *Florilus*)
2. Puerto Rico (Seiglie 1970) total (as *Florilus*)
3. Puerto Rico (Seiglie 1974) living, total (as *Florilus*)

Additional common species:

Triloculina oblonga	1
Reophax comprima	1
Rosalina globularis	1 (as *Discrobis*)
Quinqueloculina spp.	1, 3 (as *miliolids*)
Cibicides antilleanus	1
Neoeponides auberii	1 (as *Rosalina*)
Reophax caribensis	2
Ammonia beccarii	3 (as *A. tepida*)
Fursenkoina pontoni	3

***Planulina exorna* association**

Salinity: 35–36‰

Temperature: 18 to 31 °C

Substrate: sand

Depth: 31–42 m

Distribution:
1. Cedar Keys, Florida (Parker 1954, trav. VIII) total
2. Chocktawatchee, Florida (Parker 1954, trav. VI) total
3. Mobile, Alabama (Parker 1954, trav. V) total

Additional common species:
Miliolidae

***Planulina ornata* association**

Salinity: 35–36‰

Temperature: 17 to 31 °C

Substrate: sand

Depth: 25–85 m

Distribution:
1. Fort Myers, Florida (Bandy 1956) total
2. Tampa, Florida (Bandy 1956) total
3. Tarpon Springs, Florida (Bandy 1956) total

Additional common species:

Rosalina floridana	1 (as *Discorbis*)
Hanzawaia concentrica	1, 2, 3
Bigenerina irregularis	1, 2, 3
Textularia spp.	2
Gaudryina aequa	2
Elphidium poeyanum	3
Quinqueloculina spp.	3

Table 7.10 *(continued)*

***Quinqueloculina* spp. association**

Salinity: 35–40‰

Temperature: 15 to 31 °C

Substrate: fine shelly quartz sand

Depth: 1–46 m

Distribution:
1. Cape Romano, Florida (Benda and Puri 1962) total
2. Fort Myers, Florida (Bandy 1956) total
3. Panama City, Florida (Bandy 1956) total
4. Mississippi (Phleger 1954) total (as Miliolidae)
5. Louisiana (Bandy 1954) total
6. Brownsville, Texas (Phleger 1951, trav. I) total
7. Campeche Bank, Mexico (Davis 1964) total
8. Belize (Wantland 1975) living, total
9. Golfo de Santa Fe, Venezuela (Sellier de Civrieux and Bermudez 1973) total
10. Puerto Rico (Seiglie 1974) living, total (as Miliolidae)

Additional common species:

Ammonia beccarii	1, 2 (as *Streblus*), 4 (as *'Rotalia'*), 8, 10 (as *A. tepida*)
Textularia gramen	1
Elphidium spp.	1, 9
Triloculina trigonula	1
Spiroloculina dentata	1
Planulina ornata	3
Bigenerina irregularis	3, 4
Hanzawaia concentrica	4 (as *Cibicidina strattoni*), 5 (as *H. strattoni*), 6 (as *Cibicides*)
Rosalina columbiensis	4 (as *Discorbis*)
Textularia sp.	6, 9
Archaias angulatus	7
Triloculina spp.	7, 8
Laevipeneroplis proteus	7 (as *Peneroplis*)
Discorbis spp.	8
Buliminella elegantissima	8
Sagrina pulchella	8
Brizalina striatula	8
Fursenkoina pontoni	10
Cassidella complanata	10 (as *Fursenkoina*)

***Tretomphaloides concinnus* association**

Salinity: 34–36‰

Temperature: 18 to 31 °C

Substrate: sand

Depth: (1) 12–30, (2) 15, (3) 86, (4a) 20, (4b) 38–42, (5) 22–119, (6) 20–33, (7) 20–42 m

Distribution:
1. Fort Myers, Florida (Bandy 1956) total (as *Discorbis*)
2. Florida (Parker 1954, trav. VIII) total (as *Rosalina*)
3. Cape San Blas, Florida (Parker 1954, trav. VIII) total (as *Rosalina*)
4. Panama City, Florida: (a) (Bandy 1956) total (as *Discorbis*); (b) (Bock 1982) living (as *Rosalina*)
5. Chocktawatchee, Florida (Parker 1954, trav. VI) total (as *Rosalina*)
6. Mobile, Alabama (Parker 1954, trav. V) total (as *Rosalina*)
7. Mississippi (Parker 1954, trav. IV) total (as *Rosalina*)

Additional common species:

Rosalina floridana	1, 4a (as *Discorbis*)
Hanzawaia concentrica	1, 4a, 4b (as *H. strattoni*) 5, 6, 7 (as *Cibicidina strattoni*)
Bigenerina irregularis	1, 4a
Asterigerina carinata	2, 4a, 5
Miliolidae	2, 3
Peneroplidae	2
Amphistegina spp.	3
Planulina ornata	4a
Quinqueloculina spp.	4a
Ammonia beccarii	4a (as *Streblus tepidus*), 7 (as *'Rotalia'*)
Planulina exorna	4b, 5
Globocassidulina subglobosa	4b (as *Cassidulina*)
Cibicides deprimus	6
Nonionella opima	7

Table 7.10 *(continued)*

***Uvigerina peregina parvula* association**

Salinity: 35–36‰

Temperature: 14 to 20 °C

Substrate: mud

Depth: (1) 84–155, (2) 106–960, (3a) 90, (3b) 77–110, (4) 86–91 m

Distribution:
1. Louisiana (Phleger 1951, trav. X, XII) total
2. Galveston, Texas (Phleger 1951, trav. VI, IX) total
3. Matagorda, Texas: (a) (Phleger 1951, trav. IV) total; (b) (Phleger 1956, trav. V) total
4. Brownsville, Texas (Phleger 1951, trav. II) total

Additional common species:

Uvigerina laevis	1, 2
Cibicidoides aff. *floridanus*	1, 2, 3b (as *Cibicides*)
Brizalina striatula spinata	1 (as *Bolivina*)
Brizalina subaenariensis mexicana	1
Brizalina ordinaria	2 (as *Bolivina simplex*)
Osangularia culter	2 (as *Parrella culter*)
Brizalina albatrossi	2 (as *Bolivina*)
Bulimina alazanensis	2
Brizalina fragilis	4 (as *Bolivina*)
Brizalina lowmani	4 (as *Bolivina*)

***Brizalina albatrossi* association**

Salinity: 35‰

Temperature: 12 to 13 °C

Substrate: mud

Depth: (1a) 265–1280, (1b) 646–1164, (2) 185–1152, (3) 833 m

Distribution:
1. Louisiana: (a) (Phleger 1951, trav. X, XII) total (as *Bolivina*); (b) (Pflum and Frerichs 1976, trav. II) total
2. Galveston, Texas (Phleger 1951, trav. VII–IX) total (as *Bolivina*)
3. Texas (Pflum and Frerichs 1976, trav. III) total (as *Bolivina*)

Additional common species:

Brizalina subaenariensis mexicana	1a (as *Bolivina*)
Uvigerina peregrina	1a, 2
Bulimina alazanensis	1a, b, 2, 3
Bulimina aculeata	1b, 2
Epistominella exigua	1b, 2 (as *Pseudoparrella*)
Uvigerina peregrina dirupta	1b, 3
Brizalina ordinaria	1b (as *Bolivina*)
Osangularia culter	3

***Brizalina ordinaria* association**

Salinity: 35‰

Temperature: 5 to 15 °C

Substrate: mud

Depth: (1) 430–615, (2) 353–481, (3) 205–366 m

Distribution: (all as *Bolivina simplex*)
1. Louisiana (Phleger 1951, trav. XII) total
2. Matagorda, Texas (Phleger 1951, trav. IV–V) total
3. Brownsville, Texas (Phleger 1951, trav. I) total

Additional common species:

Brizalina albatrossi	1 (as *Bolivina*)
Uvigerina peregrina	2
Bulimina striata mexicana	2
Brizalina lowmani	3 (as *Bolivina*)

***Bulimina aculeata* association**

Salinity: 35‰

Temperature: 5 to 8 °C

Substrate: mud

Depth: (1) 1317, (2a) 631–1481, (2b) 598–1682, (3) 1286–1374, (4) 917–1178 m

Distribution:
1. Florida (Parker 1954, trav. X) total
2. Mississippi Delta: (a) (Parker 1954, trav. II–III) total; (b) (Pflum and Frerichs 1976, trav. I) total
3. Louisiana (Pflum and Frerichs 1976, trav. II) total
4. Texas (Pflum and Frerichs 1976, trav. III) total

Additional common species:

Brizalina ordinaria	2a (as *Bolivina*)
Uvigerina peregrina	2a
Bólivinellina tessellata	2a
Trochulina translucens	2b (as *Rotorbinella*)
Sphaeroidina bulloides	2b
Cassidulina neocarinata	2b
Uvigerina peregrina dirupta	2b
Gyroidina orbicularis	2b
Osangularia culter	2b
Fontbotia wuellerstorfi	2b (as *Cibicides*)
Ammoglobigerina globulosa	2b (as *Trochammina*)
Repmanina charoides	2b (as *Glomospira*)
Brizalina albatrossi	3 (as *Bolivina*)
Bulimina alazanensis	3, 4

Table 7.10 *(continued)*

Bulimina alazanensis **association**

Salinity: 35‰

Temperature: 5 °C

Substrate: mud

Depth: (1) 1051, (2a) 914–951, (2b) 710–939, (3) 768–932 m

Distribution:
1. Florida (Parker 1954, trav. X) total
2. Louisiana: (a) (Phleger 1951, trav. X, XII) total; (b) (Pflum and Frerichs 1976, trav. II) total
3. Galveston, Texas: (a) (Phleger 1951, trav. VII, IX) total; (b) (Pflum and Frerichs 1976, trav. III) total

Additional common species:

Globocassidulina subglobosa	1, 2a (as *Cassidulina*)
Uvigerina peregrina	2a, 3
Brizalina albatrossi	2a, b, 3a (as *Bolivina*)
Uvigerina peregrina dirupta	2b, 3b
Epistominella exigua	2b, 3a (as *Pseudoparrella*), 3b
Cassidulina crassa	2b (as *Globocassidulina*)
Epistomaroides mexicana	2b (as *Anomalina*)
Bulimina aculeata	3a
Gyroidina orbicularis	3a
Osangularia culter	3a (as *Parrella*)

Epistominella exigua **association**

Salinity: 35‰

Temperature: 7 to 13 °C

Substrate: mud

Depth: (1) 914, (2) 878, (3) 479–939 m

Distribution:
1. Florida (Parker 1954, trav. IX) total
2. Cape San Blas, Florida (Parker 1954, trav. VII) total
3. Louisiana (Pflum and Frerichs 1976, trav. II) total

Additional common species:

Sphaeroidina bulloides	3
Uvigerina peregrina	3
Brizalina albatrossi	3 (as *Bolivina*)
Cassidulina crassa	3 (as *Globocassidulina*)
Bulimina alazanensis	3
Epistomaroides mexicana	3 (as *Anomalina*)

Eponides turgidus **association**

Salinity: 35‰

Temperature: 5 °C

Substrate: mud

Depth: (1) 2697, (2) 2213, (3) 2388 m

Distribution:
1. Chocktawatchee, Florida (Parker 1954, trav. VI) total
2. Mobile, Alabama (Parker 1954, trav. V) total
3. Mississippi Delta (Parker 1954, trav. III) total

Additional common species:

Nuttallides decoratus	1, 3 (as *Epistominella*)

Fontbotia wuellerstorfi **association**

Salinity: 35‰

Temperature: 5 °C

Substrate: mud

Depth: (1) 3072–3283, (2) 2999, (3a) 2468–2972, (3b) 1883–3252, (4) 2836–3367 m

Distribution:
1. Florida (Parker 1954, trav. IX, X, XI) total (as *Cibicides*)
2. Cape San Blas, Florida (Parker 1954, trav. VII) total (as *Cibicides*)
3. Mississippi Delta: (a) (Parker 1954, trav. II) total; (b) (Pflum and Frerichs 1976, trav. I) total (as *Cibicides*)
4. Louisiana (Phleger 1951, trav. XII) total (as *Planulina*)

Additional common species:

Nuttallides decoratus	1, 2 (as *Epistominella*), 3b (as *Alabamina*)
Hoeglundina elegans	3b
Globocassidulina subglobosa	3b
Laticarinina pauperata	3b
Cibicidoides bradyi	3b (as *Cibicides*)
Eponides turgidus	3b
Melonis pompilioides	3b
Eponides tumidulus	4
Pullenia quinqueloba	4
Nuttallides rugosus	4 (as *Pseudoparrella*)

Table 7.10 *(continued)*

***Nuttallides decoratus* association**

Salinity: 35‰

Temperature: 5 °C

Substrate: mud

Depth: (1) 1920–3614, (2) 3017, (3) 1573, (4) 1417, (5a) 1875–3017, (5b) 1883–2093, (6) 1528–3517, (7) 1920 m

Distribution:

1. Florida (Parker 1954, trav. VIII, IX, X) total (as *Epistominella*)
2. Cape San Blas, Florida (Parker 1954, trav. VII) total (as *Epistominella*)
3. Chocktawatchee, Florida (Parker 1954, trav. VI) total (as *Epistominella*)
4. Mobile, Alabama (Parker 1954, trav. V) total (as *Epistominella*)
5. Mississippi Delta: (a) (Parker 1954, trav. II) total (as *Epistominella*); (b) (Pflum and Frerichs 1976, trav I) total
6. Louisiana (Pflum and Frerichs 1976, trav. II) total (as *Alabamina*)
7. Galveston, Texas (Phleger 1951, trav. VIII) total (as *Pseudoparrella*)

Additional common species:

Fontbotia wuellwestorfi	1, 2, 5a, b, (as *Cibicides*)
Globocassidulina subglobosa	1, 5a, b (as *Cassidulina*)
Eponides turgidus	1
Bulimina aculeata	4, 6
Gyroidina orbicularis	5b
Ehrenbergina pupa	6
Pullenia subsphaerica	6
Osangularia culter	6
Hoeglundina elegans	6
Eponides tumidulus	7
Repmanina charoides	7 (as *Glomospira*)
Adercotryma glomerata	7 (as *Haplophragmoides*)

***Sphaeroidina bulloides* association**

Salinity: 35‰

Temperature: 6 to 13 °C

Substrate: mud

Depth: (1) 997–1109, (2) 468, (3) 556 m

Distribution:

1. Mississippi (Pflum and Frerichs 1976, trav. I) total
2. Louisiana (Pflum and Frerichs 1976, trav. II) total
3. Texas (Pflum and Frerichs 1976, trav. III) total

Additional common species:

Uvigerina peregrina dirupta	1
Bulimina aculeata	1
Bulimina alazanensis	1
Globocassidulina subglobosa	1
Epistominella exigua	1
Trochulina translucens	2 (as *Rotorbinella*)
Brizalina albatrossi	3 (as *Bolivina*)

***Trochulina translucens* association**

Salinity: 35–36‰

Temperature: 8 to 13 °C

Substrate: mud

Depth: (1) 320–585, (2) 366–585, (3) 446–555, (4) 530–549 m

Distribution:

1. Florida (Parker 1954, trav. VIII, IX) total (as '*Rotalia*')
2. Cape San Blas, Florida (Parker 1954, trav. VII) total (as '*Rotalia*')
3. Chocktawatchee, Florida (Parker 1954, trav. VI) total (as '*Rotalia*')
4. Mobile, Alabama (Parker 1954, trav. V) total (as '*Rotalia*')

Additional common species:

Brizalina minima	1 (as *Bolivina*)
Globocassidulina subglobosa	1, 4 (as *Cassidulina*)
Nuttallides rugosus	1 (as *Epistominella*)
Brizalina albatrossi	2, 3 (as *Bolivina*)
Brizalina ordinaria	2 (as *Bolivina*)

Tables 7.10 *(continued)*

***Uvigerina peregrina* association**

Salinity: 35–36‰

Temperature: 5 to 15 °C

Substrate: mud, sandy mud

Depth: (1) 914, (2) 950–1144, (3) 530–713, (4a) 298–1417, (4b) 300–375, (5a) 229–749, (5b) 181–375, (6) 398–914, (7a) 1005, (7b) 373–459, (8) 68–100, (9) 35–140 m

Distribution:
1. Cedar Keys, Florida (Parker 1954, trav. VII) total
2. Chocktawatchee, Florida (Parker 1954, trav. VI) total
3. Mobile, Alabama (Parker 1954, trav. V) total
4. Mississippi Delta: (a) (Parker 1954, trav. II) total; (b) (Pflum and Frerichs 1976, trav. I) total
5. Louisiana: (a) (Phleger 1951, trav. XI, XII) total; (b) (Pflum and Frerichs 1976, trav. II) total
6. Galveston, Texas (Phleger 1951, trav. VII, IX) total
7. Matagorda, Texas: (Phleger 1951, trav. V) total
8. Golfo de Santa Fe, Venezuela (Sellier de Civrieux and Bermudez 1973) total
9. Araya-Los Testigos, Venezuela (Seiglie 1966) total

Additional common species:	
Bulimina alazanensis	2, 4a, 5a, 6
Globocassidulina subglobosa	2, 6 (as *Cassidulina*)
Bulimina aculeata	3a, 5a, 6
Uvigerina hispidocostata	3a
Bulimina striata mexicana	4a
Cassidulina neocarinata	4b
Sphaeroidina bulloides	4b, 7b
Valvulineria complanata	4b
Brizalina ordinaria	5a, 6 (as *Bolivina simplex*)
Brizalina albatrossi	4b, 5a (as *Bolivina*)
Brizalina subaenariensis mexicana	4b, 5a (as *Bolivina*)
Osangularia culter	5a (as *Parrella culter*)
Trochulina translucens	5b (as *Rotorbinella*)
Bolivina goësii	5b
Cassidulina curvata	5b
Cibicides umbonatus	5b
Epistominella exigua	6 (as *Pseudoparrella*)
Uvigerina spp.	8
Nonionoides sp.	8
Brizalina spp.	8, 9
Nonionella opima	9
Nonionoides grateloupi	9 (as *Florilus*)

***Uvigerina peregrina dirupta* association**

Salinity: 35‰

Temperature: 7 °C

Substrate: mud

Depth: (1) 805–904, (2) 761–1024 m

Distribution:
1. Mississippi (Pflum and Frerichs 1976, trav. I) total
2. Texas (Pflum and Frerichs 1976, trav. III) total

Additional common species:	
Bulimina aculeata	1, 2
Brizalina albatrossi	1, 2 (as *Bolivina*)
Bulimina striata mexicana	1
Sphaeroidina bulloides	1, 2
Cassidulina neocarinata	1
Osangularia culter	1
Bulimina alazanensis	2
Bulimina spicata	2

	Area \ Association	*Quinqueloculina* spp.	*Ammonia beccarii*	*Hanzawaia concentrica*	*Bigenerina irregularis*	*Globocassidulina subglobosa*	*Cibicidoides floridanus*	*Elphidium discoidale*	*Fursenkoina pontoni*	*Brizalina lowmani*	*Buliminella morgani*	*Uvigerina peregrina parvula*	*Nonionella opima*	*Asterigerina carinata*	*Tretomphaloides concinnus*	*Planulina ornata*	*Planulina exorna*
Florida	Cape Romano	x															
	Fort Myers	x	x	x	x		x							x	x	x	
	Tampa		x			x	x							x		x	
	Tarpon Springs		x	x										x		x	
	Cedar Keys					x											x
	Cape San Blas		x	x		x								x	x		
	Panama City	x	x	x		x									x		x
	Chocktawatchee					x								x	x		x
Al.	Mobile		x	x		x	x								x		
Mississ.	Mississippi Sound	x	x	x									x		x		
	Mississippi Delta		x										x				
La.	Louisiana	x	x	x	x	x	x	x	x	x	x	x	x	x			
Texas	Galveston		x	x	x	x	x	x	x	x	x	x					
	Matagorda		x	x	x			x	x	x		x	x				
	Brownsville	x		x				x	x	x		x					
	<100 m inner shelf	x	x	x	x	x	x	x	x	x	x	x	x	x	x	x	x
	100–200 m outer shelf					x				x		x			x		
	>200 m slope					x				x		x					

Fig. 7.10 Distribution of principal associations in the northern Gulf of Mexico shelf and slope

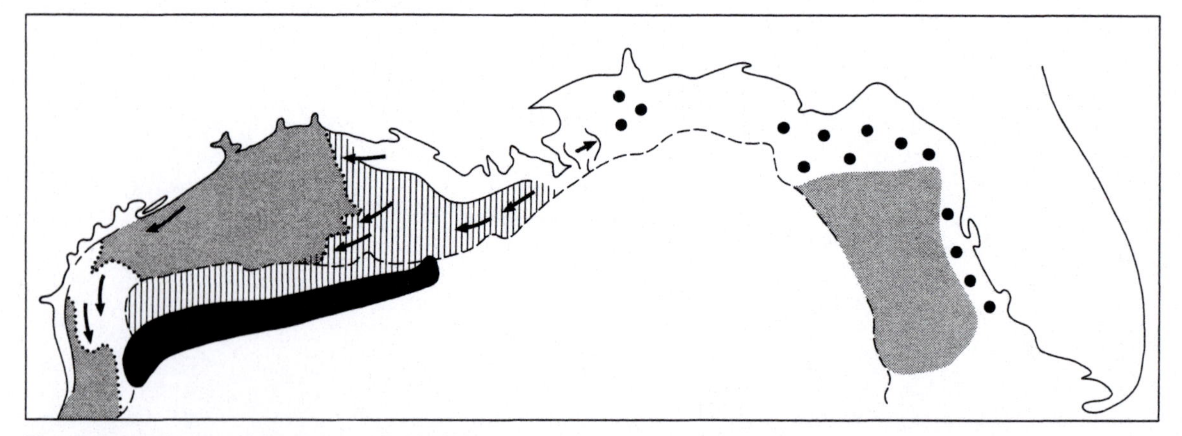

Fig. 7.11 Sediments of the northern Gulf of Mexico. Stipple: areas of slow or non-deposition, mud–sand; vertical lines: active deposition of mud and silt – arrows show dispersion paths; large black circles and black area: quartz sand: black circles and black area: muds with > 1% organic carbon (data from Gould and Stewart 1955, Van Andel 1960, Emery and Uchupi 1972)

Table 7.11 Summary of the abiotic environmental factors for shelf to deep-sea associations, Gulf of Mexico

Association	Salinity (‰)	Temperature (°C)	Depth (m)
Mainly shelf			
Ammonia beccarii	22–37	10–32	3–48
Asterigerina carinata	35–37	12–30	13–62
Bigenerina irregularis	36	16–23	22–82
Brizalina lowmani	31–35	5–15	27–1426
Brizalina subaenariensis mexicana	35–36	5–18	105–1828
Buliminella morgani	35–36	16–22	33–77
Cibicidoides floridanus	36	16–22	71–95
Elphidium discoidale	?	14–20	18–84
Fursenkoina pontoni	35	14–18	22–70
Globocassidulina subglobosa	35–36	5–18	53–2560
Hanzawaia concentrica	34–36	17–31	8–71
Nonionella opima	31–36	16–26	5–49
Planulina exorna	35–36	18–31	31–42
Planulina ornata	35–36	17–31	25–85
Quinqueloculina spp.	35–40	14–31	1–46
Tretomphaloides concinnus	34–36	18–31	12–119
Uvigerina peregrina parvula	?	14–20	84–960
Mainly slope–deep sea			
Brizalina albatrossi	35	5–15	185–1280
Brizalina ordinaria	35	12–13	205–615
Bulimina aculeata	35	5–8	631–1682
Bulimina alazanensis	35	5	710–1051
Epistominella exigua	35	5	479–939
Eponides turgidus	35	5	2213–2697
Fontbotia wuellerstorfi	35	5	2468–3367
Nuttallides decoratus	35	5	1417–3614
Sphaeroidina bulloides	35	6–13	468–1109
Trochulina translucens	35–36	8–13	320–585
Uvigerina peregrina	35–36	5–15	181–1417
Uvigerina peregrina dirupta	35	7	716–1014

values >1% (Emery and Uchupi 1972) and most of this is detritus from phytoplankton derived from the adjacent shelf (Walsh 1983).

Shelf

Throughout most of the area salinities are normal (35–36‰) and presumably in not an important environmental control. However, the *Ammonia beccarii* association is present over a salinity range of 22–36‰ and the *Nonionella opima* association, 31–36‰. The *Brizalina lowmani* association occurs mainly in waters of normal salinity but tolerates slightly brackish conditions (31–35‰). The *Quinqueloculina* spp. association is present in normal marine to slightly hypersaline conditions (Table 7.11).

The surface waters are mixed down to a depth of ~100 m and therefore experience seasonal temperature variation (roughly 12–32 °C). All the inner shelf associations are under its influence (Fig. 7.12) but each has a reasonably distinct temperature range so this may be a significant environment control here. The same applies to outer shelf and slope associations down to a depth of 1500 m, but deeper than this temperature is constant at 5 °C and salinity at 35‰ (Nowlin 1971).

The inner shelf associations of the northern Gulf of Mexico show an interesting pattern of distribution (Fig. 7.13). Six are ubiquitous and each occupies a different depth range with considerable overlap between associations. The Mississippi Delta, a site of rapid sediment accumulation in contrast to the slow accumulation elsewhere, forms a divide between two groups of associations. In the western group, the *Elphidium discoidale, Fursenkoina pontoni* and *Buliminella morgani* associations have very similar depth ranges which overlap with the upper part of that of

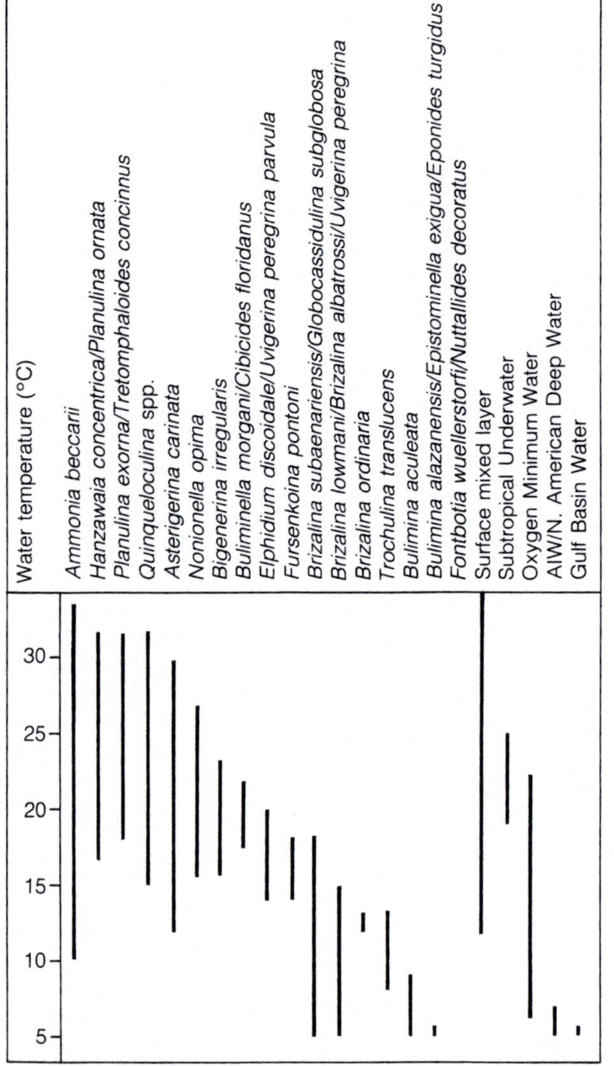

Fig. 7.12 Distribution of shelf associations with respect to water temnerature

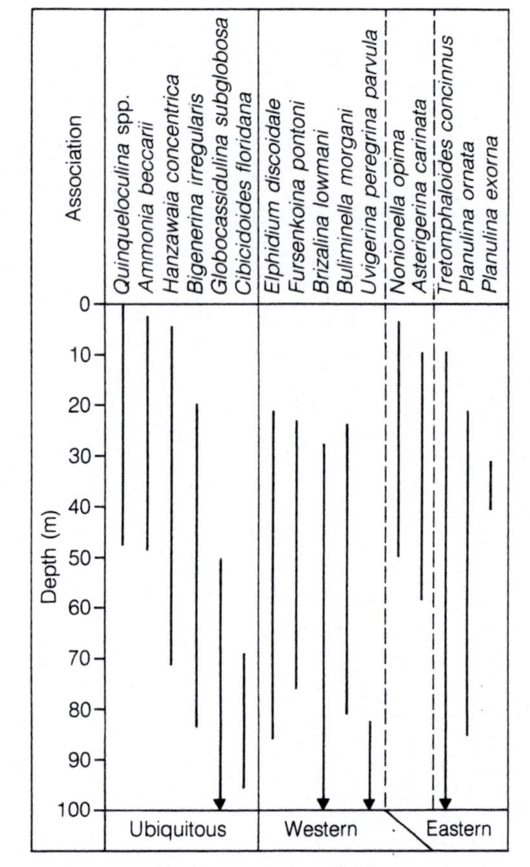

Fig. 7.13 Geographical distribution of inner shelf associations in the northern Gulf of Mexico

the *Brizalina lowmani* association. In the case of the eastern group, the *Tretomphaloides concinnus* association has a broad depth range which completely overlaps those of the other two associations. These overlaps cannot be explained by temperature differences so the control may be the substrate.

Two associations span the Mississippi Delta area. The *Nonionella opima* association is best developed around the delta (Fig. 7.13), especially in the deltaic marine environment where living assemblages have large standing crops (up to 8000 per 10 cm^2, Phleger 1955; Lankford 1959; Murray 1973). This association is also present off Matagorda, Texas. The *Asterigerina carinata* association, which is essentially characteristic of the Florida inner shelf, is also known from a single sample off Louisiana (Bandy 1954).

The *Brizalina lowmani* and *Globocassidulina subglobosa* associations extend on to the slope, and the variation in their depth distribution in different areas is shown in Fig. 7.14. There is no obvious correlation between any of the slope associations and the bottom-water masses (but note that the water-mass information is generalised). Beyond the outer shelf, the substrate is principally mud with some silt and often a large component of sand-sized planktonic foraminiferal tests.

In his study of living assemblages from the shelf off Panama City, Florida, Bock (1982) noted that although there is some seasonal change in the dominant species, the annual variation is small. The variation in standing crop size is shown in Table 7.12. The maximum is only rarely four times greater than the mini-

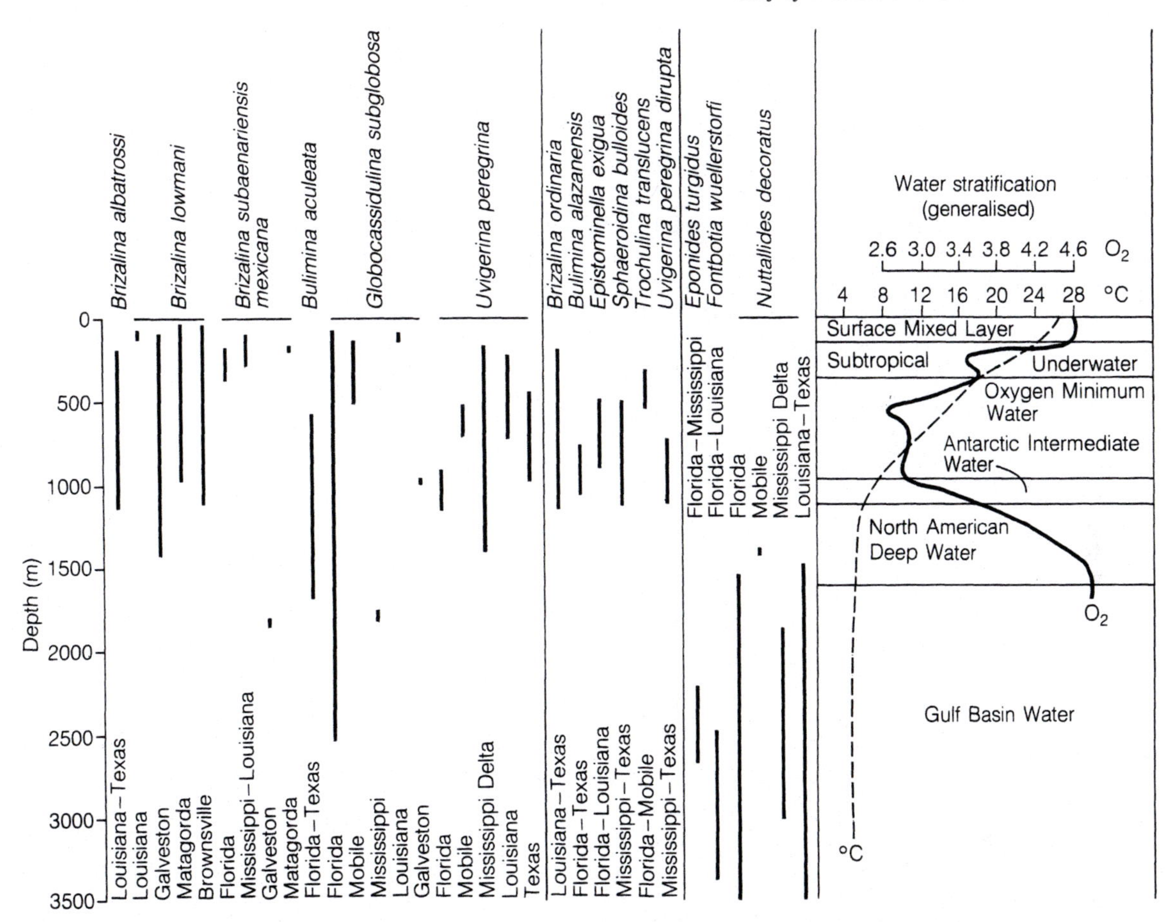

Fig. 7.14 Generalised water masses (after Nowlin 1971) and depth distribution of shelf and slope associations, northern Gulf of Mexico

Table 7.12 Variation in numbers of living individuals peer 10 cm³ sediment on the shelf off Panama City, Florida (data from Bock 1982)

Depth(m)	Minimum	Maximum	Max÷min
38	260	830	3.1
38	380	440	1.1
42	550	900	1.6
45	790	4310	5.4
53	480	1250	2.6
69	220	2960	13.4
75	460	1340	2.9
118	1250	6040	4.8
191	1980	6720	3.4

mum. Overall the higher values occur in deeper water where the sediment is finer grained. The diversity ranged from $H(S)$ 2.5–3.5 for the three seasons sampled. Bock pointed out that the total assemblages are quite different from the living ones, principally because of the high abundance of the relict fauna. *Amphistegina gibbosa* commonly makes up >50% of the total assemblage although it is present living in very small numbers. However, its tests are robust and heavy whereas those of most of the present living taxa are light and readily removed by current winnowing.

Bock accounts for the absence from this area of what he considers to be the normal inner shelf fauna of *Ammonia* and *Elphidium*, and its replacement by a mid-shelf fauna comprising *Cibicidoides floridanus, Hanzawaia concentrica, Planulina exorna* and *Tretomphaloides concinnus*, as due to the presence of the Loop Current which bring this deeper water fauna on to the inner shelf.

Reophax hispidulus was found by Parker (1954) to be common living but rare dead in the northeast gulf and she inferred that its fragile test is destroyed after death. Only seven of the samples contained >100 living individuals and of these only two show similar living and dead assemblages. Sample 208 has a *Buliminella morgani* association, but the proportion of agglutinated tests is 15% in the living and 43% in the dead assemblage (*Goësella mississippiensis* and *Textularia earlandi*). Sample 214 has a *Nonionella opima* association with a diversity of α 2 for the living and α 15 for the dead assemblage. For the remaining five samples (from the Mississippi Delta) the dead assemblages from traverses I and II show enrichment in agglutinated tests, while those of traverses III and IV show gain of calcareous tests (they are more diverse than the living). Nevertheless, despite these postmortem changes these samples give a good record of the fauna because of the rapid sedimentation (see Murray 1976a).

The Central Texas Shelf yielded 30 samples with >100 living individuals (Phleger 1956). He considered that the nearshore fauna down to a depth of 20–30 m was adapted to wave turbulence. Murray (1973) reported that the diversity at depths <21 m was α 2.5–6 while at depths of 21–110 m it was α 5–11. Phleger observed that the dead representatives were more widely distributed than the living and attributed this to downslope transport or the presence of relict tests. He recognised a series of depth boundaries based on total assemblages but Murray (1973) showed that they were not evident from the living assemblages. Buzas (1967) using canonical analysis confirmed the validity of depth zones based on total assemblages and further studies using *Q*- and *R*-mode cluster analysis showed good agreement for the depth boundaries of the living (12 and 77 m) and total (10 and 77 m) assemblages (Mello and Buzas 1968). A different approach is that of Severin (1983) who divided the 45 principal species into six morphogroups: straight–cylindrical, planoconvex, elongate–flattened, biconvex–keeled, tapered and rounded–planispiral. The living and total assemblages were analysed separately and the results are shown in Fig. 7.15. The plano-convex and rounded–planispiral groups make up the largest part of the living and total assemblages and Severin considered these morphologies to characterise generalists.

This part of the shelf is a region of relict sediments (Van Andel 1960). A detailed comparison of the living and dead assemblages showed that on the inner shelf the latter contains *Elphidium discoidale* and *E. gunteri* neither of which is common living (Murray 1976a). However, *E. gunteri* is common in the living assemblages of the nearby lagoons. This led Murray to

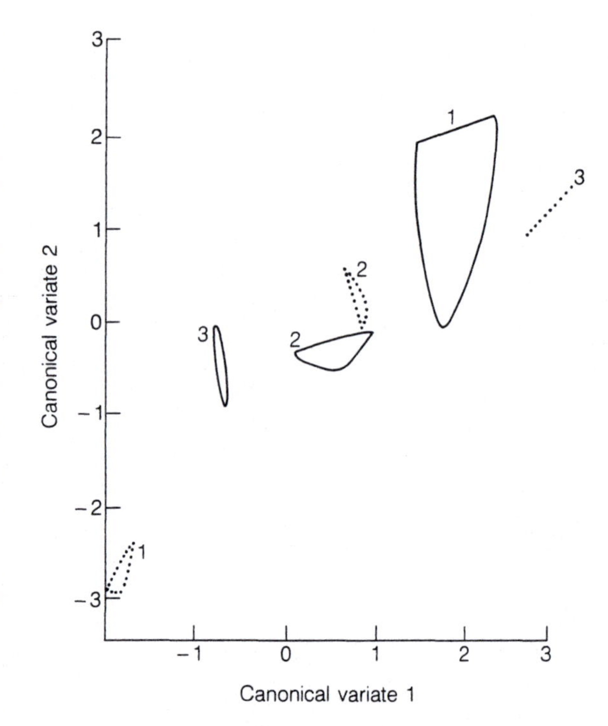

Fig. 7.15 Biofacies in the Texas area discriminated using six morphotypes (based on Severin 1983). Line = live, dots = total

suggest that the inner shelf total assemblage might be made up of three source assemblages:

1. A relict lagoonal assemblage dominated by *Ammonia beccarii* and *E. gunteri* (from a period of low sea-level).
2. A modern assemblages dominated by *A. beccarii* with locally common forms such as *Bigenerina irregularis, Ammoscalaria pseudospiralis, Fursenkoina pontoni, Brizalina lowmani* and *Nonionella opima*
3. Additional rare species not present in the living assemblages of the lagoons or inner shelf: *B. fragilis, Planulina exorna* and *Uvigerina peregrina parvula*. They are, however, found living further offshore and they may be transported landwards after death (although recent studies by Snedden *et al.* 1988 suggest that storm transport is away from the shore).

Another postmortem change might be destruction of some of the *A. pseudospiralis* tests.

Thus, from the limited data available, it appears

that the total assemblages of the shelf, away from the area of active sedimentation of the Mississippi Delta, include an element of relict fauna as well as being modified by normal postmortem processess. This fact has long been recognised for assemblages from offshore 'banks'.

According to Seiglie (1968b) the four most significant reef foraminifera in the Antillean–Caribbean region are *Amphistegina gibbosa, Archaias angulatus, Asterigerina carinata* and *Trochulina rosea* (as *Rotorbinella*). To the west of Puerto Rico *Amphistegina gibbosa* lives between 8 and 20 m water depth, but it is abundant on wave-cut terraces at 55 m and on submerged reefs at 85 m. Similar submerged reefs are known from various parts of the northern Gulf of Mexico shelf (Bandy 1956, Ludwick and Walton 1957, Poag 1972, Poag and Sweet 1972). Poag (1972) considers that all occurrences of *Amphistegina* at depths >42 m are relict from a period of lower sea-level. West Flower Garden Bank on the Texas Shelf lies at a depth of 20–38 m and supports a reef of West Indian hermatypic corals and associated organisms. The abundant red coralline algae, together with encrusting *Gypsina plana*, form nodules and crusts which constitute much of the reef (Poag and Tresslar 1979, 1981). Other taxa found living only on hard substrates include *Carpenteria utricularis, Carterina spiculotesta, Homotrema rubra, Planogypsina squamiformis, Planorbulina acervalis, P. mediterranensis*, and *Rotaliammina squamiformis*. Species living on both hard substrates and sediments are *Amphistegina gibbosa, Eponides repandus* and *Spirillina vivipara*. Poag and Tresslar noted that many of the hard substrate taxa are known only sparsely from other gulf and Caribbean reefs and they attributed this to differences of sampling technique: they used Scuba divers whereas most samples are taken with grabs.

Outer shelf to deep sea

The sediments of the outer shelf, slope and deep sea are mainly fine grained (terrigenous clay and silt) with sand-sized particles in the form of tests of planktonic foraminifera. The outer shelf and upper slope are bathed by the core of Tropical Underwater which is warm and well oxygenated, but beneath this, from ~300 to 900 m, is Oxygen Minimum Water (2.6 to 3.5 ml l^{-1} O_2) which decreases in temperature from 17 °C at the top to 7 °C at the base. These water masses have salinities of >35‰. Antarctic Intermediate Water is marked by salinities slightly below 35‰. The North American Deep Water and Gulf Basin Water have low temperatures, salinity ~35‰ and high dissolved oxygen (Nowlin 1971; Fig. 7.14)

The benthic associations do not show a clear-cut relationship with the water masses except in a general way (Table 7.13, Fig. 7.14). The clearest break is at ~1500 m. Above this, associations either span several water masses (Table 7.13, column 1) with only occasional extensions below 1500 m, e.g. *Globocassidulina subglobosa*, or they are confined to the Oxygen Minimum Water–Antarctic Intermediate Water, (Table 7.13, column 2) with its associated muddy sediments rich in carbon (Fig. 7.14). The Gulf Basin Water, which occupies depths >1500 m is characterised by three associations whose upper limits occasionally extend into the lowermost North American Deep Water (although it must be emphasised that the water-mass data are generalised for the whole gulf). These broad relationships also apply for plots of individual species present in >5% abundance (Fig. 7.16).

Because the Gulf of Mexico waters are stratified in the same way over the entire gulf, a very clear correlation between foraminiferal taxa and depth might be expected but this is not always the case (see Fig. 7.16).

Table 7.13 Associations of foraminifera and water masses in the Gulf of Mexico

Tropical Underwater to North American Deep Water	Oxygen minimum to Antarctic Intermediate Water	Gulf Basin Water
Brizalina albatrossi	*Brizalina ordinaria*	*Eponides turgidus*
Brizalina lowmani	*Bulimina alazanensis*	*Fontbotia wuellerstorfi*
Brizalina subaenariensis mexicana	*Epistominella exigua*	*Nuttallides decoratus*
Bulmina aculeata	*Sphaeroidina bulloides*	
Globocassidulina subglobosa	*Trochulina translucens*	
Uvigerina peregrina	*Uvigerina peregrina dirupta*	

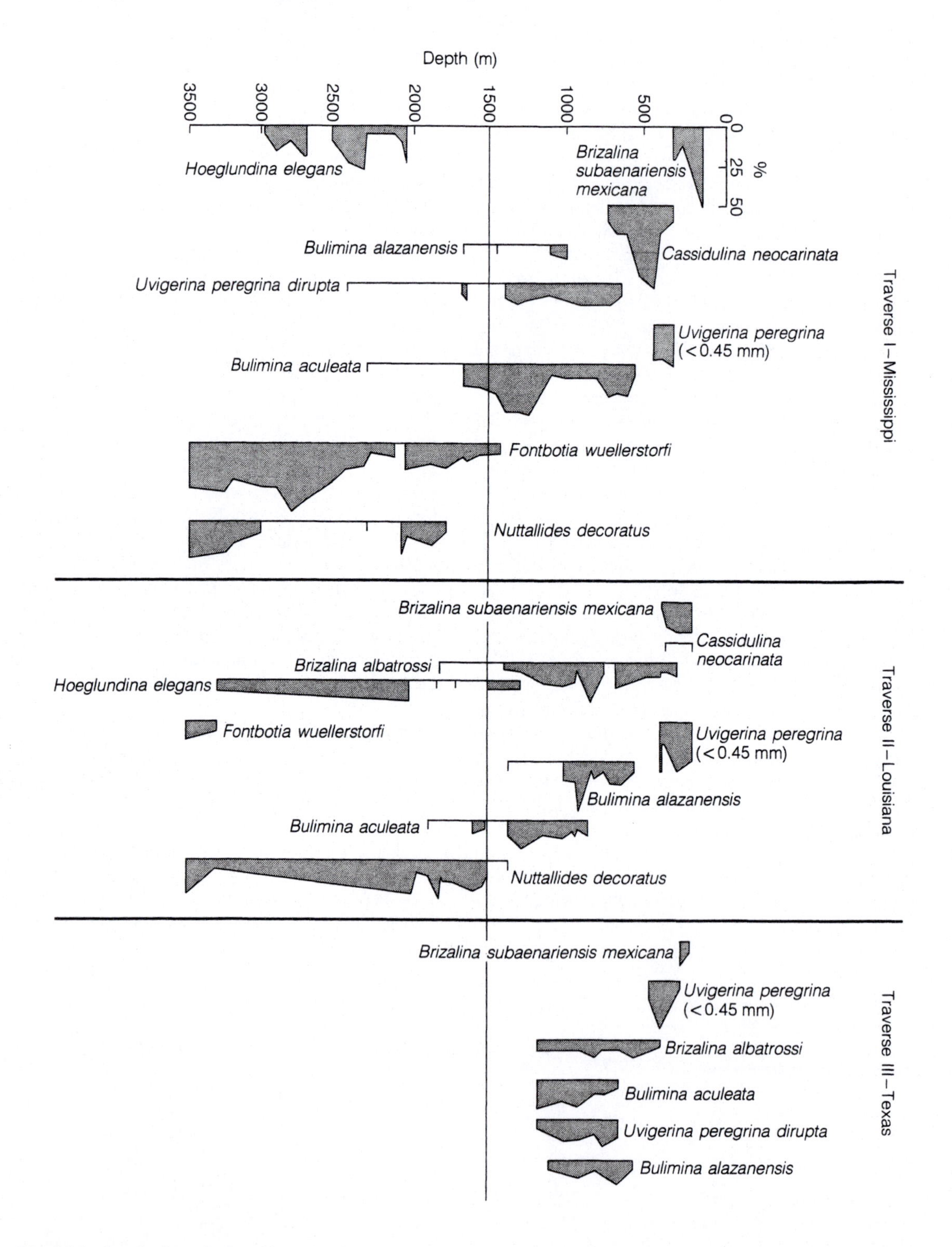

Fig. 7.16 Depth plots of selected species forming ≥ 5% of the total assemblages (data from Pflum and Frerichs 1976)

Some associations show major depth differences from one area to another (see Fig. 7.14). There are also regional differences in the distribution of associations and four (*Brizalina albatrossi, B. lowmani, B. ordinaria* and *Uvigerina peregrina dirupta*) are confined to the Louisiana–Texas area (Fig. 7.17). Notwithstanding these points, Pflum and Frerichs (1976) recognised 18 bathymetric subdivisions based on upper depth limits or abundance, and 27 isobathyal species having more or less consistent upper depth limits. Species having variable upper depth limits (heterobathyal) show changes related to the Mississippi Delta, some having a depressed and other and elevated upper limit. Species showing a marked increase in size with increasing depth include *Hoeglundina elegans, Sphaeroidina bulloides and Laticarinina pauperata.* Some species show morphological clines, e.g. *Bulimina alazanensis* from bathyal depths has notches on the distal portions of the costae and the costae are slightly irregular, whereas those from abyssal regions have heavy continuous costae that terminate in an apical spine.

An ecological–biosociological analysis of Pflum and Frerich's data enabled Hiltermann (1987) to recognise eight biocoenoses the distribution of which appears to be controlled mainly by temperature.

In the absence of any data on living assemblages it is not possible to evaluate postmortem changes in detail. Downslope transport by turbidity currents may take place but was considered to be localised and to involve only minor amounts of sediment (Phleger 1960b).

In their analysis of benthic provinces in the Gulf of Mexico, Culver and Buzas (1981a, b, d, 1982b, 1983a, b) distinguished between coastal, inner shelf, outer shelf, Mississippi mouth and slope–abyssal plain divisions, but did not recognise any faunal differences from east to west along the northern gulf. This is because although associations (marked by an abundance of given species) may be of restricted geographic distribution, individual species are commonly widely present in low abundance beyond the range of their association (see maps in Culver and Buzas 1981b).

Generic predominance Poag (1981) has published a detailed map for the entire gulf. The patterns on the continental shelf are complex as already seen from a study of associations. Because relict sediments are present on the shelf of Florida and Texas the patterns there may be misleading. Beyond the shelf the picture is more orderly (Poag 1981, 1984). On the upper continental slope *Brizalina* is generally predominant around the northern gulf but off Mexico *Uvigerina* is co-predominant (Fig. 7.18). On the lower slope *Bulimina* is predominant but off the Mississippi Delta

Association	*Trochulina translucens*	*Eponides turgidus*	*Epistominella exigua*	*Fontbotia wuellerstorfi*	*Brizalina subaenariensis mexicana*	*Bulimina aculeata*	*Bulimina alazanensis*	*Globocassidulina subglobosa*	*Nuttallides decoratus*	*Uvigerina peregrina*	*Sphaeroidina bulloides*	*Uvigerina peregrina dirupta*	*Brizalina albatrossi*	*Brizalina lowmani*	*Brizalina ordinaria*	*Uvigerina peregrina parvula*
Florida	|	|	|	|	|	|	|	|	|	|						
Alabama	|	|						|	|	|						
Mississippi		|		|	|	|			|	|	|	|				
Louisiana			|	|	|	|	|	|		|	|		|	|	|	|
Texas					|	|	|	|	|	|	|	|	|	|	|	|

Fig. 7.17 Geographical distribution of shelf and slope associations in the northern Gulf of Mexico

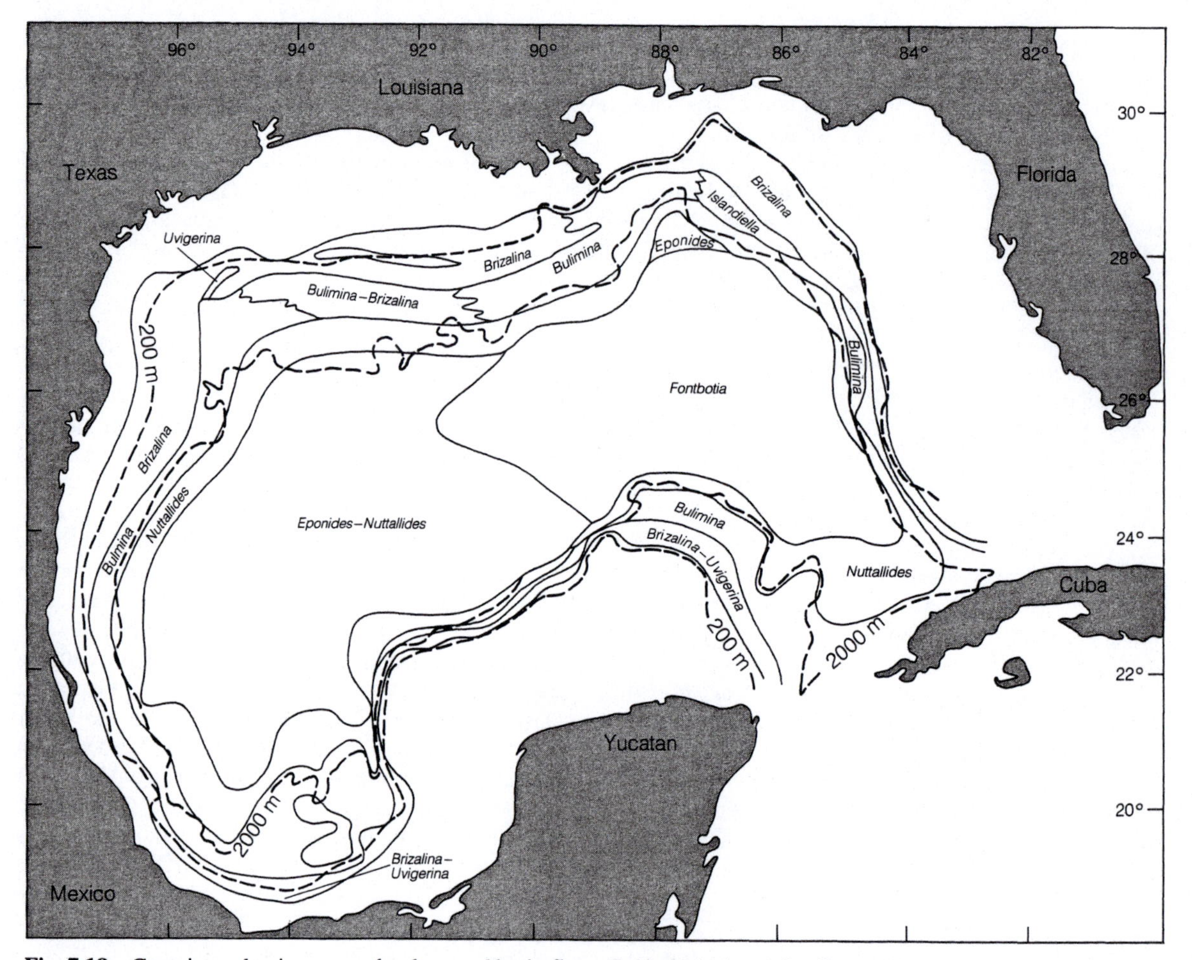

Fig. 7.18 Generic predominance on the slope and basin floor, Gulf of Mexico, (after Poag 1981, 1984)

this zone of predominance extends to the upper slope. Off Florida *Islandiella* replaces *Bulimina, Brizalina* is co-predominant off Louisiana–Texas and *Cibicidoides* replaces it off Yucatan. The upper continental rise is characterised by *Nuttallides* and the deepest parts by *Eponides–Nuttallides* and *Fontbotia* (*Cibicides* of Poag 1981, 1984).

There is no clear relationship between the distribution of generic predominance and water masses. For instance, although the *Bulimina–Brizalina, Bulimina, Islandiella* and *Cibicidoides* predominance zones occur mainly beneath the Antarctic Intermediate Water and North American Deep Water, they also overlap the upper part of the Gulf Basin Water. However, there is a general correlation of the *Brizalina, Uvigerina* and *Brizalina–Uvigerina* predominance fields with the Tropical Underwater and Oxygen Minimum Layer. In the deep gulf, the *Fontbotia* occurrence is nearly coincident with the Mississippi Fan which has a high rate of sediment accumulation and which is rich in terrigenous organic detritus. The *Eponides–Nuttallides* predominance area is one of slower sedimentation rates, higher calcium carbonate content and mainly marine organic detritus. Poag (1981, 1984) also recognised localised occurrences of *Hoeglundina* and *Repmanina* (as '*Glomospira*'), the latter agglutinated assemblages being developed through syndepositional carbonate dissolution.

Planktonic–benthic ratio Data for the northern Gulf of Mexico are summarised in Fig. 7.19. Although there is an overall increase in the percentage of planktonic tests within increasing depth, the correlation is far from perfect. The depression of the percentage contours around the Mississippi Delta may be due in part to the influence of fresh water causing a re-

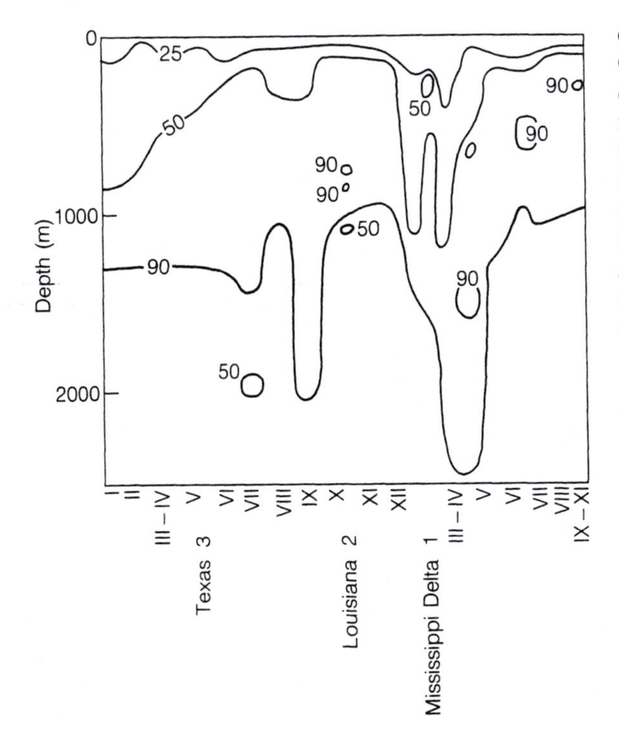

Fig. 7.19 Contours of percentage of planktonic and benthic tests for northern Gulf of Mexico. Data from Phleger and Parker (1951), Parker (1954) and Pflum and Frerichs (1976) – positions of traverse given in roman and arabic numerals (based on Pflum and Frerichs 1976)

duced abundance of living planktonic foraminifera. It is also likely that downslope movement of bottom sediment may play a role in disturbing the pattern. Local patches of <50% planktonic tests at depths >1000 m must be due to postmortem dissolution. Values >90% off Florida reflect the increased oceanic influence produced by the Loop Current (Bock 1982).

Caribbean Sea

Shelf assemblages are poorly known. The Belize Shelf has a nearshore lagoon separated from the open sea by a barrier reef complex. Most of the sediments are carbonate and *Thalassia* meadows are present in sheltered areas. In a general account of the foraminifera by Wantland (1975) 10 assemblages were described. These include typical shallow-water associations with abundant Miliolina but also the *Fursenkoina pontoni* association which was present living but not as dead assemblages. The *Globocassidulina subglobosa* association was present at 55–64 m, shallower than in the Gulf of Mexico and because of this the additional common species are different (Table 7.10). Wantland noted that there were significant differences between the living and total assemblages which are only partly accounted for by sampling just the sediments and not the *Thalassia*. In current-swept areas lag concentrations of the robust tests of *Archaias angulatus* and *Asterigerina carinata* were developed and these have also been reported from Pedro Bank to the southwest of Jamaica (Marshall 1976).

Two contrasting embayments on the Venezuela seaboard have been studied in a general way. The Golfo de Santa Fe is 11 km long, is land-locked on three sides and has its maximum depth of 70 m at the entrance. The sediment is calcareous and clastic sand at depths <55 m, and deeper than this there are silts. At depths of <35 m there is a *Quinqueloculina* spp. association with additional *Textularia* spp. in its deeper occurrences and *Elphidium* spp. in the coastal zone (bordered by mangroves). The *Hanzawaia concentrica* association is present at depths of 50–105 m on sand or muddy sand. The fine-grained substrates are dominated by various nonionids, such as *Nonionella atlantica* and *Nonionoides grateloupi* (= *Florius* spp. of Sellier de Civrieux and Bermudez 1973). At the mouth of the gulf *Uvigerina* spp. and *Brizalina* spp. associations are developed on muds.

The Golfo de Cariaco is a silled basin and is almost completely land-locked. The shallow peripheral regions are occupied by assemblages dominated by agglutinated foraminifera and miliolids, *Ammonia beccarii* (as *Streblus tepidus*), *Elphidium cariacoense*, *Buccella* spp. or *Fursenkoina pontoni* (as *Virgulina*). The sill has an agglutinated–*Hanzawaia concentrica* assemblage. The deepest part of the gulf has a *Buliminella silviae* assemblage in sediments rich in organic carbon (up to 10%) and low in dissolved oxygen (Seiglie and Bermudez 1963). A *Nonionoides grateloupi* association (as *Florilus*) with accessory *Fursenkoina pontoni* is present in waters polluted with effluent from fish industries (Seiglie 1968a).

The shelf between the gulfs of Santa Fe and Cariaco has eight widely distributed genera: *Quinqueloculina* (0–30 m) and *Textularia* (20–50 m) on sand and biogenic sand, *Elphidium* (0–15 m), *Nonionoides* (as *Florilus*) (15–90 m) and *Brizalina* (as *Bolivina* (80–180 m) on mud, and *Hanzawaia* (40–180 m), *Cibicides* (15–180 m) and *Uvigerina* (70–180 m) on mud or sand on the central to outer shelf (Sellier de Civrieux 1970, 1977). In shallow water, biogenic sand near reefs between 0 and 30 m water depth have

60–90% *Amphistegina* in total assemblages (temperature ~26 °C).

The Araya–Los Testigos Shelf has *Ammonia beccarii* and *Elphidium poeyanum* associations in the nearshore zone, the *Hanzawaia concentrica* association at depths of 40–90 m and *Bigenerina irregularis* and *Textularia calva* on skeletal sands on reefs or other topographic highs. Muddy substrates have the *Uvigerina peregrina* association at 35–140 m, and in shelf depressions either the *Nonionoides sloanii* or *Buliminella silviae* associations. The latter is present in sediments rich in organic carbon (4.35–10% organic C, Seiglie 1966).

The diverse faunas of Tobago (Radford 1976a, b) include abundant miliolids. Major faunal breaks can be recognised at 19 and 21 m in the northwest bays and at 9, 15–18 and 24–36 m in eastern bays.

Several bays around Puerto Rico have been studied by Seiglie. In Yabucoa Bay few living foraminifera were encountered (but this may be due to the use of a Shipek sampler). The total assemblages fall into five associations: *Amphistegina gibbosa, Trochulina rosea* (as *Rotorbinella*), *Ammonia beccarii, Nonionoides grateloupi* (as *Florilus*) and *Archaias angulatus*. The latter association is probably relict. Some of the foraminiferal tests are infilled with glauconite (Seiglie 1970).

On the Cabo Rojo Shelf, southwest Puerto Rico, total assemblages from the reefs and adjacent shelf represent the *Asterigerina carinata* association together with *Trochulina rosea* (as *Rotorbinella*), *Amphistegina gibbosa, Textularia* spp. and *Neoeponides auberi* (as *R. mira*) minor associations (Table 7.14). Relict *Archaias* and *Articulina* and glauconitised tests are also present (Seiglie 1971a).

Mayagüez Bay and adjacent areas are contaminated by pollution. Three species are particularly affected: *Fursenkoina pontoni, F. spinicostata* and *Nonionoides grateloupi* (as *Florilus*) (Seiglie 1971b). *Fursenkoina pontoni* is very abundant living but much less so in the total assemblages. This is normally the case elsewhere even in the absence of pollution. *Fursenkoina spinicostata* was rare at the time of sampling but has dead abundances of 5–8% at 40 cm below the sediment surface. This species may therefore have recently declined in abundance. *Nonionoides grateloupi* is associated with low oxygen water and sediments rich in organic carbon. In additional to these, the *Quinqueloculina* spp. and *Ammonia beccarii* associations are also represented (Seiglie 1974). Pyrite is present in three forms in this area: inside foraminiferal tests, replacing plant debris and as framboids (Seiglie 1973). It was claimed that living foraminifera contained pyrite. However, the cytoplasm filled only a part of the test and this suggests that these individuals had recently died (Seiglie gave no details of his preparation techniques). He interpreted the pyritisation as due to bacterial activity and to be a disease. There is no clear

Table 7.14 Minor benthic associations from the shelf seas of the Caribbean (L = living, T = total)

Association		Depth (m)	Area	Source
1. *Elphidium poeyanum*	T	0–4	Belize	Cebulski 1969, Wantland 1975
	T	10–15	Venezuela	Seiglie 1966
2. *Heterillina cribrostoma*	L, T	9	Belize	Wantland 1975
3. *Brizalina* spp.	T	95–107	Golfo de Santa Fe	Sellier de Civrieux and Bermudez 1973
4. *Buliminella silviae*	T	?	Golfo de Cariaco	Seiglie and Bermudez 1963
		5–35	Araya–Los Testigos	Seiglie 1966
5. *Nonionoides sloanii*	T	5–35	Araya–Los Testigos	Seiglie 1966
6. *Textularia calva*	T	5–40	Araya–Los Testigos	Seiglie 1966
7. *Reophax comprima*	L	19–35	Tobago	Radford 1976a, b
8. *Reussella atlantica*	L	20–50	Tobago	Radford 1976a, b
9. *Amphistegina gibbosa*	T		Puerto Rico	Seiglie 1970, 1971a
10. *Trochulina rosea*	T		Puerto Rico	Seiglie 1970, 1971a
11. *Archaias angulatus*	T		Puerto Rico	Seiglie 1970, 1971a
12. *Neoeponides auberi*	T		Puerto Rico	Seiglie 1971a
13. *Textularia* sp.	T		Puerto Rico	Seiglie 1971a
14. *Articulina mexicana*	T		Puerto Rico	Seiglie 1971a

link between foraminifera and pollution except that pyritised tests were more abundant in the most polluted area of the bay (Seiglie 1974). However, pyritisation also increased from 12% at the sediment surface to 65–96% at >80 cm below the surface and this is independent of pollution.

Assemblages from the southern shelf, defined on presence–absence criteria, have been described by Brooks (1973).

In summary, there is clearly a great diversity of faunas in the Caribbean Sea related to clear or turbid water, vegetated or non-vegetated sea floors and carbonate to clastic substrates, but as yet few detailed studies have been carried out.

CHAPTER 8

Atlantic seaboard of South America

From Trinidad (~10° N) to Cape Horn (~55° S) the South American margin changes from tropical to cool temperate. It encompasses a great range of environments including the enormous delta of the Amazon, coral reefs of I. Fernando de Noronha, mangrove swamps and tidal marshes, hypersaline and brackish lagoons as well as narrow and wide continental shelves. However, from the point of view of foraminiferal studies it is at a very early stage of knowledge. Scarcely anything is known of the faunas between Trinidad and Salvador. The coasts of southern Brazil, Uruguay and Argentina have received some attention, but many of the papers are taxonomic and give only the sparsest ecological data. Major reviews include that of Boltovskoy (1976) on zoogeography and Boltovskoy *et al.* (1980) which illustrates the fauna of Argentinian waters and summarises the ecology.

All the localities to which reference is made are listed on Fig. 8.1.

The coast of Brazil from Recife to around the area of Lagoa de Patos is bathed in warm waters of the south-flowing Brazil Current which arises from the South Equatorial Current. Off Argentina there is a belt of coastal water derived from the subantarctic together with land run-off and beyond this is the Malvin (Falkland) Current flowing north and also derived from subantarctic waters. This cool water meets the warm Brazil Current on the shelf off Patos (data from Sverdrup *et al.* 1942).

Mangrove swamps

Total assemblages from mangrove swamps at Acupe, Brazil, have been described by Zaninetti (1979), Zaninetti *et al.* (1979) and Hiltermann *et al.* (1981) and the taxonomy of allogromiine and textulariine species has been reviewed by Brönnimann *et al.* (1979). Four subenvironments were recognised: internal, subinternal, subexternal and external mangrove areas. All are brackish but lower salinities (<10‰) are found in the internal zone and higher values (up to 25‰) in the external zone. The temperature ranges from 13 to 27 °C throughout the year. *Arenoparrella mexicana* is the most abundant species in all subenvironments. Other common forms include *Miliammina fusca*, *Siphotrochammina elegans*, *S. lobata* and *Crithrionina* sp. In the internal and subinternal zones agglutinated and organic tests from >70% of the assemblage and hyaline tests <30%. In the external zone the agglutinated organic forms range from 35 to 100%, hyaline taxa 0–65% and porcellaneous up to ~10% due to the presence of near-normal salinities in the adjacent bay. Diversity is generally low (α 1–3) although 63 species were encountered altogether.

The fauna of the mangrove areas of Guaratiba, Rio de Janeiro, are very similar to those described above (Zaninetti *et al.* 1976, 1977; Brönnimann *et al.* 1981b). *Arenoparrella mexicana* is the dominant species and the majority of the assemblages have >50% agglutinated and organic tests with only rare porcellaneous forms. Diversity of these total assemblages ranges from α 1–6.5 and occasionally up to α 10. The salinities here range from 27 to 35‰ so this is much less brackish than the Acupe mangrove swamp.

Lagoons

Minor associations are listed in Table 8.1 and major ones in Table 8.2.

The Lagoa de Araruama is hypersaline with salinities ranging from 55 to 70‰. Six species were recorded by Tinoco (1958): in order of decreasing abundance –

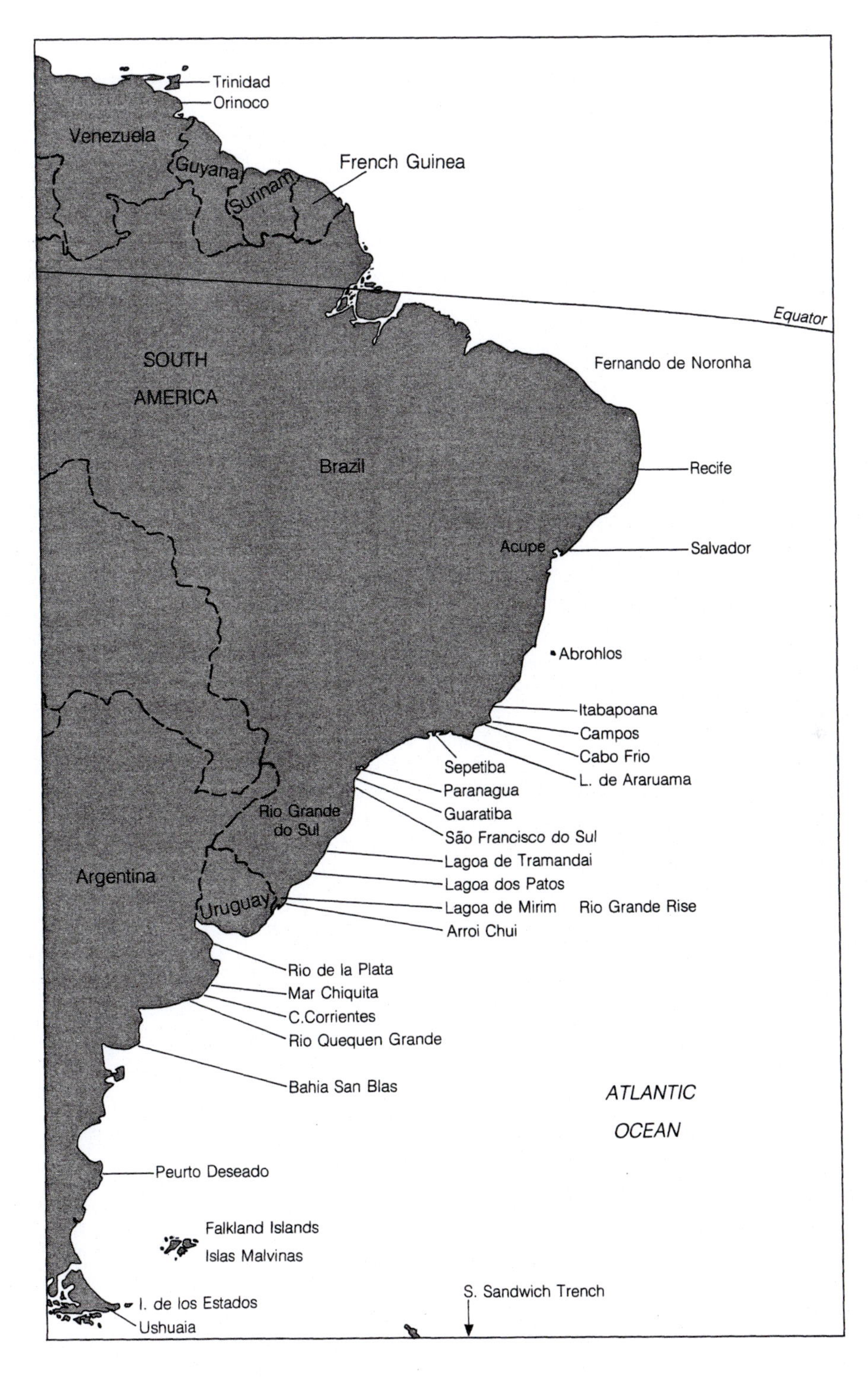

Figure 8.1 Localities mentioned in the text

Table 8.1 Minor associations in lagoons and estuaries along the South American Atlantic seaboard (T = total)

Association		Depth(m)	Area	Source
1. *Ammonia rolshauseni*	T	0–14	Baia de Sepetiba	Brönnimann *et al.* 1981a
2. *Gaudryina exilis*	T	2–9	Baia de Sepetiba	Brönnimann *et al.* 1981a
3. *Ammotium pseudocassis*	T	<6	Lagoa de Patos	Closs 1963

Elphidium morenoi, *Ammonia beccarii* (as *Streblus catesbyanus* var. *tepida*), *Triloculina oblonga*, *Quinqueloculina costata*, *Triloculina* cf. *T. insignis* and *Elphidium* cf. *E. discoidale*.

Baia de Sepetiba near Rio de Janeiro is mostly normal marine except in marginal areas where there is inflow of river water. Brönnimann and Zaninetti (1984) described 28 taxa of trochamminaceans from this lagoon and stated that they were an important element of the agglutinated fauna. Tinoco (1966a) gave a brief account of the fauna but the main ecological study is that of Brönnimann *et al.* (1981a). The *Elphidium* spp. and *Ammonia beccarii* associations were found in the brackish areas while the *Gaudryina exilis*, *Quinqueloculina* spp. and *A. rolshauseni* associations were present in the marine areas.

The brackish Baia de Paranagua is characterised by the *Elphidium* spp. association whereas the Lagoa da Tramandai, which has a salinity range of 0–29‰, has the *Miliammina fusca* association.

The largest Brazilian lagoon is the non-tidal Lagoa dos Patos which is subject to large and somewhat erratic salinity fluctuations according to the wind direction and the amount of freshwater runoff. A general account of the fauna was given by Closs (1963). The *Miliammina fusca* association is dominant in the major part of the area, with the *Elphidium* spp. association near the connection with the sea. Further sampling showed that *Haynesina germanica* (as *Nonion tisburyensis* of Closs and Madeiros 1965) is common, living in the most brackish areas. Seasonal studies of stations near the connection with the sea showed that there are changes in the living assemblages. Forti and Roettger (1967) recorded *Buliminella elegantissima* to be abundant in the first year but rare in the second. The standing crop was highest in the winter, and this was considered to be the period of growth of calcareous species and of reproduction of agglutinated ones. Closs and Madeira (1968) repeatedly sampled two stations. That nearest the sea had the *Elphidum* spp. association during the winter (salinity 2–27‰) and late autumn (salinity 15–31‰), only rare living forms during the spring (salinity <2‰) and *Trilocularena patensis* was dominant in the summer and early autumn (salinity 2–31‰). A similar picture was seen at the station further into the lagoon, except that in the first winter the standing crop was very low (salinity 3–27‰) while in the second it was higher and the *Elphidium* spp. association was present (salinity 10–20‰).

Lagoa Mirim, which is almost completely fresh water (salinity 0–2‰), was found to contain only rare dead foraminifera (Closs and Medeiros 1965). The brackish parts of Arroio Chui had living *Elphidium discoidale*, *Buccella frigida*, *Ammonia beccarii* (as *Rotalia*) and *Quinqueloculina seminula* (Closs and Madeira 1962).

The large estuary of the Rio de la Plata, Argentina, was divided into three zones on the occurrence of living individuals: 'fluvial' (salinity <10‰) with *Haynesina germanica* (as *Nonion tisburyensis*), 'fluvial marine' (salinity 10–30‰) with *Ammonia beccarii* (as *Rotalia*) and marine (salinity >30‰) with *Buliminella elegantissima* (Boltovskoy and Lena 1974). Standing crops were low but this may be due to using a small-diameter corer, snappers and dredges, for collecting the samples. Living foraminifera were present in freshwater lakes close to this estuary and also that of the Rio Quequen Grande: *Haynesina germanica*, *Miliammina fusca*, *Psammosphaera* sp. and *Trochammina* sp. (Boltovskoy 1958; Boltovskoy and Lena 1971).

Low-diversity assemblages in Mar Chiquita (salinity 1–34‰) fall into the *Ammonia beccarii* and *Elphidium* spp. associations. Living individuals were found as deep as 12 cm below the sediment surface (Lena and L'Hoste 1974). In the Rio Quequen Grande salinities are very variable. Standing crop values were highest at salinities of 2–5‰ and this was attributed to pollution (Boltovskoy and Boltovskoy 1968). The *Elphidium* spp., *Ammonia beccarii* and *Quinqueloculina* spp. associations are represented here. Because of the strong tides, there is commonly a diurnal salinity variation of 23‰. Although a few miliolids (*Q. milletti* and *Q.* aff. *seminula*) are able to tolerate these extreme conditions, most are present only where the salinity

Table 8.2 Principal associations in lagoons and estuaries along the South American Atlantic seaboard

***Ammonia beccarii* association**

Salinity: 1–34‰

Temperature: 20 to 24 °C

Substrate: mud, sandy mud

Depth: 0–16 m

Distribution:
1. Baia de Sepetiba, Brazil (Brönnimann *et. al.* 1981a) total (= biofacies 2)
2. Lagoa dos Patos, Brazil (Closs 1963) total (as *Rotalia*)
3. Rio de la Plata, Argentina (Boltovskoy and Lena 1974) living
4. Mar Chiquita, Argentina (Lena and L'Hoste 1974) living
5. Rio Quequen Grande, Argentina (Boltovskoy and Boltovskoy 1968) living

Additional common species:

Elphidium spp.	1 (as *Cribroelphidium*), 3, 4, 5
Gaudryina exilis	1

***Elphidium* spp. association**

Salinity: 1–34‰

Temperature: 14 to 27 °C

Substrate: mud, muddy sand

Depth: 0–8 m

Distribution:
1. Baia de Sepetiba, Brazil (Brönnimann *et al.* 1981a) total (= biofacies 1)
2. Baia de Paranagua, Brazil (Closs and Madeira 1966) living, dead
3. Lagoa de Tramandai, Brazil (Closs and Madeira 1967) total
4. Lagoa dos Patos, Brazil (Closs 1963) total
5. Mar Chiquita, Argentina (Lena and L'Hoste 1974) living
6. Rio Quequen Grande, Argentina (Boltovskoy and Boltovskoy 1968) living

Additional common species:

Ammonia beccarii	1, 2, 3, 4, (as *Rotalia*), 5 (as *R. parkinsoniana*)
Gaudryina exilis	1

***Miliammina fusca* association**

Salinity: 0–29‰

Temperature: 14 to 27 °C

Substrate: fine sand, silt, mud

Depth: 0–3 m

Distribution:
1. Lagoa de Tramandai, Brazil (Closs and Madeira 1967) total
2. Lagoa dos Patos, Brazil (Closs 1963) total
3. Rio Quequen Grande, Argentina (Boltovskoy and Boltovskoy 1968) living

Additional common species

Trilocularena patensis	1, 2
Ammotium salsum	1

***Quinqueloculina* spp. association**

Salinity: 31‰

Temperature: 20 to 24 °C

Substrate: sand

Depth: 0–2 m

Distribution:
1. Baia de Sepetiba, Brazil (Brönnimann *et al.* 1981a) total (= biofacies 4)
2. Rio Quequen Grande, Argentina (Boltovskoy and Boltovskoy 1968) living (as small Miliolidae)

Additional common species:

Ammonia beccarii	1, 2
Elphidium spp.	2

does not fall below 32‰ (Wright 1968). It is possible that the occurrences in the lower salinity upper part of the estuary are due to transport of tests in the powerful tidal currents. Wright considered that pollution led to more robust and more costate tests in *Q. milletti* and *Q.* aff. *seminula*.

From this review it can be seen that few authors have published numerical data. Only four principal associations appear to be widespread: *Ammonia beccarii*, *Elphidium* spp. and *Miliammina fusca* in brackish areas and *Quinqueloculina* spp. in slightly brackish to near marine. All are present from the Baia de Sepetiba (lat. 24° S) to Rio Quequen Grande (lat. 48° S) although the northern area is under the influence of the warm Brazil Current and the cool southerly derived coastal current.

Shelf

Because of the paucity of numerical data on the composition of shelf assemblages it is possible to give only a general account.

In the area between Trinidad and the Orinoco Delta, Drooger and Kaasschieter (1958) recognised three main total assemblages: *Ammonia sarmientoi* (as *Rotalia*) is dominant at depths down to 36 m (but some may be reworked Quaternary); *Nonionella atlantica*/*Uvigerina peregrina* with *Brizalina lowmani* (as *Bolivina*) and *B. striatula spinata* (as *Bolivina*) from 35 to 60 m on mud on the Orinoco Shelf; and *U. peregrina* with *U. proboscidea*, *Hoeglundina elegans* and *Eponides regularis* off Trinidad and in the Gulf of Paria on mud down to 150 m. Some of the *U. peregrina* have a rusty colour and may be relict, and relict sediment carpets the outer Orinoco Shelf.

Hofker (1983) has dealt with the taxonomy and distribution of the fauna off Surinam. Due to upwelling the water temperatures are low and the tropical fauna is absent. Relict, corroded *Amphistegina gibbosa* are present at 80–100 m.

The West Indian tropical reef fauna extends south to the Atol das Rocas (Tinoco 1966b) and Fernando de Noronha (Jindrich 1983). Bioclastic carbonate sands from the former contain common *Archaias angulatus* and *Amphistegina radiata* together with frequent *Laevipeneroplis proteus* (as *Peneroplis*), *Amphisorus hemprichii*, *Borelis pulchra*, *Heterostegina suborbicularis* and *Gypsina vesicularis* from 30 to 40 m water depth (salinity 36‰, temperature 26 °C). The sediments also contain fragments of the calcareous algae *Halimeda*, *Jania* and *Amphiroa*. On Fernando de Noronha *Homotrema rubra* is a framework organism in foraminifer–algal reefs. It forms up to 65 vol. % of the structure and is especially abundant on the undersides of low intertidal accretionary lips at the landward side of the algal ridge. These structures are undergoing syndepositional lithification with the pervasive deposition of a Mg-calcite micrite cement.

Forty-three species were recorded by Closs and Barberena (1960) from beach sands of Salvador, Brazil. The dominant species were *Archaias angulatus* and *Pyrgo subsphaerica* with miliolids and peneroplids. Only 13 species were found in the beach sands of Itabapoana; *Amphistegina radiata* and *Quinqueloculina lamarckiana* were dominant (Rodrigues 1968).

In a mainly taxonomic study, Brady *et al.* (1888) listed the occurrence of 124 taxa from 57 to 1720 m around the Abrohlos Islands. Between Campos and the mouth of the Rio de la Plata the main shelf species are *Bulimina marginata*, *Uvigerina peregrina parvula*, *Cibicides boueanum* and *Cassidulina curvata* (Boltovskoy 1961).

According to Brönnimann and Beurlen (1977a) rich assemblages of living foraminifera are present at depths of <50 m off Campos, Brazil. The agglutinated component includes trochamminids, spiroplectammininids, *Cribrostomoides* and *Haplophragmoides* (Brönnimann and Beurlen 1977b; Brönnimann 1978, 1979). *Paratrochammina simplissima* is the most common trochamminid.

A single sample from the harbour at São Francisco do Sul was dominated by *Flintina bradyana* together with *Textularia candeina*, *Ammonia rolshauseni* (as *Rolshausenia*), *Elphidium discoidale* and *Nonionoides grateloupi* (as *Nonion*, Madeira 1969).

On the shelf between Cabo Frio and the border with Uruguay the following were abundant in one or more samples: *Nonionoides grateloupi* (as *Nonion*), *Nonionella atlantica*, *Buliminella elegantissima*, *Bulimina marginata*, *Cassidella complanata* (as *Virgulina*), *Chrysalidinella dimorpha*, *Uvigerina peregrina parvula*, *Textularia lateralis*, *Quinqueloculina vulgaris*, *Rosalina floridana* (as *Discorbis*), *Poroeponides lateralis*, *Ammonia beccarii* (as *Rotalia*), *Elphidium discoidale*, *Globocassidulina subglobosa* (as *Cassidulina*), *Islandiella norcrossi australis* (as *Cassidulina*), *Planulina foveolata* and *Gypsina vesicularis* (Boltovskoy 1959). No faunal changes occurred within this 10° span of latitude. Essentially the same area was studied by Pereira (1969) and, in a sample from the slope at 2200 m, *U. peregrina* was dominant.

Beach sands from Rio Grande do Sul are dominated by *Elphidium discoidale*. Common forms include *Quinqueloculina seminula*, *Nonionella atlantica* and *Buccella frigida* (Closs and Barberena 1962). The shelf was studied by Röettger (1970) who obtained essentially the same results as Boltovskoy (1959).

Studies of littoral foraminifera from Argentina include Boltovskoy (1971) – whole coast, Boltovskoy (1963, 1964, 1966), Boltovskoy and Lena (1966, 1969b) – Puerto Deseado and Boltovskoy (1970 – a review). The Puerto Deseado studies are of particular interest because in 1963 44 species were found (28 living), another 32 (8 living) in 1966 and by 1970 the total had risen to 130 species (based on 'several thousand samples', Boltovskoy and Lena 1970). In addition, it was discovered that living individuals were present to a depth of 16 cm below the sediment surface. The annual production of *Elphidium macellum* and the entire living assemblage at this station is discussed in Chapter 4. At Ushuaia, time-series studies showed that there were blooms in the spring and summer and that the six most abundant species (trochamminids) reproduce throughout the year. The average standing crop was 53 per 10 cm^2 and the annual production 640 per 10 cm^2 (Boltovskoy *et al.* 1983a). The other abundant species in this area are *Buccella frigida*, *Cassidulina crassa*, *Cibicides acknerianus*, *E. articulatum* and *Nonion depressulus* (Lena 1966).

On the shelf off Rio de la Plata, in the area of meeting of the Brazil and Malvin currents, the coastal waters are characterised by *Epistominella* sp. (as *E. exigua* of Lena 1976), the Malvin Current by large *Cibicides acknerianus* and *Trifarina angulosa* (as *Angulogerina*) and the Brazil Current by *Bulimina patagonica* (but Boltovskoy 1976 considers this to be an indicator of the Malvin Current).

In Bahia San Blas the dominant species between 5 and 25 m is *Buccella frigida* (= *Eponides* of Boltovskoy 1954) with subsidiary *Quinqueloculina seminula*. Because the tests were small, Boltovskoy suggested that there was a shortage of food but it is possible that they were size sorted by currents.

The only quantitative results for the mainland Argentinian shelf are those of Boltovskoy and Totah (1985) for five stations along a transect off Cape Corrientes (see Table 8.3). The coastal water is subantarctic water mixed with surface runoff and has an annual temperature of 9–18 °C. The Malvin Current Water is also subantarctic water flowing north at up to 1.7 ms^{-1} (1 knot). The annual surface temperature range is ~5–11 °C. The coastal water is characterised by high diversity whereas that of the Malvin Current

Table 8.3 Summary of data for a transect off Cape Corrientes, Argentina (data from Boltovskoy and Totah 1985). Dominant species are in italic figures (L = living, D = dead). X = <1%, XX = present (standing crop <100)

Water mass	Coastal water				Malvin Current					
Depth(m)	40		69		78		87		119	
Station	1		2		3		4		5	
Living/dead	L	D	L	D	L	D	L	D	L	D
H(*S*)	4.24	3.96	3.96	3.48	2.15	2.47	2.35	1.57	2.51	1.86
α	13	10	11	11	—	—	4	4	2.5	3
Standing crop per unit area	260		565		19		794		3714	
Gavelinopsis praegeri	*19*	6	12	4	—	—	1	2	1	1
Buccella frigida	9	*30*	x	4			—	x	—	—
Epistominella sp.	2	x	*29*	*42*	—	xx	*48*	*69*	*32*	*58*
Ammonia beccarii	13	4	—	x			—	—	—	—
Bulimina patagonica	8	9	7	12	xx	—	—	x	—	—
Cibicides dispars	2	x	3	1	xx	xx	25	18	10	5
Globocassidulina subglobosa	—	—	—	x			3	x	14	7
Quinqueloculina spp.	11	16	—	x			—	—	—	—
Cuneata arctica	—	—	10	8	xx	xx	x	x	—	—
Textularia gramen	6	10	—	—			—	—	—	—
Trifarina angulosa	—	—	x	x			4	3	28	23
Trochammina advena	—	—	x	x	xx	xx	11	6	3	x

water is low. The opposite is true of standing crop. The boundary between the two water masses is around stas. 3 and 4 at 78–89 m and this may account for the poor fauna of sta. 3. The living and dead assemblages at sta. 1 are dominated by *Gavelinopsis praegeri* (= *Discorbis williamsoni* of Boltovskoy and Totah 1985) and *Buccella frigida* (as *B. peruviana* s.l.) respectively, whereas *Epistominella* sp. (as *E. exigua*) dominates stas. 2, 4 and 5. Accessory species in coastal water assemblages include *B. frigida*, *Ammonia beccarii*, *Bulimina patagonica*, *Quinqueloculina* spp., *Cuneata arctica* (as *Reophax*) and *Textularia gramen*, while for those of the Malvin Current water they are *Globocassidulina subglobosa* (as *Cassidulina*), *Trifarina angulosa* (as *Angulogerina*) and *Trochammina advena*. Similar results were obtained by Giussani de Kahn and Watanabe (1980) on total assemblages. Heron-Allen and Earland (1932) described the faunas around the Falkland Islands (Islas Malvinas) as 'monotonous' and noted that 90% of it was composed of *C. crassa*.

At the southeasternmost tip of Argentina lies Isla de los Estados in the region of powerful currents and cool temperatures (5–7 °C). Thompson (1978) studied the living assemblages and divided them into four groups. The protected intertidal assemblages are from fine-grained sediments rich in biogenic calcareous debris (bryozoa, echinoid spines, bivalve shells) and are composed of *Rosalina globularis*, *Cibicides lobatulus*, *Patellina corrugata*, *Elphidium lessonii* and *E. crispum*. Exposed intertidal assemblages are dominated by *R. globularis*, *C. lobatulus* and *E. lessonii*. Protected offshore assemblage from muddy substrates in bays are dominated by *R. globularis*, *C. lobatulus*, *C. fletcheri*, *Gavelinopsis praegeri* (as *Rotorbinella*) and *Cassidulina crassa* (as *Globocassidulina*). The exposed offshore assemblage is from depths of 10–50 m and mainly from carbonate sands. It is variable in composition with the following forms normally dominant: *Rosalina globularis*, *Cibicides lobatulus*, *C. fletcheri*, *Gavelinopsis praegeri*, *Discanomalina vermiculata* and *Cribrostomoides jeffreysii*.

Using the association approach of this book, there is a widely developed *Cibicides lobatulus* association with additional *C. fletcheri* and *Rosalina globularis*, found in both protected and exposed offshore areas (α 3.5–6.5), and a *R. globularis* association with additional *C. lobatulus* present in the exposed intertidal and exposed offshore (α 4.5–5.5). Other associations are of minor importance.

This area is of interest because the dominant species are all epifaunal attached forms which probably live on bryozoa, hydroids and bivalve shells.

The zoogeography of the inner shelf foraminifera has been reviewed by Boltovskoy (1976). Some of the salient points are summarised in Fig. 8.2. Approximately 1550 taxa have been recorded although some may be synonyms. Many species are rare, others are cosmopolitan and very few are endemic. In broad terms the faunas off Brazil are of tropical character and are assigned to the West Indian Province. They are under warm waters of the southward-flowing Brazil Current. Several larger foraminifera have their southern limit of distribution approximately at Campos. Between here and the province boundary to the south Boltovskoy distinguished a South Brazilian Subprovince in which there is a mixture of predominantly West Indian taxa and those from the Argentine Province. Along the coast of Argentina there is a coastal water mass beyond which is Malvin Current water. Both are derived from subantarctic sources and both are temperate. The Brazil Current and the Argentine Coastal Water converge off Lagoa de Patos and this is taken as the province boundary. However, the precise area of mixing is likely to vary through time and the faunal changes between the provinces are diffuse.

Deep sea

Only a few data are available. Two main associations were recognised on the Rio Grande Rise off Brazil (Lohmann 1978). The *Fontbotia wuellerstorfi* association (as *Planulina*) from 2145 to 3700 m, with additional *Globocassidulina subglobosa*, *Uvigerina pergerina*, *Pyrgo* spp., *Cibicidoides kullenbergi* and *Oridorsalis umbonatus* occurs beneath NADW. The *Nuttallides umboniferus* association (as *Epistominella*), from 3815 to 4833 m, with additional *F. wuellerstorfi* (as *Planulina*), *U. peregrina*, *Pyrgo* spp., and *O. umbonatus* characterises Antarctic Bottom Water.

In the Polar Front region of Marrice Ewing Bank three associations were described by Mead and Kennett (1987, see Mead 1985 for description of the foraminifera). The *Bulimina aculeata* association with additional *Trifarina angulosa* occurs beneath Circumpolar Deep Water (CDW) (temperature 0.5 to 1.71 °C, salinity 34.70–34.73‰, 1493–2557 m). The *Uvigerina peregrina* association with additional *Pullenia bulloides* (temperature 0.26–0.4 °C, salinity 34.69–34.70‰, 2621–2771 m, 3091 m) and the *Nuttallides*

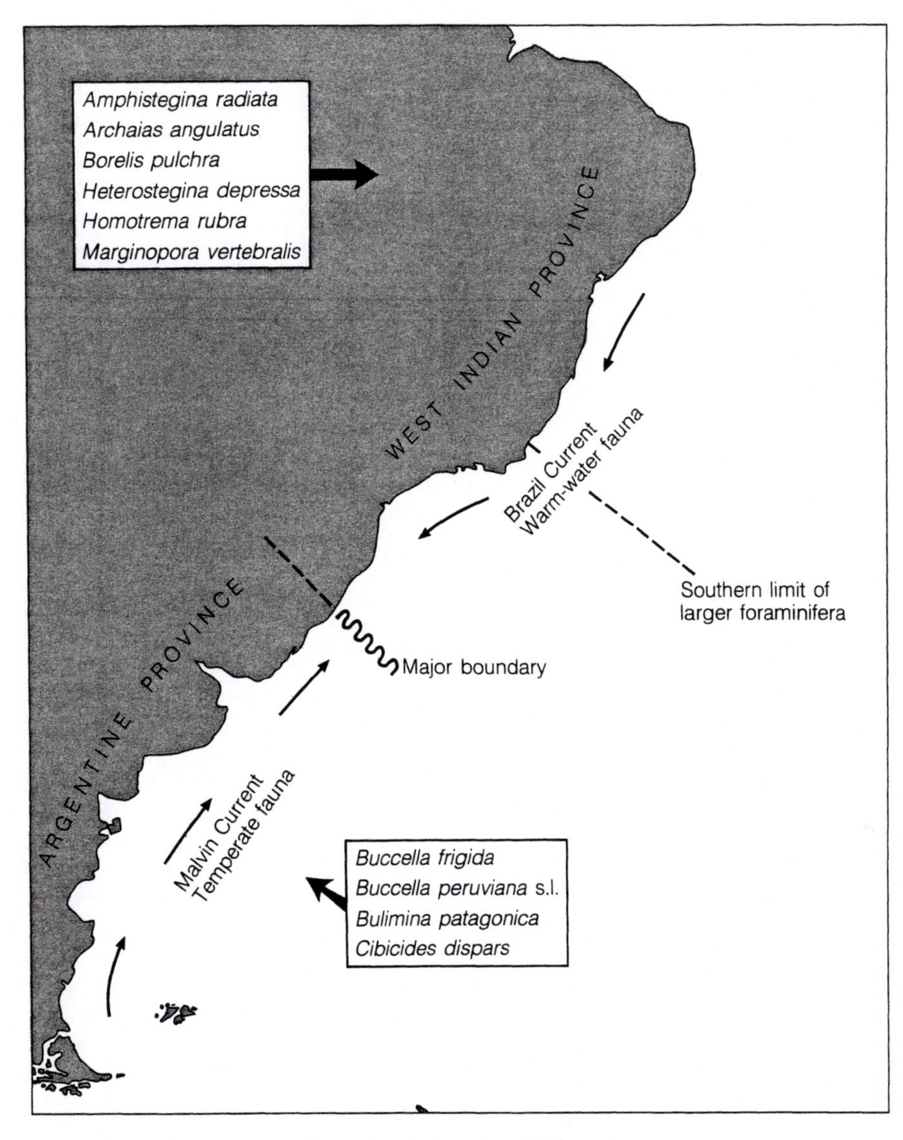

Figure 8.2 Biogeography of the inner shelf (based on Boltovskoy 1976)

umboniferus–Ehrenbergina trigona association with *Oridorsalis umbonatus* and *Pullenia bulloides* (temperature 0.14 to 0.36 °C, salinity 34.68–34.69‰, 2773–3028 m, 3091–3122 m) occur at the CDW – Antarctic Bottom Water transition.

Two samples from the South Sandwich Trench at 4900 and 7000 m and from below the Calcite Compensation Depth (CCD), had agglutinated assemblages with abundant *Tolypammina vagans* (4900 m) and *Trochammina* sp. (7000 m) (Basov 1975).

CHAPTER 9

Atlantic seaboard of Europe and Africa

Introduction

The shallow-water areas of Europe have been studied for two centuries and many taxa have been described. However, there has been much confusion because many of the early descriptions were poor and the illustrations inadequate. As elsewhere, species of *Elphidium* have presented problems. Many forms attributed to *E. excavatum* (Terquem) are now considered to be *E. williamsoni*. *Nonion depressulus* was used both for the forms that belong there and also for another group which was named *Protelphidium anglicum* (Murray 1965b) and is now considered to be *Haynesina germanica* (see Banner and Culver 1978). Species of *Cassidulina* have been discussed by Sejrup and Guibault (1980) and Mackensen and Hald (1988).

Many species are illustrated in Murray (1971a, 1979b) and Haynes (1973).

The shelf faunas off Africa are much less well known and there have been no major taxonomic reviews.

The localities mentioned in this chapter are shown in Fig. 9.1.

Marshes

Records to date are only from northern Europe. Tides vary from microtidal, e.g. Baltic (almost non-tidal) to macrotidal in the Severn Estuary (8–11 m). High marsh areas are not flushed daily; they are normally covered by water only at spring tides. Pools may evaporate during the summer, or be flooded by rainwater. Thus, the salinity ranges from very brackish to hypersaline. The temperature range is also extreme, from frozen in the winter to very warm (? 30 °C) in the summer. The low marsh is normally covered by the tide each day. The water is brackish to marine depending on the area and the temperature range is less extreme than that of the high marsh.

Most species are herbivorous and they may be epifaunal or infaunal (Table 9.1). The depth of the oxidised surface layer governs the depth to which they live. Five associations occur widely (Table 9.2). The *Jadammina macrescens* and *Trochammina inflata* associations are characteristic of high marshes. There is high dominance with one species commonly forming >80% of the living assemblage. Agglutinated tests are predominant but hyaline forms are sometimes present. Porcellaneous tests are present only in normal marine examples, e.g. *T. inflata* association of the Netherlands. Diversity is low (α <2). Low marshes have the *Ammonia beccarii*, *Elphidium williamsoni* or *Haynesina germanica* associations depending on the salinity range. The most euryhaline species is *H. germanica*. Such assemblages are predominantly calcareous (hyaline) and porcellaneous forms are present only when salinities are normal marine. Diversity is

Table 9.1 Typical marsh species (feeding strategy from Ch. 2)

Species	Mode of life	Feeding strategy
Ammonia beccarii	Infaunal, free	? Herbivore
Elphidium williamsoni	Infaunal, free	Herbivore
Haynesina germanica	Infaunal, free	? Herbivore
Jadammina macrescens	Epifaunal, free	Herbivore or detritivore
Trochammina inflata	Epifaunal–infaunal, free	Herbivore or detritivore

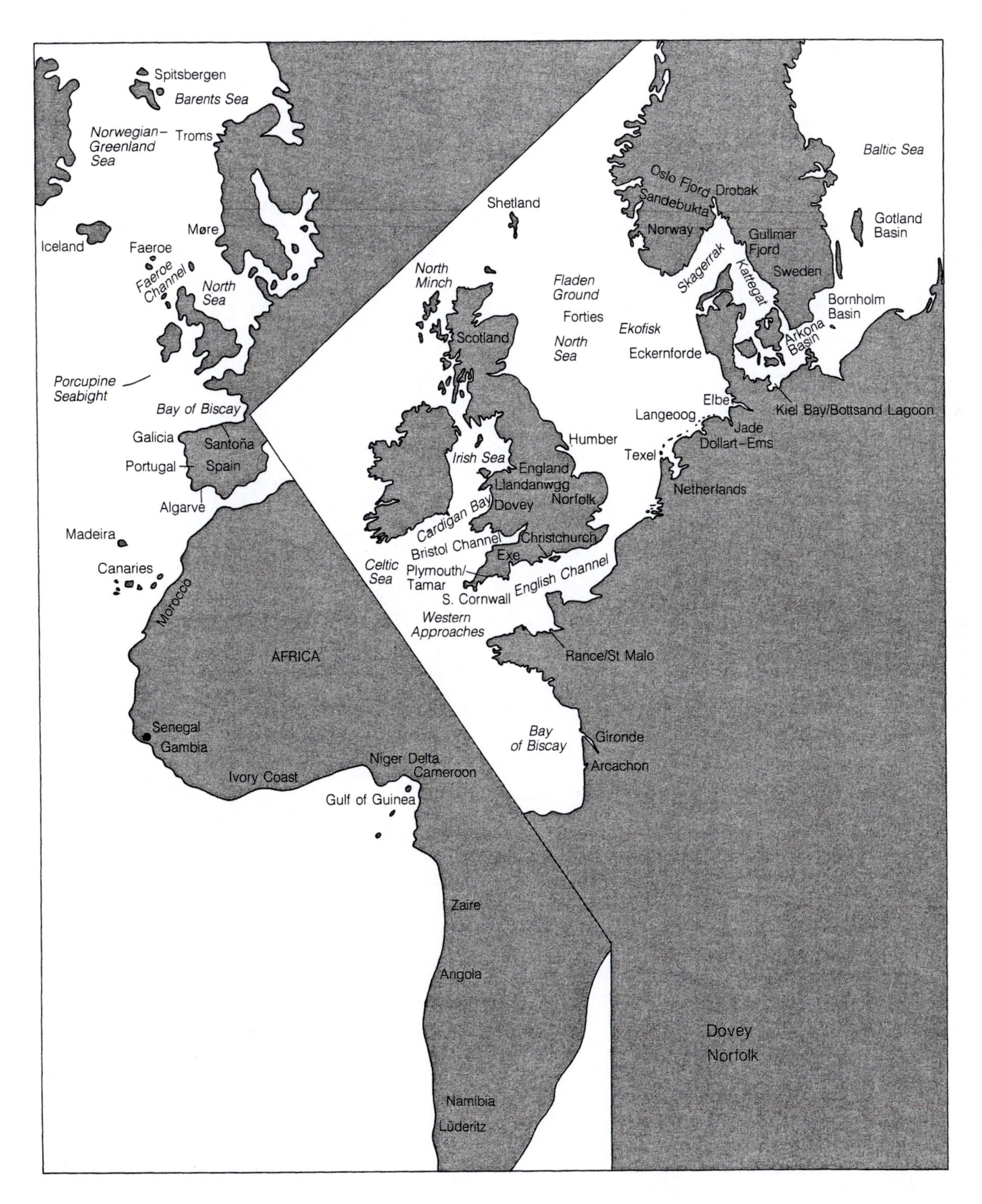

Figure 9.1 Localities

Table 9.2 Marsh associations

***Ammonia beccarii* association**

Salinity: brackish–marine

Temperature: 0 to 20+ °C

Substrate: muddy silt and sand

Depth: intertidal low marsh

Distribution:
1. Netherlands (Phleger 1970) living
2. Norfolk (Phleger 1970) living
3. Dovey Estuary (Haynes and Dobson 1969) living, dead
4. Gironde (Pujos 1976) living

Additional common species:

Haynesina germanica	1 (as *Nonion* cf. *N. tisburyensis*)
Elphidium williamsoni	1, 2 (as *E.* cf. *translucens*)
Trochammina inflata	1
Jadammina macrescens	2 (as *J. polystoma*)
miliolids	2

***Elphidium williamsoni* association**

Salinity: brackish

Temperature: 0 to 20+ °C

Substrate: muddy silt and sand

Depth: intertidal, low marsh with *Spartina*

Distribution:
1. Netherlands (Phleger 1970) living (as *Elphidium* cf. *E. translucens*)
2. Norfolk (Phleger 1970) living (as *Elphidium* cf. *E. translucens*)
3. Severn (Murray and Hawkins 1976) living
4. Dovey Estuary (Haynes and Dobson 1969) living, dead (as *E. excavatum*)

Additional common species:

Ammonia beccarii	1
Haynesina germanica	1 (as *Nonion* cf. *N. tisburyensis*), 3 (as *Protelphidium anglicum*), 4 (as *Protelphidium depressulum*)
Jadammina macrescens	1 (as *J. polystoma*), 4
miliolids	1
Trochammina inflata	1
Miliammina fusca	4

***Haynesina germanica* association**

Salinity: very brackish

Temperature: 0 to 20+ °C

Substrate: muddy silt and sand

Depth: intertidal, low marsh with *Spartina*

Distribution:
1. Netherlands (Phleger 1970) living (as *Nonion* cf. *N. tisburyensis*)
2. Humber (Brasier 1981) living
3. Norfolk (Phleger 1970) living (as *Nonion* cf. *N. tisburyensis*)
4. Exe Estuary (Murray new data) living, dead
5. Severn (Murray and Hawkins 1976) living (as *Protelphidium anglicum*)
6. Dovey Estuary (Haynes and Dobson 1969) living, dead (as *Protelphidium depressulum*)

Additional common species:

Elphidium williamsoni	1, 3 (as *Elphidium* cf. *E. translucens*), 2, 4, 5, 6
Ammonia beccarii	2, 4
Miliammina fusca	6

***Jadammina macrescens* association**

Salinity: brackish

Temperature: 0 to 30+ °C

Substrate: muddy silts rich in plant debris

Depth: intertidal, high marsh

Distribution:
1. Netherlands (Phleger 1970) living (as *J. polystoma*)
2. Norfolk (Phleger 1970) living (as *J. polystoma*)
3. Severn (Murray and Hawkins 1976) living
4. Dovey Estuary (Haynes and Dobson 1969) living, dead
5. Gironde (Pujos 1976, 1983) living (as *J. polystoma*)

Additional common species:

Ammonia beccarii	1, 2, 5
Elphidium williamsoni	1, 2 (as *E.* cf. *translucens*)
Trochammina inflata	1, 2, 3

Table 9.2 *(continued)*

***Trochammina inflata* association**

Salinity: brackish

Temperature: 0 to 30+ °C

Substrate: muddy silt rich in plant debris

Depth: intertidal, high marsh

Distribution:
1. Netherlands (Phleger 1970) living
2. Humber (Brasier 1981) living
3. Severn (Murray and Hawkins 1976) living
4. Gironde (Pujos 1976) living
5. Arcachon (Le Campion 1970) living, dead

Additional common species:	
Ammonia beccarii	1, 4
Jadammina macrescens	1 (as *J. polystoma*), 3
miliolids	1
Elphidium williamsoni	1 (as *E.* cf. *translucens*), 4 (as *E. umbilicatulum*)
Miliammina fusca	4, 5
Nonion depressulus	4
Trochammina rotaliformis	5

low (α <3). Apart from the occurrences listed in Table 9.2, Hofker (1977) has listed the common species from Texel (Netherlands) which include all the above except *A. beccarii.*

An unusual occurrence of the *Jadammina macrescens* association is in inland marshes developed where springs have brought brackish water to the surface from Permian halite deposits. Sea-birds from the Baltic and North seas are thought to have involuntarily transported living forms to establish viable populations (Haake 1982).

Localised occurrences of minor associations include a *Tiphotrocha comprimata–Miliammina fusca* association from the Bottsand Lagoon, Baltic Sea (salinity 6–14‰, temperature 0–30 °C; Lutze 1968a), *Haplophragmoides wilberti* from the Gironde (Pujos 1976), and '*Ammoscalaria pseudospiralis*'(? *Ammotium)* and *M. fusca* associations, living and dead respectively, from Arcachon (Le Campion 1970).

Postmortem changes

Because marsh associations occur in the upper part of the intertidal zone, the water has lost most of its energy by the time the marshes are covered and the plants at the seaward edge filter out much of the sediment in suspension. Hence, dead assemblages from these marshes do not contain many transported tests from adjacent environments. Postmortem dissolution of calcareous tests is likely, but there are no published observations on this for the European marshes. Most of the dead assemblages appear to be of field I type.

Mangrove marsh

Although there are estuaries and marshes along the coast of Portugal, no studies have yet been carried out. North Africa is a desert as far south as Senegal where the Senegal River has a mangrove swamp. During the dry season the waters are low, and during the wet season high. Salinities range from 0 to 58‰ and 0 to 27‰ respectively. The dry season fauna is dominated by *Ammotium salsum* and *Ammonia beccarii* (as *A. tepida*) and assemblages range from 100% hyaline to ~70% agglutinated with a single occurrence of 50% hyaline, 50% porcellaneous (normal marine). The wet season fauna is an *Ammotium salsum–Miliammina fusca* association. Most assemblages have >70% agglutinated tests although one was 100% hyaline. At both seasons diversity is low (α <1) (Ausseil-Badie 1983).

Lagoons and estuaries

In addition to lagoons and estuaries there is the Baltic Sea, an epicontinental brackish body of water, and fjords, over-deepened valleys having more or less normal marine conditions in their deeper parts but brackish conditions in the shallow areas.

Minor associations of restricted geographic occurrence are listed in Table 9.3. Details of the principal species and associations are given in Tables 9.4 and 9.5.

Table 9.3 Minor associations from European lagoons and estuaries (L = living, D = dead, T = total)

Association		Depth (m)	Area	Source
1. *Subreophax aduncus*	L	20	Baltic	Lutze 1965
2. *Elphidium clavatum*	L, D	>23	Baltic	Lutze 1965, Wefer 1976b
3. *Elphidium gerthi*	L, D	<10	Baltic	Lutze 1965, 1974a, Wefer 1976b
4. *Elphidium incertum*	L	20–25	Baltic	Lutze 1974a, Wefer 1976b
	L	9	Oslo Fjord	Alve and Nagy 1986
5. *Nodulina dentaliniformis*	D		Baltic	Lutze 1965
6. *Brizalina skagerrakensis*	T	110–330	Oslo Fjord	Risdal 1964, Thiede *et al.* 1981
7. *Cassidulina laevigata*	T	100–140	Oslo Fjord	Risdal 1964, Thiede *et al.* 1981
8. *Elphidium albiumbilicatulum*	T	10	Oslo Fjord	Thiede *et al.* 1981
9. *Nonionellina labradorica*	T	50	Oslo Fjord	Thiede *et al.* 1981
10. *Eggerelloides medius*	T	60–200	Oslo Fjord	Thiede *et al.* 1981,
	L, T			Alve and Nagy 1986
11. *Spiroplectammina biformis*	L	9–13	Oslo Fjord	Alve and Nagy 1986
12. *Adercotryma glomerata*	L	35	Oslo Fjord	Alve and Nagy 1986
13. *Ammoscalaria pseudospiralis*	L, T	1–18	Oslo Fjord	Alve and Nagy 1986
	L	0	Arcachon	Le Campion 1970
14. *Recurvoides trochamminiforme*	L, T	24–66	Oslo Fjord	Alve and Nagy 1986
15. *Bulimina marginata*	T	30–120	Oslo Fjord	Risdal 1964, Thiede *et al.* 1981
	T		Gullmar Fjord	Qvale *et al.* 1984
16. *Elphidium gunteri*	L, D		Dollart	Van Voorthuysen 1960
17. *Elphidium* sp.	L, D	0–4	Christchurch	Murray 1968a
18. *Reophax moniliformis*	D	2	Christchurch	Murray 1968a
19. *Planorbulina mediterranensis*	L, D		Arcachon	Le Campion 1970
20. *Triloculina inflata*	L		Arcachon	Le Campion 1970
21. *Trochammina inflata*	D		Arcachon	Le Campion 1970
22. *Elphidium lidoense*	L		Gironde	Pujos 1976
23. *Elphidium macellum*	L		Gironde	Pujos 1976
24. *Cibicides* cf. *ungerianus*	L		Gironde	Pujos 1976

Table 9.4 Typical lagoon species (feeding strategies from Ch. 2)

Species	Mode of life	Feeding strategy
Ammonia beccarii	Infaunal, free	? Herbivore
Ammotium cassis	Infaunal, free	Detritivore
Eggerelloides scabrus	Infaunal, free	Detritivore
Elphidium excavatum	Infaunal, free	? Herbivore
Elphidium williamsoni	Infaunal, free	Herbivore
Haynesina germanica	Infaunal, free	? Herbivore
Miliammina fusca	Infaunal, free	Detritivore
Nonion depressulus	Infaunal, free	Herbivore

Table 9.5 Principal associations in European lagoons and estuaries

***Ammonia beccarii* association**

Salinity: 0–35, optimum >10‰

Temperature: 0 to 20, optimum 15 to 20 °C

Substrate: muddy silt

Depth: <10 m

Distribution:
1. Oslo Fjord: (a) (Risdal 1964, Thiede *et al.* 1981) total (as *A. batavus*); (b) (Alve and Nagy 1986) living, total (as *A. batavus*)
2. Elbe Estuary (Wang 1983) total
3. Dollart Estuary (Van Voorthuysen 1960) living, dead (as *Streblus batavus*)
4. Christchurch Harbour (Murray 1968a) living
5. Tamar Estuary (Murray 1965c) living, dead
6. Gironde (Pujos 1976) living
7. Arcachon (Le Campion 1970) living
8. Santona (Cearreta 1988b) living, dead

Additional common species:

Elphidium albiumbilicatulum	1a (as *Nonion depressulum asterotuberculatum*)
Elphidium incertum	1a, b
Elphidium excavatum	2
Haynesina germanica	2 (as *Protelphidium*), 3 (as *Nonion depressulus*), 4 (as *Protelphidium anglicum*), 5 (as *Protelphidium* sp. nov.), 6 (as *Protelphidium paralium*), 7 (as *Nonion depressulus*) 8
Ephidium williamsoni	2, 3 (as *E. excavatum*)
Nonion depressulus	2 (as *Protelphidium*), 6
Miliammina fusca	4
Elphidium oceanense	4
Elphidium sp.	4
Cibicides lobatulus	6
Triloculina inflata	7

***Ammotium cassis* association**

Salinity: 17–29, optimum 20–22‰

Temperature: 0 to 19, optimum <3 °C

Substrate: mud, silt, sand, generally rich in organic matter

Depth: 4–30 m (at discontinuity levels)

Distribution:
1. Baltic: (a) (Lutze 1965) living, dead; (b) (Lutze 1974a) living, dead; (c) (Wefer 1976b) living
2. Swedish estuaries: (a) (Olssen 1976) living; (b) (Nyholm *et al.* 1977) living
3. Oslo Fjord: (a) (Risdal 1964, Thiede *et al.* 1981) total; (b) (Alve and Nagy 1986) living, total

Additional common species:

Elphidium clavatum	1a, b (as *Cribrononion*)
Elphidium excavatum	1a, b (as *Cribrononion*), 2a, b
Elphidium incertum	1a, b, 2a (as *Cribrononion*), 3b
Subreophax aduncus	1a (as *Reophax*)
Nodulina dentaliniformis	1a (as *Reophax*)
Eggerelloides scabrus	1a, 2b, 3b (as *Eggerella*)
Thurammina faerleensis	2a
Haynesina germanica	2a (as *Protelphidium germanicum*)
Ammonia beccarii	2a, b, 3a (as *A. batavus*)
Eggerelloides medius	2b (as *Verneuilina*)
Ammoscalaria runiana	3a
Miliammina fusca	3a, b

Table 9.5 *(continued)*

***Eggerelloides scabrus* association**

Salinity: 17–29‰, minimum 24‰ for most of year

Temperature: 1 to 20 °C

Substrate: sand, mud

Depth: 1–30 m

Distribution: (all records as *Eggerella*)
1. Baltic: (a) (Lutze 1965) dead; (b) (Lutze 1974a) dead; (c) (Haake 1967) total; (d) (Lutze *et al.* 1983) living, dead
2. Oslo Fjord: (a) (Risdal 1964, Thiede *et al.* 1981) total; (b) (Alve and Nagy 1986) living, total
3. Gullmar Fjord (Qvale *et al.* 1984) total
4. Christchurch (Murray 1968a) dead
5. Arcachon (Le Campion 1970) living, dead

Additional common species:

Elphidium clavatum	1a (as *Cribrononion*)
Ammotium cassis	1a, 2a
Reophax subfusiformis	2a
Bulimina marginata	2a
Ammoscalaria pseudospiralis	2a
Ammoscalaria runiana	2a
Reophax scorpiurus	2a
Adercotryma glomerata	2b
Spiroplectammina biformis	2b
Reophax scottii	2b
Ammonia beccarii	4
Elphidium incertum	4
Haynesina germanica	4 (as *Protelphidium anglicum*)
Quinqueloculina dimidiata	4

***Elphidium excavatum* association**

Salinity: 12–21‰

Temperature: 4 to 17 °C

Substrate: silt, sand, gravel, algae

Depth: <16 m

Distribution:
1. Baltic Sea: (a) (Lutze 1965) living, dead (as *Cribrononion*); (b) (Lutze 1974a), living, dead; (c) (Haake 1967) living, (d) (Wefer 1976b) living
2. Elbe Estuary (Wang 1983) living, total
3. Jade (Richter 1967) living, dead (as *E. selseyense*)
4. Langeoog (Haake 1962) living, total (as *E. selseyense*)
5. Dollart Estuary (Van Voorthuysen 1960) living, dead (as *E. selseyensis*)

Additional common species:

Elphidium incertum	1c (as *Cribrononion*)
Ammonia beccarii	2, 3, 4, 5 (as *Streblus batavus*)
Haynesina germanica	2 (as *Protelphidium*), 3, 4, 5 (as *Nonion depressulus*)
Nonion depressulus	2 (as *Protelphidium*)

***Elphidium williamsoni* association**

Salinity: 2–35‰

Temperature: 0 to 32 °C

Substrate: sand, muddy sand, *Zostera*

Depth: <2 m

Distribution:
1. Bottsand Lagoon, Baltic (Lutze 1968a) living, dead (as *Cribrononion articulatum*)
2. Jade (Richter 1967) living, dead (as *E. excavatum*)
3. Dollart Estuary (Van Voorthuysen 1960) living, dead (as *E. excavatum*)
4. Christchurch Harbour (Murray 1968a) living, dead (as *E. excavatum*)
5. Exe Estuary (Murray 1980) living (as *E. excavatum*)
6. Tamar Estuary (Murray 1965c) living (as *E. excavatum*)
7. Llandanwg Lagoon (Haman 1969) living (as *E. excavatum*)
8. Gironde (Pujos 1976) living (as *E. umbilicatulum*)
9. Arcachon Lagoon (Le Campion 1970) living, dead (as *E. excavatum*)

Additional common species:

Miliammina fusca	1, 4
Haynesina germanica	2, 3, 8 (*as Nonion depressulus*), 4 (as *Protelphidium anglicum*), 5, 6 (as *Protelphidium anglicum*)
Ammonia beccarii	2, 3 (as *Streblus batavus*), 4, 6, 8, 9
Elphidium excavatum	3 (as *E. selseyensis*)
Elphidium oceanense	4
Elphidium lidoense	8 (as *Cribrononion*)
Cibicides cf. *ungerianus*	8
Ammoscalaria pseudospiralis	9
Eggerelloides scabrus	9 (as *Eggerella*)
Reophax moniliformis	9 (as *R. findens*)

Table 9.5 *(continued)*

***Haynesina germanica* association**

Salinity: 0–35‰

Temperature: 0 to 20 °C

Substrate: mud, muddy sand

Depth: 0–20 m

Distribution:
1. Elbe Estuary (Wang 1983) total
2. Jade Bay (Richter 1964a, b, 1967) living (as *Nonion depressulus*)
3. Langeoog (Haake 1962) living, total (as *Nonion depressulus*)
4. Dollart Estuary (Van Voorthuysen 1960) living, dead (as *Nonion depressulus*)
5. Humber Estuary (Brasier 1981) living, total
6. Christchurch Harbour (Murray 1968a) living, dead (as *Protelphidium anglicum*)
7. Exe Estuary (Murray 1980) living, dead (as *Protelphidium*)
8. Tamar Estuary: (a) (Murray 1965c) living, dead (as *Protelphidium* sp. nov.); (b) (Ellison 1984) living
9. Cale de Dourduff (Vénec-Peyré 1981, 1983a) living, dead (as *Protelphidium paralium*)
10. Arcachon (Le Campion 1970) living, dead (as *Nonion depressulum*)

Additional common species:

Elphidium excavatum	1, 2, 3 (as *E. selseyense*)
Ammonia beccarii	1, 3 (as *Streblus batavus*), 6, 8a
Elphidium williamsoni	2, 6 (as *E. excavatum*),7
Elphidium gunteri	3, 4
Nonion depressulus	3 (as *N. umbilicatulus*), 6, 7
Eggerelloides scabrus	6 (as *Eggerella*)
Miliammina fusca	6
Elphidium sp.	6
Quinqueloculina dimidiata	6
Reophax moniliformis	6, 10 (as *R. findens*)
Elphidium oceanense	6
Bulimina cf. *elegans*	8a
Ammoscalaria pseudospiralis	10

***Miliammina fusca* association**

Salinity: 0–35‰

Temperature: 0 to 25 °C

Substrate: silt

Depth: 0–2 m

Distribution:
1. Baltic: (a) (Lutze 1965) dead; (b) (Wefer 1976b) living
2. Bottsand Lagoon, Baltic (Lutze 1968a) living, dead
3. Oslo Fjord (Alve and Nagy 1986) living, total
4. Christchurch Harbour (Murray 1968a) living
5. Arcachon (Le Campion 1970) dead

Additional common species:

Elphidium clavatum	1a (as *Cribrononion*)
Elphidium williamsoni	2 (as *Cribrononion articulatum*)
Ammotium aff. *cassis*	2
Ammonia beccarii	4
Elphidium sp.	4
Haynesina germanica	4 (as *Protelphidium anglicum*)
Trochammina inflata	5
Reophax moniliformis	5 (as *R. findens*)

***Nonion depressulus* association**

Salinity: 19–35‰

Temperature: 6 to 20 °C

Substrate: muddy sand

Depth: 0–2 m

Distribution:
1. Langeoog (Haake 1962) living (as *N. umbilicatulum*)
2. Christchurch Harbour (Murray 1968a) living
3. Exe Estuary (Murray 1980, 1983) living
4. Gironde (Pujos 1976) living

Additional common species:

Haynesina germanica	1 (as *Nonion depressulus*), 2 (as *Protelphidium anglicum*), 3 (as *Protelphidium*), 4 (as *Protelphidium paralium*)
Elphidium williamsoni	1, 2 (as *E. excavatum*), 3
Ammonia beccarii	4
Elphidium lidoense	4 (as *Cribroelphidium*)
Cibicides cf. *ungerianus*	4

Baltic Sea

The Baltic Sea is landlocked and its connection with the North Sea is via the Skagerrak–Kattegat where the water has a salinity of 32‰. Because precipitation greatly exceeds evaporation and the area is non-tidal, the waters are stratified. The surface layer is brackish (salinity 7–10‰) and seasonally variable in temperature. The lowest temperatures are found in the lower part of this layer. Beneath is more saline water which flows in from the North Sea. The salinity decreases with passage into the Baltic. A number of basins are present and each sill acts as a threshold for the bottom water to cross. The salinity in successive basins is 11–22‰ (Arkona Basin), 15–21‰ (Bornholm Basin) and 11–14‰ (Gotland Basin). Renewal of basin water is sometimes infrequent (i.e. decades) and under these conditions anoxia develops. The Baltic Sea reaches a maximum depth of 459 m although most is <200 m.

The general characteristics of the foraminiferal assemblages have been described by Brodniewicz (1965) and Lutze (1965). Both found that fragile agglutinated taxa are present in places as a high proportion of the living assemblages, but they are seen only in wet samples. Once samples have been dried these tests are destroyed, but the residual assemblages more closely resemble those that would be fossilised. Although Lutze stained his material only a small number of samples had ≥100 living individuals. As might be expected, the living forms show a patchy distribution (Lutze 1968b). The surface-water layer is characterised by the *Elphidium excavatum* association. This species reproduces throughout the year and the life cycle lasts at least 20 days (Haake 1967). It is infaunal in sand but also lives epiphytically on algae and tolerates a broad range of salinities (12–21‰) and temperatures (4–17 °C) (Wefer 1976b).

The *Miliammina fusca* association was found to be widespread dead but absent living (Lutze 1965). However, Wefer (1976b) recorded it living shallower than 13 m both on sediment and on algae.

The *Elphidium gerthi* association was said by Lutze (1965) to be the key assemblage in shallow water but none of his samples had ≥100 living individuals. In the 'Hausgarten' area of Kiel Bay it was present living (Lutze 1974a). This species lives mainly attached to algae (Wefer 1976b).

The *Ammotium cassis* association is present in the transition zone at the junction of the surface and deep water at a depth of 17–23 m in Kiel Bay. Its presence seems to be a recent event because it was not recorded by Rhumbler (1935, 1936) and only a few specimens were found by Rottgardt (1952). According to Wefer (1976b), *A. cassis* reproduces at >85% O_2 saturation, temperatures below 3 °C and during periods of high food availability.

The *Elphidium clavatum* association is present in the more saline deep-water layer and its optimum is at 20–24‰ (Lutze 1974a). This species appears to reproduce principally during the deepest extension of the euphotic zone (Wefer 1976a). It is epifaunal (Wefer 1976b) and is opportunistic for it was the first and principal species to colonise artificial substrates in an experiment in Eckernförder Bay (Wefer and Richter 1976).

The *Elphidium incertum* association is found around the basin margins on sandy sediments beneath the topmost part of the deep water (Lutze 1974a). It is infaunal and reproduces during the winter months (Wefer 1976b).

In the very shallow (<1 m) Bottsand Lagoon, the *Elphidium williamsoni* and *Miliammina fusca* associations are present. The daily temperature range in summer is sometimes 10 °C. *Elphidium williamsoni* reproduces in April and reaches its adult size by the winter. Because the environment is extreme, diversity is very low (α <1).

The *Eggerelloides scabrus* association was found only as dead assemblages by Lutze (1965) and he considered the species to be related to the incoming North Sea water. This has been confirmed by a subsequent study in Kiel Bay (Lutze *et al.* 1983) where this species forms >50% of some dead assemblages. At present it lives only in those areas where the salinity is at least 24‰ for a major part of the year.

Similarly, the *Nodulina dentaliniformis* association was recorded only as dead assemblages although rare living individuals are present (as *Reophax*, Lutze 1965). The taxonomy of Baltic reophacids is confused (compare Rhumbler 1936, Rottgardt 1952, Lutze 1965, 1974a and Brodniewicz 1965). Biometric studies led Hermelin (1983) to recognise two ecophenotypes with different distributions. Forms from the west are more elongate than those from elsewhere. However, the holotype, which comes from >1800 m in the Atlantic (Brady 1884) is quite distinct from the Baltic examples. Are these all morphotypes of one species or is the Baltic form a separate species?

The diversity of all the living assemblages from the Baltic is low (α <2, Murray 1976a).

Swedish estuaries

In estuaries on the Swedish west coast, the *Ammotium cassis* association is well developed. This species lives horizontally within the top 5 cm of sediment, ingesting organic particles. It peaks at discontinuity levels such as the halocline (Olsson 1976) or breaking internal waves (Olsson 1977). As in the Baltic, it favours transitional water masses with salinities only rarely lower than 16‰, and it is eurythermal. It can withstand low oxygen concentrations. The test serves as a barrier to adverse physiochemical conditions. Most specimens have an apertural plug of 10–24 μm grains of quartz and mica which may assist this function but may also serve as a source of material for test growth.

Oslo Fjord

Oslo Fjord is divided into a number of sub-basins (maximum depth >300 m) separated from one another by thresholds, the shallowest (Drøbak Sill) being <20 m deep. The water is stratified with a surface brackish layer subject to seasonal temperature variation of ~2 to 20 °C. It extends to 40–50 m in the inner part and 80–90 m in the outer part. Below this, the deep water remains fairly uniform with salinities of 33–34‰ and temperatures of 7–8 °C. Water renewal in the deep basins takes place in the late winter.

From the data on total assemblages collected by Risdal (1964) and reinterpreted by Thiede *et al.* (1981) it can be seen that there are three principal associations: *Eggerelloides scabrus* beneath the surface water at 10–30 m, *Bulimina marginata* beneath the surface water and the uppermost deep water at 30–120 m, and *Brizalina skagerrakensis* (as *Bolivina* cf. *robusta*) beneath the well-oxygenated deep water (110–330 m). Minor associations include *Elphidium albiumbilicatulum* and *Nonionellina labradorica* beneath the surface water (10, 50 m) and *Cassidulina laevigata* in deeper water (100–140 m). More detailed studies of a shallow arm of the fjord (Sandebukta, see Alve and Nagy 1986) has shown a greater diversity of associations. The *Miliammina fusca* association is present on non-vegetated mudflats, the *Ammotium cassis* association at 4–7 m at the halocline, the *Ammonia beccarii* association at depths of <10 m together with the *E. incertum* association. Four minor associations are confined to Sandebukta: *Adercotryma glomerata*, *Ammoscalaria pseudospiralis*, *Recurvoides trochamminiforme* and *Spiroplectammina biformis*. The *Eggerelloides medius* association (as *Verneuilina*) is present at depths of 60–200 m throughout Oslo Fjord. The living agglutinated assemblages of Dröbak Sound are mainly dominated by *Reophax fusiformis* (= *Proteonina* of Christiansen 1958; see also Hiltermann 1973).

Sandebukta is polluted with effluent from a paper mill, domestic sewage and from adjacent farmland. The sediments contain 0.7–2.6% total organic carbon and pH values range from 6.5 to 7.2. The *Eggerelloides scabrus* association is best developed on sediments rich in wood fibres where the salinity is 24–34‰ and temperatures 3 to 12 °C (Alve and Nagy 1986). From a study of short sediment cores, it appears that during the past 100 years there has been an increase in nutrients through pollution and this is matched by an increase in the agglutinated component. In cores from the deeper parts of Sandebukta, the pre-pollution assemblages are the *Bulimina marginata* association, but the modern assemblages are the *Adercotryma glomerata* and *E. medius* associations (= *Verneuilina* of Nagy and Alve 1987).

Nearby Gullmar Fjord was studied by Höglund (1947) who concentrated on the agglutinated taxa. Some information on total assemblages is given by Qvale *et al.* (1984). The *Bulimina marginata* association may be present in more marine waters flowing in from the Skagerrak. The *Eggerelloides scabrus* association is dominant at depths 5–25 m under the surface water (salinity 25–31‰).

Elbe Estuary

There is a single living association with *Elphidium excavatum* forming >80% in all but one sample. The average test size increases from 125 to ~200 μm from landward to seaward. The total assemblages fall into the *Ammonia beccarii*, *Elphidium excavatum* and *Haynesina germanica* associations. They are size-sorted in relation to the sediment grain size (Wang and Murray 1983; Wang 1983).

Jade Bay

Jade Bay is mesotidal and has almost normal marine salinities in its outer part (32‰) but is brackish in its inner part (24‰). Richter (1964a) recognised the following distribution pattern:

High-water mark: maximum standing crop *Haynesina germanica* association with *Elphidium williamsoni*
Open tidal flats: small standing crop.
Lower water mark and channels: intermediate standing crop. *Haynesina germanica* association with *E. excavatum*.

The additional species were thought to be controlled in their distribution by currents; *Elphidium williamsoni* favours areas of little water movement while *E. excavatum* tolerates stronger currents and mobile sediments for it lives at depths of 0.5–6 cm below the surface (Richter 1964b). *Haynesina germanica* lives in the topmost layer (down to ~3 cm). Living individuals (as well as dead tests) may be transported by currents and by wave attack. However, at high or low water the tidal currents and waves are weakest (at high water the waves are weak because of the large extent of shallow water to be traversed). Large-scale transport of *H. germanica* and *E. williamsoni* takes place on the flood tide (60–240 individuals per 10 cm^3 sediment). In culture, these two species adhere to the underside of the water surface with their pseudopodia and under calm conditions they could be transported landwards on the rising tide (Richter 1965).

During a hurricane and floods in 1962, a 3–5 cm layer of sediment was deposited on reclaimed land. This sediment contained 71 living individuals per 40 cm^3. Ten days after the flood, individuals that had been buried deeper than 4 cm had managed to migrate up to the top 4 cm layer. Ice also transports foraminifera by the repeated stranding and floating off of ice slabs. The included sediments contain up to 600 foraminifera per 10 cm^3. Richter estimated that at high water near Hooksiel in the winter of 1962–63, 3.5–4 million living foraminifera were present in the 2 m thick ice accumulation. Thus, the Watt area of the southern North Sea is one where physical environmental controls exert the major influence (see Richter 1967 for a review). Apart from currents and waves, the annual temperature range is from frozen to ~31 °C and salinity from 14 to 38‰ (Vangerow 1965).

Southern North Sea coast

The tidal flats between Langeoog and the mainland are sands and muddy sand. The area is mesotidal (~2.4 m) with current velocities averaging 0.15 and a maximum of 0.33 m s^{-1}. The dominant association is of *Haynesina germanica* and there is a correlation between the abundance of this species and diatom growths. The *Elphidium excavatum* association lives mainly on sand flats but the total assemblages are more widely distributed. There is a small occurrence of living *Nonion depressulus* association on muddy sand (Table 9.5, Haake 1962). Postmortem bed-load transport is active and the common taxa have settling velocities equivalent to fine sand.

The Ems–Dollart Estuary has both living and dead *Ammonia beccarii, Elphidium williamsoni, Haynesina germanica* (Table 9.5) and *E. gunteri* (Table 9.3) associations but only the latter is widespread (Van Voorthuysen 1960). Muds from high water have the same taxa but only *A. beccarii* and *H. germanica* are common (Reinhöhl-Kompa 1985). Similar faunas from the Dutch coast were described by Hofker (1977).

British Isles

Seasonal samples from the brackish estuarine lagoon, Christchurch Harbour, England, were dominated by two associations, *Elphidium* sp. and *Haynesina germanica*. The former was dominant in the autumn, occasionally present during the winter and spring and absent during the summer. The *H. germanica* association was dominant during the winter, spring and summer and, at some stations, during the autumn too. The *E. williamsoni* association was present at six stations in the spring and one during the autumn. Although *Miliammina fusca* and *Ammonia beccarii* were widely present as additional common species, they were rarely present as associations. The *Nonion depressulus* association were encountered at the mouth of the estuary in the winter.

In the lower estuary, diversity was highest in the summer (α 1–5), and autumn (α 2–4), while in the spring (α 1–4) and winter (α 1–2) generally lower values were attained. In the upper estuary, with more extreme salinity variation, the annual range was α 1–3.

Although numerous environmental parameters were measured no relationship was found with the abundance of any of the two principal species, *Elphidium* sp. and *Haynesina germanica* (Murray 1968a). However, *E. williamsoni* appears to have a lower optimum temperature range (10–15 °C) than *Elphidium* sp. (15–20 °C) and this is reflected in the pattern of occurrence of the associations. Multivariate analysis of the summer data only showed that *H. germanica* (as *Protelphidium anglicum*) was present in the higher salinity and higher calcium waters (Howarth and Murray 1969).

The dead assemblages are mainly *Haynesina germanica* and *Elphidium* sp. associations and this suggests that postmortem modification is not important. Some of the hyaline tests are etched but dissolution appears to be minor. In the lagoon mouth there are nearshore species transported in even though the area is microtidal. At a couple of stations the dead assemblage is the *Eggerelloides scabrus* association, although this species is only found living during the periods of higher salinity so many of these tests may have been transported in. The abundance of exotic tests ranges from 0 to 22% (Wang and Murray 1983).

The most abundant living form in the lower intertidal zone of the more marine part of the Exe Estuary is *Nonion depressulus* (Murray 1980) and its population dynamics have been studied over a $2\frac{1}{2}$-year period (Murray 1983). Standing crop values are highest in spring and autumn and coincide with those of the diatoms. During these periods, salinity was 19–35‰, therefore this species is euryhaline but less so than *Haynesina germanica* and *Elphidium williamsoni* both of which are present as associations in the more brackish parts. This estuary is mesotidal and exotic tests form up to 36% of the dead assemblages (Wang and Murray 1983).

Even though the Tamar Estuary is only moderately brackish (salinity 25–35‰), the *Haynesina germanica* association is still the principal one throughout the year, both living and dead and from intertidal (Ellison 1984) to 20 m depth (Murray 1965c). The *Ammonia beccarii* and *Elphidium williamsoni* associations are both present although the latter was found only living.

France

The Gironde Estuary has the *Ammonia beccarii*, *Elphidium williamsoni* and *Nonion depressulus* associations in the brackish parts. Near the mouth there are more marine associations: *E. lidoense*, *E. macellum*, *Cibicides* cf. *ungerianus* (Table 9.3). *Jadammina macrescens* (as *J. polystoma*) which elsewhere is confined to high marshes, is said to live under the turbidity maximum and to migrate seasonally within the estuary (Pujos 1976, 1983).

Arcachon Lagoon is mainly normal marine but in the inner parts, salinities fall to 29‰. It is mesotidal and much of the bottom is exposed at low water. Le Campion (1970) distinguished a number of subenvironments of which the most important are:

1. Subtidal sands with *Abra* or *Venus*. These have living and dead assemblages of a *Planorbulina mediterranensis* association and dead *Eggerelloides scabrus* association. This is an area of powerful currents.
2. Intertidal *Zostera*. *Triloculina inflata* association living, *Elphidium williamsoni* association living and dead, *Ammonia beccarii* association living, *Eggerelloides scabrus* association living and dead, *Ammoscalaria pseudospiralis* association living, *Haynesina germanica* association living and dead, *Reophax moniliformis* association dead.
3. *Scrobicularia* mud. *Ammoscalaria pseudospiralis* association living and dead together with dead *Eggerelloides scabrus*, *Miliammina fusca* and *Trochammina inflata* associations.

It is clear that postmortem transport takes place due to the powerful tidal currents, hence the abundance of *Trochammina inflata* away from the marsh.

The Santona Estuary in north Spain has two distinct living (and dead) assemblages. In the marine lower estuary is the *Ammonia beccarii* association while in the brackish upper estuary is the *Haynesina germanica* association (Cearreta 1988b).

In Senegal, the Casamance Estuary experiences an annual salinity range of 0–110‰ because of the extremes of high rainfall and drought. Three principal associations represented by living and total assemblages are present. At the mouth, in near-marine conditions is an *Eggerelloides scabrus* association with *Elphidium gunteri* and *Ammonia* sp. The major part is occupied by an *Ammonia* spp. association with *E. gunteri* and *Ammotium salsum* where the salinity is 35–55‰ or an almost monospecific *A. salsum* association where salinities are >60‰. *Ammonia beccarii tepida* reaches its maximum development during the wet months when salinities are between 35 and 50‰. *Ammotium salsum* is more euryhaline than *Ammonia beccarii tepida* especially in the hypersaline range (Debenay and Pages 1987). Similar relationships are seen in other estuaries in Senegal (Debenay *et al.* 1987).

Total assemblages from lagoons on the Niger Delta are *Ammonia beccarii* associations with *A. beccarii* often forming >90% (Adegoke *et al.* 1976).

Pollution and the influence of man

The effects of pollution on faunal succession in Sandebukta, Oslo Fjord, have already been discussed. In Swedish estuaries recovering from sewage and heavy metal pollution, there is also faunal succession. *Elphidium excavatum* was the only foraminifer in Valen Estuary while it was polluted, but after 2 years of recovery two additional species were established, namely *Ammoscalaria runiana* and *Miliammina-fusca*. The Gota Estuary was devoid of foraminifera until recovery commenced. The pioneer taxa were

Ammonia beccarii and three allogromiids followed by a more diverse fauna including *Psammosphaera bowmani*, *Haynesina germanica* (as *Protelphidium anglicum*), *E. excavatum* and *Ammotium cassis*. The pioneer species of the two estuaries were said to be able to tolerate oxygen concentrations below 0.5 mg l^{-1} (Cato *et al.* 1980).

In the Cale du Dourduff (Brittany), following major pollution from a wrecked oil tanker, *Haynesina germanica* (= *Protelphidium paralium* of Vénec-Peyré 1981) showed deformities of shell growth and calcification but after 1 year the incidence of this had fallen.

The restricted and often oxygen-depleted waters of the dock basins of St Malo have a very low diversity living fauna dominated by *Elphidium oceanensis* (as *Cribrononion lidoense*), *Haynesina germanica* (as *Protelphidium paralium*) and *E. excavatum* (as *Cribrononion*) together with *Eggerelloides scabrus* (as *Eggerella*) and *Ammonia beccarii* (Rouvillois 1972a). This is a normal brackish fauna.

Comparative studies of the Rance Estuary, before the construction of the barrage, during its construction and the final conditions when it was in operation for generating electricity through tidal power, have been made by Rouvillois (1967, 1972b). Originally *Ammonia beccarii* was dominant but after the construction of the barrage (and before it became operational) the salinity became much reduced. *Haynesina germanica* (as *Protelphidium paralium*) and *Elphidium oceanensis* (as *Cribroelphidium lidoense)* became dominant. Once electricity generation commenced, seawater was able to enter the estuary again. *Elphidium oceanensis* became rare and the dominant species were *A. beccarii* and *H. germanica*.

Postmortem changes

Some of the differences between the living and dead assemblages in the Baltic Sea have already been mentioned: the recent introduction of *Ammotium cassis* and the presence of relict *Eggerelloides scabrus*. At depths shallower than 13 m, wave disturbance of the sea floor causes fragmentation of foraminifera and the loss of fine material to give a coarse lag sediment (Flemming and Wefer 1973). This was termed the abrasion zone by Grobe and Fütterer (1981) who also recognised two further zones. From 13 to 26 m is a disintegration zone where the sand fraction from the abrasion zone is deposited. Here, the organic matrix of tests is destroyed. Finally, at depths >26 m, there is the corrosion zone where the bottom waters undersaturated with respect to $CaCO_3$ cause dissolution of calcareous tests. This may be partial, leading to broken chambers and thinned walls, or total (Jarke 1961a). The effects of these processes were quantified by Wefer and Lutze (1978). At depths of 5–18 m, annual carbonate production by foraminifera ranged from 10 to 133 mg m^{-2} yet only 0–0.27 mg m^2 was preserved. Even at 27–28 m where annual production was 3117 mg m^{-2} only 4.3% was preserved.

In Sandebukta, Oslo Fjord, the pH of the muddy sediments ranges from 6.5 to 7.2. Calcareous taxa normally form only a minor part of the living assemblages and postmortem dissolution is evident from the presence of partially dissolved tests and organic linings. Dissolution takes place between April and June (Alve and Nagy 1986).

Dissolution of calcareous tests has not been recognised as a major postmortem process in most other European estuaries and lagoons and the presence of calcareous assemblages in cores from subrecent deposits confirms this. Postmortem transport operates with an intensity which in general varies with the magnitude of the tidal range (Wang and Murray 1983). In the microtidal Christchurch Harbour it is only significant near the mouth. The mesotidal Exe and Dovey have modest numbers of size-sorted exotic tests (<40%) while the macrotidal examples have up to 70% exotic tests (Humber, Severn, Rance–Rouvillois 1974). In addition, bed-load transport must take place on the intertidal sand flats between the Friesian Islands and the mainland. This area is also affected by ice transport.

The nature of the dead assemblages is summarised in Table 9.6.

Continental shelf and slope

Because of the varied nature of continental shelf and slope environments over the geographic range from the Barents Sea to southern Africa, a large number of minor associations, present in a very restricted area, have been recognised (Table 9.7).

Typical shelf and slope species are listed in Table 9.8 and the principal associations in Tables 9.9 and 9.10 respectively.

Barents Sea

In shallow, cold waters off Spitsbergen five minor associations have been recognised: *Cassidulina reni-*

Table 9.6 Classification of dead assemblages from European lagoons and estuaries

Field I

Oslo Fjord deeper than ? 60 m
Swedish estuaries
Christchurch Harbour – most
Santona Estuary – upper

Field I-C to I-CC

Baltic Sea basins

Field II

Jade Bay – part
Christchurch Harbour – mouth
Tamar Estuary
Santona Estuary – lower

Field III

Baltic Sea <13 m
Elbe Estuary
Jade Bay – part
North Sea coast
Arcachon

forme, *Buccella frigida*, *Spiroplectammina biformis*, *Pateoris hauerinoides* and *Astrononion gallowayi* and spanning the outer shelf–upper slope, the *Thurammina sphaerica* association (Table 9.7). The three major associations are *Cibicides lobatulus* (0–105 m), *Elphidium clavatum* (0–35 m) and *Cassidulina laevigata* (100–300 m) (Table 9.9).

Norwegian Sea

The continental shelf off Norway has irregular topography with banks and troughs that extend down to 500 m. This topography is a product of Plio-Pleistocene glacial activity which caused both erosion and deposition of till. Subsequently there has been marine erosion and reworking of the till by powerful bottom currents. In places the continental slope is steep and has been subject to large-scale slumping and sliding.

The waters of the Norwegian–Greenland Sea are stratified (Fig. 9.2, Table 9.11). Within about 100 km of the coast, there is the lower-salinity, variable-temperature Norwegian Coastal Current water mass. Beneath this is North Atlantic Water (NAW), warm, highly saline water which enters the Norwegian Sea

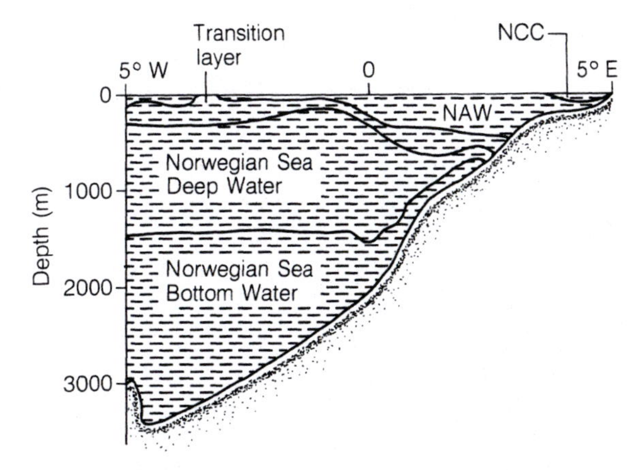

Figure 9.2 Water masses along the Norwegian margin (modified from Mackensen *et al.* 1985). NCC = Norwegian Coastal Current, NAW = North Atlantic Water

Table 9.11 Water masses of the Norwegian–Greenland Sea (based on Sejrup *et al.* 1981, and Qvale 1986a)

	Salinity (‰)	Bottom temperature (°C)	Approximate boundary (m)
Norwegian Coastal Current	<35		
			—
North Atlantic Water	>35	5 to 7	
			500
Transition Layer			
			650
Norwegian Sea Deep Water	34.91	−1 to 0	
			750
Norwegian Sea Bottom Water	34.91	−1	

via the Faeroe–Shetland Channel. The north-flowing Norwegian Current, which bathes much of the shelf and upper slope down to ~500 m, is formed by NAW. The Transition Layer separates this from the underlying Norwegian Sea Deep Water which forms a much thicker layer in the open sea (Fig. 9.2). Finally, there is Norwegian Sea Bottom Water, the upper surface of which is marked by a pycnocline at ~−1 °C. This water mass covers the bottom from around 700 m to the maximum depth at around 3900 m (data from Sejrup *et al.* 1981, Mackensen *et al.* 1985 and Qvale 1986a).

Table 9.7 Minor associations from the shelf and slope of the European and African margins (L = living, D = dead, T = total)

Association		Depth (m)	Area	Source
1. *Thurammina sphaerica*	T	120–320	Barents Sea	Jarke 1960
2. *Cassidulina reniforme*	T	44	Spitsbergen	Nagy 1965
3. *Spiroplectammina biformis*	L, D	5–15	Spitsbergen	Rouvillois 1966
4. *Pateoris hauerinoides*	D	5	Spitsbergen	Rouvillois 1966
5. *Buccella frigida*	L, D	25	Spitsbergen	Rouvillois 1966
6. *Astrononion gallowayi*	D	15	Spitsbergen	Rouvillois 1966, Nagy 1965
7. *Cibicides refulgens*	T	215–255	Norway	Qvale 1986a
	T	230	Biscay	Caralp *et al.* 1970
8. *Cibicidoides pseudoungerianus*	T	225	Norway	Qvale 1986a
	D	950	Porcupine Seabight	Weston 1985
	D	744–996	NW Africa	Haake 1980a
	T	1400	Lüderitz	Martin 1981
9. *Cribrostomoides subglobosum*	L	1100–2005	Norway	Mackensen *et al.* 1985
10. *Reophax scorpiurus*	D	3004	Norway	Mackensen *et al.* 1985
11. *Robertinoides normani*	D	499	Norway	Mackensen *et al.* 1985
12. *Fissurina semimarginata*	D	1600–2683	Norway	Mackensen *et al.* 1985
13. *Recurvoides turbinatus*	T	1515	Norway	Belanger and Streeter 1980
14. *Islandiella norcrossi*	T	635	Norway	Belanger and Streeter 1980
15. *Cassidulina teretis*	T	954	Norway	Belanger and Streeter 1980
	D	510–1140	Porcupine Seabight	Weston 1985
16. *Islandiella islandica*	T	—	North Sea	Jarke 1961b
17. *Rosalina anomala*	L	110	W. Scotland	Murray 1985
18. *Acervulina inhaerens*	L, D	110	W. Scotland	Murray 1985
19. *Eponides repandus*	D	115	W. Scotland	Murray 1985
	T	56	North Minch	Edwards 1982
20. *Massilina secans*	L, T	0	Irish Sea	Atkinson 1969
21. *Cribrostomoides jeffreysii*	L	—	Irish Sea	Haward and Haynes 1976
22. *Miliolinella subrotunda*	L	~100	Irish Sea	Dobson and Haynes 1973
23. *Nonionella turgida*	L, D	95–135	Celtic Sea	Murray 1979a
24. *Cancris auricula*	L	92	Celtic Sea	Murray 1979a
	L, D	0–550	NW Africa	Lutze and Coulbourn 1984, Haake 1980a
	L	39–768	NW Africa	Lutze 1980
25. *Epistominella vitrea*	L, D	99–100	Celtic Sea	Murray 1979a
	L, D	618–3905	Portugal	Seiler 1975
26. *Adercotryma glomerata*	L	100	Celtic Sea	Murray 1979a
27. *Gavelinopsis praegeri*	D	108–119	Celtic Sea	Murray 1979a
	T	115	N. Biscay	Schnitker 1969
	D	1456	Portugal	Seiler 1975
	D	43–62	NW Africa	Lutze 1980
28. *Nonion depressulus*	T	—	Brittany	Rosset-Moulinier 1972
	L	17–18	S. Cornwall	Murray 1970c
29. *Ammoglobigerina globigeriniformis*	L	84–100	English Channel	Murray 1970c
30. *Bulimina gibba/elongata*	L, D	77–95	Bristol Channel	Murray 1979a
	D	45	Portugal	Seiler 1975
31. *Asterigerinata mamilla*	D	18	S. Cornwall	Murray 1970c
	D	—	Douarnenez	Rosset-Moulinier 1972
32. *Uvigerina pygmaea*	D	960–1380	Porcupine Seabight	Weston 1985
33. *Bulimina striata/inflata*	D	1310–1500	W. Approaches	Weston 1985
34. *Bolivinellina pseudopunctata*	L	10	Plymouth	Murray 1965c

Table 9.7 *(continued)*

Association		Depth (m)	Area	Source
35. *Quinqueloculina lata*	D	50	Lyme Bay	Murray 1986a
36. *Textularia truncata*	L, D	50	Lyme Bay	Murray 1986a
37. *Elphidium magellanicum*	L		Baie de Seine	Dupeuble *et al.* 1971
38. *Cibicides boueanum*	T	60	N. Biscay	Rouvillois 1970
39. *Elphidium lidoense*	T	20–60	Biscay	Pujos 1972
40. *Cibicides* cf. *C. ungerianus*	T	15–70	Biscay	Pujos 1972, Caralp *et al.* 1970
41. *Nonionoides boueanum*	T	27–44	Galicia	Van Voorthuysen 1973
	T	118	Lüderitz	Martin 1981
42. *Brizalina striatula*	L	119–146	Portugal	Seiler 1975
	L	151	NW Africa	Lutze 1980
43. *Trifarina fornasini*	L	211	Portugal	Seiler 1975
	L, D	140–290	NW Africa	Haake 1980a
44. *Brizalina subspinescens*	L	287	Portugal	Seiler 1975
45. *Brizalina spathulata*	L	377	Portugal	Seiler 1975
	T	800	Lüderitz	Martin 1981
46. *Trifarina bradyi*	L	432–816	Portugal	Seiler 1975
47. *Rhabdaminella cylindrica*	L	1010	Portugal	Seiler 1975
48. *Eilohedra levicula*	L	2217	Portugal	Seiler 1975
	L, D	1517–3075	NW Africa	Lutze 1980
49. *Hoeglundina elegans*	T, D	1948–2761	N. Atlantic	Phleger *et al.* 1953, Weston and Murray 1984, Lutze 1980
50. *Cibicidoides kullenbergi*	L, T	1400–5400	N. Atlantic	Lutze and Coulbourn 1984, Phleger *et al.* 1953
51. *Globobulimina* spp.	L, D	50–450	NW Africa	Lutze and Coulbourn 1984, Haake 1980a
52. *Uvigerina finisterensis*	L, D	450–1800	NW Africa	Lutze and Coulbourn 1984
53. *Hanzawaia concentrica*	L	32	NW Africa	Haake 1980a
	L	9–35	Niger Delta	Adegoke *et al.* 1976
54. *Rectuvigerina phlegeri*	L	41	NW Africa	Haake 1980a
	L, D	39–70	NW Africa	Lutze 1980
55. *Suggrunda eckisi*	L, D	60	NW Africa	Haake 1980a
56. *Reophax calcareus*	L	94	NW Africa	Haake 1980a
57. *Textularia pseudogramen*	L, D	38–148	NW Africa	Haake 1980a, Lutze 1980
58. *Textularia panamensis*	D	41	NW Africa	Haake 1980a
	T	25–80	Zaire	Kouyoumontzakis 1981
59. *Bolivina dilatata dilatissima*	L	299–744	NW Africa	Haake 1980a
60. *Brizalina subaenariensis*	L, D	497	NW Africa	Haake 1980a
61. *'Nonion' asterizans*	L, D	81–103	NW Africa	Haake 1980a, Lutze 1980
	T	12–50	Zaire	Kouyoumontzakis 1981
62. *Bulimina alazanensis*	L, D	2090	NW Africa	Lutze 1980
63. *Rectuvigerina arquatensis*	L	342–514	NW Africa	Lutze 1980
64. *Cassidulina crassa*	L, D	157–3008	NW Africa	Lutze 1980
65. *Uvigerina bononiensis*	L	46–109	NW Africa	Lutze 1980
66. *Brizalina variabilis*	L, D	33–70	NW Africa	Lutze 1980
67. *Bolivina pseudoplicata*	L, D	38	NW Africa	Lutze 1980
68. *Bolivina dilatata*	L, D	103–966	NW Africa	Lutze 1980
69. *Bolivina cincta*	D	1024–1854	NW Africa	Lutze 1980
70. *Brizalina difformis*	D	109–316	NW Africa	Lutze 1980
71. *Bolivina compacta*	D	3488	NW Africa	Lutze 1980
72. *Uvigerina auberiana*	T	1000	SW of Azores	Hermelin and Scott 1985

Table 9.7 *(continued)*

Association		Depth (m)	Area	Source
73. *Rectuvigerina nicoli*	T	35–87	Niger Delta	Adegoke *et al.* 1976
74. *Amphicoryna scalaris*	T	50–100	Cameroon	Berthois *et al.* 1968
75. *Cancris congolensis*	T	25–28	Zaire	Kouyoumontzakis 1982a
76. *Ammobaculites americanus*	T	100–180	Lüderitz	Martin 1981
77. *Textularia millettii*	T	1256	Lüderitz	Martin 1981
78. *Bulimina elongata*	T	117	Lüderitz	Martin 1981
79. *Cibicides praecinctus*	T	170	Lüderitz	Martin 1981
80. *Elphidium advenum*	T	51	Lüderitz	Martin 1981
81. *Uvigerina cushmani*	T	195	Lüderitz	Martin 1981

Table 9.8 Typical shelf and slope species

Species	Mode of life	Feeding strategy
Shelf to upper slope		
Ammonia beccarii	Infaunal, free	? Herbivore
Bulimina marginata	Epifaunal, infaunal, free	
Cassidulina laevigata	Infaunal, free	
Cassidulina obtusa	Infaunal, free	
Cibicides lobatulus	Epifaunal, attached	? Herbivore, detritivore
Cibicides refulgens	Epifaunal, attached	Herbivore, suspension feeder, parasitic
Eggerelloides scabrus	Epifaunal, free	Detritivore
Elphidium clavatum	Epifaunal, free	Detritivore
Elphidium crispum	Epifaunal, free or clinging	Herbivore
Elphidium excavatum	Infaunal, epifaunal free	? Herbivore
Fursenkoina fusiformis	Infaunal, free	
Gaudryina rudis	Epifaunal, clinging	
Hyalinea balthica	Epifaunal, free	
Quinqueloculina seminula	Epifaunal, free or clinging	
Reophax fusiformis	Infaunal, free	Detritivore
Rosalina anomala	Epifaunal, clinging or attached	
Textularia sagittula	Epifaunal, clinging	
Trifarina angulosa	Infaunal, free	
Slope		
Brizalina skagerrakensis	Infaunal, free	
Epistominella exigua	Epifaunal, free	
Fontbotia wuellerstorfi	Epifaunal, free	
Globocassidulina subglobosa	Infaunal, free	
Melonis barleeanum	Infaunal, free	
Nuttallides umboniferus	Epifaunal, free	
Ordorsalis umbonatus	Epifaunal, free	
Uvigerina peregrina	Infaunal, epifaunal, free	

Table 9.9 Principal shelf associations of the Atlantic seaboard of Europe and Africa

***Ammonia beccarii* association**

Salinity: 22–35‰

Temperature: 0.7 to 29 °C

Substrate: muddy sand

Depth: 0–60 m

Distribution:
1. Southern North Sea (Jarke 1961b) total (as *Streblus batavus*)
2. Cardigan Bay: (a) (Haman 1971) total; (b) (Atkinson 1971) total
3. S. Cornwall (Murray 1970c) living
4. Plymouth (Murray 1965c) living, dead
5. Lyme Bay (Murray 1986a) living, dead
6. Baie de Seine (Moulinier 1967) dead
7. North Brittany (Dupeuble *et al.* 1971) living, dead
8. Galicia (Van Voorthuysen 1973) total (as *A. inflata*)
9. Niger Delta (Adegoke *et al.* 1976) total
10. Cameroon (Berthois *et al.* 1968) total
11. Lüderitz, Namibia (Martin 1981) total

Additional common species:

Quinqueloculina seminula	1, 2a, b, 7a
Elphidium excavatum	2a, b (as *E. selseyense*), 5
Eggerelloides scabrus	3 (as *Eggerella*), 5
Bulimina gibba/elongata	3, 5, 11
Textularia sagittula	4
Reophax fusiformis	4
Cibicides lobatulus	4, 5
Bolivinellina pseudopunctata	5 (as *Brizalina*)
Fursenkoina fusiformis	5
Textularia panamensis	9 (as *Poritextularia*)
Elphidium gunteri	9 (as *Cribroelphidium*), 10
Nonionella atlantica	9 (as *Florilus*)
Elphidium advenum	10, 11
Textularia mexicana	10
Ammobaculites americanus	11
Pararotalia cf. *nipponica*	1

***Bulimina marginata* association**

Salinity: (2) 25–35, others 35‰

Temperature: 5.5 to 13 °C

Substrate: muddy sand

Depth: (1) 900, (2) 70–99, (3) 70, (4) 20–176, (5) 95, (6) 105, (7) 1400 m

Distribution:
1. Møre, Norway, (Mackensen *et al.* 1985) dead
2. Skagerrak (Qvale and Van Weering 1985) total
3. Forties, North Sea (Murray 1985) living
4. North Minch (Edwards 1982) total
5. Celtic Sea (Murray 1979a) living
6. N. Biscay (Rouvillois 1970) total
7. Lüderitz, Namibia (Martin 1981) total

Additional common species:

Melonis barleeanum	1 (as *M. zaandami*)
Cassidulina laevigata	1, 6
Uvigerina peregrina	1
Hyalinea balthica	2, 4
Fursenkoina fusiformis	3
Cibicides lobatulus	4
Rosalina globularis	4
Textularia sagittula	4, 5
Bulimina aculeata	6
Brizalina difformis	6 (as *Bolivina*)
Fissurina orbignyana	6
Fursenkoina rotundata	7
Praeglobobulimina spinescens	7

Table 9.9 *(continued)*

Cassidulina laevigata **association**

Salinity: 34.9–35.11‰

Temperature: –1 to 17 °C

Substrate: sand, mud

Depth: 100–2500 m

Distribution:

1. Barents Sea (Østby and Nagy 1981, 1982) total
2. Møre, Norway: (a) (Sejrup *et al.* 1981) total; (b) Mackensen *et al.* 1985) living, dead
3. Norway (Qvale 1986a) total
4. Norwegian Channel (Qvale and Van Weering 1985) total
5. Skagerrak (Qvale and Van Weering 1985) total
6. Northern North Sea: (a) (Jarke 1961b) total (as *C. norcrossi*); (b) (Mackensen and Hald 1988)
7. Western Approaches: (a) (Murray 1970c) dead; (b) (Weston 1985) dead
8. N. Biscay: (a) (Rouvillois 1970) total; (b) (Schnitker 1969) total
9. Biscay: (a) (Pujos 1972) total; (b) (Caralp *et al.* 1970) total
10. Portugal (Seiler 1975) dead (as *C. carinata*)
11. Morocco (Mathieu 1971) total (as *C. carinata*)
12. NW Africa: (a) (Haake 1980a) living, dead (as *C. carinata*); (b) (Lutze 1980) living
13. Lüderitz, Namibia (Martin 1981) total (as *C. carinata*)

Additional common species:

Species	Distribution
Cassidulina ? reniforme	1 (as *C. crassa*), 2
Elphidium excavatum	1
Melonis barleeanum	2a (as *Nonion*), 2b, 12b (as *M. zaandami*)
Trifarina angulosa	2a, b, 3, 4, 5, 7b, 9b
Cassidulina obtusa	2a, 7a (as *C. cassa*), 7b, 8a, b (as *Islandiella crassa*), 9b (as *C. crassa*), 10
Cibicides lobatulus	2a, 8b
Uvigerina peregrina	2a, 3, 6a, 9b, 13
Fontbotia wuellerstorfi	2a, 13 (as *Cibicides*)
Epistominella exigua	2a
Elphidium clavatum	2a (as *E. excavatum*)
Pullenia bulloides	2b
Planulina ariminensis	2b
Bulimina marginata	2b, 6a, 8a, 10, 11
Lagena apiopleura	2b
Fissurina semimarginata	2b
Reophax scorpiurus	2b
Bulimina cf. *B. alazanensis*	7a
Brizalina spathulata	7a, b, 10 (as *Bolivina*)
Cassidulina teretis	7b
Brizalina difformis	8a, 12b (as *Bolivina*)
Gavelinopsis praegeri	8b, 12b
Spiroplectinella wrightii	8b (as *Spiroplectammina*)
Eggerelloides scabrus	9a (as *Eggerella*)
Cibicides cf. *C. ungerianus*	9a, b
Globocassidulina sp.	9b
Elphidium lidoense	9b
Hyalinea balthica	11
Uvigerina peregrina dirupta	12a
Brizalina subaenariensis	12a
Bolivina dilatata dilitatissima	12a
Trifarina fornasinii	12a
Textularia sagittula	12a
Cassidulina crassa	12b
Globocassidulina subglobosa	12b
Bolivina dilatata	12b
Bolivina pseudoplicata	12b
Chilostomella ovoidea	12b
Rectuvigerina arquatensis	12b
Uvigerina cushmani	13
Cibicidoides pseudoungerianus	13 (as *Cibicides*)
Hoeglundina elegans	13
Oridorsalis umbonatus	13
Fursenkoina cf. *davisi*	13

Table 9.9 *(continued)*

***Cassidulina obtusa* association**

Salinity: 35‰

Temperature: 7 to 13 °C

Substrate: sandy and silty mud

Depth: 177–3299 m

Distribution:
1. Troms, Norway (Hald and Vorren 1984) total
2. Celtic Sea, (Murray 1979a) dead
3. Western Approaches: (a) (Murray 1970c) living; (b) (Weston 1985) dead
4. N. Biscay (Schnitker 1969) total
5. Biscay (Caralp *et al.* 1970) total (as *C. crassa*)
6. Portugal (Seiler 1975) dead

Additional common species:	
Cibicides lobatulus	1, 4
Trifarina angulosa	1
Gavelinopsis praegeri	2, 4
Fursenkoina fusiformis	2
Epistominella vitrea	2
Textularia sagittula	3
Nonionella turgida	3
Brizalina spathulata	3
Bolivinellina pseudopunctata	3 (as *Brizalina*)
Cassidulina laevigata	3b, 6 (as *C. carinata*), 4, 5
Brizalina difformis	4 (as *Bolivina*)
Elphidium lidoense	5
Cibicides cf. *ungerianus*	5
Hansenisca soldanii	5 (as *Gyroidina*)
Uvigerina peregrina	5
Bulimina marginata	6
Globocassidulina subglobosa	6

***Cibicides lobatulus* association**

Salinity: 32–36.5‰

Temperature: 0 to 30 °C

Substrate: gravel, sand, mud, seaweed

Depth: 0–900 m

Distribution:
1. Spitsbergen: (a) (Rouvillois 1966) dead; (b) (Nagy 1965) total
2. Barents Sea: (a) (Østby and Nagy 1981, 1982) total; (b) (Jarke 1960)
3. Troms, Norway (Hald and Vorren 1984) total
4. Møre, Norway: (a) (Sejrup *et al.* 1981) total; (b) (Mackensen *et al.* 1985) living, dead
5. Norway (Qvale 1986a) total (including *Cibicides* spp.)
6. Iceland–Faeroe Ridge (Jarke 1958) total
7. Orkney–Lincolnshire (Jarke 1961b) total
8. W. of Scotland (Murray 1985) living, dead
9. North Minch (Edwards 1982) total
10. W. Ireland (Heron-Allen and Earland 1913) total
11. Celtic Sea (Murray 1979a) dead
12. Channel: (a) (Murray 1970c) living, dead; (b) (Sturrock and Murray 1981), living, dead (groups 1–4)
13. N. Brittany (Rosset-Moulinier 1972) dead
14. Brest (Rosset–Moulinier 1972) dead
15. Douarnenez (Rosset–Moulinier 1972) dead
16. N. Biscay (Schnitker 1969) total
17. Galicia (Van Voorthuysen 1973) total
18. Portugal (Seiler 1975) living, dead (? *Cibicides* div. sp.)
19. NW Africa (Lutze 1980) dead

Additional common species:	
Elphidium clavatum	1b
Haynesina orbiculare	1b (*Elphidium*)
Cassidulina ? *reniforme*	1b, 2a (as *C. crassa*), 4a
Astrononion gallowayi	1b, 2a, b
Trifarina angulosa	2b (as *Angulogerina*), 3, 5
Melonis barleeanum	2b (as *Nonion zaandamae*)
Cibicidoides pseudoungerianus	2b, 5, 6, 7 (as *Cibicides*)
Islandiella norcrossi	2b, 6 (as *Cassidulina*)
Cassidulina laevigata	4, 5, 8, 16
Cassidulina obtusa	4a, 16 (as *Islandiella crassa*)
Cibicides refulgens	4b, 5
Discanomalina semipunctata	4b, 5 (as D. *pseudopunctata*)
Planorbulina mediterranensis	7
Rosalina anomala	8, 12b
Textularia sagittula	8, 9 11, 12a, b
Rosalina globularis	9
Eponides repandus	9
Gaudryina rudis	9
Epistominella vitrea	11
Gavelinopsis praegeri	11, 12a, b, 16, 18, 19
Ammonia beccarii	12a, 13
Quinqueloculina seminula	12a, 14
Spirillina vivipara	12b
Plascopsilina confusa	12b

Table 9.9 *(continued)*

Cribrostomoides jeffreysii	12b, 18
Ammoglobigerina globigeriniformis pygmaea	12b
Rosalina williamsoni	12b
Elphidium gerthi	13 (as *Cribononion*)
Elphidium crispum	14
Miliolinella subrotunda	14
Asterigerinata mamilla	14, 15
'Nonion asterizans'	19
Textularia pseudogramen	19

Eggerelloides scabrus association

Salinity: 35‰

Temperature: 8 to 14 °C

Substrate: sand, muddy sand

Depth: 0–97 m

Distribution: (all as *Eggerella*)
1. Southern North Sea (a) (Richter 1967) living, dead, (b) (Jarke 1961b) total
2. Celtic Sea (Murray 1979a) living
3. S. Cornwall (Murray 1970c) living
4. Galicia (Colom 1974) total

Additional common species:

Fursenkoina fusiformis	1a (as *Virgulina*), 3
Ammonia beccarii	1a (as *Streblus*), 3
Haynesina germanica	1a (as *Nonion depressulus*)
Bulimina marginata	2
Cribrostomoides kosterensis	2
Textularia sagittula	2
Bulimina gibba/elongata	3
Reophax fusiformis	3
Quinqueloculina sp.	3
Bolivinellina pseudopunctata	3 (as *Brizalina*)

Elphidium clavatum association

Salinity: 10–35‰

Temperature: 0 to 7 °C

Substrate: muddy gravel, sand

Depth: 0–35 m, except (2b) 160 m and (3a) 285 m

Distribution:
1. Spitsbergen: (a) (Rouvillois 1966) dead; (b) (Nagy 1965) total
2. Barents Sea (Jarke 1960) total
3. Møre, Norway (Sejrup *et al.* 1981) total (as *E. excavatum*)
4. Skagerrak: (a) (Hansen 1965) living; (b) (Van Weering and Qvale 1983) total (as *E. excavatum*)

Additional common species:

Buccella frigida	1a
Tholosina bulla	1a, b
Elphidium subarcticum	1b
Elphidium bartletti	1b
Cibicides lobatulus	1b, 2
Cassidulina ? reniforme	1b (as *C. crassa*)
Nonionellina labradorica	1b (as *Nonion*)
Elphidium sp.	1b
Trifarina angulosa	2 (as *Angulogerina*), 3
Ammonia beccarii	4 (as *A. batava*)

Elphidium crispum association

Salinity: 30–35‰

Temperature: 8 to 18 °C

Substrate: seaweed, muddy sand

Depth: 0–25 m

Distribution:
1. Cardigan Bay (Atkinson 1969) living, total
2. Brest (Rosset-Moulinier 1972) dead
3. Galicia (Van Voorthuysen 1973) total

Additional common species:

Elphidium macellum	1
Rosalina globularis	1
Cibicides lobatulus	2

Elphidium excavatum assemblage

Salinity: 34.8–35.0‰

Temperature: 7 to 14 °C

Substrate: sand, muddy sand

Depth: 0–40 m

Distribution:
1. Southern North Sea: (a) (Richter 1967) living, dead (as *E. selseyense*); (b) (Jarke 1961b) total (as *E. incertum*)
2. Lyme Bay (Murray 1986a) dead

Additional common species:

Ammonia beccarii	1a (as *Streblus*), 1b (*S. batavus*), 2
Fursenkoina fusiformis	1a (as *Virgulina*)
Haynesina germanica	1a (as *Nonion depressulus*)

Table 9.9 *(continued)*

***Fursenkoina fusiformis* association**

Salinity: 34–35‰

Temperature: 5 to 15 °C

Substrate: fine sand, muddy sand

Depth: 20–130 m

Distribution:
1. Central North Sea: (a) (Collison 1980) living; (b) (Murray 1985) living, dead; (c) (Jarke 1961b) total (as *Bulimina*)
2. Celtic Sea (Murray 1979a) living, dead
3. Bristol Channel (Murray 1970c) living
4. S. Cornwall (Murray 1970c) living
5. Plymouth (Murray 1965c) living
6. Lyme Bay (Murray 1986a) living
7. NW Africa (Lutze 1980) living, dead

Additional common species:

Bulimina marginata	1a, b, c
Hippocrepina pusilla	1a
Reophax subfusiformis	1a
Ammonia beccarii	1a (as *A. batavus*), 4, 5, 6
Trifarina angulosa	1b
Hyalinea balthica	1b, c (as *Anomalina*)
Cassidulina obtusa	1b
Epistominella vitrea	1b
Reophax scottii	1b
Nonionella sp.	1b
Nonionella turgida	2
Adercotryma glomerata	2
Gavelinopsis praegeri	2
Textularia sagittula	2
Bulimina gibba/elongata	3, 4, 6
Eggerelloides scabrus	4 (as *Eggerella*), 6
Bolivinellina pseudopunctata	6 (as *Brizalina*)
Brizalina variabilis	7 (as *Bolivina*)
Bolivina pacifica	7

***Gaudryina rudis* association**

Salinity: 35–36.5‰

Temperature: 7 to 14 °C

Substrate: biogenic sand and gravel

Depth: 62–140 m

Distribution:
1. W. Scotland (Murray 1985) dead
2. North Minch (Edwards 1982) total
3. W. English Channel (Murray 1970c) dead
4. N. Biscay (Schnitker 1969) total
5. Biscay (Caralp *et al.* 1970) total (as *G. pseudoturris*)
6. NW Africa (Haake 1980a) dead

Additional common species:

Eponides repandus	1
Cibicides lobatulus	1, 2, 3, 4, 5
Cibicides refulgens	1
Textularia sagittula	3, 5
Cassidulina laevigata	4, 6 (as *C. carinata*)
Cibicidoides pseudoungerianus	4, 5 (as *Cibicides*)
Quinqueloculina seminula	4
Spiroplectinella wrightii	4 (as *Spiroplectammina*)
Cibicides cf. *ungerianus*	5
Textularia pseudogramen	6
Trifarina fornasinii	6

***Hyalinea balthica* association**

Salinity: 35‰

Temperature: 9 to 13 °C

Substrate: sand, muddy sand

Depth: 100–400 m

Distribution:
1. North Sea (Jarke 1961b) total (as *Anomalina*)
2. North Minch (Edwards 1982) total
3. Celtic Sea (Murray 1970c, 1979a) living, dead
4. N. Biscay (Rouvillois 1970) total
5. Galicia (Colom 1974) total

Additional common species:

Bulimina marginata	2, 4
Fursenkoina fusiformis	4
Nonionella turgida	4
Cassidulina laevigata	4

Table 9.9 *(continued)*

***Quinqueloculina seminula* association**

Salinity: 34–36.5‰

Temperature: 8 to 18 °C

Substrate: sand

Depth: 20–120 m

Distribution:
1. Bristol Channel (Murray 1970c) dead
2. S. Cornwall (Murray 1970c) dead
3. N. Biscay: (a) (Dupeuble *et al.* 1971) living, dead; (b) (Rosset-Moulinier 1972) dead
4. N. Biscay (Schnitker 1969) total
5. Biscay (Pujos 1972) total
6. Morocco (Mathieu 1971) total

Additional common species:

Ammonia beccarii	1, 2, 3a, b, 5
Eggerelloides scabrus	1 (as *Eggerella*)
Textularia sagittula	1, 2, 5
Quinqueloculina dunkerquiana	3a
Quinqueloculina lata	3b
Cibicides lobatulus	3b, 4, 6
Cassidulina laevigata	4
Cibicidoides pseudoungerianus	4 (as *Cibicides*)
Gaudryina rudis	4

***Reophax fusiformis* association**

Salinity: 35‰

Temperature: (1) –1, (2) and (3) 7 to 13, (4) 4 °C

Substrate: mud, muddy sand

Depth: (1) 2702–3260, (2) 100, (3) 40–50, (4) 2032 m

Distribution:
1. Norway (Mackensen *et al.* 1985) dead
2. Celtic Sea (Murray 1979a) living
3. Plymouth (Murray 1965c) living
4. NW Africa (Lutze 1980) living

Additional common species:

Fissurina semimarginata	1
Brizalina skagerrakensis	1 (as *Bolivina* cf. *B. robusta*)
Reophax scorpiurus	1
Haplophragmoides bradyi	1
Bulimina marginata	2
Nonionella turgida	2
Triloculina spp.	3
Fursenkoina fusiformis	3
Ammonia beccarii	3

***Textularia sagittula* association**

Salinity: 35–36.5‰

Temperature: 7 to 16 °C

Substrate: sand

Depth: 52–145 m

Distribution:
1. W. Scotland (Murray 1985) dead
2. Celtic Sea (Murray 1979a) living, dead
3. W. English Channel (Murray 1970c) dead
4. Plymouth (Murray 1965c) dead
5. NW Africa (Haake 1980a) dead

Additional common species:

Cibicides lobatulus	1, 2, 3, 4
Cassidulina laevigata	1
Cibicides refulgens	1
Trifarina angulosa	1
Fursenkoina fusiformis	2
Gavelinopsis praegeri	2, 4
Quinqueloculina seminula	2
Gaudryina rudis	3
Ammonia beccarii	4
Textularia pseudogramen	5
Spiroloculina angulosa	5
Bulimina gibba	5
'Nonion asterizans'	5

Table 9.9 *(continued)*

***Trifarina angulosa* association**

Salinity: 35‰

Temperature: –1 to 10 °C

Substrate: gravel, sand, silty mud

Depth: 128–801 m

Distribution:
1. Barents Sea (Jarke 1960) total
2. Møre, Norway: (a) (Sejrup *et al.* 1981) total; (b) (Mackensen *et al.* 1985) living, dead
3. Norway (Qvale 1986a) total
4. Norwegian–Greenland Sea (Belanger and Streeter 1980) total
5. Porcupine Seabight (Weston 1985) dead
6. Western Approaches (Weston 1985) dead
7. Biscay (Caralp *et al.* 1970) total
8. NW Africa (Lutze 1980) dead

Additional common species:

Elphidium clavatum	1
Cibicides lobatulus	1a, b, 2
Uvigerina peregrina	1a, 2
Cassidulina laevigata	2a, b, 3, 5 (as *C. carinata*)
Cassidulina obtusa	2a, b, 3
Textularia sagittula	2a
Melonis barleeanum	2a (as *Nonion*), 2b (as *M. zaadami*), 3, 4
Elphidium excavatum	2a, 6
Astrononion gallowayi	2b
Robertinoides normani	2b
Bulimina marginata	2b, 5
Cibicides refulgens	3
Cibicides spp.	3
Cassidulina teretis	4
Recurvoides turbinatus	4
Globocassidulina subglobosa	5
Cassidulina obtusa	7 (as *C. crassa*)
Cassidulina crassa	8

Table 9.10 Principal slope associations of the Atlantic seaboard of Europe and Africa

***Brizalina skagerrakensis* association**

Salinity: 3–35‰

Temperature: –0.81 to > 6 °C

Substrate: mud with 2–3‰ organic carbon

Depth: (1) 3490, (2) 75–700 m

Distribution:
1. Møre, Norway (Mackensen *et al.* 1985) dead (as *Bolivina* cf. *robusta*)
2. Skagerrak (Qvale and Van Weering 1985) total (as *Bolivina*)

Additional common species:

Species	
Reophax fusiformis	1
Bulimina marginata	2
Hyalinea balthica	2
Cassidulina laevigata	2

***Epistominella exigua* association**

Salinity: 34.90–34.97‰ = NEADW

Temperature: (1) –1, (2) 3 to 4, (4b) 3 to 10 °C

Substrate: mud

Depth: 501–7500 m

Distribution:
1. Møre, Norway: (a) (Sejrup *et al.* 1981) total; (b) (Mackensen *et al.* 1985) living
2. NE Atlantic (Weston and Murray 1984) total
3. Biscay: (a) (Caralp *et al.* 1970) total; (b) (Pujos-Lamy 1973a) total
4. NW Africa: (a) (Phleger *et al.* 1953) total; (b) (Lutze 1980) living, dead
5. SW of Azores (Hermelin and Scott 1985) total
6. Sierra Leone Rise (Sen Gupta *et al.* 1987) total

Additional common species:

Species	
Fontbotia wuellerstorfi	1a (as *Cibicides*), 1b (as *Cibicidoides*), 2, 3a, b, 4a (as *Planulina*)
Oridorsalis umbonatus	1b, 2
Triloculina frigida	1b
Cassidulina teretis	2
Nuttallides rugosus	2 (as *Osangularia*)
Bolivina aff. *B. thalmanni*	2
Hoeglundina elegans	2
Reophax spp.	3
Pullenia bulloides	4a
Nuttallides umboniferus	4a (as *Epistominella*), 6
Ammoglobigerina globigeriniformis	4b (as *Trochammina*)
Eilohedra levicula	4b (as *Epistominella*)
Trifarina bradyi	4b
Globocassidulina subglobosa	4b, 6 (as *Cassidulina*)
Pullenia quinqueloba	4b
Planulina ariminensis	5
Uvigerina auberiana	5
Bolivina pseudoplicata	5
Eponides turgidus	6

***Fontbotia wuellerstorfi* association**

Salinity: 34.9–35.0‰ = NADW

Temperature: (1a) –2, (3b) 2.5 to 4 °C

Substrate: mud

Depth: 1360–4280 m

Distribution:
1. Møre, Norway: (a) (Sejrup *et al.* 1981) total (as *Cibicides*); (b) (Mackensen *et al.* 1985) living (as *Cibicidoides*)
2. Norwegian–Greenland Sea (Belanger and Streeter 1980) total (as *Cibicidoides*)
3. NE Atlantic: (a) (Phleger *et al.* 1953) total; (b) (Weston and Murray 1984) total (both as *Planulina*)

Additional common species:

Species	
Epistominella exigua	1a, b, 2, 3a, b
Elphidium clavatum	1a (as *E. excavatum*)
Oridorsalis umbonatus	1a (as *Eponides*), 1b, 2, (as *O. tener*), 3b
Bulimina marginata	1a
Triloculina frigida	1b
Hormosinella distans	1b (as *Reophax*)
Bulimina alazanensis	3a
Pullenia quinqueloba	3a
Globocassidulina subglobosa	3b
Cibicidoides kullenbergi	3b

Table 9.10 *(continued)*

***Globocassidulina subglobosa* association**

Salinity: >35‰ = Mediterranean Water

Temperature: 4 to 9 °C

Substrate: mud

Depth: 377–4095 m

Distribution:
1. Western Approaches (Weston 1985) dead
2. Portugal (Seiler 1975) dead
3. NE Atlantic (Weston and Murray 1984) total
4. NW Africa (Lutze 1980) dead (as *Cassidulina*)

Additional common species:

Cassidulina laevigata	2 (as *C. carinata*)
Cassidulina obtusa	2
Cassidulina crassa	4

***Melonis barleeanum* association**

Salinity: 34.93–35.6‰

Temperature: –0.4 to 9 °C

Substrate: silty mud

Depth: 280–2710 m

Distribution:
1. Barents Sea (Jarke 1960) total (as *Nonion zaandamae*)
2. Møre, Norway: (a) (Qvale 1986a) total; (b) (Mackensen *et al.* 1985) living (as *M. zaandami*)
3. Norwegian–Greenland Sea (Belanger and Streeter 1980) total
4. Western Approaches (Weston 1985) dead
5. Portugal (Seiler 1975) living
6. NW Africa (Lutze 1980) living (as *M. zaandami*)

Additional common species:

Cibicidoides pseudoungerianus	1 (as *Cibicides*)
Buccella frigida	1 (as *Eponides*)
Islandiella islandica	1 (as *Cassidulina*)
Marsupulina schultzei	1
Cassidulina laevigata	2a, b
Cassidulina reniforme	2a
Pullenia bulloides	2b, 3
Trifarina angulosa	2b
Cribrostomoides subglobosum	2b
Cibicides lobatulus	1, 2b
Recurvoides turbinatus	3
Islandiella norcrossi	1 (as *Cassidulina*), 3
Cassidulina teretis	3
Nuttallides rugosus	4 (as *Osangularia*)
Rhabdamminella cylindrica	5 (as *Marsipella*)
Globobulimina spp.	6

***Nuttallides umboniferus* association**

Salinity: <34.9‰ = AABW

Temperature: <2.0 °C

Substrate: mud

Depth: 4001–5502 m

Distribution:
1. N. Atlantic (Phleger *et al.* 1953) total (as *Epistominella*)
2. NE Atlantic (Weston and Murray 1984) total (as *Osangularia*)
3. Sierra Leone Rise (Sen Gupta *et al.* 1987)

Additional common species:

Epistominella exigua	1, 2, 3
Fontbotia wuellerstorfi	1 (as *Planulina*)
Globocassidulina subglobosa	1 (as *Cassidulina*), 2
Pullenia bulloides	1, 2
Oridorsalis umbonatus	2, 3
Eponides turgidus	3

***Oridorsalis umbonatus* association**

Salinity: 34.84–35‰

Temperature: (1) –0.8, (3) 2.5 to 4.0 °C

Substrate: mud

Depth: 1734–3877 m

Distribution:
1. Møre, Norway (Mackensen *et al.* 1985) living
2. Norwegian–Greenland Sea (Belanger and Streeter 1980) total (as *O. tener*)
3. NE Atlantic (Weston and Murray 1984) total

Additional common species:

Epistominella exigua	1, 2
Eponides tumidulus	1, 2
Triloculina frigida	1
Recurvoides turbinatus	2
Fontbotia wuellerstorfi	2 (as *Cibicidoides*), 3 (as *Planulina*)
Globocassidulina subglobosa	3

Table 9.10 *(continued)*

***Uvigerina peregrina* association**

Salinity: 34.9–35.1‰

Temperature: 6 to 15 °C

Substrate: sand, mud rich in organic carbon

Depth: 201–3000 m

Distribution:

1. Møre, Norway: (a) (Sejrup *et al.* 1981) total; (b) (Mackensen *et al.* 1985) living (as var. *finisterrensis*)
2. Norway (Qvale 1986a) total
3. Norwegian Channel (Qvale and Van Weering 1985) total
4. Biscay (Caralp *et al.* 1970) total
5. Portugal (Seiler 1975) living
6. NW Africa: (a) (Lutze and Coulbourn 1984) living; (b) (Lutze 1980) living, dead
7. Niger Delta (Adegoke *et al.* 1976) total (as *Eouvigerina*)
8. Lüderitz, Namibia (Martin 1981) total

Additional common species:

Cassidulina laevigata	1a, b, 2, 3, 6b, 8 (as *C. carinata*)
Cibicides lobatulus	1a
Trifarina angulosa	1a, b, 2, 3
Melonis barleeanum	1b (as *M. zaandami*)
Textularia saggittula	1b
Hyalinea balthica	2
Bulimina marginata	2, 3, 7, 8
Uvigerina peregrina dirupta	4
Cassidulina obtusa	5
Hoeglundina elegans	5, 6b
Trifarina fornasinii	5
Bolivina dilatata	6b
Cancris auricula	6b
Bulimina striata	6b
Cassidulinoides bradyi	8
Bulimina cf. *alazanensis*	8
Praeglobobulimina spinescens	8
Cibicidoides pseudoungerianus	8 (as *Cibicides*)
Cibicides praecinctus	8

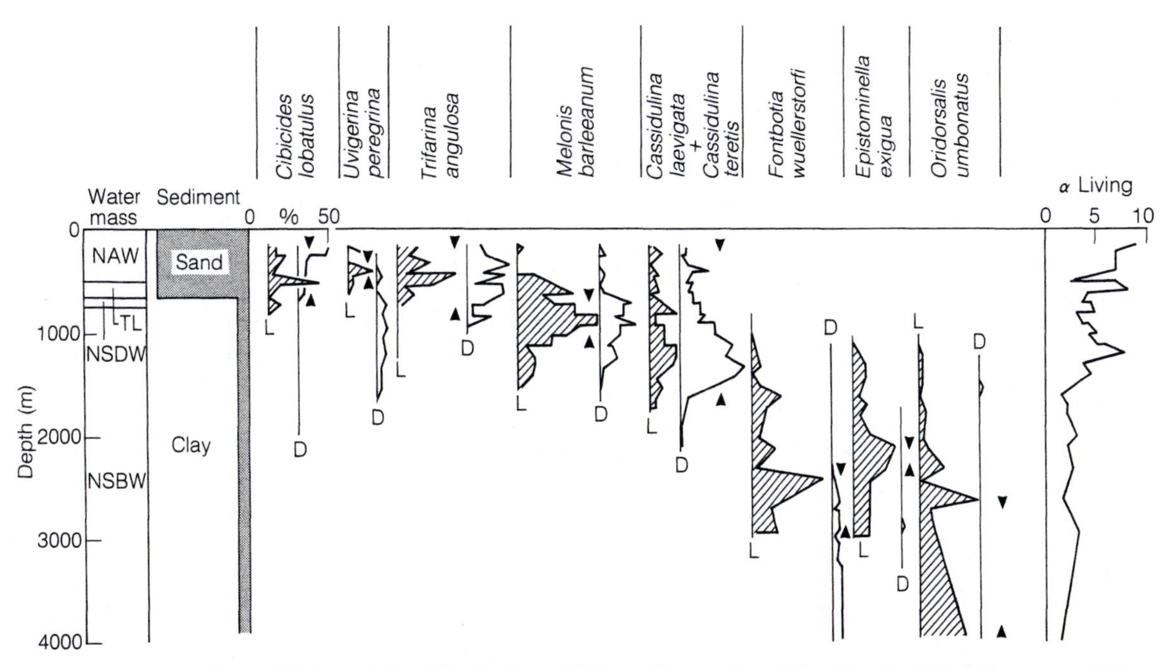

Figure 9.3 Comparison of living and dead distributions off Møre, Norway. L = living, D = dead, limits between which each species is dominant as an association data from Mackensen *et al.* 1985)

The sediments are predominantly terrigenous down to ~1200 m and pelagic below this. The terrigenous sediments range from gravels and sands to clays depending on the current regime and topography. Their thickness is highly variable from almost zero to >3 m. The pelagic sediments have >30% $CaCO_3$ and this is composed of planktonic foraminifera. The terrigenous component consists of clay and silt. The Holocene pelagic sediments (i.e. last 10 000 years) are usually less than 4 cm thick (Sejrup *et al.* 1981).

Living and dead tests were differentiated in traverses off Møre, Norway (Mackensen *et al.* 1985). Living *Cibicides lobatulus*, *Uvigerina peregrina* and *Trifarina angulosa* correlate well with sandy substrates and NAW (Fig. 9.3). The dead distributions of *C. lobatulus* and *U. peregrina* are considerably extended downslope, but that of *T. angulosa* agrees well with the living. Mackensen *et al.* (1985) correlated the occurrence of *Melonis barleeanum* (as *M. zaandami*) with muddy sediments containing 15–30% $CaCO_3$ on the slope at depths of 600–1200 m. *Cassidulina laevigata* and *teretis* were not separated. They are more abundant dead than living but their depth distributions are similar. Norwegian Sea Bottom Water is characterised by *Fontbotia wuellerstorfi*, *Epistominella exigua*, *Oridorsalis umbonatus* (Fig. 9.3) and, in its upper part, by *Cribrostomoides subglobosum* (Table 9.7). There are significant abundance and depth differences in the distribution of *F. wuellerstorfi* but the reasons for this are not known. *Elphidium excavatum* and *Cassidulina reniforme* in the dead assemblages are considered to be reworked from glacial marine sediments. However, *C. reniforme* lives in low abundance between ~500 and ~1300 m. It was considered to be an Arctic species by Sejrup and Guibault (1980).

The diversity of the living assemblages is highest at depths of <1200 m (α 2–9) and consistently low at greater depths (α 1.5–4.5, Fig. 9.3). Values lower than 5 indicate an ecologically stressed environment. As the area is noted for its stable water stratification, environmental variability, as in shallow waters, cannot be the cause. Perhaps the very low water temperatures (<0 °C) are important in this respect.

Studies of total assemblages include Sejrup *et al.* (1981), Qvale (1981, 1986a), and Hiltermann (1984). Corliss and Chen (1988) reanalysed the data of Mackensen *et al.* (1985) in terms of morphotypes and mode of life. Infaunal morphotypes peak at 500–1000 m where the sediments are richer in mud and organic carbon.

Norwegian Channel and Skagerrak

The Norwegian Channel is dominated by the *Uvigerina peregrina* association at depths of 240–430 m.

The sediment is mud with >1% organic carbon. *Uvigerina peregrina* is infaunal/epifaunal, free, and the overlying waters are possibly somewhat depleted in oxygen although definitely not anoxic (Qvale and Van Weering 1985). Along the eastern margin of the channel, on sandy substrates under the inflowing Atlantic water, is the *Cassidulina laevigata* association. It is also present in the southern Skagerrak along the boundary between the variable surface waters and the stable deeper water.

The principal association of the Skagerrak is that of *Brizalina skagerrakensis*. It is present on muddy substrates with 1 to >3% organic carbon at depths of 75–700 m. Most of its range is in the stable bottom water (salinity 33–35‰, temperature 6–7 °C, well oxygenated) (Van Weering and Qvale 1983). Close to the Swedish shore the *Bulimina marginata* association is present between 70 and 99 m in waters of variable salinity. Qvale *et al.* (1984) considered that it might be able to tolerate high concentrations of organic matter and low oxygen content of the water. However, there is no correlation between this species and the organic content of the sediment (Qvale and Van Weering 1985). This association is common in adjacent fjords as discussed above. Near the Danish coast, the sandy sediments at depths of <32 m support and *Elphidium clavatum* association. The water is brackish (salinity 10–30‰, Hansen 1965).

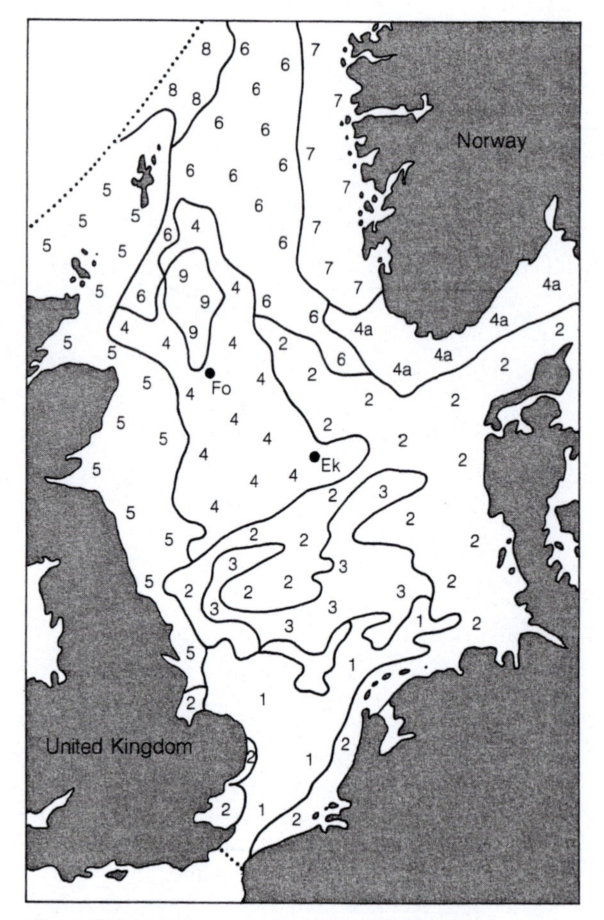

Figure 9.4 North Sea foraminiferal associations (modified from Jarke 1961b). Ek = Ekofisk, Fo = Forties. (1) *Ammonia beccarii* with *Quinequeloculina seminula*; (2) *Elphidium excavatum*; (3) *Eggerelloides scabrus*, (4) *Fursenkoina fusiformis*; (4a) *F. fusiformis* Jarke, based on > 63 μm); *Brizalina skagerrakensis* (Qvale and Van Weering 1985, based on > 125 μm); (5) *Cibicides lobatulus* with *Planorbulina mediterranensis*; (6) *Cassidulina laevigata*; (7) *Uvigerina peregrina*; (8) *Islandiella islandica*; (9) *Hyalinea balthica*

North Sea

In a general study of North Sea total assemblages, Jarke (1961b) recognised 12 facies. His co-worker Gabel (1971) published the same distribution map, illustrated many more species and updated the taxonomy. Uncertainties over the identification of the elphidiids and nonionids makes it difficult to relate these data to the associations discussed here, but the overall pattern has been summarised in Fig. 9.4.

Living assemblages from the coastal region off Germany are either *Elphidium excavatum* or *Eggerelloides scabrus* associations with additional *Fursekoina fusiformis* and *Ammonia beccarii* (Richter 1967).

In the central latitudes of the North Sea the *Fursenkoina fusiformis* association is widespread. *Fursenkoina fusiformis* commonly makes up >80% of the living assemblages. From experimental tank observations, Richter (1967) concluded that it lived just beneath the sediment surface. However, from cores taken off Northumberland, Collison (1980) recovered living individuals down to the limit of sampling (11 cm). Whereas it represented only 21% of the assemblage in the top 1 cm, it formed 57% of the overall living assemblage in 11 cm. It is possible that it lives in open burrows made by the macrofauna. In the Forties and Ekofisk areas *F. fusiformis* associations have low diversity (α 1.5–5) but for the dead the range is α 3.5–8 (Murray 1985). These areas lie within Jarke's (1961b) group IV characterised by '*Bulimina fusiformis*, *B. marginata* and *Anomalina balthica*' (= *Hyalinea balthica*). This part of the North

Sea develops thermohaline stratification during the summer so the waters at the bottom are cool throughout the year.

Biologists carrying out macrofaunal surveys have reported certain larger agglutinated foraminifera from size fractions larger than 1.3–1.5 mm. For example, *Astrorhiza arenarea*, *Saccammina spherica* and *Psammosphaera fusca* at depths >100 m (Stephens 1923), *S. spherica*, *Cribrostomoides crassimargo* (as *Alveolophragmium*) and *Reophax scorpiurus* on Fladen Ground (McIntyre 1961), and *Astrorhiza limicola* off Northumberland (Buchanan and Hedley 1960).

The flexible tests of *Reophax scottii* were present in plankton samples taken at 10 m below the water surface during the winter months at a time of vertical mixing (John 1987).

West Scotland

The Fair Isle Channel, between Shetland and Scotland and the western shelf of Scotland, are areas of temperate carbonate accumulation (biogenic sands; Wilson 1979, 1982) under vigorous wave activity. Fair Isle Channel is also influenced by powerful tidal currents and here the *Cibicides lobatulus* association is present living on hydroids. On the shelf west of Scotland it is present living on bivalve shells and the tubes of *Ditrupa*. A living *Rosalina anomala* association occurs on hydroids. The sediment dead assemblages fall into the *Cibicides lobatulus*, *Eponides repandus*, *Gaudryina rudis* and *Textularia sagittula* associations. All these species are epifaunal and live attached to firm substrates. In additional *Acervulina inhaerens* commonly encrusts bivalve shells (Table 9.7). The diversity of the living assemblages is low (α 1–2) (Murray 1985). Similar total assemblages were described from North Minch Channel coarse sands and gravels. In deeper less disturbed areas with muddy sediments, the *Bulimina marginata* and *Hyalinea balthica* associations are present (Edwards 1982).

Irish Sea

Cardigan Bay in the Irish Sea is <50 m deep and has a salinity of 34.5–35‰. Phytal and attached species have been reported from intertidal rock pools (Atkinson 1969). The *Elphidium crispum* association is present together with a *Massilina secans* association (on *Laminaria*) with *Ammonia beccarii*, *E. crispum*, *Cibicides lobatulus*, *Asterigerinata mamilla* (as *Eoeponidella*) and *Bulimina gibba*. Subtidal epifaunal occurrences on hydroids and shells of *Chlamys* include the *Cibicides lobatulus* (with *Rosalina anomala*), *Miliolinella subrotunda* and *Cribrostomoides jeffreysii* associations (Dobson and Haynes 1973, Haward and Haynes 1976). The subtidal sediments have an *Ammonia beccarii* association (Haman 1971, Atkinson 1971).

Celtic Sea and English Channel

The Celtic Sea is stratified from April until December with a thermocline between 30 and 50 m. The bottom waters experience only a small seasonal temperature variation, from 9 to 13 °C. Because of its exposed position, it is subject to large waves (up to 21 m) and oscillatory currents disturb the sea bed at water depths of up to 180 m. The sediments range from coarse sand and gravel on banks, through poorly sorted muddy fine sands to very muddy, silty fine sand in the deeper parts.

Both the living and dead assemblages consist of mixtures of agglutinated and hyaline forms with only a few samples having >10% porcellaneous tests. Diversity is moderate, α 4–12 for the living and α 4–15 for the dead assemblages. Standing crop values are low to moderate, 10–550 per 10 cm^2, and show an inverse relationship with sediment grain size.

Major associations found only as living include (Table 9.9) *Bulimina marginata*, *Eggerelloides scabrus* and *Reophax fusiformis* and minor associations *Cancris auricula* and *Adercotryma glomerata*. Major associations present both living and dead are *Fursenkoina fusiformis*, *Hyalinea balthica* and *Textularia sagittula* plus minor *Epistominella vitrea* and *Nonionella turgida*. Major associations present only as dead assemblages are *Cassidulina obtusa*, *Cibicides lobatulus*, *Elphidium excavatum* and, in the Bristol Channel, *Quinqueloculina seminula* (data from Murray 1970c, 1979a).

At depths shallower than ~90 m, the sediments are mobile sands influenced by storm waves and possibly by tidal currents. This may be less hospitable to epifaunal and semi-infaunal free-living species, certainly during periods of sea-floor disturbance. Epifaunal clinging (*Textularia sagittula*) and epifaunal attached (*Cibicides lobatulus*) species may be better adapted for these conditions, although they may also flourish under more stable regimes. Free-living taxa are common in the muddy fine sands of the lower-energy areas (Sturrock and Murray 1981).

The English Channel and its Western Approaches are meso- to macrotidal. There are no major rivers flowing into these seas so the bottom waters have salinities of 34.8–35‰ throughout the year. Only in the western English Channel and the Western Approaches is a thermocline developed during the summer months. Elsewhere the powerful tidal currents mix the waters from top to bottom. Thus, most of the area is a high-energy environment with coarse sand to gravel sediment, some of it biogenic (mainly bivalve shells with some bryozoa). Finer-grained sediments containing mud are confined to sheltered bays, mainly along the English coast.

General summaries of the faunas (Rosset-Moulinier 1981, 1986, 1988) show that most of the English Channel is occupied by a *Cibicides lobatulus–Textularia truncata* assemblage (*T. truncata* is here included within a broad *T. sagittula* group). Associated species from west to east are *Ammoglobigerina globigeriniformis* var. *pygmaea* (as *Trochammina*), *Ammonia beccarii* and *C. pseudoungerianus* (as *Heterolepa*). In the coastal regions there are several different assemblages including those dominated by *Lepidodeuterammina ochracea*, *Remaneica plicata*, *Eggerelloides scabrus* (as *Eggerella*), and *Nonion depressulus*. However, insufficient data were given to place these in the associations recognised in this book.

The results of more detailed studies of restricted areas are summarised in Tables 9.7 and 9.9. The living assemblages from the mid shelf (48–95 m) of the western English Channel are the *Ammoglobigerina globigeriniformis* var. *pygmaea* association (as *Trochammina*) with *Cribrostomoides jeffreysii*, *Gavelinopsis praegeri*, *Spirillina vivipara* and *Lepidodeuterammina ochracea* (as *Trochammina*). All these species are known to be epifaunal, clinging or attached (Murray 1970c). Further samples showed epifaunal attached ('attached immobile' of Sturrock and Murray 1981) species such as *Cibicides lobatulus*, *Acervulina inhaerens*, *Placopsilina confusa*, *Nubecularia* sp. and *Remaneica* sp. on shells, hydroids and pebbles. In addition, *Rosalina anomala* and *Textularia sagittula* are epifaunal, clinging on hydroids and shells respectively. This area is swept by powerful tidal currents and the sea floor consists of a shell pavement with a thin, discontinuous cover of biogenic sand which is formed into ripples. Television surveys have shown that the hydroids are attached to larger shells and pebbles and that the ripples change their orientation with the tide. In this highly mobile environment the microfauna needs to hold on to something firm. However, on death small tests are transported away so the dead assemblages are the *T. sagittula* association (= *C. lobatulus–T. truncata* assemblage of Rosset-Moulinier 1986) or *Gaundryina rudis* association.

In the coastal regions of southwest England, living and dead assemblages of the *Ammonia beccarii* and *Cibicides lobatulus* associations are found on sandy substrates together with living *Eggerelloides scabrus*, *Nonion depressulus* and dead *Elphidium excavatum*, *Quinqueloculina seminula* and *Asterigerinata mamilla* associations. Sediments with some mud support the *Fursenkoina fusiformis* association. With the exception of *C. lobatulus*, all these species are free-living.

Along the French coast there are living and dead *Ammonia beccarii*, *Nonion depressulus* and *Quinqueloculina seminula* associations together with dead *Elphidium crispum*, *Cibicides lobatulus* and *Asterigerinata mamilla* associations (Moulinier 1967, Dupeuble *et al.* 1971, Rosset-Moulinier 1972, Rosset-Moulinier and Roux 1977).

Bay of Biscay – Portugal

The continental shelf of the Bay of Biscay slopes gently seaward as far as the 100 m isobath, and between there and the top of the slope at ~200 m the topography is more irregular with relief of ~20 m. The slope is dissected by submarine canyons, the largest being the Gouf de Cap Breton in the southern part. Bottom salinities are 35‰ and the bottom temperature below the thermocline is in the range 10–12 °C. In the northern part, the sediments are sands and gravels down to ~90 m, a mud zone between 90 and 100 m and sands with very little mud >110 m (Rouvillois 1970). In the southern part, sand and fine sand span the whole shelf except off the Gironde (Pujos 1972).

Three major associations are widely developed. The *Quinqueloculina seminula* association characterises the inner shelf down to ~60 m where it gives way to the *Cassidulina laevigata* association which extends down the slope to 2500 m (Caralp *et al.* 1970). On the outer shelf and slope (130–3200 m) the latter overlaps with the *C. obtusa* association. The *Bulimina marginata* and *Hyalinea balthica* associations are developed over a short depth range on the northern mud zone (Rouvillois 1970). The *Cibicides lobatulus* association is present from 135 to 140 m in the north (Schnitker 1969).

The western seaboard of Spain and Portugal has been studied by Braga (1942), Van Voorthuysen (1973) and Colom (1974), total assemblages, and Seiler (1975), living and dead assemblages. Shallow-water associations include *Ammonia beccarii*, *Eggerelloides scabrus* and *E. crispum* with *Hyalinea balthica* at 100–400 m (all based on total assemblages). Living assemblages are represented by the *Cibicides lobatulus* (45 m, as *Cibicides* spp.), *Brizalina striatula* (119–146 m, as *Bolivina*) and *Uvigerina peregrina* (182 m) associations. Dead assemblages fall into the *Bulimina gibba/elongata* (45 m), *Cassidulina laevigata* (146–1730 m, as *C. carinata*) and *Cibicides lobatulus* (119 m, as *Cibicides* spp.) associations.

Off the Algarve, Galhano (1963) distinguished an inner shelf zone down to 20 m with sand and shell debris having an assemblage rich in miliolids (mainly *Quinqueloculina seminula*) and *Elphidium* spp. and a mud zone >20 m with *Nonionoides boueanum* (as *Nonion*) and *Uvigerina peregrina bradyana*. In the nearshore zone, miliolids are more abundant off Portugal than off Spain.

Some comments of the distribution of 276 species around Madeira are given by Braga and Galhano (1965). This area represents the most northerly occurrence of peneroplids (*Peneroplis pertusus* as *P. carinatus* and *P. planatus* as *P. proteus*) and of *Amphistergina lessoni*.

West Africa

Off Morocco, the *Quinqueloculina seminula* association is present on the inner shelf sands and the *Cassidulina laevigata* association on mid-shelf muds (Mathieu 1971, 1988). *Cibicides lobatulus* is present as an accessary species down to 80 m. The water is saline (salinity 36‰) but the temperatures are modest (13–18 °C) due to coastal upwelling.

The middle and outer shelf off central and southern Morocco is a region of biogenic carbonate sediments with a high abundance of *Miniacina miniacea* at depths of 95–115 m. Milliman (1976) argued that these sediments must be forming now, but the common presence of associated coralline algae suggests accumulation at a lower sea-level.

The shelf off Senegal and Gambia has three associations based on the >250 μm fraction (Lutze and Coulbourn 1984), *Cibicides lobatulus*, *Cancris auricula* and *Globobulimina* spp., and each is represented by both living and dead assemblages. The first two are present on sand while the third is on mud with a low organic carbon content. However, Haake (1980a) studied the >125 μm fraction and found eight living associations none of which is distributed elsewhere (see Table 9.7). Major associations represented by dead assemblages are *Cassidulina laevigata* (as *C. carinata*, 148 m), *Textularia sagittula* (52 m) and *Gaudryina rudis* (as *Dorothia*, 140 m). Lutze (1980), on the basis of the >63 μm fraction, recorded living and dead *Fursenkoina fusiformis*, living *Cassidulina laevigata* and dead *Cibicides lobatulus* associations.

Off the Ivory Coast, at depths of 40–80 m, the dominant forms are large agglutinated *Jullienella foetida* and *Schizammina* spp. (Buchanan 1960; Le Calvez 1963; Berthois and Le Calvez 1966). Otherwise shelf faunas are diverse but not readily grouped into associations on the limited data available (Mathieu 1970).

In the northern Gulf of Guinea, of 13 total assemblages described by Basov (1976) only 2 show dominance of one species – *Trifarina carinata* (as *Angulogerina*, 140, 225 m). The Niger Delta shows a progression from *Hanzawaia concentrica* and *Rectuvigerina nicoli* associations on the inner shelf to *Uvigerina peregrina* association from mid shelf to bathyal (Adegoke *et al.* 1976). Off Cameroon, the *Ammonia beccarii* association is present in the slightly brackish coastal waters while from 50 to 100 m there is an *Amphicoryna scalaris* (as *Nodosaria*) association (Berthois *et al.* 1968). On muddy sediments in the seasonally variable environmental conditions of the mouth of the Congo, Zaire, there are '*Nonion asterizans*' and *Cancris congolensis* associations (Kouyoumontzakis 1982a). The latter replaces the normal inner shelf *Textularia panamensis* association present beneath the turbid waters of the Zaire shelf (Kouyoumontzakis 1981, 1982b).

Larsen (1982) recorded the depth and latitudinal distribution of 233 benthic species from the seaboard between Senegal and Angola but, as he gave no abundance data, it has been impossible to recognise associations. Similarly, Mikhalevich (1983) carried out a mainly systematic review of tropical African faunas.

Between Luderitz and the Orange River the total assemblages include the *Ammonia beccarii*, *Bulimina marginata* and *Cassidulina laevigata* associations together with a number of minor associations. However, some assemblages contain many relict or reworked individuals characterised by infillings of glauconite or phosphorite or by damage to the test. These occurrences have been omitted from Table 9.9.

Geographical distribution of shelf associations

The geographical distribution of major shelf associations (Fig. 9.5) shows only one that is confined to the northernmost latitudes, namely *Elphidium clavatum*. Two are more or less ubiquitous: *Cibicides lobatulus* and *Cassidulina laevigata*. The remainder show varying degrees of restriction. The only porcellaneous-dominated association (*Quinqueloculina seminula*) does not extend north of the Celtic Sea.

The total depth range of the major associations along the European seaboard is given in Fig. 9.6 but, of course, the ranges of individual species are greater. There are some notable differences in the depth occurrence of the same association in different areas (Fig. 9.7). For instance, the *Cibicides lobatulus* association has the following ranges: Barents Sea–Norway 10–499 m, Iceland–Faeroe Ridge 400–900 m, Celtic Sea 100–104 m, English Channel 13–150 m, Bay of Biscay 135–148 m. Although the salinity in each is ~35‰, the temperature varies from 0–3 °C in the north to 8–18 °C off Brittany. As already discussed, this association is substrate controlled.

It seems likely that substrate is the dominant control for the following associations: *Fursenkoina fusiformis* (muddy sand), *Gaudryina rudis* and *Textularia sagittula* (shell debris–biogenic sediment). Substrate together with a temperature critical for reproduction seems to control the following associations: *Ammonia beccarii* (sand, ~15 °C for reproduc-

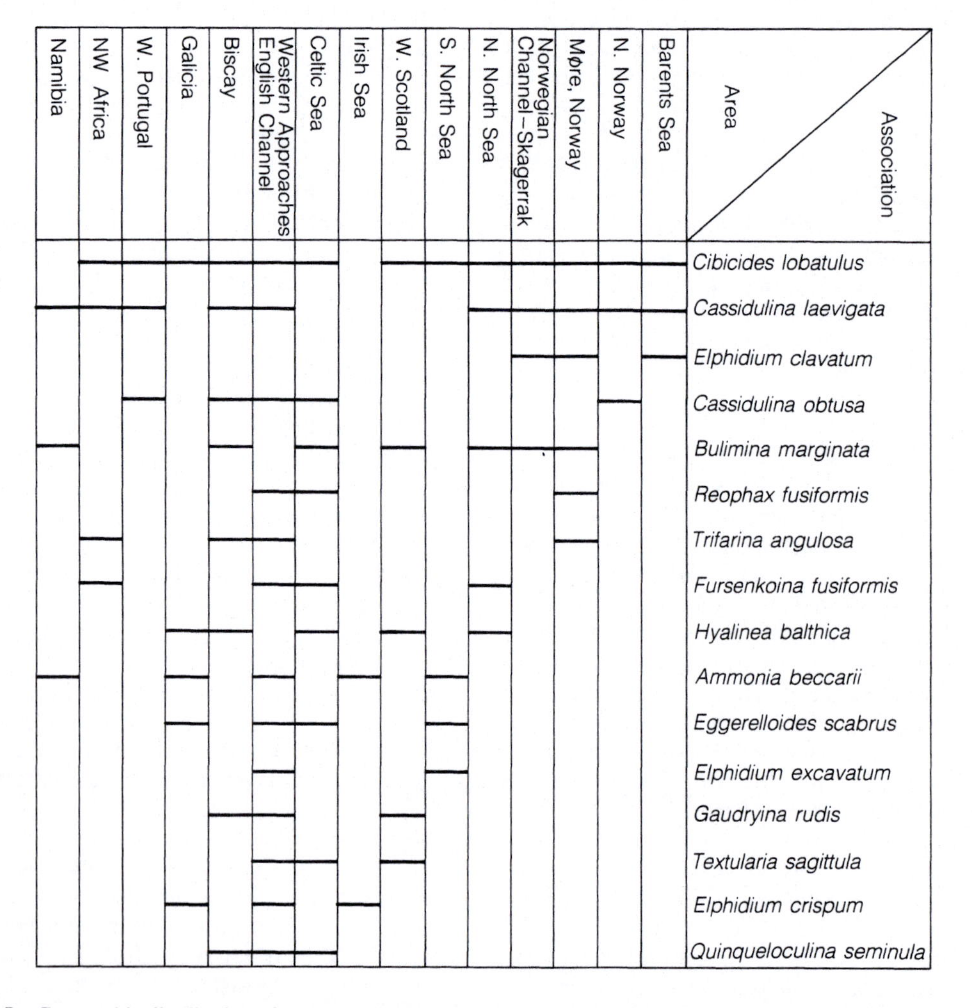

Figure 9.5 Geographic distribution of major shelf and upper slope associations along the Atlantic seaboard of Europe and north Africa

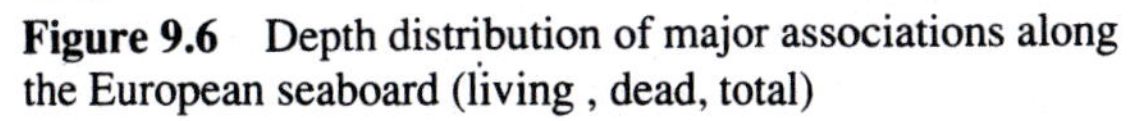

Figure 9.6 Depth distribution of major associations along the European seaboard (living , dead, total)

tion), *Eggerelloides scabrus* (sand, ~13 °C for reproduction), *Elphidium clavatum* (sand, maximum temperature 7 °C ?). *Elphidium excavatum* (sand, minimum temperature for reproduction >10 °C ?), *Hyalinea balthica* (muddy sediment, 9–13 °C), and *Quinqueloculina seminula* (sand, minimum temperature for reproduction >15 °C ?). The *E. crispum* association is probably confined to very shallow water because only there are temperatures suitable for reproduction (Fig. 9.8).

Although temperature gradients are broad on the eastern sides of oceans, it is nevertheless possible to recognise faunal boundaries and some examples are given in Table 9.12. Blanc-Vernet *et al.* (1983) consider that although the English Channel belongs to the same faunal province as the Bay of Biscay, it is poorer in species. However, taking all the seas around the British Isles into consideration, the converse is true.

Attempts to define faunal provinces have been made by Mikhalevich (1972) and Boltovskoy and Wright (1976) but there is little agreement on the province boundaries for the North Atlantic. A study of only two samples from off Guinea showed the presence of several species previously not recorded off West Africa (Debenay and Konate 1987). Our knowledge of modern foraminifera in this area is too incomplete to draw meaningful conclusions about geographic distribution.

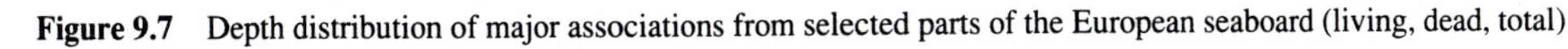

Figure 9.7 Depth distribution of major associations from selected parts of the European seaboard (living, dead, total)

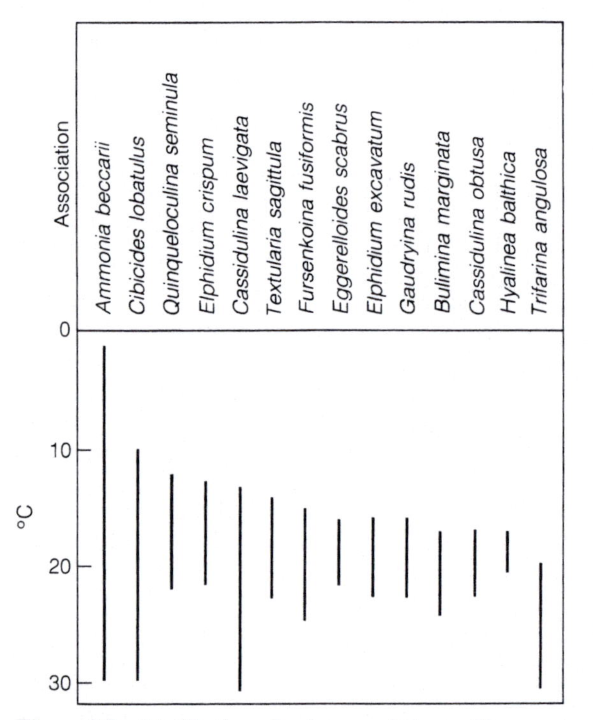

Figure 9.8 Distribution of major associations with respect to temperature

Table 9.12 Species close to their limit of distribution in the area of the British Isles (based on Murray 1971a)

Species of southern origin reaching their northern limit of distribution in the North Sea:
Ammonia beccarii
Asterigerinata mamilla
Bolivina pseudoplicata
Bulimina gibba/elongata
Eggerelloides scabrus
Elphidium crispum
Elphidium excavatum
Eponides repandus
Gavelinopsis praegeri
Haynesina germanica
Massilina secans
Nonion depressulus
Quinqueloculina spp.
Rosalina globularis
Textularia sagittula

Species of northern origin reaching their southern limit of distribution in the northern North Sea–Baltic Sea:
Ammotium cassis
Elphidium clavatum
Nonionellina labradorica

Species of northern origin reaching their southern limit of distribution in the English Channel:
Brizalina skagerrakensis
Buccella frigida
Pateoris hauerinoides

Pollution and the influence of man

At present there is severe pollution from industrial wastes along the southern margin of the North Sea. Oil exploration must have had some effect although nothing was evident in the foraminiferal faunas (Murray 1985). As far as can be seen, pollution has not seriously modified the foraminiferal assemblages on the shelf or slope.

Slope and deep sea

There have been few studies of living assemblages because of the difficulties in searching for them in the overwhelming numbers of planktonic tests in deep-sea sediments. The most detailed analyses are those of Gooday in the Porcupine Seabight. In Gooday (1986a), he described a diverse assemblage of living allogromiids and saccamminids >45 μm in length which made up 9–27% of the living assemblage in seven subsamples from a single station from 1330 m. From the same locality, Gooday (1986b) reported standing crops of 1246–2324 per 10 cm^2 which he attributed to three causes: the undisturbed sediment–water interface collected by the multiple corer, the abundance of individuals in the 45–62 μm size fraction and the care taken to retrieve the tectinous and agglutinated forms. The 45–62 μm fraction yielded 27% of the living individuals but only 5–12% of the total number of species. The diversity of the >45 μm living assemblage was high (α 40–63, $H(S)$ 3.17–4.09). Most species are represented by single individuals and the most dominant species formed only 6.81% of the assemblage. Within the sediment, >50% live in the 0–1 cm interval and >74% in the 0–2 cm interval. The latter includes *Nonionella iridea*, *Cassidulina teretis* and *Eilohedra levicula*. *Haplophragmoides bradyi* and *Melonis barleeanum* extend from 0 to 5 cm (the limit of sampling), the former peaking at 1–3 cm and the latter at 2–3 cm.

Further studies in Porcupine Seabight have shown that there is a seasonal flux of phytodetritus (i.e. dead plant plankton) at the sampled depths of 4483–4539 m

Table 9.13 Standing crop per 10 cm²

	Porcupine Seabight		W. Approaches–Portugal		NW Africa	
	Gooday (1986b)	Weston (1985)	Weston (1985)	Seiler (1975)	Haake (1980a)	Lutze (1980)
Depth (m)	>45 μm	>125 μm	>125 μm	>63 μm	>63 μm	>63 μm
200–1000		2–5	3–34	Av. 69	300–400	15 to <300
1000–2000	1246–2324	2–3	4	Av. 13	200	49–63
2000				Av. 11		20–82

and that the benthic foraminifera respond to its arrival (Gooday 1988). The phytodetritus settles on to the sediment surface and develops a low-diversity assemblage ($H(S)$ 1.50–1.69) in which *Epistominella exigua* and *Alabaminella weddellensis* (as *Eponides*) predominate together with *Fontbotia wuellerstorfi* (as *Planulina*) and *Pyrgoella* sp. None of these was common in the top 2 mm of the underlying sediment but *Adercotryma glomerata* was fairly common in both. The sediment assemblage was more diverse ($H(S)$ 2.83–3.62). These observations are of great biological and ecological interest but it would be impractical to examine samples routinely in this way. The results for this single station were obtained after months of picking (A J Gooday 1988, pers. comm.).

Standing crop values using conventional techniques are much lower than those reported by Gooday (see Table 9.13), even though the margins off Portugal and northwest Africa are regions of upwelling. Some epibenthic taxa such as *Fontbotia wuellerstorfi* (as *Cibicidoides* of Lutze and Thiel 1989) live attached to pebbles, sponges and polychaete tubes. Densities of 100–280 per 10 cm² are observed on the most elevated part of pebbles to take advantage of water movement. Similarly, *Planulina ariminensis* attaches itself to sponges which extend 10–15 cm above the sea floor. Living individuals are under-represented in sediment samples. The diversity of the living assemblages are α 9–22, but mainly >13 off Portugal and $H(S)$ 0.8–1.2 and α 4–11 off northwest Africa.

Some of the associations described from the shelf extend into the bathyal zone (Table 9.9). These include *Hyalinea balthica* (to 400 m), *Cibicides lobatulus* (900 m), *Trifarina angulosa* (1011 m), *Cassidulina laevigata* (1730 m), *C. obtusa* (3200 m) and *Reophax fusiformis* (3260 m). Others do not generally extend on to the shelf (Table 9.10).

The *Epistominella exigua* association is widely developed and commonly has *Fontbotia wuellerstorfi* as an additional species (Table 9.10). It is present in waters with a salinity fractionally less than 35‰ and at temperatures of –1 to 10 °C. The depth range is 501–7500 m. Where details of the water mass are known, the dominance of this species has been related to NEADW (Schnitker 1974b; Weston and Murray 1984).

The *Fontbotia wuellerstorfi* association in the North Atlantic is generally correlated with NADW (Schnitker 1974; Weston and Murray 1984; Lutze and Coulbourn 1984), but it is much better developed in the Norwegian Sea Bottom Water which is significantly cooler (–2 °C compared with 2.5–4.0 °C). Its total depth range is 1360–4280 m. The *Oridorsalis umbonatus* association is similar (depth range 1734–3877 m) and may tolerate a low food supply (Mackensen *et al.* 1985).

Nuttallides umboniferus, like *Epistominella exigua*, *Fontbotia wuellerstorfi*, and *Oridalis umbonatus*, is epifaunal. This association is found only in the deeper part of the northeastern Atlantic (>4000 m) south of the Azores Ridge. It is limited to bottom waters containing at least 5% AABW (Murray *et al.* 1986).

The other major associations are dominated by infaunal species and may therefore be less influenced by bottom water. The *Uvigerina peregrina* association correlates well with sediments rich in organic carbon (Lutze 1980; Lutze and Coulbourn 1984; Coulbourn and Lutze 1988) although it is commonly believed to prefer waters of low oxygen content. However, organic-rich sediments are likely to have oxygen-depleted pore waters even if the overlying bottom water is not oxygen deficient.

The *Globocassidulina subglobosa* association may show a correlation with MW (Weston 1985) as the temperature (4–9 °C) and salinity (>35‰) suggest but it might also be related to oxygen minimum water (as off northwest Africa, Lutze 1980) although the overall depth range is large (377–4095 m).

The *Melonis barleeanum* association extends from the Norwegian Sea to northwest Africa over a temperature range of –0.5 to 9 °C depth range of 466–2710 m. Mackensen *et al.* (1985) consider substrate to be the dominant control and Weston (in Murray 1984b) noted that it was abundant in areas of active bottom currents.

In addition to those papers with data tables that could be used to recognise the associations described above there are others which give general data for the Faeroe Channel, Bay of Biscay and the Gulf of Guinea.

Dredge samples from the Faeroe Channel were examined for larger agglutinated foraminifera by Carpenter and Pearcey during the last century and their results have been re-examined by Murray and Taplin (1984). The warm NAW flowing northwards has a diverse fauna including *Astrorhiza arenaria*, *Bathysiphon* spp., *Hyperammina* spp., *Marsipella* spp., *Rhabdammina* spp., *Rhizammina* spp., *Cyclammina cancellata*, *Hormosina* spp., *Reophax* spp. and *Vulvulina pennatula*. No large agglutinated taxa were found beneath the cold Norwegian Sea Overflow Water and, with the exception of *B. crassatina*, the few agglutinated taxa recorded also occurred in the warm NAW, i.e. *Botellina labyrinthica*, *Saccammina spherica*, *Saccorhiza ramosa*, *Thurammina papillata*, *Cribrostomoides subglobosum* (as *Alveolophragmium*), *Reophax* spp. and *Ammoglobigerina globigeriniformis*.

Brizalina subaenariensis (as *Bolivina*) was described by Pujos-Lamy (1973b) as living under inferred conditions of low dissolved oxygen in the water and presence of H_2S in the sediment of a Bay of Biscay submarine canyon. At the head of the canyon adults were rare and the abundance of small individuals was interpreted as evidence of high infant mortality. In the middle part many adult tests were deformed and this was taken as indicating unfavourable conditions. Adult tests with strong ornament plus the rarity of juveniles and abnormal tests was taken as evidence of a favourable environment. Weston (1985) also found this species on the canyon-cut slope of the Western Approaches. Noting that Kontrovitz *et al.* (1978) had determined from flume studies that this species had a high traction velocity, she suggested that *B. subaenariensis* and others with heavy tests (e.g. *B. dilatata*) would be well suited to life in a current-swept environment.

Associations of epifaunal, attached foraminifera and ferro-manganese crusts have been reported by Schaaf *et al.* (1977), Bignot and Lamboy (1980) and Janin (1984). These come from depths of 2000–4000 m from the Bay of Biscay and the Iberian margin. Species present include *Cibicides lobatulus*, and *Ammotrochoides bignoti*, the latter using coccoliths in the construction of its wall. Bignot and Lamboy consider that the polymetallic crusts are no longer forming.

General accounts of the faunas of the Gulf of Guinea have been given by Salami (1982, to ~900 m) and Levy *et al.* (1982a, 2475–4331 m). Subrecent assemblages have been decided by Brouwer (1973).

Postmortem changes

The northern European shelf seas are meso- to macrotidal and subject to severe wave attack during the winter months. In areas where living and dead assemblages have been distinguished they are commonly quite different (e.g. Celtic Sea and English Channel). Possible explanations for this are (a) seasonal changes in the composition of the living assemblages, (b) epifaunal living forms (on hyroids, shells, etc.) undersampled, (c) postmortem transport leading to loss or gain and (d) mixing with relict faunas.

At present there is no reliable information on seasonal changes of living assemblages because most areas have been sampled only once. There is no doubt that epifaunal taxa have been undersampled so that our understanding of the living assemblages is far from complete. Postmortem winnowing of small tests from the inner shelf has been widely recognised and these are transported in suspension into low-energy areas which included estuaries and outer shelf or slope areas of mud accumulation. Such tests are valuable guides to net sediment transport directions and can be used to quantify the amounts involved (Murray 1987b). Bed-load transport takes place in regions of powerful current or waves such as the southern North Sea, English Channel and Bay of Biscay. Such tests become abraded and fragmented and they are normally found only on the inner shelf.

Relict faunas are of two kinds. In some areas fossil forms are reworked into modern sediments, e.g. Palaeogene larger foraminifera in the English Channel (Le Calvez and Boillot 1967). These are easy to recognise because, apart from the obvious age and environment differences, the tests show diagenetic effects and are damaged through abrasion. The second type is a shallow-water fauna commonly observed in mid to outer shelf sediments. Sometimes the species are extant but clearly not in the correct environment, e.g. *Elphidium crispum* at 100–112 m in the Celtic Sea (Murray 1979a). In other cases they are extant but not anywhere near the relict occurrences, e.g. *Amphistegina* at 100–130 m on the shelf from Senegal to Angola (Kouyoumontzakis 1984) and *Ammonia beccarii* extending down to 350 m off Namibia (Martin 1981). In the former example, they have been dated at 11 000–12 000 BP. Such occurrences relate to periods of lower sea-level. The presence of relict tests indicates slow or non-deposition of modern sediment.

Dissolution of calcareous tests has not so far been reported from the shelf environments although it is known to take place in marginal marine environments such as the Baltic Sea. Postmortem destruction of fragile agglutinated tests has been reported from the Norwegian Slope (*Cribrostomoides subglobosum* between 1000 and 2000 m, Mackensen *et al.* 1985).

Infilling of modern tests with glauconie/glauconite has been reported from the Norwegian Shelf (Bjerkli and Östmo-Saeter 1973) and the Celtic Sea–English Channel (Murray 1965c; George and Murray 1977). Sometimes tests are broken through the swelling caused by glauconite growth.

Downslope transport has been universally recognised for the upper part of the continental slope and in the past major slides may have transported material into deeper water. Submarine canyons provide a route by which shallower water sediments and faunas can be transported into depths >4000 m and laterally by distances of up to 120 km (Pujos-Lamy 1973c).

Many authors have emphasised the bathymetric distribution of species. While within a limited area depth distributions may be consistent and predictable (see Haake *et al.* 1982) they can be disturbed by downslope transport, and comparisons between different areas invariably show major depth differences for individual species.

Shelf and slope environments are classified in Table 9.14.

Table 9.14 Classification of postmortem assemblages from the Atlantic seaboard of Europe and Africa

Field I
Slope, basin, abyssal plain
Field I–A
Norwegian Sea, 1000–2000 m (loss of *Cribrostomoides subglobosum*)
Field II
Upper slope
Norwegian Channel
Skagerrak Basin
Northern North Sea
Celtic Sea >90 m
Bays marginal to English Channel
Outer shelf, Bay of Biscay
Fields III–IV
Norwegian shelf and slope (to ~750 m)
Southern North Sea
Fair Isle Channel and W. Scotland
Celtic Sea <90 m
English Channel

Planktonic:benthic ratio

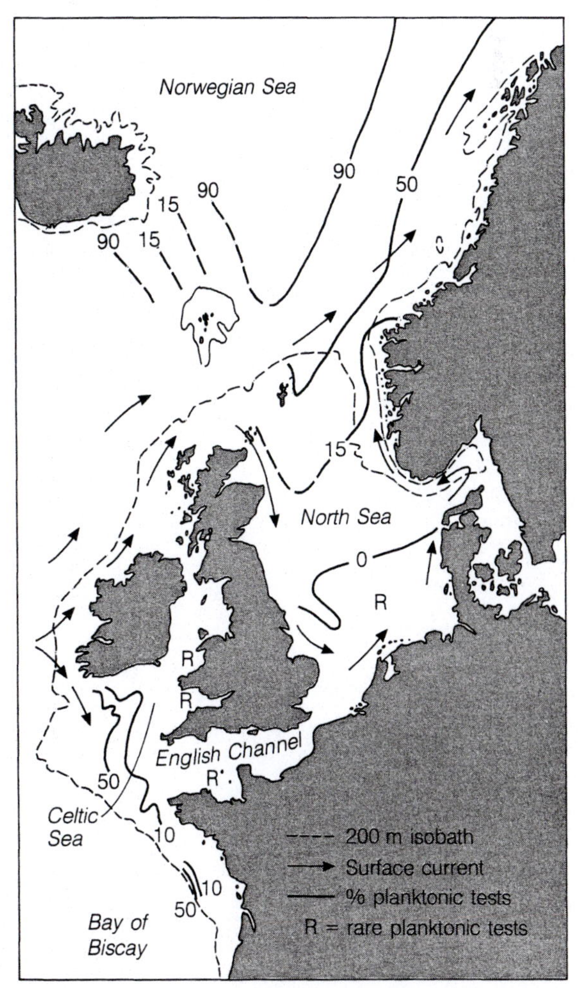

Figure 9.9 Percentage of planktonic and benthic tests, plus surface currents (data from Jarke 1958, 1961b; Qvale 1981; Schnitker 1969; Murray 1976b, 1988; Murray *et al.* 1982)

The data for northern Europe are summarised in Fig. 9.9. The decrease in plankton abundance from ocean to shore is complicated around the British Isles because of the epicontinental seas. In the North Sea there is a rapid decrease from north to south and planktonic tests are absent from the southern part (data from Jarke 1961b). Similarly, they are absent from much of the English Channel. Values of >50% are recorded on the outer shelf in the Celtic Sea in a region of mud deposition.

Apart from the decrease in abundance from ocean on to the shelf, the planktonic foraminifera show a decrease in size so that in areas with 10% or fewer planktonic tests they are juveniles rather than adults (Norway, Hald and Vorren 1984; North Sea, Murray 1985; Celtic Sea, Murray 1976a; Channel, Sturrock *et al.* 1981). This is because adults do not live in shelf waters so tests are transported in from the adjacent ocean.

CHAPTER 10

Atlantic Ocean – summary and comparison

A review of the data on the Atlantic Ocean has already been presented in Chapters 6, 8 and 9. It was decided not to choose an arbitrary depth to separate off the data on the continental margins from those of the deep sea because it would have been artificial and would have obscured the progressive changes in the faunas. The aim of this chapter is to summarise some of the important features and to draw comparisons. Where necessary, reference is made to the seas adjacent to the Atlantic, especially the Gulf of Mexico–Caribbean (Ch. 7) and the Mediterranean (Ch. 11).

Figure 10.1 shows the surface-water isotherms for winter in both hemispheres together with the principal currents. In the North Atlantic the effects of the cold Labrador Current influences the shallow areas as far south as Cape Hatteras (lat. 35° N), whereas the North Atlantic Current takes relatively warm water as far north as the British Isles (~lat. 50° N) and this can readily be seen from the position of the 5 and 10 °C isotherms. However, the clockwise circulation also causes the 20 °C isotherms to be deflected south off northwest Africa. Thus, the latitudinal distribution of shallow-water faunas, if influenced by temperature, should differ on the two sides of the North Atlantic. The same should also be true for the South Atlantic due to the anticlockwise circulation.

Care has been taken in this book to use a consistent set of names so that false differences between faunas are not introduced by inconsistent nomenclature of species.

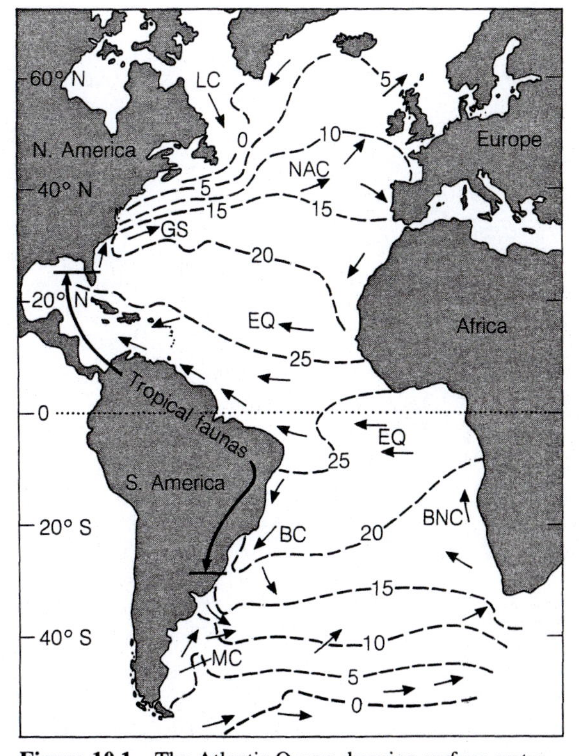

Figure 10.1 The Atlantic Ocean showing surface-water isotherms for Northern and Southern Hemisphere winters and surface currents (based on Sverdrup *et al.* 1942). C = current, LC = Labrador, NA = North Atlantic, GS = Gulf Stream, EQ = equatorial, BC = Brazil, BNC = Benguela, MC = Malvin (or Falkland). Also shown are the limits of tropical–subtropical faunas along the inner shelf of the Americas

Marshes

On both sides of the North Atlantic high to mid marshes are characterised by *Jadammina macrescens* and *Trochammina inflata* associations and low marshes by *Elphidium williamsoni*, *Ammonia beccarii* and *Haynesina germanica* associations. However, in North America *Tiphotrocha comprimata* on high marshes and the *Miliammina fusca* association on low marshes are more widely developed than in Europe. Furthermore, only the *J. macrescens* association has been recorded north of 46° N in Canada, while all are present at ~56° N in Europe.

Mangrove swamps

The Brazilian examples have an *Arenoparrella mexicana* association (which also characterises brackish to hypersaline marshes in the Gulf of Mexico) whereas the one example from northwest Africa has an *Ammotium salsum–Ammonia beccarii* association in the dry season and an *Ammotium salsum–Miliammina fusca* association in the wet season.

Brackish lagoons and estuaries

There are major differences in the composition of associations from different sectors of the Atlantic. Only the *Ammonia beccarii* and *Miliammina fusca* associations are known from North America, South America and Europe. Other major associations in North America and Europe include *Elphidium clavatum* (rare in Europe), *E. williamsoni*, *Ammotium cassis* (rare in North America) and *Haynesina germanica* (rare in North America). The *H. orbiculare*, *Eggerella advena* and *Ammobaculites* spp. associations are confined to North America and the *E. excavatum*, *Eggerelloides scabrus* and *Nonion depressulus* associations to Europe. It is not easy to account for these differences because all the species involved are euryhaline and eurythermal to a greater or lesser extent. On the whole, North American lagoons and estuaries are microtidal whereas most of those in Europe are meso- or macrotidal and therefore experience greater diurnal salinity variations. This might affect species which are less euryhaline. Also there is some geographical isolation of species: *E. scabrus* and *N. depressulus* are absent from the American seaboard (although the latter name has often mistakenly been used for other species there).

Shelf seas

There are major contrasts between the two margins of the North Atlantic (Table 10.1). Some of these are because there is a coastal zone of slightly brackish water along the North American margin from Canada to near Cape Hatteras whereas along the Europe–African margin fully marine conditions prevail. Some are due to limited geographic distribution. For instance, the following species are not known to occur along the North American margin: *Eggerelloides scabrus*, *Gaudryina rudis*, *Hyalinea balthica* and *Textularia sagittula* while *Eggerella advena* is absent from the European shelves.

Table 10.1 Distribution of shelf associations off N. America and Europe–NW Africa (P = principal, M = minor)

	N. America	Europe
Adercotryma glomerata	P	M
Saccammina atlantica	P	
Eggerella advena	P	
Islandiella helenae	P	
Islandiella islandica	P	M
Cibicides lobatulus	P	P
Elphidium clavatum	P	P
Bulimina marginata	P	P
Trifarina angulosa	P	P
Fursenkoina fusiformis	P	P
Ammonia beccarii	P	P
Cassidulina laevigata		P
Cassidulina obtusa		P
Eggerelloides scabrus		P
Elphidium crispum		P
Elphidium excavatum		P
Gaudryina rudis		P
Hyalinea balthica		P
Quinqueloculina seminula	M	P
Reophax fusiformis		P
Textularia sagittula		P
Elphidium spp.	P	
Quinqueloculina spp.	P	
Cibicides mollis	P	
Hanzawaia concentrica	P	M
Textularia secasensis	P	
Globocassidulina subglobosa	P	

The seas around the British Isles present a great diversity of environments especially with respect to seasonal variation in bottom temperature, substrate type and the effects of tidal amplitude and waves. Indeed the western margin of the British Isles (and the Bay of Biscay) has the highest internal tidal energy flux in the world (Baines 1982). Because of the negligible flux of terrigenous sediment along this western margin (most river flow into the North Sea), the region is one of temperate carbonate accumulation. The sediment consists principally of whole and broken bivalve mollusc shells with some echinoderm and barnacle debris. This provides habitats for the epifaunal *Gaudryina rudis* and *Textularia sagittula* associations. Such habitats do not exist along the North American margin. The only widespread epifaunal assemblage is the *Cibicides lobatulus* association which occurs north of 37° N (Maryland, USA.), north of 22° N (Dakar northwest Africa) and south of ~52° S (Tierra del Fuego, Argentina).

Parts of the shelf to the north of Cape Hatteras to Canada have corrosive bottom waters which lead to the development of agglutinated assemblages such as the *Saccammina atlantica* association, but no analogous environment exists along the European margin.

From these examples, it can be seen that there are large differences in the range of environments available on the shelves on the two sides of the North Atlantic, and these are undoubtedly the principal cause of the disparate development of the benthic associations.

The tropical/subtropical shallow-water fauna extends from southern Florida (26° N) to Brazil (31° S) but is less clearly defined along the European–African margin. Peneroplids and *Amphistegina* are present on Madeira (33° N) but the record of *Amphistegina* along the African shelf is complicated by the relict occurrences. The best development of the fauna is in the Mediterranean which is warm because it is surrounded by land having a warm climate. In terms of latitude, it is equivalent to Georgia to Nova Scotia (see Fig. 10.1).

The Greenland–Scotland Ridge

The Greenland–Scotland Ridge can be regarded both as a barrier to faunal exchange between the deep waters of the Norwegian Sea and the North Atlantic and as a corridor along which shelf and slope assemblages may migrate across the ocean.

From the sea around Iceland, Nørvang (1945) recorded 87 species from depths down to ~400 m. Only 7 were Arctic forms and 20 from warmer water, while 34 occurred in both cool and warm waters. Thus, in this depth range, the area is not a major faunal barrier. Lukashina (1987) looked at the depth range 100–1250 m and concluded that the faunas are tied to water mass rather than depth. Mackensen (1987) studied one profile across the ridge to the east of Iceland and another across the Wyville–Thomson Ridge. Like Lukashina, he also found that the faunas relate to water masses, including bottom currents, and to surface sediment type. The ridge faunal resemble those of the European seaboard and it seems that the ridge serves as a corridor.

Mackensen noted that on the south side of the ridge the distribution of the living fauna is quite unlike that of the dead due to the combined effects of downslope transport and redeposition of Pleistocene material. He cautioned against the use of dead assemblages in ecological studies.

Slope and deep sea

Although the deep sea is generally considered to have very diverse faunas, the number of species which are dominant is relatively small especially at depths greater than ~1000 m. This is perhaps because there is greater uniformity of water mass and substrate over large areas at great depth.

The bottom waters of the Atlantic Ocean are derived from two main sources, the Norwegian Sea, which gives rise to NADW, and the Weddella Sea, which gives rise to AABW. The Mediterranean supplies warm saline MW which helps in the formation of NADW after its passage along the European margin into the Norwegian Sea. The slightly below normal salinity Labrador Sea supplies water which mixes with NADW to form NEADW otherwise known as upper NADW. Although individual water masses retain their integrity over great distances, they slowly mix with adjacent water masses and change their characteristics. Thus, although AABW can be recognised in the northeastern Atlantic it is much modified from that which started out from the Weddell Sea.

The water masses have their individual temperature, salinity and therefore density differences, apart from varying in their content of dissolved oxygen, silicate, degree of corrosiveness with respect to $CaCO_3$, etc. The stratification of water masses is density dependent with the densest at the bottom. The boundaries between water masses in the Atlantic Ocean are commonly depth parallel over great distances but some boundaries transgress different depths.

The recognition of deep-sea foraminiferal groupings has almost invariably been achieved through multivariate analysis (principal component or varimax factor). However, they are just as readily recognised by the association approach followed here. The use of different sediment size fractions for foraminiferal analysis introduces inconsistencies because one of the key species, *Epistominella exigua*, is not satisfactorily sampled if >150 m μm or >250 μm fractions are studied. (see, for instance, Schröder *et al.* 1987, Sen Gupta *et al.* 1987, Pawlowski and Lapierre 1988). Associations dominated by epifaunal taxa show a reasonable correlation with water mass. Thus, the *E. exigua* association correlates with upper NADW (= NEADW) while the *Fontbotia wuellerstorfi* association correlates with true NADW. The *Oridorsalis umbonatus* and *Hoeglundina elegans* associations also occur under NADW although the latter species, being aragonitic, is confined to modest depths (2500–3800 m). The *Nuttallides umboniferus* association occurs in the deepest parts under the influence of AABW. At the Polar Front there is a variant, the *N. umboniferus–Ehrenbergina trigona* association, which is present in waters transitional from AABW to CDW. It has been argued that *N. umboniferus* is related to the carbonate undersaturation of the seawater rather than AABW *per se* (Bremer and Lohmann 1982).

Associations characterised by infaunal taxa are less clearly related to water mass. The *Globocassidulina subglobosa*, *Melonis barleeanum* and *Uvigerina peregrina* associations are all dependent on food availability as indicated by elevated organic carbon contents of the sediments in which they live. *Uvigerina peregrina* is abundant where the flux of organic carbon exceeds 2–3 g m^{-2} year (Altenbach 1988).

Data on ~250 foraminiferal assemblage analyses from the slope and abyssal plain from North Africa to the Norwegian Sea are summarised in Fig. 10.2. *Epistominella exigua* is clearly not depth dependent nor is it tied exclusively to NEADW for it is abundant in the southern part of the Norwegian Sea. *Fontbotia wuellerstorfi* is far more abundant in the Norwegian Sea than elsewhere and it too is not depth dependent. It is well adapted to low annual flux rates of organic carbon (Altenbach 1988). *Nuttallides umboniferus* is confined to depths >4000 m.

Among the infaunal taxa, *Melonis barleeanum* is bimodally abundant in the Norwegian Sea–west of Britain and off northwest Africa. *Uvigerina* spp. are abundant off northwest Africa at depths down to 3000 m and locally off Europe and in the Norwegian Sea down to 1500 m. Finally, *Trifarina angulosa* is very abundant in the Norwegian Sea and off Britain but is of limited occurrence elsewhere. *Uvigerina* spp. and *T. angulosa* show a more or less antipathetic relationship. These distribution patterns reveal the limitations of using benthic foraminifera for precise depth determination.

Data on the stable isotope signature of benthic foraminifera and their ambient water masses are discussed in Chapter 3 (see Fig. 3.3).

Conclusions

It will be apparent that there are large gaps in our knowledge of Atlantic foraminifera and yet it remains our most intensely studied ocean. Nevertheless, data on the South Atlantic are sparse and large areas of the North Atlantic have received little attention. Many studies are little more than reconnaissance. We are far from achieving an understanding of the dynamics of foraminiferal processes even in the shallow, more accessible parts of the marginal seas.

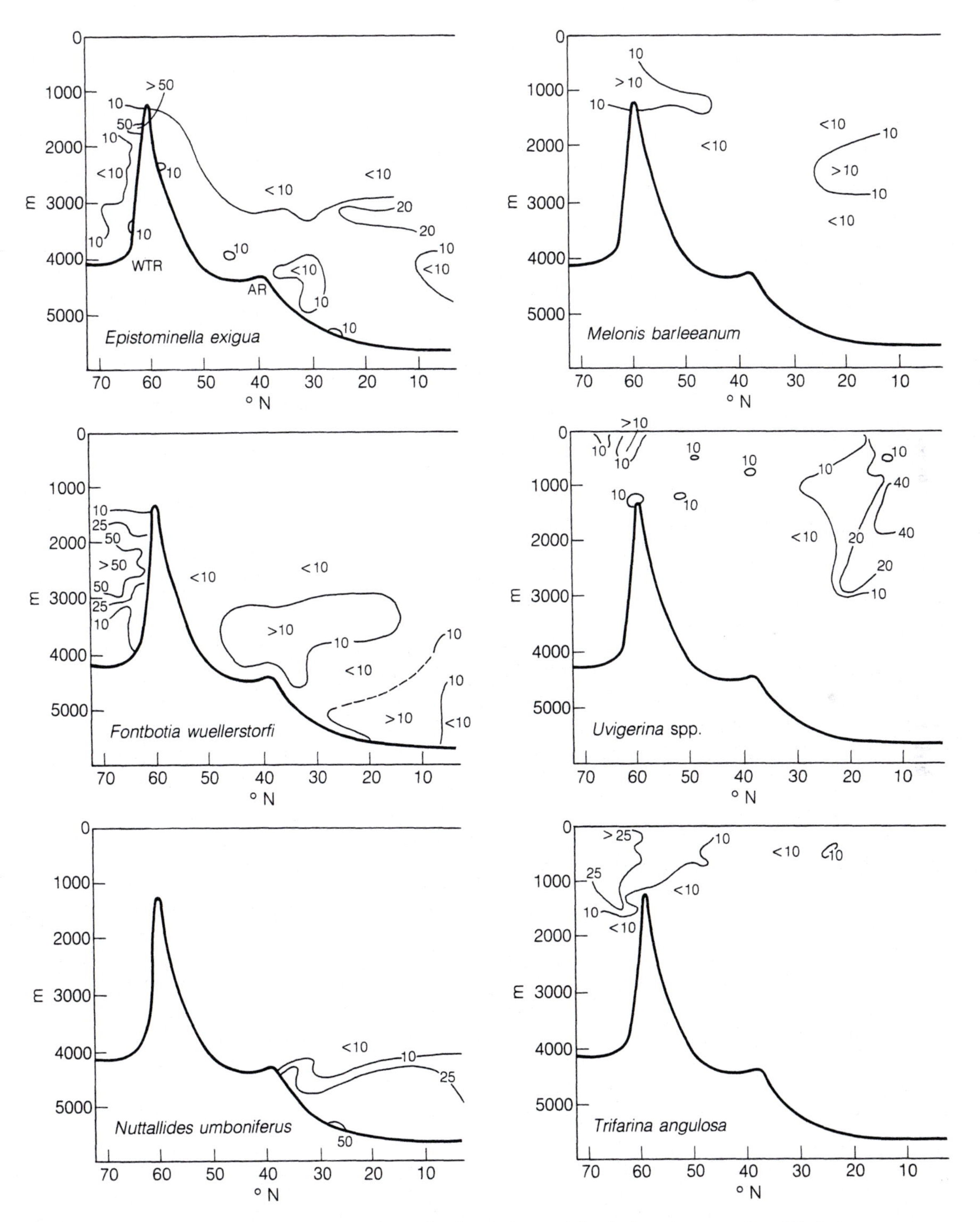

Figure 10.2 Latitude–depth plots of selected taxa for the northwest Atlantic (based on own data, Weston pers. comm. 1988, and sources quoted in Tables 9.9 and 9.10). WTR = Wyville–Thomson Ridge, AR = Azores Ridge. Bathymetric curve generalised

CHAPTER 11

The Mediterranean

Introduction

The Mediterranean is a land-locked sea connected to the Atlantic Ocean via the Strait of Gibraltar. There are two main basins, eastern and western, and communication between them is across sills in the Strait of Sicily and Strait of Messina. The annual rainfall and the contribution of fresh water via rivers is less than the water lost by evaporation from the sea surface. Consequently, water is drawn in from the North Atlantic and it flows along the coast of North Africa towards the east. Surface salinities increase from ~35‰ at the Strait of Gibraltar to ~39‰ in the eastern Mediterranean. During the winter, the more saline surface waters cool and therefore become more dense, and they sink into the deep basins. The latter are filled with relatively warm (~13 °C) more saline waters (>38‰). Circulation in the eastern basin is sluggish and the waters are somewhat depleted in oxygen although not dysaerobic (~4.5 ml l^{-1}). However, in the past anoxic conditions have developed. There is an outflow of MW at a depth >300 m through the Strait of Gibraltar into the North Atlantic. The water in the Mediterranean is renewed in about 75 years (Sverdrup *et al.* 1942). For a general review of the sediments and geology see Stanley (1972). Localities are shown in Fig. 11.1 and surface-water temperature in Fig. 11.2.

Although there is a fair amount of literature on the taxonomy of Mediterranean foraminifera and some general accounts of their distribution, there are rela-

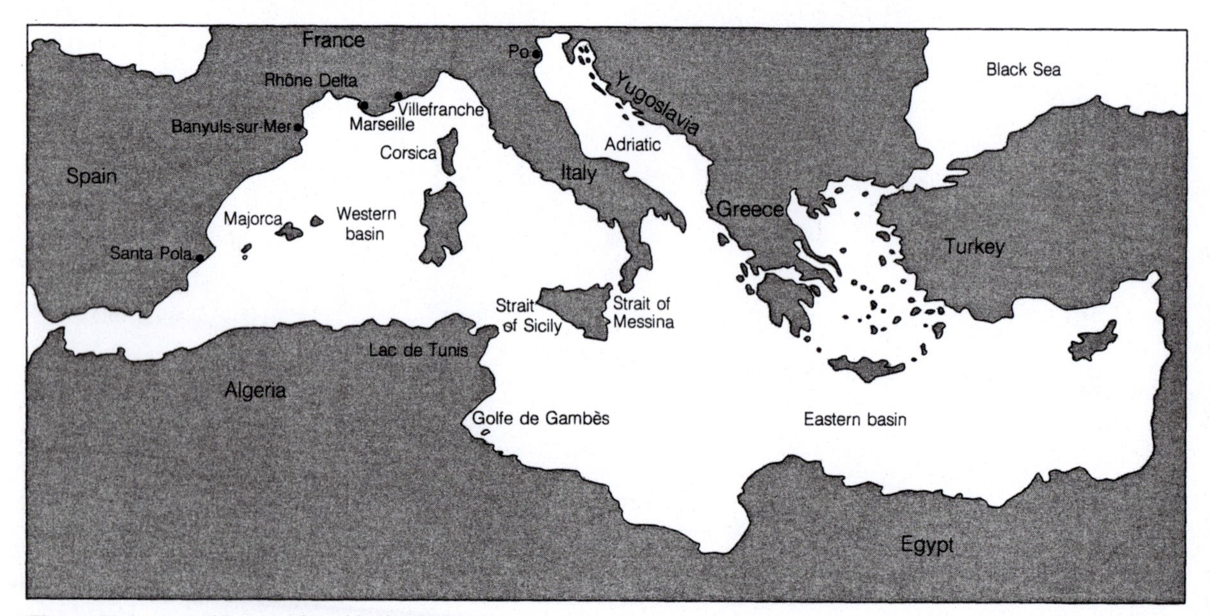

Figure 11.1 Localities mentioned in the text

tively few studies with numerical data on the composition of assemblages. Therefore, the ecology of this area is rather poorly known.

Because the water is clear and in many places it is possible to see shoals of little fish and meadows of seagrass, the impression is given of fertile conditions. However, this is misleading for the nutrient content of the water is low. In shallow waters the nutrients are more rapidly recycled and some of the foraminifera may carry symbionts to aid in this process. In deeper waters the density of life is low.

Marshes and lagoons

In the microtidal Venice Lagoon an *Ammonia beccarii* associations with *Haynesina germanica* (= *Nonion pauciloculum* of Albani *et al.* 1984) is present on all marshes except the higher salt marsh which has a *Trochammina inflata* association with *Haplophragmoides* sp., *Quinqueloculina seminula* and *A. beccarii*.

Lagoons

Living assemblages from salt-pans are *Ammonia beccarii* associations with *Nonion depressulus*, *Haynesina germanica* and small, smooth or striate miliolids. Assemblages from Spain have a diversity of α <1 in comparison with the adjacent marine area of α 10–12 (Zaninetti 1984), while those from the western part of the Rhone Delta have α <1–3.5 compared with α 8 in the adjacent sea (Zaninetti 1982). Deformed specimens having imperfect coiling form 1–10% of the common species. Malmgren (1984) analysed the morphological variability of *A. beccarii* in relation to measured environmental variables, but was unable to demonstrate any statistically valid relationships between morphotypes and salinity, temperature, pH or O_2. However, none of the morphotypes appears to have been deformed. Zaninetti believes that the salt-pan faunas are brackish lagoonal faunas which have adapted to hypersaline conditions up to a maximum of ~80‰ salinity. The *A. beccarii* association is characteristic of lagoons along the French coast (see review by Levy 1971).

Mudflat and marsh areas in Greece also have the *Ammonia beccarii* association as the principal assemblage even though the salinity range is 26–89‰ (Scott *et al.* 1979). The principal additional species is *Haynesina germanica* and locally the *H. germanica* association is present. *Quinqueloculina seminula*, *Discorinopsis aguayoi*, *Elphidium translucens* (as *Cribroelphidium*), *Jadammina macrescens* (as *J. polystoma*) and *Trochammina inflata* are all present living. In view of the small development and discontinous extent of marshes in the Mediterranean, it is remarkable that typical marsh species such as *J. macrescens* and *T. inflata* should be able to colonise them.

The *Haynesina germanica* association is known from the Po Delta where it occurs in brackish water (17– 29‰, d'Onofrio *et al.* 1976). *Miliammina fusca*, common in lagoons elsewhere, is present in Mediterranean lagoons only in small numbers, e.g. the Rhône Delta (Vangerow 1972), marsh/mudflats in Greece (Scott *et al.* 1979).

The Black Sea has an oxygenated surface brackish layer down to ~50 m, a dysaerobic layer down to ~150–200 and deeper than this conditions are anoxic (Fairbridge 1966). The benthic foraminifera are represented by the *Ammonia beccarii* association and only *Nonion depressulus*, *Haynesina germanica*, *Elphidium poeyanum*, *Discorbis villardeboanus* and *Jadammina macrescens* are common in addition (see Table 11.1). The fauna is impoverished and only 51 benthic species are known to be present (Tufescu 1973). The variability of *Ammonia tepida*, including deformities, has been discussed by Tufescu (1968).

The Caspian Sea is totally isolated from other seas and is very brackish. Extremely low-diversity *Ammonia beccarii* dominated assemblages are known from a bordering lagoon (Yassini and Ghahreman 1977).

Shelf and basin

Around most of the Mediteranean the shelf is narrow and its most extensive development is seen in the Adriatic Sea. Ten major associations can be recognised (Table 11.2) and 34 minor ones (Table 11.3).

Widely distributed associations include *Ammonia beccarii* on the inner shelf, *Cassidulina laevigata* and *Uvigerina mediterranean* on the outer shelf and slope. In addition, where seagrasses are present in the nearshore zone, there are abundant miliolids.

Seagrass assemblages

Seagrasses recorded from the Mediterranean include *Posidonia*, *Cymodocea* and *Halophila*. *Posidonia* meadows are present discontinuously along the coast of Spain to southern France at depths of 2–40 m (Colom 1964, 1974; Vénec-Peyré and Le Calvez 1981; Vénec-Peyré 1983a, 1984). Attached foraminifera include *Nubecularia lucifuga*, *Planorbulina mediterranensis*, *P. acervalis*, *Cibicides lobatulus*, *Rosalina posidonicola*, *Hemisphaerammina bradyi* and *Iridia* spp. which live on the blades. At Banyuls-sur-Mer, *R. posidonicola*, with *C. lobatulus* and *P. mediterranensis*, is dominant at depths of 15–20 m while *Iridia* spp. and *H. bradyi* are common at 10 m. The rhizoids support a different attached fauna comprising *Miniacina miniacea*. In addition there are forms which cling to the *Posidonia* and these include abundant and diverse *Quinqueloculina* spp. and *Triloculina* spp. together with *R. bradyi*, R. *globularis*, *Ammonia beccarii*, *Elphidium crispum* and *E. macellum*. At Banyuls-sur-Mer, the miliolids are less diverse, less robust and less ornamented than those from Villefranche (Vénéc-Peyré and Le Calvez 1981). Along the Spanish coast *Peneroplis* and *Vertebralina* are found (Colom 1974).

The distribution of *Posidonia* is interrupted by the Rhône Delta and commences again to the east. The miliolids at Villefranche were examined in detail by Le Calvez and Le Calvez (1958). They collected *Posidonia* blades and washed the foraminifera from them. The average composition of the fauna was *Quinqueloculina* spp. 50% and *Triloculina* spp. 28%.

Typical seagrass faunas with attached and clinging species from near Marseille include, during the summer months, *Vertebralina striata*, *Peneroplis pertusus*, *P. planatus*, *Spirolina arietina* and *Sorites variabilis* (Blanc-Vernet 1969). In the winter, when the vegetation has died down, these species are sparsely present in the sediment.

Langer (1988) made comparative studies of different types of epiphytic assemblages from off Italy. He recognised four morphotypes:

1. Permanently attached, encrusting forms with flat attachment surfaces, dominant on *Posidonia* leaves, e.g. *Cibicides lobatulus*, *C. refulgens*, *Cyclocibicides vermiculatus*, *Miniacina miniacea*, *Nubecularia lucifuga*, *Planorbulina acervalis* and *P. mediterranensis*. They are attached by a secreted organic glue and they are herbivorous (on diatoms).
2. Attached forms which are sometimes mobile, e.g. *Asterigerinata*, *Cymbaloporetta* and *Rosalina*.
3. Free-living forms with a canal system allowing rhizopods to be developed all over the test so that the whole surface acts as an apertural face.

Table 11.1 Lagoon association

***Ammonia beccarii* association**

Salinity: 5–89‰

Temperature: 4 to 37 °C

Substrate: muddy

Depth: shallow except in Caspian Sea, 0–170 m

Distribution:
1. Santa Pola, Spain (Zaninetti 1984) living (salt works)
2. SE France (Le Calvez and Le Calvez 1951) total
3. W. Rhône Delta, France (Zaninetti 1982) living (salt pans)
4. E. Rhône Delta, France (Vangerow 1972, 1974) living
5. Venice Lagoon, Italy: (a) (Albani *et al.* 1982) total; (b) (Olivieri 1963) total
6. Greece (Scott *et al.* 1979) living, total
7. Black Sea (a) (Macarovici and Cehan-Ionesi 1962) total (as *Rotalia*), (b) (Tufescu 1973) total
8. Caspian Sea (Yassini and Ghahreman 1977) total
9. Lac de Tunis (Carbonel and Pujos 1982) total

Additional common species:

Nonion depressulus	1 (as *Haynesina*, 7a (as *N. stelligerum*), 7b (as *Protelphidium tuberculatum*), 9
Haynesina germanica	1, 3, 6a, b (as *Nonion depressulum*), 5a (as *Nonion paucilocum*), 6 (as *Protelphidium depressulum*), 9 (as *Protelphidium paralium*)
Miliolinella sp.	1
miliolids	1, 3, 9
Trochammina inflata	3
Jadammina macrescens	3, 6a, b (as *J. polystoma*)
Elphidium spp.	4
Miliammina fusca	4
Elphidium lidoensis	5a (as *Cribrononion granosum*)
Quinqueloculina seminula	6
Elphidium translucens	6 (as *Cribroelphidium*)
Discorbis villardeboanus	7a
Elphidium poeyanum	7b (as *Cribroelphidium*)
Elphidium iranicum	8 (as *Cribroelphidium*)

Table 11.2 Principal associations of benthic foraminifera in the Mediterranean Sea

***Ammonia beccarii* association**

Salinity: 35–38‰

Temperature: 10 to 25 °C

Substrate: muddy sand

Depth: 0–100 m

Distribution:
1. Adriatic: (a) (Daniels 1970) living, dead; (b) (Haake 1977) living; (c) (Chierici *et al.* 1962) total; (d) (Jorissen 1988) total (including *A. parkinsoniana* and *Pseudoeponides falsobeccarii*)
2. S. Italy: (a) (Sgarrella *et al.* 1985) dead; (b) (Iaccarino 1969) total
3. Rhône Delta, France (Kruit 1955) total (as *Rotalia*)
4. Banyuls-sur-Mer, France (Vénec-Peyré 1984), living
5. Algeria (Levy *et al.* 1980) total

Additional common species:

Textularia agglutinans	1a, b (as *Textilina pseudogramen*), 1d
Eggerelloides scabrus	1a, b, 2a, b (as *Eggerella*), 3 (as *Verneuilina*)
Quinqueloculina schlumbergeri	1a
Brizalina striatula	1a
Elphidium sp.	1a (as *Cribrononion excavatum*), 3
Nonionella opima	1b, d (as *N. turgida*)
Bulimina spp.	1b
Buccella granulata	1c (as *Eponides*), 1d
Ammonia perlucida	1d
Elphidium poeyanum	1d
Cassidulina laevigata	1d
Bulimina marginata	1d
Elphidium granosum	1d
Elphidium crispum	1d, 3, 5
Elphidium advenum	1d
Quinqueloculina seminula	1d, 3, 5
Cibicides lobatulus	1d, 5
Adelosina spp.	1d
Elphidium cuvilieri	2a (as *Cribrononion*)
Triloculina trigonula	2b, 3
'Nonion asterizans'	3
Buccella tenerrima	3 (as *Rotalia faramanensis*)
Quinqueloculina rugosa	3
Nonion commune	4
Massilina secans	5
Valvulineria complanata	5

***Buccella granulata* association**

Salinity: 37–38‰

Temperature: 10 to 25 °C

Substrate: sand, mud

Depth: 25–42 m

Distribution:
1. Adriatic: (a) (Chierici *et al.* 1962) total; (b) (Jorissen 1988) total

Additional common species:

Ammonia beccarii	1a, b
Asterigerinata mamilla	1a, b
Elphidium decipiens	1a
Rosalina globularis	1a (as *Discorbis*)
Elphidium granosum	1b
Textularia agglutinans	1b
Bulimina marginata	1b
Cibicides lobatulus	1b

***Bulimina marginata* association**

Salinity: 37–38‰

Temperature: 10 to 2 °C

Substrate: mud

Depth: 32–396 m

Distribution:
1. Adriatic (Jorissen 1988)

Additional common species:

Valvulineria complanata	1 (as *V. bradyana*)
Bigenerina nodosaria	1
Uvigerina mediterranea	1
Globocassidulina oblonga	1 (as *Cassidulina*)
Textularia agglutinans	1
Elphidium poeyanum	1
Elphidium granosum	1
Ammonia beccarii	1
Nonionella opima	1 (as *N. turgida*)
Ammonia perlucida	1
Melonis barleeanum	1 (as *Nonion*)
Textularia sagittula	1
Hyalinea balthica	1

Table 11.2 *(continued)*

***Cassidulina laevigata* association**

Salinity: 37–39‰

Temperature: 13 to 20 °C

Substrate: mud

Depth: 25–853 m

Distribution:
1. Adriatic: (a) (Chierici *et. al.* 1962) total; (b) (Jorissen 1988) total
2. E. Mediterranean (Parker 1958) total (as *C. carinata*)
3. W. Mediterranean: (a) (Moncharmont–Zei 1964) total, (b) (Fierro 1961) total; (c) (Sgarrella *et al.* 1985) dead

Additional common species:

Brizalina catanensis	1a (as *Bolivina*)
Uvigerina peregrina	1a
Cibicidoides pseudoungerianus	1a (as *Cibicides*)
Bulimina marginata	1b, 3c
Textularia sagittula	1b
Trifarina angulosa	1b
Bulimina aculeata	3b
Globocassidulina subglobosa	3b (as *Cassidulina*), 3c
Valvulineria complanata	3c (as *V. bradyana*)
Bolivina alata	3c

***Gyroidina–Repmanina* association**

Salinity: 39‰

Temperature: 13 to 14 °C

Substrate: mud

Depth: 746–2900 m

Distribution:
1. E. Mediterranean: (a) (Parker 1958) total; (b) (Massiota *et al.* 1976) total; (c) (Cita and Zocchi 1978) total (*Repmanina* = *Glomospira charoides*)
2. W. Mediterranean (Cita and Zocchi 1978) total

Additional common species:

Melonis barleeanum	1a (as *Nonion*)
Uvigerina mediterranea	1a
Glomospira gordialis	1a
Articulina tubulosa	1b, 2
Robertina translucens	2

***Nonionella opima* association**

Salinity: 35–38‰

Temperature: 10 to 25 °C

Substrate: muddy silt

Depth: 13–70 m

Distribution:
1. Adriatic Sea: (a) (Daniels 1970) living; (b) (Haake 1977) living, (c) (Jorissen 1988) total (as *N. turgida*)

Additional common species:

Adercotryma glomerata	1a (as *A.* sp.)
Ammonia beccarii	1a, c
Epistominella vitrea	1a
Brizalina striatula	1a
Textularia agglutinans	1a,b (as *Textilina pseudogramen*), c
Nouria polymorphinoides	1a
Eggerelloides scabrus	1a, b (as *Eggerella*)
Reophax scorpiurus	1a
Quinqueloculina spp.	1a
Astrononion sidebottomi	1b
Elphidium lidoense	1b
Cibicides boueanum	1b
Bulimina spp.	1b
Buccella granulata	1c
Cibicides lobatulus	1c
Asterigerinata mamilla	1c
Elphidium poeyanum	1c
Bulimina marginata	1c
Valvulineria complanata	1c (as *V. bradyana*)

***Quinqueloculina* spp. association**

Salinity: 37–39‰

Temperature: 10 to 25 °C

Substrate: (1,2) seagrass, (3) sand

Depth: (1,2) 2–40, (3) 8–65 m

Distribution:
1. See text – seagrass assemblages of whole Mediterranean
2. E. Mediterranean only: (a) (Blanc-Vernet 1969 living; (b) (Cherif *et al.* 1988) total
3. Rhône Delta (Kruit 1955) total

Table 11.2 *(continued)*

Additional common species:

Triloculina spp.	1, 2a
Peneroplis spp.	1, 2a, b
Spirolina spp.	1, 2a
Amphistegina madagascariensis	2a, b
Rotalinoides gaimardi	2a (as *Ammonia*)
Ammonia beccarii	3 (as *Rotalia*)
Buccella tenerrima	3 (as *Rotalia faramanensis*)
'Nonion asterizans'	3

Textularia agglutinans **association**

Salinity: 35–38‰

Temperature: 10 to 25 °C

Substrate: muddy silt

Depth: 20–132 m

Distribution:
1. Adriatic: (a) (Daniels 1970) living, dead; (b) (Jorissen 1988) total

Additional common species:

Nonionella opima	1a
Ammonia beccarii	1a, b
Adercotryma glomerata	1a (as *A.* sp.)
Epistominella vitrea	1a
Brizalina striatula	1a
Hopkinsina sp.	1a
Elphidium translucens	1a (as *Cribrononion*)
Bulimina marginata	1b
Elphidium poeyanum	1b
Buccella granulata	1b
Reussella spinulosa	1b
Elphidium granosum	1b
Elphidium advenum	1b

Uvigerina mediterranea **association**

Salinity: 37–39‰

Temperature: 13 to 16 °C

Substrate: mud

Depth: 95–2500 m

Distribution:
1. Adriatic (Jorissen 1988) total
2. E. Mediterranean (Parker 1958) total
3. W. Mediterranean: (a) (Cita and Zocchi 1978) total; (b) (Blanc-Vernet 1969) ?total; (c) (Blanc-Vernet *et al.* 1979) total; (d) (Levy *et al.* 1980) total; (e) (Colom 1970) total; (f) (Sgarrella *et al.* 1985) dead

Additional common species:

Trifarina angulosa	1
Cibicidoides pseudoungerianus	1 (as *C. pachydermus*), 3e
Cassidulina laevigata	1
Bulimina marginata	1
Textularia agglutinans	1
Uvigerina peregrina	1, 2, 3d
Bulimina costata	1, 2
Gyroidina altiformis	1, 2
Gyroidina orbicularis	1
Melonis barleeanum	2 (as *Nonion*), 3b (as *Nonion parkeri*)
Spirophthalmidium acutimargo	2
Gyroidina neosoldanii	2
Hoeglundina elegans	3a
Globobulimina turgida	3d
Amphicoryna scalaris	3d
Sigmoilopsis schlumbergeri	3e (as *Sigmoilina*)

Valvulineria complanata **association**

Salinity: 36–38‰

Temperature: 10 to 25 °C

Substrate: muddy sand

Depth: 30–100 m

Distribution:
1. Rhône Delta (Kruit 1955) total (as *V. mediterranensis*?)
2. Adriatic (Jorissen 1988) total (as *V. bradyana*)
3. S. Italy (a) (Sgarrella *et al.* 1985) dead (as *V. bradyana*), (b) (Iaccarino 1967, 1969) total (as *V. bradyana*)
4. Algeria (Levy *et al.* 1980) total

Additional common species:

Reophax cylindrica	1
Lagenammina difflugiformis	1
Ammonia beccarii	1 (as *Rotalia*), 4
Eggerelloides scabrus	1 (as *Verneuilina*)
Textularia agglutinans	1, 2
Bulimina marginata	2
Nonionella opima	1, 3b (as *N. turgida*)
Melonis barleeanum	2
Cassidulina neocarinata	3b
Brizalina oceanica	3b (as *Bolivina*)
Brizalina alata	3b

Table 11.3 Minor benthic associations in the Mediterranean Sea (L =living, D = dead, T = total)

	Association		Depth (m)	Area	Source
1.	*'Nonion asterizans'*	T	12	Rhône Delta	Kruit 1955
2.	*Triloculina trigonula*	T	8–30	Rhône Delta	Kruit 1955
3.	*Elphidium crispum*	T	12–38	Rhône Delta	Kruit 1955
4.	*Lagenammina difflugiformis*	T	31	Rhône Delta	Kruit 1955
5.	*Triloculina longirostra*	T	7–18	Rhône Delta	Kruit 1955
6.	*Adercotryma glomerata*	L	35	Adriatic	Daniels 1970
7.	*Eggerelloides scabrus*	L	35	Adriatic	Daniels 1970
		T	19–21	S. Italy	Sgarrella *et al.* 1985
8.	*Miliolinella subrotunda*	L	5	Adriatic	Daniels 1970
9.	*Affinetrina planciana*	L	5	Adriatic	Daniels 1970
10.	*Brizalina striatula*	L	5	Adriatic	Daniels 1970
11.	*Elphidium advenum*	L	30	Adriatic	Haake 1977
12.	*Bulimina* spp.	L	10–50	Adriatic	Haake 1977
13.	*Brizalina catanensis*	T	166	Adriatic	Chierici *et al.* 1962
14.	*Adelosina* sp.	T	12	Adriatic	Jorissen 1988
15.	*Quinqueloculina aspera*	T	23–45	Adriatic	Jorissen 1988
16.	*Asterigerinata mamilla*	T	35–70	Adriatic	Jorissen 1988
		T	51–117	E. Mediterranean	Parker 1958
17.	*Cibicides lobatulus*	T	52–99	Adriatic	Jorissen 1988
		T	143	E. Mediterranean	Parker 1958
18.	*Quinqueloculina seminula*	T	32	Adriatic	Jorissen 1988
19.	*Reussella spinulosa*	T	53–60	Adriatic	Jorissen 1988
20.	*Textularia sagittula*	T	68–101	Adriatic	Jorissen 1988
21.	*Melonis barleeanun*	T	102–843	Adriatic	Jorissen 1988
		T	201–1016	E. Mediterranean	Parker 1958
22.	*Uvigerina peregrina*	T	152–957	Adriatic	Jorissen 1988
		T	105–107	Algeria	Levy *et al.* 1980
23.	*Bulimina costata*	T	355–868	Adriatic	Jorissen 1988
		T	287–567	E. Mediterranean	Parker 1958
24.	*Globocassidulina oblonga*	T	101	Adriatic	Jorissen 1988
25.	*Trifarina angulosa*	T	249	Adriatic	Jorissen 1988
26.	*Massilina secans*	T	30–35	Algeria	Levy *et al.* 1980
27.	*Amphicoryna scalaris*	T	67–170	Algeria	Levy *et al.* 1980
28.	*Textularia bocki*	T	88	Algeria	Levy *et al.* 1980
29.	*Neoconorbina terquemi*	D	55	S. Italy	Sgarrella *et al.* 1985
30.	*Bolivina alata*	D	99–188	S. Italy	Sgarrella *et al.* 1985
31.	*Rosalina obtusa*	D	22	S. Italy	Sgarrella *et al.* 1985
32.	*Cassidulina neocarinata*	T	120–300	S. Italy	Iaccarino 1967, 1969
33.	*Triloculina trigonula*	T	12	S. Italy	Iaccarino 1969
34.	*Elphidium granosum*	T	29	S. Italy	Iaccarino 1969

Elphidiids are predominant within dense vegetation where they are suspended by rhizopods. *Sorites orbiculus* also has this mode of life.

4. Free-living forms with a terminal aperture which hold their test erect when moving. They are probably omnivorous and they reproduce in the sediment rather than on the plants, e.g. *Anomalina*, *Globulina*, *Miliolinella*, *Nummoloculina*, *Peneroplis*, *Quinqueloculina*, *Spiroloculina*, *Textularia* and *Triloculina*.

Langer notes that the size of the plant substrates and their length of life controls the diversity of the epi-

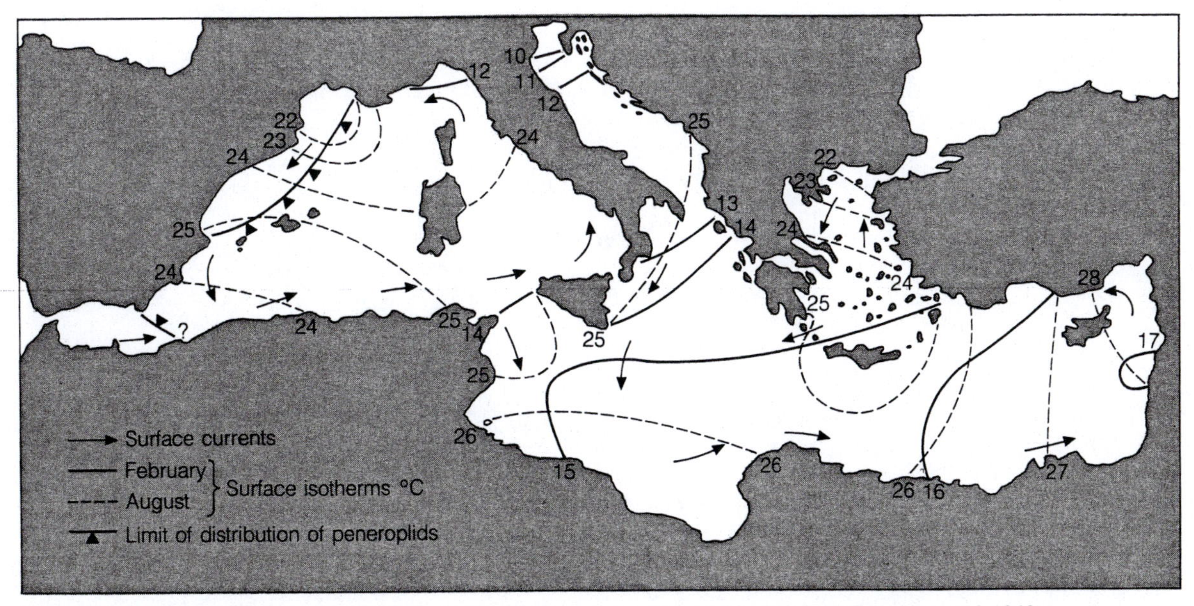

Figure 11.2 Distribution of peneroplids in the Mediterranean (oceanographic data from Sverdrup *et al.* 1942)

phytic foraminifera; larger plants support more species than smaller ones.

Seagrass assemblages from the eastern Mediterranean are similar to those of the western part, but with the addition of the subtropical species *Amphistegina madagascariensis* and *Rotalinoides gaimardi* (as *Ammonia*, Blanc-Vernet 1969).

Miliolid-rich total assemblages with peneroplids are well developed on the north African coast in the Golfe de Gabès (Blanc-Vernet 1974, Blanc-Vernet *et al.* 1979). Although Levy *et al.* (1980) reported miliolid-rich sediments from ~10 m off Africa, they did not record any peneroplids.

The distribution of peneroplids is summarised in Fig. 11.2. Their absence from the northwest part of the western Mediterranean may be due to the influence of the Rhône discharge, and from southernmost Spain (Mateu 1971) due to the incoming relatively cool Atlantic water. Although present along the Catalan coast, only *Peneroplis pertusus*, *P. planatus* and *Amphisorus hemprichii* are present and *Spirolina* spp. are absent (Mateu 1970).

Rhône Delta

Because of the non-tidal nature of the area the fresh water flows out over the seawater so most of the offshore delta is fully marine. The principal total assemblage at depths of <32 m is the *Ammonia beccarii* association and this gives way to the *Valvulineria complanata* association from 30 to 90 m (Table 11.2). In addition, there are five minor associations (Table 11.3) (data from Kruit 1955, >250 μm fraction only). However, somewhat different results were obtained by Bizon and Bizon (1984a) who studied total assemblages from the >63 μm fraction. Although no tables of data were given, from their distribution maps it can be seen that there is an offshore increase in the abundance of tests associated with a decrease in grain size. *Ammonia beccarii* is common close to the shore away from the river mouth. '*Bolivina*' and *Nonionella turgida* are common at depths of 30–80 m and beyond this *Cassidulina carinata* and *Bulimina* spp. (~50–100 m).

Adriatic Sea

The Adriatic Sea has the most extensive development of continental shelf in the Mediterranean. There is a basin extending down to 270 m in the central part and a larger one 1225 m deep in the southern part. The Adriatic Sea is connected with the Mediterranean by the Strait of Otranto with a sill at 750 m depth. In general salinities are 37–38‰ and even near the Po Delta brackish conditions do not affect the shelf because the freshwater discharge forms a surface layer. The temperature shows an annual variation from ~9 to 25 ° C in shallow water and is ~13 °C in the deepest parts. There is a clockwise pattern of surface circulation and only very small tides.

Living assemblages have been studied by Daniels (1970, 1971) and Haake (1977) and total assemblages by Chierici *et al.* (1962) and Jorissen (1987, 1988). Daniels made a detailed study of the population dynamics of *Nonionella opima* and seasonal changes in the assemblages of a small inlet on the Yugoslavia coast.

On the basis of principal component and cluster analysis, Jorissen recognised four biofacies: II, 7.5–25 m along the Italian coast on sands rich in $CaCO_3$ but poor in organic matter; III, sand platform rich in $CaCO_3$ but poor in organic matter; IV, 20–100 m, along the Italian coast, muddy and rich in organic matter; I, 50–1225 m, clayey with intermediate amounts of organic matter. He concluded that there was a major faunal break at 100 m which was probably determined by the depth of the photic zone.

The Po introduces nutrients into the sea and these feed blooms of phytoplankton. The surface circulation carries this organic matter and fine-grained sediment to the south where it is deposited as a narrow band along the Italian coast. During the summer, when the waters are stratified, there may be almost total consumption of the available oxygen in the area of organic-rich mud deposition and this has a profound effect on the fauna. Jorissen considered that *Bulimina marginata*, *Textularia agglutinans* and *Nonionella opima* (as *N. turgida*) were correlated with a high organic carbon content, as were *Melonis barleeanum* (as *Nonion*) and *Valvulineria bradyina* but to a lesser degree. He also recognised ecophenotypes of *Ammonia beccarii* and *A. parkinsoniana* (here treated as *A. beccarii s.l.*), *B. marginata*, *Elphidium* spp. and related them to known (e.g. organic carbon) or supposed (e.g. mode of life, stress from Po discharge) causes.

Using the association approach of this book, it is possible to recognise 7 major associations and 21 minor associations (each represented by only a few samples) (see Tables 11.2 and 11.3). In Fig. 11.3 these have been plotted against depth and in crude sediment groups. In sands and muddy sands, there is a depth progression from the *Ammonia beccarii* association to the *Cassidulina laevigata* association and similarly on mud from the *Nonionella opima* to the *Uvigerina mediterranea* association. Three minor associations from deep water have also been plotted as they are probably undersampled. The faunal change at 100 m, recognised by Jorissen, is somewhat diffuse (90–140 m).

The only diversity data available are for the Lim Fjord, Yugoslavia (Murray 1973 using tables from Daniels 1970). In this sheltered inlet the living assemblages showed a range of α 11–19 near the mouth to 7–14 at the head.

The low oxygen environment of another embayment on the coast of Yugoslavia supports an assem-

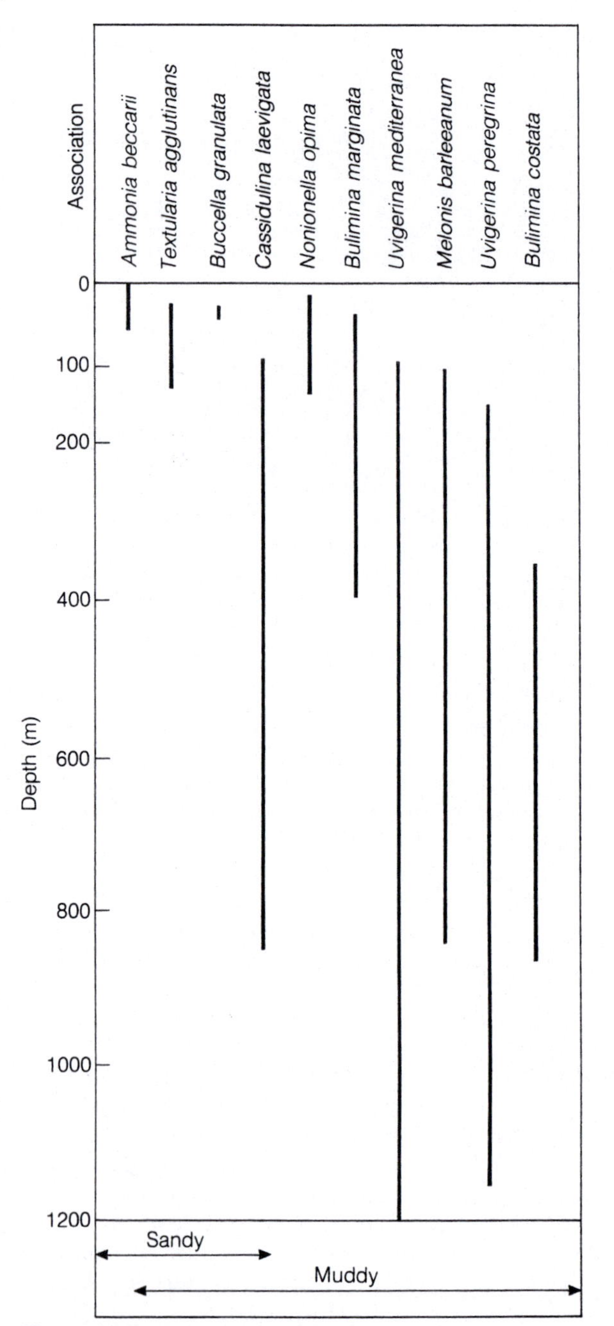

Figure 11.3 Depth/substrate distribution of major associations in the Adriatic Sea

blage of *Bolivina pseudoplicata*, *Brizalina spathulata*, *B. variabilis*, *Bulimina aculeata*, *B. elongata*, *B. marginata* and *Trifarina angulosa* (Cimmerman *et al.* 1988 – no quantitative data). This may be a modified *B. marginata* association.

A comparative study of epifaunal Lituolacea and Miliolacea on two submarine cliffs off the Yugoslavian coast showed that in the open sea miliolids (*Quinqueloculina ungeriana*, *Q. undulata* and *Affinetrina planciana*) dominate above the thermocline at depths of 4–20 m and temperatures of 20–23 °C, while below the thermocline from 20 to 70 m and temperatures ~15 °C *Textularia bocki*, *T. pseudorugosa* and *Gaudryina rudis* are the principal species. Although the water characteristics are the same in the more sheltered example, *Peneroplis pertusus* and *P. planatus* together with some *T. bocki* are dominant above the thermocline, while *T. bocki* is by far the most important species below the thermocline (Drobne and Cimmerman 1984).

Other shelf areas

The faunas of the shelf off Motril and Nerja near Malaga, Spain, have been listed by Sanchez-Ariza (1983, 1984) but because of the curious way of presenting the data it is impossible to determine associations. The diversity $H(S)$ varies from 2.72 at 5 m to 4.42 at 200 m (total assemblages).

Four ecophenotypes of *Ammonia beccarii* have been recognised through a biometric analysis of material from 35 to 90 m off Banyuls-sur-Mer (Vénec-Peyré 1983b). The life span was determined as 5–7 months. Reproduction was thought to take place four times a year but it is likely to be almost continuous with overlapping generations.

The general characteristics of the shelf faunas near Marseille have been given by Blanc-Vernet (1969). Close to the shore, seagrass is common and these faunas have been described above. Nearshore sediment away from seagrass supports only a sparse foraminiferal assemblage. *Ammonia beccarii* and *Eggerelloides scabrus* (as *Eggerella*) are present living throughout the year although both are rare in January and February. *Elphidium macellum* occurs from March to October.

A belt of detrital sands extends from ~40 to 100 m. The living assemblage is poor although dead tests are common. Typical living species are *Cibicides lobatulus*, *C. refulgens*, *Planorbulina variabilis* (as *Cibicidella*), *Quinqueloculina* spp. *Triloculina* spp. and various discorbids. Where mud is present *Cassidulina*, *Bulimina* and *Bolivina* are found. The dead assemblage includes *remanié* tests and Blanc-Vernet considers these to represent a period of lower sea-level.

In regions of powerful currents there are coarse sands and gravels at depths of a few meters to 100 m. The fauna is similar to that of the detrital sands with many species living attached to stones and shells.

Between 100 and ~300 m there are shelly sands with mud which support only a sparse living assemblage while the dead assemblage is largely relict Quaternary material.

At Banyuls-sur-Mer there is a coastal seagrass fauna dominated by miliolids and beyond this, at depths of 35–90 m is mud with a living *Ammonia beccarii* association. The only other abundant species is *Nonion commune* (35–50 m). The standing crop varies from 30 to 40 per 10 cm^2 (Vénec-Peyré 1983a).

In the Gulf of Policastro, Southern Italy, Sgarrella *et al.* (1985) related different assemblages to substrate and inflow of river water. However, no salinity data were given and it seems likely that none of the assemblages was affected by brackish water because the river outflow spreads over the sea surface due to its lower density. The main dead assemblages fall into the *Ammonia beccarii* (10–21 m on muddy substrates), *Quinqueloculina* spp. (6–35 m, seagrass and muddy sand substrates), *Valvulineria complanata* (45–84 m on mud and sandy mud) and *Cassidulina laevigata* associations (97–207 m, mud).

Western Mediterranean Basin

Taxonomic notes and illustrations of Neogene species, many of which are still extant, are given by Wright (1978). Very little is known of the deeper-water assemblages. The diversity is less than that of the adjacent Atlantic Ocean but higher than that of the Eastern Mediterranean Basin (Cita and Zocchi 1975). Deep-sea forms which appear to be absent from the Mediterranean include *Cibicidoides kullenbergi*, *Epistominella exigua*, *Nuttallides umboniferus* and *Oridorsalis umbonatus*.

Around Majorca and Minorca Colom (1964) related total assemblages to substrate type. On gravel and shelly bottoms the principal forms are *Sphaerogypsina globula* (as *Gypsina*), *Elphidium crispum*, *Textularia pseudoturris*, *T. pseudorugosa* and *Elphidium complanatum* down to a depth of ~400 m. From 550 and 650 m on muddy sediment the main form is *Martinottiella gymnesica* (as *Listerella*).

Off Provence and Corsica, the total assemblages on muddy substrates are dominated by hyaline taxa. Genera present >10% include '*Bolivina*' (600–2400 m) *Uvigerina* (600–1600 m), *Cassidulina* (600–2400 m, >30% ~600 m), *Melonis* (700–2400 m) and species *Bulimina costata–inflata* (1000–1600 m), *Eponides pusillus* (1200–2600 m) and *Anomalinoides minimus* (1500–2600 m, >30% 2000–2600 m) (Bizon and Bizon 1984b).

In the Balearic Basin, Cita and Zocchi (1978) recognised a *Uvigerina peregrina–U. mediterranea* association with *Hoeglundina elegans* and *Gyroidina* at 1800–2500 m, and a calcareous association with no obvious dominant species but including *Gyroidina*, *Robertina translucens*, *Articulina tubulosa* and *Repmanina charoides* (as *Glomospira*) at 2650–2900 m.

Eastern Mediterranean Basin

The Eastern Mediterranean Basin, which is mostly between 2000 and 3500 m deep, is separated from the western part of the Mediterranean by a sill only ~450 m deep in the Strait of Sicily. Below ~800 the salinity is 38.6‰ and the temperature 13.7 °C.

There are limited data on total assemblages only (Parker 1958; Massiota *et al.* 1976; Cita and Zocchi 1978). Hooper (1969) reinterpreted Parker's data. Parker's data.

Three major associations can be recognised: *Cassidulina laevigata* (71–179 m), *Uvigerina mediterranean* (201–1016 m) and *Gyroidina – Repmanina* association (746–2852 m). The latter is particularly characteristic of depths >1900 m according to Cita and Zocchi. The typical *Gyroidina* species are *G. altiformis* and *G. neosoldanii*. *Repmanina charoides* and *Glomospira gordialis* make up the '*Glomospira*' element originally described by the authors listed above. An additional component at depths >2500 m is *Articulina tubulosa*. The diversity in terms of numbers of species present is 11–64 at 1000–1800 m, 4–8 at 1800–2500 m and <8 at 2500–4000 m (Cita and Zocchi 1975). The number of individuals is also low at depth and it appears that this low-nutrient sluggishly circulating deep water is not favourable for benthic life and, indeed, during the Quaternary underwent several stagnation episodes (see for instance Oggioni and Zandini 1987).

Pollution

Comparison of an unpolluted bay with another affected by sewage and industrial pollution showed that although the foraminifera were affected to a limited degree, no profound changes were observed (Menorca, Mateu 1974).

Postmortem changes

Because there have been few studies of both living and dead assemblages from the same stations it is not possible to determine postmortem changes in detail.

Jorissen (1988) considered that in the Adriatic Sea some agglutinated taxa would be rapidly destroyed after death, e.g. *Reophax* spp. and *Eggerelloides scabrus* (as *Eggerella*). However, the latter species is normally sufficiently robust to survive burial elsewhere.

Dissolution of calcareous tests has been reported off the Po Delta (Jorissen 1988) and in the Black Sea (Macarovici and Cehan-Ionesi 1962) but is not a widespread phenomenon.

There are major differences between the living and dead assemblages in Lim Fjord although on a ternary plot of wall structure the dead assemblages lie within the field for living assemblages from all seasons (Murray 1976a). *Nonionella opima*, which is dominant living, is much reduced in abundance in the dead assemblages. Daniels (1970) attributed this to destruction of the fragile tests by burrowing elements of the *Schizaster–Turritella* community. There is no evidence of loss through transport.

Displaced faunas have been reported from the Eastern Mediterranean Basin (Parker 1958; Massiota *et al.* 1976; Cita and Zocchi 1978) and reworked fossil material is present at a variety of depths from intertidal to the basins. Structureless muds (unifites) of gravity-flow origin are widely distributed throughout the Mediterranean basins (Stanley 1981).

Comment

Overall, the fauna of the Mediterranean is impoverished especially in deeper water. This is partly due to its geographic isolation from both the Atlantic and Indian oceans. Also, the deep waters are more saline and warmer than those of normal oceanic areas. Thus, the deep waters of the Eastern Mediterranean Basin, which are both geographically and environmentally furthest removed from those of the Atlantic, have very low diversity assemblages. The brackish environment of the Caspian Sea has lower diversity than that of the Black Sea.

CHAPTER 12

Indian Ocean

Introduction

The Indian Ocean is here taken to include the Red Sea, Arabian Gulf and the Andaman Sea. To the south it passes into the Southern Ocean, the boundary between the two being at approximately 40° S.

Although there has been a long history of research on the foraminifera (see Bandy *et al.* 1972 for a review and bibliography up to 1971) there have been few detailed ecological studies and large areas remain completely unknown. The exceptions are the Arabian Gulf, studied by researchers from Britain, West Germany and the Netherlands, and the Gulf of Aqaba in the Red Sea, studied by researchers from Israel and several European countries. More is known about the deep sea than of the shelves with the exception of those around India.

All the places to which reference is made in this chapter are shown on Fig. 12.1.

Summary data on associations are given in Tables 12.1–12.8.

Red Sea

The Red Sea is almost completely land-locked and is connected with the Indian Ocean only through the Strait of Bab-el-Mandeb which has a sill depth of 125 m. The climate is hot and arid and there is an absence of rivers. Evaporative loss is balanced by an inflow of Indian Ocean surface water. Salinities are hypersaline and at depths of more than 200 m the water is isothermal (21.7 °C) and isohaline (40.6‰). The nutrient content is low so conditions are oligotrophic. The sediments are almost entirely biogenic with very little terrigenous clastic material (Fairbridge 1966).

Studies of the benthic foraminifera are patchy. Taxonomic reviews include those of Said (1949, 1950a), Perelis and Reiss (1975), Zweig-Strykowski and Reiss (1975), Halicz and Reiss (1979) and Larsen (1976). Ecological studies have been mainly on individual taxa or groups of larger foraminifera. Gabrié and Montaggioni (1982) examined sediment samples from the Jordanian side of the Gulf of Aqaba and recorded a foraminiferal distribution pattern similar to that of the Israeli side (see Reiss and Hottinger 1984).

Said (1950b) examined 3 samples from the Gulf of Suez (59–64 m), 7 from the Gulf of Aqaba which he described as 'the most desolate sea in the world' (265–1161 m) and 40 from the northern Red Sea (24–1148 m). He noted that the deeper parts of the Gulf of Aqaba are dominated by species of *Uvigerina* together with *Cibicides mabahethi* and *Fontbotia wuellerstorfi* (as *Cibicides*). The presence of the latter is of interest bearing in mind that the sill depth in the Bab-el-Mandeb Strait is a mere 125 m.

A mangrove lagoon on the margin of the Gulf of Aqaba (Elat) (salinity 47‰) yield living assemblages dominated mainly by *Sorites* although most of the dead assemblages were dominated by miliolids or *Spirolina*/*Peneroplis*. Halicz *et al.* (1984) suggest that living and dead assemblages are very similar if *Sorites* is ignored. This form is exclusively epiphytic and tests may be partly resorbed and destroyed during reproduction.

The open Gulf of Aqaba is hypersaline (40.3–41.6‰) and isothermal and isohaline from ~200 m down to its maximum depth of 1830 m. The seasonal temperature variation at the surface is 20.5–27.3 °C. The foraminiferal faunas have been reviewed by Reiss and Hottinger (1984). Agglutinated forms occur on all types of substrate and at all depths. Some are attached, e.g. saccamminids, tolypamminids and some textulariids, and are found on large shell debris and other hard substrates. Others occur in association with filamentous green algae. Many of the textulariids have a calcareous

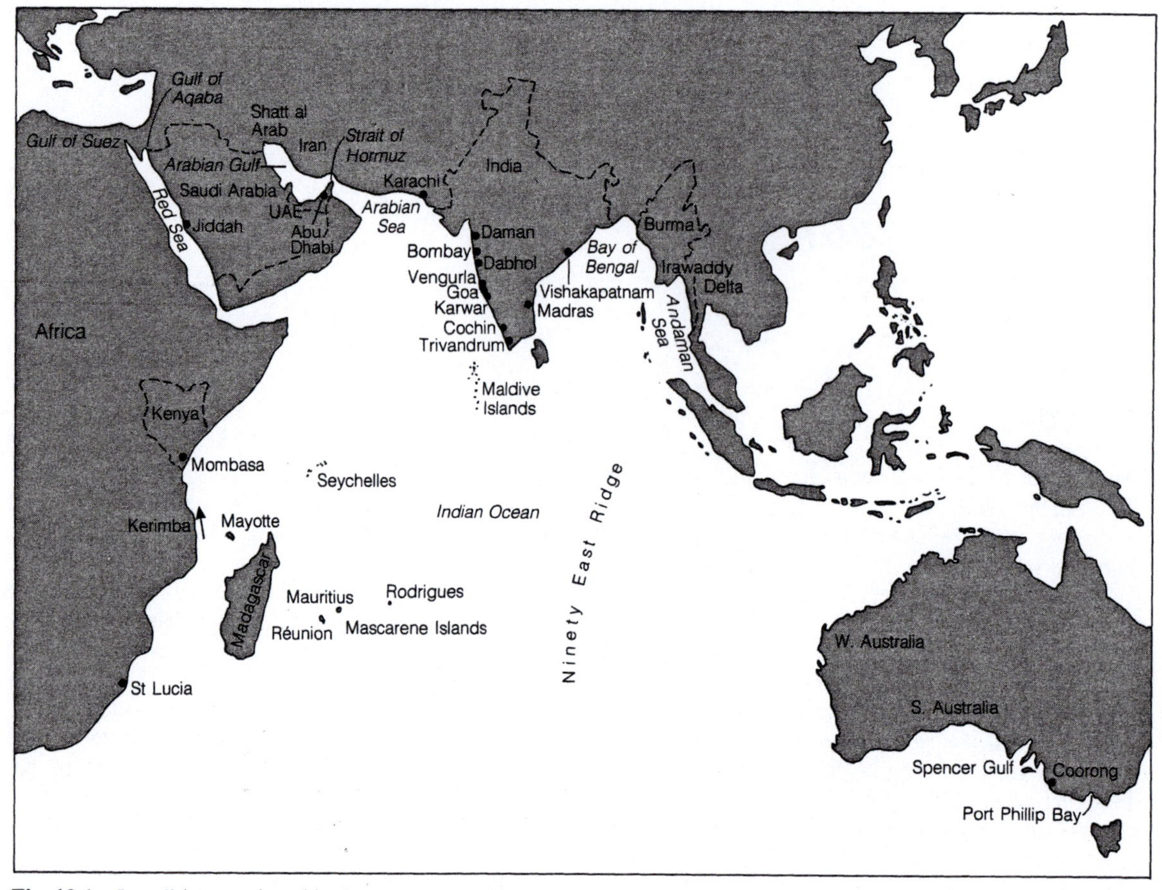

Fig. 12.1 Localities mentioned in the text

cement. Although porcellaneous foraminifera are abundant most of the studies have concerned peneroplids and larger species. Miliolids form ~50% of the assemblage in shallow water but their abundance drops to ~10% at 50 m.

Shallow-water *Brizalina* spp. are ornamented with spines. Buliminids and uvigerinids become abundant below the shelf edge. Cibicidids are abundant at depths shallower than 200 m. Many are attached to hard substrates (e.g. *Cibicides lobatulus*, *Caribeanella elatensis*) but other are free-living. Encrusting taxa include *Planorbulina mediterranensis*, *P. acervalis* and *Acervulina inhaerens*, the latter playing an important role in cementing coral and other coarse debris below the reef front. Arborescent genera (*Miniacina*, *Homotrema*, *Sporadotrema*) are frequent on hard substrates at depths >90 m. They form a pioneer assemblage on fresh surfaces and they become overgrown by organisms of the successor community which includes *A. inhaerens* at depths <130 m where light is present. *Elphidium* spp. commonly live in dense mats of the alga *Cystoseira* and their dead tests are most abundant between 20 and 50 m depth.

The larger foraminifera can be divided according to substrate preference (Fig. 12.2). *Sorites orbiculus* and *Amphisorus hemprichii* live epiphytically on the seagrass *Halophila*. *Amphisorus* has the ability to crawl back on to its host plant if it is dislodged during a storm. It is also negatively phototrophic at high light intensities so that it can moderate the light reaching its dinophycean symbionts (see Fig. 12.2). *Sorites orbiculus* is most abundant in *Diplantheria* meadows. It has a thicker disc than *S. variabilis* which lives in deeper water where it moulds its test to the form of the hard substrate.

On soft sediment, between *Halophila* plants and elsewhere, *Operculina ammonoides* is present. The tests of the deeper forms are thinner than those from shallow water and there is a change from involute above 70 m to evolute below. Forms living on hard or

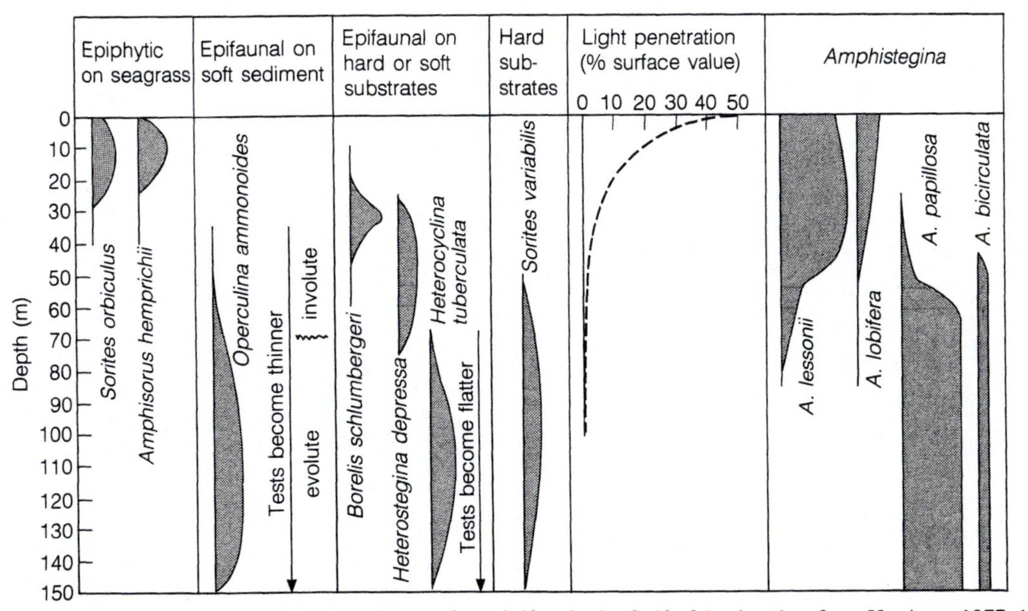

Fig. 12.2 Diagrammatic depth distribution of larger foraminifera in the Gulf of Aqaba (data from Hottinger 1977, 1980; Reiss and Hottinger 1984; Hansen and Buchardt 1977)

soft substrates include *Borelis schlumbergeri*, *Heterostegina depressa* (particularly in high-energy areas) and *Heterocyclina tuberculata* (endemic to the Red Sea–East African Province). These are all free-living species.

Of the four species of *Amphistegina* present, the two thick-walled subglobular forms, *A. lessonii* and *A. lobifera*, occur in shallow water, whereas at depths of 40–80 m they are replaced by thinner-walled, lenticular *A. bicirculata* and *A. papillosa*. *Amphistegina lessonii* favours lagoons and channels while *A. lobifera* prefers higher-energy habitats such as below the front of fringing reefs. There is seasonal control on the reproduction of *Amphisorus* (April–May) and *Amphistegina lobifera* (January and July) but not on the nummulitids.

From three summaries of depth traverses published by Reiss (1977) it can be seen that at depths of 0–50 m assemblages are dominated by miliolids including *Spirolina*, *Peneroplis*, *Sorites*, *Amphisorus* and *Borelis*. Between ~50 and ~150 m *Amphistegina* and *Operculina* are dominant, and deeper than 150 m larger foraminifera are absent. In view of the rather homogeneous water mass in the area, the only variable which could account for the depth distribution is light penetration. Many of the shallower-water species, including all the larger foraminifera, contain algal symbionts which require light. However, other depth-distributed species do not. Those with symbionts can undertake nutrient recycling which is of great benefit in oligotrophic conditions. The *Amphistegina* group is most successful since it has symbionts and also feeds actively on the benthic primary producers (diatoms, etc.) (Reiss and Hottinger 1984).

The stable isotope data for the larger foraminifera (reviewed in Ch. 3) show 2.1‰ depletion in ^{18}O through vital effects associated with the uptake of metabolic CO_2 and O_2 from the symbionts. This may reflect productivity rather than temperature. The $\delta^{13}C$ values are probably controlled by the food consumed.

Postmortem changes are said to be small. There is transport by winnowing in shallow areas exposed to waves and currents, and also transport of foraminiferal tests in clouds of filamentous algae which are moved along shore by currents (Reiss 1977; Hottinger 1977).

Midway along the length of the Red Sea is Jiddah Bay, Saudi Arabia. The area is sheltered with a series of coral reefs running parallel with the coast and intervening troughs up to ~30 m deep floored with carbonate muds and sands and with beds of seagrass. Bottom temperatures range from 20 to 34 °C and salinities 37–48‰ according to season and position within the bay.

The living assemblages have moderate to high diversities ranging from α <5 to 20 overall. Nearshore values were α <5 in winter, α 5–15 in spring and α 5–10 in summer. Offshore the values were α 5–15, 15–20 and 15–20 respectively. The highest diversities

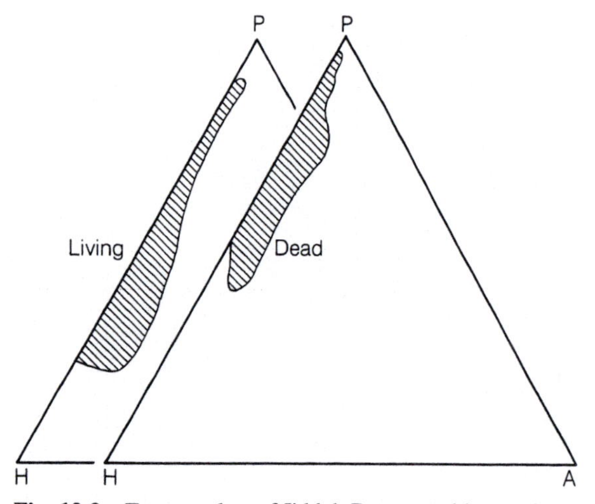

Fig. 12.3 Ternary plots of Jiddah Bay assemblages (data from Bahafzallah 1979). A = agglutinated, H = hyaline, P = porcellaneous

were in the troughs where the hypersalinity was less pronounced than in the nearshore (Bahafzallah 1979). In terms of wall structure, the living assemblages are dominated by porcellaneous and hyaline types and they fall within the field for hypersaline environments defined by Murray (1973) (see Fig. 12.3). Most assemblages appear to be the *Quinqueloculina* spp. association (Table 12.3). On the Egyptian coast Anan (1984) has recorded a *Peneroplis* spp. association from beach sands.

Arabian Gulf

Like the Red Sea, the Arabian Gulf is almost completely land-locked and it has a single connection with the Indian Ocean via the Strait of Hormuz. However, the gulf is shallow throughout and its maximum depth of ~100 m is at the entrance. The Arabian shelf is in part bordered by lagoons and it slopes gently towards the two deeper 'basins' which are close to the Iranian shore. This simple topographic pattern is much modified by the local development of rocky highs and islands sitting atop salt domes.

Fresh water enters the gulf via the Shatt al Arab Delta and ephemeral streams along the Iranian shore, but the climate is arid and very hot in summer so there is net loss of water through evaporation. Bottom salinities are generally in excess of 38‰ and they become increasingly hypersaline in the lagoons and bays along the Arabian shore. Bottom temperatures are around 18–22 °C in the deeper parts but the seasonal range in the lagoons is roughly 15–40 °C.

Most of the gulf is floored with biogenic carbonate sediments with aragonite mud being common only in the low-energy areas such as the lagoons and the deeper 'basins'. The waters are exceptionally clear and all parts of the gulf lie within the photic zone. Submarine vegetation is widely distributed in shallow water and algae, including endolithic forms, extend to considerable depths.

The sedimentary processes operating in the area have been described by various authors in Purser (1973) and by Bathurst (1971) and the marine biology of the western part is beautifully illustrated in Basson *et al.* (1977). A general regional study of the foraminifera has been carried out by Hughes-Clarke and Keij (in Purser 1973), while more detailed studies of the Iranian side are those of Lutze *et al.* (1971), Lutze (1974b), Haake (1970, 1975) and for the United Arab Emirates shore Murray (1965a, 1966a, b, c, 1970a, 1970b) and Evans *et al.* (in Purser 1973).

Hypersaline lagoons

The extensive lagoon complex along the coast of the United Arab Emirates is protected from the open gulf by barrier islands. Except in the channels, depths are generally <3 m. Subtidal rocky areas support seaweeds while subtidal sediments are sometimes clothed in seagrass meadows. Blue-green 'algae' (cyanobacteria) form mucilaginous mats on intertidal to subtidal sediment surfaces especially in the inner lagoon. Mangroves and other halophytes form thickets in the upper intertidal and supratidal zones.

Although dead benthic foraminifera are abundantly present in lagoon sediments, the standing crop of the living assemblage is low. The majority of living forms are found living epiphytically and only on death are their tests contributed to the sediment. Standing crop values range from 1 to 47 per 10 cm^2, the high values coming from plants which also yield high biomass values (max. 4 mm^3 per 10 cm^2) where *Peneroplis planatus* is abundant (Murray 1970b). Living assemblages generally have diversities of α 1–6 with one value of α 7.5.

The principal associations are those of *Peneroplis* spp. and *Quinqueloculina* spp./*Triloculina* spp. (Table 12.3). These are widely developed in most of the lagoon subenvironments. A *Rosalina adhaerens* association is present on coarse sands and shell gravels in areas of powerful tidal currents. This small species lives attached to shells, shell debris and ooliths. Other

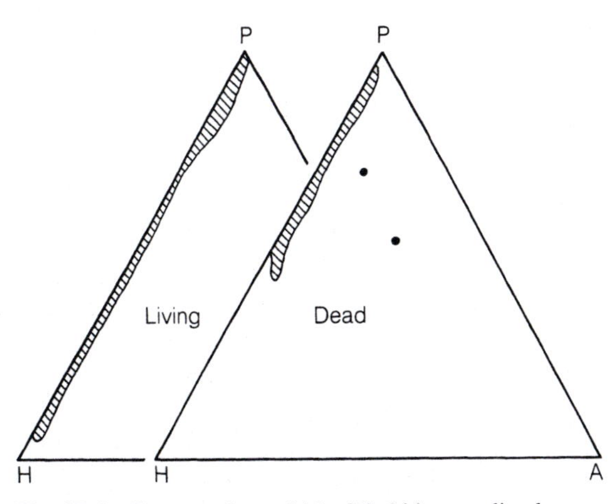

Fig. 12.4 Ternary plots of Abu Dhabi hypersaline lagoon assemblages (data from Murray 1970b)

Table 12.2 Typical lagoon species (feeding strategies from Ch. 2)

Species	Mode of life	Feeding strategy
Ammonia beccarii	Infaunal, free	? Herbivore
Miliammina fusca	Infaunal, free	Detritivore
Peneroplis spp.	Epifaunal, epiphytic, clinging	Herbivore
Quinqueloculina spp.	Epifaunal, epiphytic, free	Herbivore
Triloculina spp.	Epifaunal, epiphytic, free	Herbivore

minor assemblages form the *Elphidium* aff. *advenum* and *Ammonia beccarii* associations.

The outstanding feature of most of the assemblages is the dominance of Miliolina (Fig. 12.4). In the very hypersaline lagoons adjacent to Abu Dhabi the morphotypes of *Quinqueloculina* and *Triloculina* are smooth and striate and generally <500 μm in length. In the less hypersaline Khor al Bazam more highly ornamented morphotypes are present and the size range extends up to 1.2 mm or higher. However, the peneroplids are of similar size range throughout (Murray 1966b).

With the exception of peneroplids and the local occurrence of *Sorites marginalis* in the Halat al Bahrani Lagoon (to the west of Abu Dhabi) larger foraminifera are conspicuously absent. This must be due to the extreme salinity and temperature ranges experienced here.

Hypersaline shelf

The Arabian Gulf is the most extensive hypersaline shelf on earth. Salinities range from 38 to 40‰ throughout much of the area but they increase to 42‰ along the Arabian shore. In the Iranian part, the standing crop of the living assemblages ranges from 15 to 1054 per 10 cm^2 with most values less than 200 per 10 cm^2 (Lutze 1974b). The diversity varies from α 3–16 but is mostly α 5–10 and for the dead assemblages the values are α 3.5–19, mostly α 5–13.

The living assemblages fall mainly into the *Bulimina marginata biserialis* and *Nonionoides boueanum* associations (Table 12.6) with other samples representing the *Ammobaculites persicus*, *Ammonia* spp., *Bolivina pacifica*, *Brizalina striatula*, *Cancris auricula*, *Cassidulina* sp., *Elphidium advenum*, *E.* aff. *discoidale*, *Nonionella opima* and *Reophax calcareus* associations (Tables 12.4 and 12.6). Although the *Bulimina marginata biserialis* association is also common dead, the *Nonionoides boueanum* association is rare. The *Quinqueloculina–Textularia* association, not recorded living, is abundant as dead assemblages. It is probable that the principal elements are epifaunal on hard substrates, such as mollusc shells (see Table 12.5), and that the latter have not been adequately sampled to yield living assemblages. Off the United Arab Emirates, where the salinity is ~42‰, *Textularia* is absent and a dead *Quinqueloculina* association is present. Other minor associations represented only by dead assemblages are listed in Table 12.4.

Detailed studies of the Miliolina component of the

Table 12.1 Minor associations from Indian Ocean lagoons (L = living, D = dead, T = total)

	Association		Depth (m)	Area	Source
1.	*Elphidium* aff. *advenum*	L	0–2	Abu Dhabi, UAR	Murray 1970b
2.	*Rosalina adhaerens*	L	3–5	Abu Dhabi, UAR	Murray 1970b
3.	*Trochammina inflata*	L, T	1.5	Vembanad, W. India	Antony 1980b

Table 12.3 Principal associations in Indian Ocean lagoons and estuaries

***Ammonia beccarii* association**

Salinity: 1–50‰

Temperature: 20 to 34 °C

Substrate: silty mud, sand

Depth: 0–8 m

Distribution:
1. Abu Dhabi, UAE (Murray 1965a, 1966a) dead
2. Vembanad, W. India (Antony 1980b) living, total
3. Kayamkulam Lake, W. India (Antony 1975) living, total (as *Rotalia*)
4. Cochin, W. India (Seibold 1981) living, dead (as *A. sobrina*)
5. Vellar, E. India (Ramanathan 1970) living (as *A. beccarii sobrina*)
6. Pennar, E. India (Reddy *et al.* 1975; Reddy and Rao 1983) living, total
7. Godavari, E. India (Narappa *et al.* 1981) living
8. Suddageda, E. India (Rao and Rao 1974) living, total
9. Bendi Lagoon, E. India (Naidu and Rao, 1988) living

Additional common species:

Quinqueloculina spp.	1
Triloculina spp.	1
Peneroplis pertusus	1
Elphidium spp.	1
Quinqueloculina seminula	2, 6, 8, 9
Trochammina hadai	4
Textularia earlandi	4
Nonionoides boueanum	4 (as '*Nonion asterizans*')
Quinqueloculina cf. *milletti*	4
Murrayinella murrayi	4 (as *Pararotalia*? cf. *globosa*)
Elphidium matagordanum	6 (as *Nonion*)
Quinqueloculina lata	8
Ammobaculites exiguus	8
Elphidium simplex	8
Quinqueloculina subcuneata	8
Haynesina sp.	9

***Miliammina fusca* association**

Salinity: 0–39‰, mainly 0–20‰

Temperature: 27 to 33 °C

Substrate: silty mud or sand

Depth: <5 m

Distribution:
1. Vembanad, W. India (Antony 1980b) living, total
2. Kayamkulam Lake, W. India (Antony 1975) living, total (as *M. oblonga*)
3. Bendi Lagoon, E. India (Naidu and Rao 1988) living, total

Additional common species:

Ammonia beccarii	1
Saccammina spherica	2

***Peneroplis* spp. association**

Salinity: 42–70‰

Temperature: 16 to 40 °C

Substrate: muddy, carbonate sand; seagrass

Depth: 0–15 m

Distribution:
1. Egypt, Red Sea (Anan 1984) total
2. Abu Dhabi, Arabian Gulf (Murray 1965a, 1966a) dead (Murray 1970b) living
3. Khor Al Bazam, Arabian Gulf (Murray 1966b) dead

Additional common species:

Quinqueloculina spp.	1, 2, 3
Amphisorus hemprichii	1
Amphistegina radiata	1
Ammonia beccarii	2, 3
Triloculina spp.	2, 3
Spirolina arietina	2, 3
Spirolina acicularis	2, 3

Table 12.3 *(continued)*

***Quinqueloculina* spp. /*Triloculina* spp. association**

Salinity: 37–70‰

Temperature: 16 to 40 °C

Substrate: carbonate sand and mud; seagrass

Depth: 0–30 m

Distribution:
1. Jiddah, Red Sea (Bahafzallah 1979) living, dead
2. Abu Dhabi, Arabian Gulf (Murray 1965a, 1966a) dead, (Murray 1970b) living
3. Khor al Bazam, Arabian Gulf (Murray 1966b) dead

Additional common species:	
Peneroplis planatus	1, 2, 3
Elphidium advenum	1
Rosalina adhaerens	1
Peneroplis pertusus	2, 3
Ammonia beccarii	2, 3
Elphidium spp.	2, 3
Miliolinella spp.	2

assemblages has shown that some species are depth related. For example, *Quinqueloculina crassicarinata* ranges from 8 to 75 m but peaks at 20–60 m (Haake 1975). Furthermore, ternary plots of three groups of taxa show a progressive change in relation to depth (Fig. 12.5).

A ternary plot of the living assemblages from the Iranian part of the gulf (Fig. 12.6) shows a

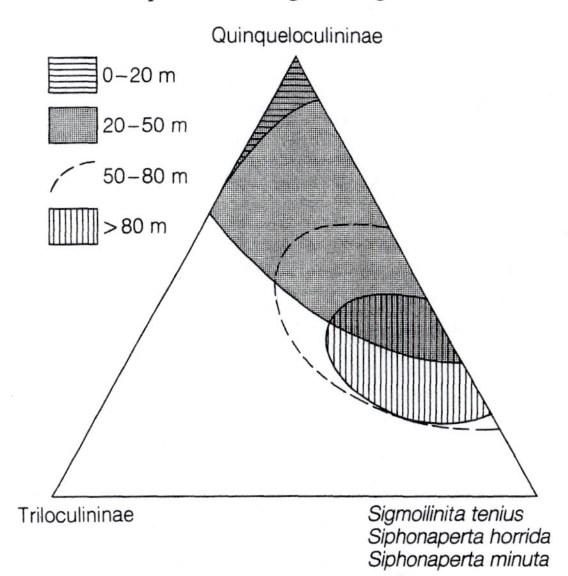

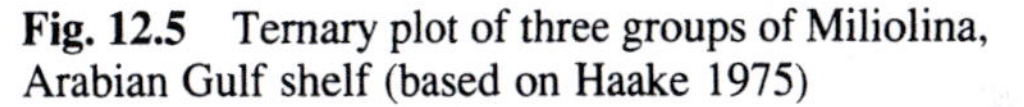

Fig. 12.5 Ternary plot of three groups of Miliolina, Arabian Gulf shelf (based on Haake 1975)

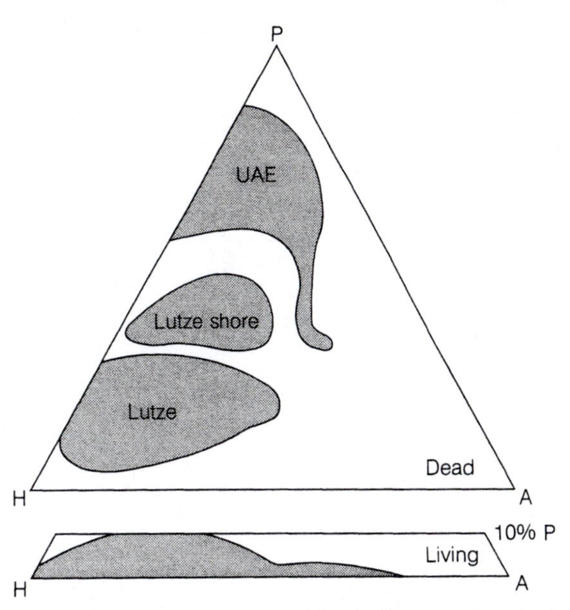

Fig. 12.6 Ternary plots of Arabian Gulf hypersaline shelf assemblages (based on data in Murray 1966c and Lutze 1974b)

Table 12.4 Minor associations from Indian Ocean shelf seas (L = living, D = dead, T = total)

Association		Depth (m)	Area	Source
1. *Rosalina adhaerens*	L, D	3–10	Abu Dhabi, UAR	Murray 1965a, 1966a
2. *Rotaliammina mayori*	L	3–5	Abu Dhabi, UAR	Murray 1965a
3. *Brizalina striatula*	L, D	16–41	Arabian Gulf	Lutze 1974b
	D	34	W. India	Setty and Nigam 1980
4. *Bolivina pacifica*	L	61–94	Arabian Gulf	Lutze 1974b
5. *Reophax calcareus*	L	68–98	Arabian Gulf	Lutze 1974b
6. *Nonionella opima*	L, D	21–45	Arabian Gulf	Lutze 1974b
7. *Cancris auricula*	L	28–50	Arabian Gulf	Lutze 1974b
	L	25–28	W. India	Seibold and Seibold 1981
8. *Elphidium* aff. *discoidale*	L	31	Arabian Gulf	Lutze 1974b
9. *Elphidium advenum*	L	64	Arabian Gulf	Lutze 1974b
	T	0–14	Albany, W. Australia	McKenzie 1962
10. *Cassidulina* sp.	L	204	Strait of Hormuz	Lutze 1974b
11. *Bolivina persiensis*	D	66–99	Arabian Gulf	Lutze 1974b
12. *Haynesina* aff. *schmitti*	D	15–25	Arabian Gulf	Lutze 1974b
	D	20	W. India	Siebold and Siebold 1981
13. *Ammonia* spp.	L, D	8–29	Arabian Gulf	Lutze 1974b
14. *Trochammina inflata*	D	15–32	W. India	Setty and Nigam 1980
15. *Nonionoides elongatum*	D	37	W. India	Setty and Nigam 1980
16. *Bulimina exilis*	D	45–51	W. India	Setty and Nigam 1980
17. *Trochammina pacifica*	D	34	W. India	Setty and Nigam 1980
18. *Hanzawaia nitidula*	D	30	W. India	Seibold and Seibold 1981
19. *Murrayinella erinacea*	L	14	W. India	Seibold and Seibold 1981
20. *Operculina ammonoides*	T	20	W. India	Rao *et al.* 1985
21. *Amphistegina radiata*	T	20	W. India	Rao *et al.* 1985
22. *Nonionoides scaphum*	T	20	Andaman Sea	Frerichs 1970
23. *Amphistegina papillosa*	T	43	Andaman Sea	Frerichs 1970
24. *Hanzawaia nipponica*	T	25	Andaman Sea	Frerichs 1970
25. *Bolivina subreticulata*	T	49	Andaman Sea	Frerichs 1970
26. *Gavelinopsis praegeri*	T	63–77	Andaman Sea	Frerichs 1970
27. *Gaudryina wrightiana*	T	61	Andaman Sea	Frerichs 1970

Table 12.5 Typical shelf taxa from the Indian Ocean

Species	Mode of life	Feeding strategy
Ammobaculites persicus	? Infaunal	? Detritivore
Ammonia	Infaunal, free	? Herbivore
Asterorotalia	? Epifaunal	?
Brizalina spathulata	? Infaunal	? Detritivore
Bulimina marginata	Infaunal or epifaunal, free	?
Cassidulina laevigata	Infaunal, free	? Detritivore
Nonionoides boueanum	? Infaunal	?
Quinqueloculina spp.	Epifaunal, free	Herbivore
Textularia	Epifaunal,	?

preponderance of hyaline and agglutinated wall types. However, as noted above, porcellaneous taxa are more abundant in the dead assemblages and especially in those from close to shores. This trend is also apparent in the shallow shelf off the United Arab Emirates. The agglutinated taxa predominantly have a calcareous cement and in this respect these assemblages differ from most other agglutinated-rich examples from non-hypersaline environments.

Hyaline larger foraminifera are know from the open gulf where they occur mainly on sandy sediments on topographic highs (Hughes-Clarke and Keij in Purser 1973). They do not extend into the shallow shelf are along the Arabian coast presumably because the waters are too hypersaline (salinity >42‰). Taxa recorded include *Operculina*, *Amphistegina* and *Heterostegina*. In the Iranian part, Lutze *et al.* (1971)

Table 12.6 Principal associations from Indian Ocean shelves

***Ammobaculites persicus* association**

Salinity: 34–40‰

Temperature: 27 to 30 °C

Substrate: mud

Depth: 6–43 m

Distribution:
1. Arabian Gulf (Lutze 1974b) living, dead
2. Daman–Bombay (Nigam 1984) living (Nigam 1986b) dead
3. Cochin (Seibold and Seibold 1981) living

Additional common species:

Species	Distribution
Brizalina striatula	1 (as *Bolivina*)
Ammonia spp.	1
Nonionella opima	1
Reophax nana	1
Miliolinae	1
Haynesina schmitti	1
Ammonia tepida	3
Asterorotalia dentata	3
Nonionoides boueanum	3 (as *'Nonion asterizans'*)
Nouria polymorphinoides	3

***Ammonia–Asterorotalia* association**

Salinity: 30–36‰

Temperature: 21 to 30 °C

Substrate: mud, often organic-rich

Depth: 3–55 m

Distribution:
1. Karachi to Cochin (Zobel 1973) dead (as *Ammonia–Florilus* and *Ammonia–Cancris* facies)
2. Daman–Bombay (Nigam 1984) living (Nigam 1986b) dead
3. Dabhol–Vengurla (Nigam *et al.* 1979; Setty and Nigam 1980) dead
4. Binge (Bhatia and Kumar 1976) total
5. Cochin (Seibold and Seibold 1981) living, dead
6. Trivandrum (Rao *et al.* 1985) total
7. Palk Bay, E. India (Rasheed and Ragothaman 1978) living, total
8. Andaman Sea (Frerichs 1970) total

Additional common species:

Species	Distribution
Nonionoides boueanum	1 (as Nonion), 3 (as *Florilus*) 4, 5 (as *Nonion asterizans*), 7 (as *Florilus*)
Quinqueloculina spp.	1, 5, 6, 7
Ammobaculites persicus	1 (as *A. agglutinans*), 5
Trochammina inflata	1 (doubtful)
Nonionoides elongatum	3 (as *Nonion*)
Nonionoides scaphum	3, 4, 8 (as *Florilus*)
Nonionella basispinata	3 (as *Florilus*)
Spiroloculina scita	4
Quinqueloculina curta	4
Murrayinella erinacea	5
Elphidium somaense	5
Hanzawaia nitidula	5
Brizalina striatula	5 (as *Bolivina*)
Cassidella panikkari	5
Elphidium advenum	6
Amphistegina radiata	6
Hanzawaia concentrica	6
Elphidium translucens	7
Bolivina pseudoplicata	7
Miliolinella circularis	7
Arenoparrella mexicana asiatica	8
Bolivina subreticulata	8
Cibicides dorsopustulosus	8
Elphidium macellum	8

***Brizalina spathulata* association**

Salinity: 32–35‰

Temperature: 17 to 25 °C

Substrate: muddy sand

Depth: 27–82 m

Distribution:
1. Andaman Sea (Frerichs 1970) total (as *Bolivina*)

Additional common species:

Species	Distribution
Nonionoides scaphum	1 (as *Florilus*)
Hanzawaia nipponica	1
Gyroidina cushmani	1
Cassidulina minuta	1
Globocassidulina subglobosa	1 (as *Cassidulina*)

Table 12.6 *(continued)*

Bulimina marginata biserialis **association**

Salinity: 38–40‰

Temperature: ?

Substrate: mud

Depth: 21–77 m

Distribution:
1. Arabian Gulf (Lutze 1974b) living, dead

Additional common species:

Nonionoides boueanum	1 (as *Nonion asterizans*)
Nonionella opima	1
Nonionoides cf. *grateloupi*	1 (as *Nonionella*)
Bolivina pacifica	1
Bolivina persiensis	1
Reophax calcareus	1
Cancris auricula	1
Ammonia spp.	1
Textularia spp.	1
Miliolinae	1
Recurvoides trochamminiforme	1 (as *Cribrostomoides*)
Aubignyna cf. *hamanakoensis*	1

Cassidulina laevigata **association**

Salinity: ?

Temperature: ?

Substrate: ?

Depth: 50–110 m

Distribution:
1. Karachi–Cochin (Zobel 1973) dead (as *Cassidulina–Cibicides* facies)

Additional common species:

Cibicides spp.	1
Hanzawaia concentrica	1

Nonionoides boueanum **association**

Salinity: 38–40‰

Temperature: ?

Substrate: muddy sand

Depth: 9–143 m

Distribution:
1. Arabian Gulf (Lutze 1974b) living, dead (as *Nonion asterizans*)
2. Dabhol–Vengurla (Setty and Nigam 1980) dead
3. Cochin (Seibold and Seibold 1981) living, dead (as '*Nonion asterizans*')

Additional common species:

Brizalina striatula	1, 2 (as *Bolivina*)
Bolivina dilatata	1
Cancris auricula	1
Nonionella opima	1
Bolivina pacifica	1
Bolivina persiensis	1
Ammobaculites persicus	1, 2
Reophax calcareus	1
Bulimina marginata biserialis	1
Miliolinae	1
Ammonia spp.	1, 2
Nonionoides elongatum	2 (as *Nonion*)
Bulimina exilis	2
Asterorotalia dentata	2
Cassidella panikkari	2
Nonionella cf. *monicana*	2

Quinqueloculina **spp. association**

Salinity: 42‰

Temperature: 20 to 33 °C

Substrate: fine sand

Depth: 5–15 m

Distribution:
1. United Arab Emirates (Murray 1965a, 1966c) dead

Additional common species:

Triloculina spp.	1
Peneroplis pertusus	1
Peneroplis planatus	1
Murrayinella murrayi	1 (as *Eponides*)
Ammonia beccarii	1
Rosalina adhaerens	1 (as *R.* n. sp.)
Articulina spp.	1

Table 12.6 *(continued)*

***Quinqueloculina-Textularia* association**

Salinity: 38–42‰

Temperature: 20 to 33 °C

Substrate: muddy carbonate sand

Depth: 5–204 m

Distribution:
1. United Arab Emirates (Murray 1966c) dead
2. Iranian part, Arabian Gulf (Lutze 1974b) dead

Additional common species:

Rosalina adhaerens	1
Articulina spp.	1
Triloculina spp.	2
Cassidulina minuta	2
Bolivina persiensis	2
Bulimina marginata biserialis	2
Asterorotalia dentata	2
Asterorotalia inflata	2
Ammonia spp.	2
Rosalina/Discorbis spp.	2
Nonionoides boueanum	2 (as *Nonion asterizans*)
Haynesina schmitti	2 (as *Protelphidium*)

***Uvigerina peregrina* association**

Salinity: ?

Temperature: ?

Substrate: ?

Depth: 106–233 m

Distribution:
1. Karachi–Karwar (Zobel 1973) dead (as *Uvigerina–Cassidulina* facies)

Additional common species:

Cassidulina laevigata	1

recorded *O. ammonoides*, *A. madagascariensis* and *H. depressa* with standing crop values of 1–4, 162–248 and 1–96 per 10 cm^2 respectively. Dead tests are widely distributed. Some examples of *Amphistegina* bear 'tooth marks' which the authors attribute to attack by echinoids. For details of the stable isotopic composition of larger foraminifera see Chapter 2.

Postmortem changes

Although early studies of the lagoonal assemblages pointed to the likelihood of major transport, subsequent sampling has shown that the living and dead distributions are very similar (Murray 1970b). Some bed-load transport takes place in the channels due to tidal currents but these are bi-directional and may not cause much net displacement. Under very calm conditions, the rising tide sometimes floats off empty tests to form a scum which is transported shorewards. Some destruction of tests, particularly of peneroplids, is caused by endolithic algae but this is common only in the Khor al Bazam (Murray 1966b).

Along the barrier island coast of the United Arab Emirates the powerful onshore wind transports foraminiferal tests from the beach on to adjacent dunes (Evans *et al.*, in Purser 1973). The preservation is good and it would not be easy to recognise that deposition had taken place subaerially. On the shelf some transport is caused by tidal currents and waves. Hughes-Clarke and Keij (in Purser 1973) suggested that the high abundance of tests at ~36 m depth may be due to the offshore transport of small forms into this muddy depositional area.

In the nearshore area off United Arab Emirates the dead assemblages contain blackened foraminiferal tests, and off Abu Dhabi these form up to 44% of the whole. They have a lower diversity and a greater proportion of broken tests than the unblackened assemblage. Black tests have been observed in lithoclasts so they may be somewhat older or relict (Murray 1966c).

Planktonic:benthic ratio

The Arabian Gulf is an excellent example of an area into which dead planktonic tests are transported. The evaporative loss from the gulf is replenished by the inflow of surface waters from the Gulf of Oman. The planktonic foraminifera rapidly die in the entrance to the gulf presumably due to the adverse salinities. A tongue of planktonic tests extends into the eastern gulf and the planktonic : benthic ratio declines from 20 to 10 to 0% with passage to the west or south. This pattern shows no correlation with depth (Hughes-Clarke and Keij in Purser 1973; Murray 1976b).

Comparison between the Red Sea and Arabian Gulf

There are major oceanographic differences between the two areas. The Red Sea is deep and hypersaline conditions are less extreme than in the Arabian Gulf. The fauna in both areas is of Indo-Pacific type but that of the gulf is more impoverished. This can be seen very clearly from the low diversity of larger foraminifera in the Gulf. It is less clear for the smaller foraminifera because not enough is known of their occurrence in the Red Sea.

East Africa

Little is known of the assemblages from the African margin. The taxonomy has been discussed by Möbius (1880) and Heron-Allen and Earland (1914–15) and the latter have beautifully illustrated the taxa.

Lagoon and marsh

The only lagoon which has been examined is that of St Lucia, South Africa (Phleger 1976). This is large, shallow (<2 m) and hypersaline to brackish depending on season, year and position within the lagoon. Most living assemblages are an *Ammonia beccarii* association (commonly >70%) with additional *Quinqueloculina seminula* in a few cases (Table 12.3). Diversity is very low (α <1). Occasionally, *Q. seminula* is dominant over *A. beccarii*. In the channel connecting the lagoon to the Indian Ocean *A. beccarii* is joined by *Bolivina* spp., *Q. seminula* and *Reophax* cf. *R. nana* (living and dead assemblages). A few dead assemblages from marshes include *Trochammina inflata, Miliammina fusca* and *Arenoparrella* cf. *mexicana*. The standing crop was 50–150 per 10 cm^2 except for some unusually high values of 2700+ and 4000+ at two localities receiving nutrients from rivers. There is no significant postmortem alteration of assemblages.

The estuary of the Mgeni River near Durban is microtidal and has a well-developed salt wedge. There is high freshwater discharge in the summer and low in the winter. Cooper and McMillan (1987) described the general features of five assemblages but gave no numerical data. The mangrove mudflats are dominated by *Haplophragmoides wilberti* with some *Helenina andersoni* and non-mangrove mudflats by *Trochammina inflata*. The estuary proper is dominated by *Ammonia parkinsoniana* with a few *Miliammina fusca*. Near the estuary mouth is a shallow marine assemblage together with planktonic tests transported in from the adjacent shelf. Finally, there is a littoral/sublittoral assemblage comprising *Cibicides lobatulus, C. refulgens, Planorbulina mediterranensis* and *Glabratella* sp. These are typical of high-energy environments and are transported in from the inner shelf. On the sandy estuarine sediments, foraminiferal tests are absent 50 cm below the surface and therefore postmortem loss must take place.

Shelf

There is little information on shelf assemblages from east Africa. Levy *et al.* (1982b) described 48 species, including diverse miliolids and larger foraminifera, from a single sample of beach sand from 100 km north of Mombasa, Kenya. Along this coast a fringing reef is separated from the shore by a 'lagoon'. Banner and Pereira (1981) recorded 206 taxa (compared with their revised figure of 350 from the Kerimba Islands based on Heron-Allen and Earland 1914–15. The living assemblages have a diversity of $H(S)$ 1.2–2.8 with the lowest values from a lagoon channel and the highest from the inner and outer reef platform. The standing crop was very variable on a local scale and ranged from 10 000 to 11 000 per 10 cm^2. Some information on the composition of the assemblages was given by Chasens (1981). The dominant species are *Quinqueloculina lamarckiana, Amphistegina lessonii, A. radiata, Calcarina calcar* (given as *Pararotalia calcar* and *P. ozawai*) and *Ammonia beccarii* (from depths of < 20 m). A similar, miliolid-dominated fauna from 25 m off Tanzania was recorded by Neagu (1982).

Heron-Allen and Earland (1914–15) noted that the tropical fauna of the Kerimba Archipelago was influenced by the presence of some cold-water taxa and this was found to be true of the Ilha da Inhaca (Mozambique) by Moura (1965).

Unspecified sessile foraminifera form up to 28% of the entire fauna in dark reef tunnels off Madagascar (Harmelin *et al.* 1985). A single sample from an area of seagrass on the inner reef had an assemblage dominated by *Ammonia convexa* together with miliolids (Monier 1973). A more detailed study of a

Madagascar reef (Battistini *et al.* 1976) revealed the presence of 141 species including a number of larger foraminifera (no numerical data).

Pakistan–West India

The northern Arabian Sea is meso- to macrotidal and the region from Karachi to Goa has a high internal tidal energy flux (Baines 1982). To the south, the tidal range decreases. Although sandy sediments are present on the shore, mud is widely distributed on the shelf and it is often rich in organic carbon (Setty and Nigam 1982). Salinities are normal except close to the shore where they may become brackish.

Lists of foraminifera from beach sands along the west coast of India have been given by Rocha and Ubaldo (1964), Bhalla and Nigam (1979), Antony (1980a) and Bhatia (1956).

It has been suggested that the regional change in mean proloculus size in *Rotalidium annectens* (= *Cavarotalia* of Nigam and Rao 1987) from 45 to 103 μm could be a palaeoclimatic index. This species is common on beaches in western India. The smallest mean proloculus size correlates with highest salinities and highest temperatures.

The only review of west coast microfaunas appears to be that of Setty (1982a).

Lagoons and estuaries

The few areas studied are all brackish to a greater or lesser extent as all are affected by the monsoon. The *Ammonia beccarii* association is the most widely developed and is probably present in Goa (Setty 1984a) as well as the localities listed in Table 12.3. It is characteristic of the lower estuary where salinities range from ~18 to 33‰. The substrate is invariably muddy and often contains more than 1% organic carbon. Because of the dramatic decrease in salinity, often to almost fresh water, during the monsoon (June–September) *A. beccarii* may cease to be able to live in the estuaries (Antony and Kurian 1975) and this implies that recolonisation must take place annually from the adjacent inner shelf. Antony (1980b) and Seibold and Seibold (1981) reported that *A. beccarii* makes a cyst of fine detrital grains and suggested that these may be a protection against the environment. It is also possible that they are normal feeding cysts.

Because of the microtidal nature of the Vembanad and Cochin estuaries, during the pre-monsoon season water enters them as a salt wedge moving upstream along the bottom. This may transport shelf foraminifera into the lower estuary. Antony (1980b) has tabulated the lowest salinities tolerated by such species, e.g. 30‰ for *Amphistegina lessonii*, 28‰ for *Operculina complanata* and *Nummulites cumingii*. However, these taxa do not colonise the estuary as the quoted salinities must be close to the lower limit for their survival.

In the upper estuary *Miliammina fusca* and *Trochammina inflata* associations are present in very brackish waters (Table 12.3).

Shelf

Rao (1969–71) described 84 species from the Gulf of Cambay, near Bombay, and noted that their distribution was related to sediment type.

Antony (1980a) recorded 17 species living in the intertidal area near Cochin and, of these, *Ammonia beccarii*, *Elphidium craticulatus*, *E. crispum* and *Quinqueloculina seminula* were dominant. He also noted a seasonal change in the abundance of dead tests with highest values in the pre-monsoon period (July) at a time when the standing crop for the living assemblages is highest. The size of the dead tests correlates with the grain size of the host sediment.

Only the inner shelf has received much attention. The most widely distributed association is that of *Ammonia–Asterotalia* which is present from the shore to ~55 m (Table 12.6). The typical component species are *Ammonia beccarii* s.l. (including *A. tepida* and *Rotalidium annectens*), *A. papillosa* and *Asterorotalia dentatata*. Where living and dead assemblages have been distinguished, the *Ammonia–Asterorotalia* association is commonly present dead even where the living association is different, e.g. off Cochin (Seibold 1975; Seibold and Seibold 1981). Setty and Nigam (1982) found a positive correlation between the abundance of *Ammonia* and that of organic carbon. Megalospheric tests form ~80% of the assemblages and ~90% of all tests are sinistral (Venkatachalapathy and Shareef 1975).

Locally the *Ammobaculites persicus* association is developed in shallow water. Often *A. persicus* has very high faunal dominance (>60%). It too shows a positive correlation with the abundance of organic carbon (Setty and Nigam 1982).

A number of other associations have been recorded from the shelf (Tables 12.4, 12.6) and of these the *Nonionoides boueanum*, *Cassidulina laevigata* and

Uvigerina peregrina associations are perhaps the most significant.

The high energy of the environments on the shallow parts of the shelf has been claimed to provide microenvironments and to account for the patchy distribution patterns (Setty *et al.* 1979; Setty and Nigam 1980) but further analysis of the data revealed nothing abnormal about the distributions (Nigam and Sarupria 1981; Nigam and Thiede 1983).

A regional study of the shelf by Antony (1968) unfortunately lacks numerical data on the composition of the assemblages but it does provide good illustrations of the taxa.

Beaches on the Laccadive Islands yield only *Calcarina* and *Amphistegina* but the number of species increases into the lagoon. Assemblages are dominated by hyaline forms with subsidiary porcellaneous and aggutinated taxa (Srivastava *et al.* 1985).

Postmortem changes

Although the estuaries of southwest India are microtidal, they have narrow mouths and powerful tidal currents are developed. Transport of inner shelf tests into lower estuary has been reported from Vembanad (Antony 1980b) and Cochin (Seibold and Seibold 1981). The smallest tests are transported furthest, probably as suspended load. In the Vembanad estuary most calcareous tests are etched, indicating some dissolution (the sediment is rich in organic carbon), but Antony considers that there is a general similarity between the living and dead assemblages.

Because of the northern part of the west coast of India experiences large tides, postmortem transport is widespread on the shelf. This affects both distribution patterns and diversity (Setty and Nigam 1986). In the southern part, the tidal range is smaller and although some transport takes place in shallow water near the shore, on muddy substrates it is small (Seibold and Seibold 1981). No authors have reported postmortem dissolution although this might be expected in the organic-rich muds. Relict foraminifera are locally present (Setty 1984b).

Pollution

Discharge of industrial wastes and sewage has been reported from most of the estuaries studied (Setty 1984a, Goa; Antony 1980b, Vembanad; Seibold and Seibold 1981, Cochin), but the effects of pollution on the benthic foraminifera have not been assessed.

Thana Creek, Bombay, has salinities of 35.9–37.7‰ and pH values of 6.5–6.7. Setty (1982b) found living foraminifera to be absent and even the exotic calcareous dead tests transported in by the powerful tidal currents undergo partial or total dissolution. He attributed this to the high level of industrial pollutants together with the organic-rich nature of the muddy sediments. However, seawards the pollutants are diluted, the pH becomes alkaline and here foraminifera become abundant and better preserved (Setty and Nigam 1984).

Acidic discharge from a titanium processing plant near Trivandrum has a profound effect on the foraminifera (Rao and Rao 1979; Setty and Nigam 1984). The effluent has pH 0.8–1.0 and a temperature of 45–50 °C. In the vicinity of discharge foraminifera are absent, and even where the effluent is diluted with seawater it is still sufficiently acidic to cause dissolution of calcareous tests.

By contrast, Cola Bay, Goa, receives hot alkaline discharge from a fertiliser factory and this appears to promote the growth and density of foraminifera (Setty 1976; Setty and Nigam 1984). *Rotalidium annectens* (as *Ammonia*), *Nonionoides scaphum*, *N. boueanum* (both as *Florilus*) and *Elphidium crispum* are abundant near the discharge. *Ammonia beccarii* is present as high-spired megalospheric tests, some of which show deformities, especially in the last whorl.

East India–Bay of Bengal

Few detailed studies have been carried out. The faunas of beach sands have been described by Hamsa (1972) and Bhalla (1967, 1970).

Estuaries

All are subject to extreme environmental conditions due to the effects of the monsoon, which introduces almost freshwater conditions, while at other times of the year salinities may be almost normal. The *Ammonia beccarii* association is widely distributed (Table 12.3). Standing crop values are very low during periods of low salinity (Rao and Rao 1974) because the substrate is mobilised and may be transported away (Ramanathan 1970; Narappa *et al.* 1981). Each estuary is recolonised after the monsoon waters recede, i.e. from November onwards. The diversity of the living assemblages is low (α <1–3.4, Rao and Rao 1974, Reddy and Rao 1983; $H(S)$ 0–2.1, Naidu and Rao 1988) and *A. beccarii* commonly shows high dominance (>70%). The living *Miliammina fusca* association is present in Bendi Lagoon especially in the inner part where summer salinities become hypersaline (Naidu and Rao 1988).

Shelf

No detailed studies have yet been carried out on the east coast of India, but from the sparse data available it seems that the inner shelf has *Ammonia beccarii* and miliolids with some larger foraminifera on clastic substrates. The outer shelf has relict carbonate sediments with robust larger foraminifera which lived in the area at a time of lower sea-level (Gulf of Mannar: Ragothaman and Manivannan 1985; Palk Bay: Ragaothman and Kumar 1985; Madras: Setty 1978a, b; Vishakhapatnam: Ganapati and Satyavati 1958, Ganapati and Sarojini 1959, Vedantam and Rao 1970). Agglutinated foraminifera are rare in these assemblages (Almeida and Setty 1972). Silty sediments of the inner shelf off Burma have *Asterorotalia* spp. total assemblages (Saing 1971).

Rasheed and Ragothaman (1978) examined seasonal changes in the living assemblages of the *Ammonia–Asterorotalia* association (Table 12.6). They found that whereas all the common hyaline species reproduced most actively from March to June, *Asterorotalia dentata* also reproduced during the monsoon months (August–September). Standing crop values were highest prior to the monsoon. Estuarine taxa such as *Ammobaculites exiguus* and *Miliammina fusca* were also living in small numbers and these are no doubt able to recolonise the adjacent estuaries following the end of the monsoon.

Postmortem changes

The extreme conditions in the estuaries with powerful outflow of fresh water from July to September/October causes the fine sediment which had accumulated during the remainder of the year to be flushed out and replaced by sand. Under these circumstances, preservation of estuarine assemblages must be a rare event.

Pollution

Most rivers in India appear to be polluted. Severe pollution involving domestic and industrial sewage, with high concentrations of nitrite, nitrate, ammonia and phosphorus, together with heavy metals (particularly copper, maximum 512 mg l^{-1}, lead, zinc, nickel, chromium and arsenic) and fluoride (maximum 3.5 mg l^{-1}), is present in Visakhapatnam Harbour. Foraminiferal test ornamentation and wall thickness are reduced in the most polluted areas. The most successful species are *Reophax nana*, *Textularia earlandi*, *Ammonia beccarii*, *Brizalina striatula* (as *Bolivina*), and '*Nonion asterizans*' (Naidu *et al.* 1985). Calcareous tests show corrosion but whether this is due to pollution is uncertain.

Andaman Sea

The Andaman Sea, adjacent to the Bay of Bengal, is a silled basin with a sill depth of 1800 m and a maximum depth of ~3800 m. The Irrawaddy Delta forms the northern margin of the area. The sea surface waters are affected by the monsoonal climate, the southwest monsoon (June–August) being wet and causing high runoff from the Irrawaddy Delta, and the northeast monsoon (December–January) being a period of minimum rainfall. Frerichs (1970) analysed the total assemblages of 86 samples from this large and varied area. More than 300 species are present but only abundance of ≥3% were reported in the data tables. Many of the assemblages are diverse so faunal dominance is low.

Only two associations are widely present in the shelf: *Ammonia–Asterorotalia* on the inner shelf and *Brizalina spathulata* association on the outer shelf (Table 12.6). In both cases the substrate is muddy sand and the bottom waters are low in dissolved oxygen (0.8–1.3 and 0.9–2.1 ml l^{-1} respectively). Six minor assemblages were recognised from seven other shelf samples (Table 12.4). Frerichs noted that close to the delta the waters are turbid and the seasonal salinity range is great (20–32‰) and considered that foraminifera are unable to live there. Alternatively, it may be that the rate of sediment build-up and low oxygen availability are the main causes of their absence.

West and south Australia

In a general study of the west shelf, Betjeman (1969) recorded nearly 400 species most of which were represented by <1% abundance. Unfortunately he gave no data tables so associations cannot be recognised. Because this shelf spans latitudes 18–34° S there are marked faunal changes from north to south. Tropical species extend as far south as 28° S where they give way to a warm temperate fauna which includes *Marginopora vertebralis* and *Peneroplis planatus*.

The inner shelf near Albany, southwest Australia, has total assemblages dominated by *Elphidium advenum* sometimes with common *Peneroplis pertusus* and *P. planatus* (McKenzie 1962). On the south coast, in Spencer Gulf, miliolid-rich assemblages (including *Peneroplis*) together with *Discorbis* are present in carbonate sediments with seagrass (Gostin *et al.* 1984). Parr (1945) has listed comments on 136 species from adjacent to Port Phillip Bay, Collins (1974) 279 species from the bay itself, and Chapman (1941) 189 species from the southeast shelf.

Marshes from the Coorong have normal inner shelf living assemblages dominated by *Ammonia beccarii* and *Elphidium* spp. The only true marsh species present is *Trochammina inflata* and this is rare (Phleger 1970).

Non-marine, saline lakes in southern Australia have yielded foraminiferal faunas (Cann and De Deckker 1981). One lake, with a salinity of < 60‰ had *Elphidium* sp. and *Trochammina* sp. Another (salinity 24‰) had *Ammonia beccarii*, ?*Triloculina rotunda* and *Trichohyalus tropicus*. Sediment taken from beneath the dried-up crust was placed in distilled water and within a short time dissolution of salt had raised the salinity to ~70‰. After 2 days, adult *Elphidium* were seen to be moving and extending their pseudopodia. From this it may be inferred that certain taxa are able to survive very hypersaline conditions in moist sediment beneath a dried crust and be able to resume normal activity when conditions become favourable.

Atolls and oceanic islands

These areas are geographically isolated from other shallow-water seas by intervening deep ocean. In the tropical parts of the Indian Ocean they are sites of coral reef growth and carbonate sediment accumulation.

The larger foraminifera of the Maldive Islands show a zonation according to the physical energy of the environment and the nature of the substrate (Hottinger 1980). From 0 to 20 m, on the higher parts of the reefs, *Calcarina calcar* is dominant with some *Heterostegina depressa* and soritids. From 20 to 40 m *H. depressa* forms >50% of the assemblage with accessory *Amphisorus* sp., *Sorites* sp., *Operculina ammonoides* and *Borelis* sp. *Heterostegina depressa* continues to be dominant from 40 to 70 m, but with increasing depth porcellaneous forms become less frequent and hyaline *O. ammonoides*, *Operculina* sp., and *Nummulites cummingii* become more abundant. Deeper than 70 m, *Cycloclypeus carpenteri* and *H. operculinoides* appear. *Amphistegina* spp. are present but no details of their occurrence were given. On muddy substrates only two zones are recognised. In a few metres of water *Marginopora* aff. *vertebralis* forms almost monospecific assemblages. At depths of 40–50 m in the lagoon *O. ammonoides* is dominant.

The faunas of the Seychelles and Mauritius were studied by Möbius (1880) but no ecological details were given. In the Mascarene Islands (Réunion, Mauritius, Rodrigues) benthic foraminifera constitute up to 58% of the biogenic particles in the fore-reef sands. *Amphistegina lessonii* and *lobifera* are abundant in fore-reef and back-reef sediments between meadows of brown algae in areas of fairly high energy and at depths of <30 m. They are especially abundant on Réunion where they form 60–85% of the total assemblage. On this island the reef itself is dominated by epifaunal cemented forms such as *Carpenteria monticularis*, *Homotrema rubra* and *Miniacina miniacea*. These forms live in cavities and under overhangs of the reef (Montaggioni 1981). Because the sediments on Réunion are so coarse and seagrass is absent, miliolids are rare, but on Mauritius and Rodrigues they are abundant especially in the more sheltered back-reef areas where they form up to 90% of the total assemblages. Altogether the fauna is diverse (130 species in the Mascarene Islands) and includes many larger foraminifera, but apart from those mentioned they are not very abundant.

Le Calvez (1965) examined sediments only from Mayotte so epifaunal attached species were not sampled. As in the Mascarene Islands, *Amphistegina* spp. are very abundant together with *Nummulites cumingii* (as *Operculinella*) and *Operculina* spp. Miliolids (*Triloculina* and *Quinqueloculina*) are abundant together with *Discorbis mira* and other attached species in sediments between pinnacles of coral.

Deep sea

Unlike the Atlantic and Pacific oceans, the Indian Ocean is land-locked to the north. The bottom topography is complex with ridges forming boundaries and, to a certain extent, barriers between a series of deep basins. The bottom waters form outside the Indian Ocean and their passage into and around the area is influenced by the bottom topography. The source

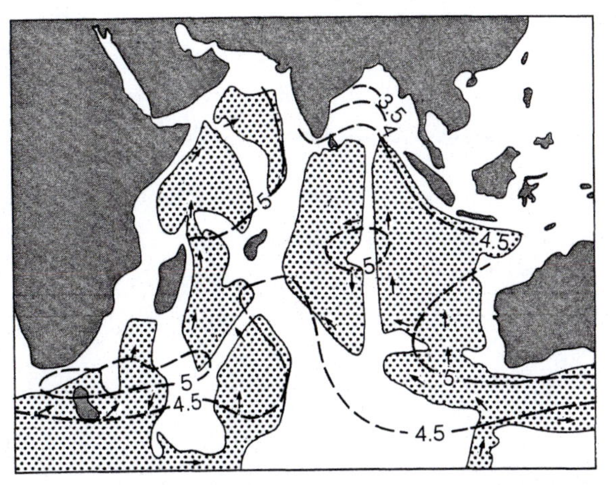

Fig. 12.7 Major basins (dotted) and ridges (unshaded) of the Indian Ocean defined by the 4000 m isobath. Generalised bottom-water movement indicated by arrows. Dashed lines show the depth (m) of the calcite compensation depth (based on Peterson 1984 and Berger and Winterer 1974)

waters are NADW (salinity 34.86–34.72‰, temperature 1.0–2.5 °C) and AABW (salinity 34.66–34.69‰, temperature –0.9 to 0.0 °C). Within the Indian Ocean these mix to form Indian Deep Water (IDW) and Indian Bottom Water (IBW) with subtle differences of temperature and salinity (Fairbridge 1966; Peterson 1984).

Figure 12.7 shows the generalised topography, bottom-water circulation and the depth of the CCD beneath which all carbonate is dissolved. The AABW is undersaturated with respect to calcium carbonate and the deeper areas undergo partial dissolution even where they lie above the CCD.

Very little is known of upper slope assemblages. Zobel (1973) distinguished a Buliminacea facies, which is found along the eastern margin of India at depths of 230–980 m but mainly at 320–485 m. This correlates with oxygen-minimum waters. At depths of 1595–2487 m there is a modified *Fontbotia wuellerstorfi* association with *Bulimina aculeata* and another at 1735–3670 m with nonionids.

At greater depths than 2000 m there are five minor and five major associations (Tables 12.7 and 12.8). Corliss (1976, 1978, 1979a, b, 1983) and Peterson (1984) used multivariate analysis to define their assemblages which are essentially the same as the associations defined here. The deepest parts are under the influence of AABW, or its derivative, IBW. These areas have a *Nuttallides umboniferus* association and this species often forms a very high percentage of the assemblages. The AABW is undersaturated with respect to calcite so postmortem dissolution of calcareous tests takes place (Corliss 1976; Peterson 1984). *Nuttallides umboniferus* is more resistant than other species to the effects of corrosive waters.

In the Southwest Indian Ocean, NADW overlies AABW (boundary at ~3800 m) and this is characterised by the *Fontbotia wuellerstorfi* association (Corliss 1983). A single sample from the east of the ocean (MD 77–148 of Peterson 1984) is the only other record of this association.

The *Epistominella exigua* association appears to span the diffuse boundary between IBW and IDW. At the Ninety East Ridge, Peterson (1984) found it to be best developed beneath the core of a cold and well-oxygenated northward-flowing geostrophic current. The *Globocassidulina subglobosa* association is found bebeath IDW in the east and southwest Indian Ocean. This species is also common in the southwest although Corliss (1979a) did not recognise a factor assemblage dominated by it. He observed that *G. subglobosa* falls into two groups, uniformly small

Table 12.7 Minor association from the deep-sea Indian Ocean (T = total)

	Association		Depth (m)	Area	Source
1.	*Pullenia bulloides*	T	1980–2226	E	Peterson 1984
		T	2674	SW	Corliss 1983
2.	*Uvigerina peregrina*	T	2010	E	Peterson 1984
3.	*Pyrgo* spp.	T	2160–3254	E	Peterson 1984
4.	*Ehrenbergina trigona*	T	2572	E	Peterson 1984
5.	*Melonis barleeanum*	T	2947	SW	Corliss 1983
		T	3240	Andaman Sea	Frerichs 1970
6.	*Karrerulina apicularis*	T	3420–3778	Andaman Sea	Frerichs 1970
7.	*Repmanina charoides*	T	2151–3618	Andaman Sea	Frerichs 1970

Table 12.8 Principal deep-sea associations from the Indian Ocean

***Epistominella exigua* association**

Salinity: 34.76–34.98‰

Temperature: (2) 1.6, (3) 4.6 to 8.3 °C

Substrate: biogenic ooze

Depth: (1,2) 2985–4200 m, (3) 1080–1600 m

Distribution:
1. North (Burmistrova 1976) total (as *Alabaminoides*)
2. Southwest (Corliss 1983) total
3. Andaman Sea (Frerichs 1970) total

Additional common species:

Alabaminella weddellensis	1
Fontbotia wuellerstorfi	2 (as *Planulina*)
Nuttallides umboniferus	2 (as *Epistominella*)

***Fontbotia wuellerstorfi* association**

Salinity: 34.72–34.73‰

Temperature: 1.1 to 1.9 °C

Substrate: biogenic ooze

Depth: 2187–3781 m

Distribution:
1. Karachi–Cochin (Zobel 1973) dead (as deep-water facies)
2. Southwest (Corliss 19833) total
3. East (Peterson 1984) total

Additional common species:

Epistominella exigua	1
Oridorsalis tener	1, 2
Pullenia bulloides	1, 2
Melonis pompilioides	1
Bulimina aculeata	1
Globocassidulina subglobosa	2
Nuttallides umboniferus	2 (as *Epistominella*)
Eggerella bradyi	3
Pyrgo spp.	3
Uvigerina peregrina	3

***Globocassidulina subglobosa* association**

Salinity: (1,2) 34.72–34.74‰, (3) 34.8–35.0‰

Temperature: (1,2) 1.3 to 2.0 °C, (3) 4.6 to 8.3 °C

Substrate: biogenic ooze

Depth: (1, 2) 2453–3656, (3) 512–1350 m

Distribution:
1. Southwest (Corliss 1983) total
2. East (Peterson 1984) total
3. Andaman Sea (Frerichs 1970) total (as *Cassidulina*)

Additional common species:

Epistominella exigua	1, 2
Oridorsalis tener	1
Hansenisca soldanii	1 (as *Gyroidinoides*)
Pyrgo spp.	2
Uvigerina peregrina	2
Pullenia bulloides	2

***Nuttallides umboniferus* association**

Salinity: 34.6–34.72‰

Temperature: 0.0 to 1.2 °C

Substrate: biogenic ooze

Depth: 2897–4700 m

Distribution:
1. North (Burmistrova 1976) total (as *Osangulariella bradyi*)
2. Southwest (Corliss 1983) total (as *Epistominella*)
3. Southwest (Corliss 1976, 1978, 1979a) total (as *Epistominella*)

Additional common species:

Epistominella exigua	1 (as *Alabaminoides*), 2
Alabaminella weddellensis	1
Fontbotia wuellerstorfi	2, 3 (as *Planulina*)
Globocassidulina subglobosa	2, 3
Pullenia bulloides	2, 3
Oridorsalis tener	2, 3

***Uvigerina* spp. association**

Salinity: 34.71–34.74‰

Temperature: 0.9 to 1.4 °C

Substrate: biogenic ooze

Depth: 2524–4561 m

Distribution:
1. Southwest (Corliss 1976, 1978, 1979a) total

Additional common species:

Epistominella exigua	1

comprising *Brizalina*, *Globobulimina* and *Uvigerina*, individuals deeper than 3500 m and small to large individuals shallower than 3500 m (Corliss 1979c).

The final group, the *Uvigerina* spp. association is confined to the southeast part of the ocean. *Uvigerina* is infaunal whereas the common additional species, *Epistominella exigua*, is epifaunal. The *Uvigerina* spp. association occurs with IBW (Corliss 1979a).

According to Burmistrova (1969) calcareous foraminifera are not found deeper than 4700 m in the Arabian Sea. At greater depths there are agglutinated faunas as would be expected from the level of the CCD.

The Andaman Sea also extends to bathyal depths (~3800 m) although it is isolated from the Indian Ocean with a sill depth of 1800 m. Two of the major associations are present, *Epistominella exigua* and *Globocassidulina subglobosa*, in waters which are marginally more saline but significantly warmer than those of the Indian Ocean (Table 12.8). In addition there are a number of minor associations represented by only one or two samples. In the deepest parts there are agglutinated assemblages of the *Karrerulina apicularis* (as *Karreriella*) and *Repmanina charoides* (as *Glomospira*) associations (Table 12.7). At these depths (>2000 m) dissolution is active (Frerichs 1970). Thus, although the Andaman Sea reaches oceanic depths, it does not have typical deep-sea assemblages because its water-mass structure is more saline and warmer than that of the Indian Ocean.

Summary

The shelf seas and lagoons of the Indian Ocean present some interesting contrasts. The Arabian Gulf, Red Sea and Kenyan coast are all regions of tropical/subtropical arid climate. The waters are hypersaline to a greater or lesser extent because there is high evaporation and no land runoff. They are also low in nutrients and very clear. They support epiphytic miliolid-rich assemblages together with symbiont-bearing larger foraminifera capable of conserving and recycling nutrients. The situation is similar in west and south Australia.

India, Burma and the Andaman Sea have the tropical humid climate with large-scale input of runoff during the summer monsoon. The waters are locally brackish, even fresh at times in the estuaries, and they are turbid. Nutrients abound and the sediments are rich in organic carbon. Most of the important species are infaunal and take advantage of the rich food supply. Porcellaneous forms are rare. *Ammonia* dominates estuaries and nearshore areas, the latter also having *Asterorotalia*. *Nonionoides* is dominant offshore. The deep sea has cold-water assemblages because the bottom waters are derived from the Antarctic (AABW) and the North Atlantic (NADW).

CHAPTER 13

Western margin of the Pacific Ocean

Introduction

This chapter considers the area from the Bering Sea in the north via the Asian seaboard and the tropical west Pacific to eastern Australia and New Zealand. It spans a great spectrum of shallow-water environments from cold to tropical, brackish to marine, turbid to clear water. From a practical point of view, it has to be recognised that the great variety of languages, the sometimes parochial attitude towards taxonomy and

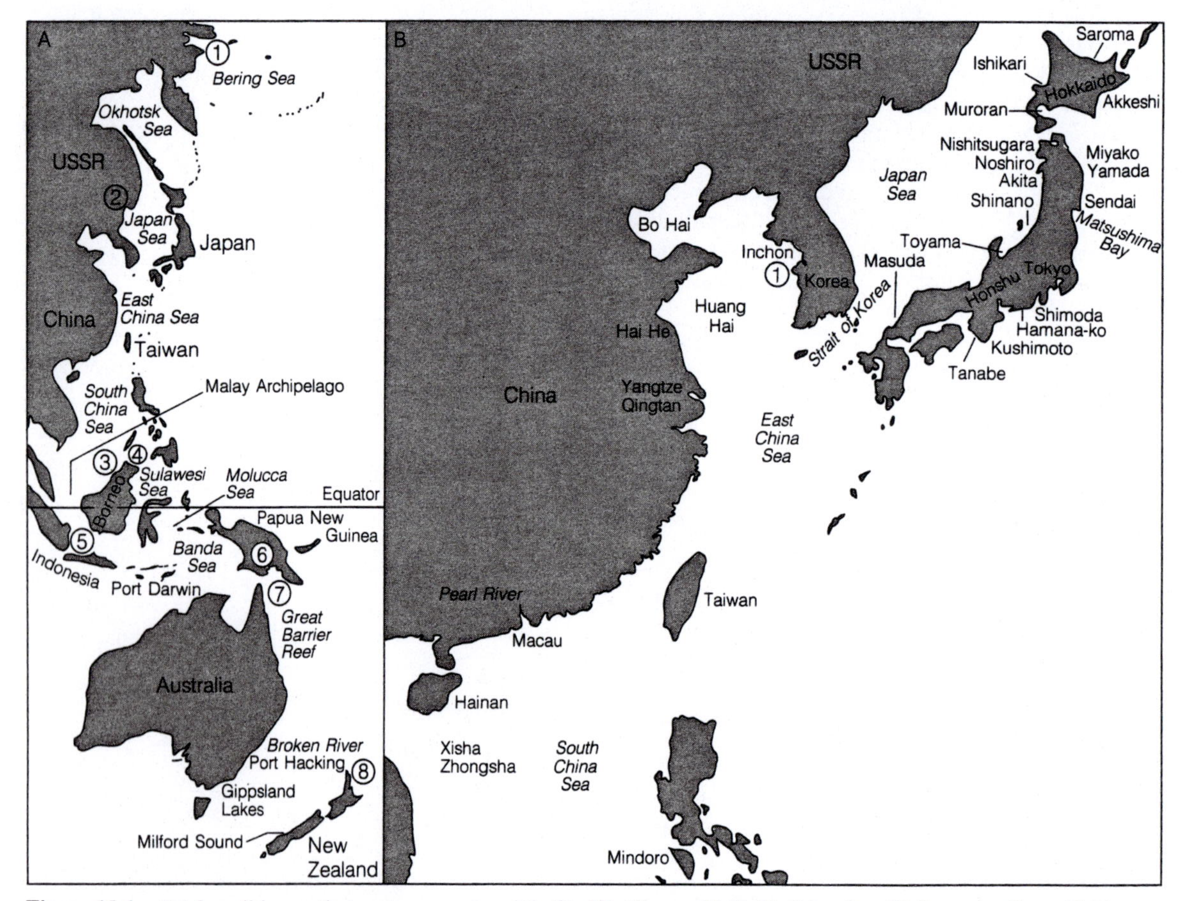

Figure 13.1 (A) Localities on the western margin of the Pacific Ocean: (1) Gulf of Anadyr; (2) Peter the Great Gulf; (3) Brunei; (4) Labuk Sabah; (5) Jakarta Bay; (6) Purari River; (7) Papuan lagoon; (8) Cavalli Islands, Bay of Islands, Great Barrier Island. (B) Detail of Southeast Asia (1) Guangyang, Asan and Gyunggi bays, Korea

the lack of published numerical data on assemblages had made this one of the most difficult areas to present as a synthesis.

Localities mentioned in this chapter are shown in Fig. 13.1

Marshes and mangals

There is very little information of these environments. Brönnimann *et al.* (1983) and Brönnimann and Keij (1986) have described agglutinated taxa from brackish mangrove creeks on Brunei where they form a major part of the fauna. Mangrove marshes from northern New Zealand have very low standing crop values (0–2090 per 10 cm^2, generally ~100, Phleger 1970). The dominant species are *Trochammina inflata*, *Haplophragmoides* sp. and *Ammonia beccarii*. Diversity is α <1. However, Gregory (1973) found *Miliammina petila* to be the dominant species on the muddy sediments of another northern New Zealand mangrove swamp. *Trochammina inflata* formed almost monospecific total assemblages in two samples from a northern New Zealand supratidal marsh (Hayward 1981).

Lagoons and estuaries

Most of the examples for which there is detailed information are from Japan and Korea but the latitudinal spread is from the cold temperate Busse Lagoon in the north to the cold-temperate Gippsland lakes in the south spanning the equatorial region in between. Minor associations are listed in Table 13.1 and major ones in Table 13.2.

Minor cool-water assemblages found only in Busse Lagoon, Russia, include *Elphidium excavatum* and

Table 13.1 Minor associations from lagoons on the western margin of the Pacific Ocean (L = living, D = dead, T = total)

Species		Depth (m)	Area	Source
1. *Ammobaculites exiguus*	T		Busse Lagoon	Fursenko and Fursenko 1973
2. *Elphidium excavatum*	T		Busse Lagoon	Fursenko and Fursenko 1973
3. *Elphidium frigidum*	L, T		Busse Lagoon	Fursenko and Fursenko 1973
4. *Elphidium etigoensis*	T		Saroma, Japan	Yoshida 1954
	L, D	0	Asan Bay, Korea	Chang and Lee 1984
5. *Elphidium jenseni*	T		Saroma, Japan	Yoshida 1954
6. *Elphidium* cf. *subincertum*	T		Saroma, Japan	Yoshida 1954
7. *Ammotium cassis*	T		Saroma, Japan	Yoshida 1954
8. *Ammoglobigerina globigeriniformis*	T		Akkeshi, Japan	Morishima and Chiji 1952
	T		Hamana-ko, Japan	Ishiwada 1958
9. *Trochammina hadai*	T	1–2	Matsushima Bay	Matoba 1970
10. *Elphidium subarcticum*	T	1–9	Matsushima Bay	Matoba 1970
11. *Pararotalia nipponica*	T	1–5	Matsushima Bay	Matoba 1970
12. *Murrayinella minuta*	T	2–4	Matsushima Bay	Matoba 1970
13. *Elphidium gerthi*	T	0	Taiwan	Haake 1980b
14. *Jadammina macrescens*	T	0	Taiwan	Haake 1980b
15. *Elphidium subincertum*	L, D	0	Asan Bay, Korea	Chang and Lee 1984
16. *Nonion nicobarensis*	L	0	Asan Bay, Korea	Chang and Lee 1984
17. *Arenoparrella mexicana*	T		Labuk estuary, Sabah	Dhillon 1968
18. *Haynesina germanica*	T		Hawaii	Coulbourn and Resig 1975
	T		Gippsland, Australia	Apthorpe 1980
19. *Triloculina trigonula*	T		Gippsland, Australia	Apthorpe 1980
20. *Elphidium macellum*	T		Gippsland, Australia	Apthorpe 1980
21. *Elphidium poeyanum*	T		Gippsland, Australia	Apthorpe 1980
22. *Quinqueloculina seminula*	T		Gippsland, Australia	Apthorpe 1980
23. *Ammobaculites barwonensis*	T		Gippsland, Australia	Apthorpe 1980
24. *Trochamminita irregularis*	T		Gippsland, Australia	Apthorpe 1980
25. *Martinottiella communis*	T		Gippsland, Australia	Apthorpe 1980
26. *Ammotium salsum*	T		Gippsland, Australia	Apthorpe 1980
27. *Elphidium williamsoni*	T		Gippsland, Australia	Apthorpe 1980

Table 13.2 Principal associations from lagoons on the western margin of the Pacific Ocean

***Ammonia beccarii* association**

Salinity: 12–35‰

Temperature: 1 to 29 °C

Substrate: mud

Depth: 0–15 m

Distribution:

1. Lake Saroma, Japan (Yoshida 1954) total (as *Rotalia*)
2. Matsushima Bay, Japan (Matoba 1970) total
3. Hamana-ko, Japan (Ishiwada 1958) total (as *Rotalia*)
4. Gwangyang Bay, Korea (Chang 1984a, 1986) living, total
5. Asan Bay, Korea (Chang 1983; Chang and Lee 1984) living, dead
6. Inchon, Korea (Chang and Lee 1982) living, dead
7. Gynggi Bay, Korea (Chang and Lee 1983) living, dead
8. Haihe and Jiyunhe, China (Li 1986) total
9. Bo Hai, China (Wang and Bian in Wang 1985) total
10. Taiwan (Haake 1980b) total
11. Hawaii (Coulbourn and Resig 1975) total (as *A. tepida*)
12. Brisbane River, Australia (Palmieri 1976a) total
13. Gippsland Lakes, Australia (Apthorpe 1980) total (as *A. aoteanus*)

Additional common species:

Trochammina japonica	1, 2, 7
Buccella frigida	1, 2
Elphidium etigoensis	1, 5, 6, 7
Eggerella advena	1 (as *Verneuilina polystroph*), 13
Elphidium subarcticum	2
Trochammina hadai	2
Ammoglobigerina globigeriniformis	3 (as *Trochammina*)
Haplophragmoides canariensis	3
Quinqueloculina ackneriana	4
Ammobaculites sp.	4
Elphidium advenum	4, 13
Elphidium subincertum	5, 6, 7, 9 (as *Cribrononion*)
Nonion nicobarensis	5, 6, 7
Nonion akitaensis	8
Elphidium hughesi	9
Elphidium simplex	9
Elphidium porisuturalis	9 (as *Cribrononion*)
Elphidium gerthi	10
Elphidium incertum	10
Haynesina germanica	11 (as *Protelphidium tisburyense*), 13 (as *Nonion depressulum*)
Elphidium williamsoni	13 (as *E. articulatum*)
Quinqueloculina seminula	13
Miliammina fusca	13
Trochammina inflata	13
Elphidium oceanesis	13
Elphidium poeyanum	13 (as *Cribroelphidium*)
Martinottiella communis	13
Ammotium salsum	13

***Buccella frigida* association**

Salinity: 31‰

Temperature: 11 to 19 °C

Substrate: mud, fine sand

Depth: 3–5 m

Distribution:

1. Busse Lagoon, Russia (Fursenko and Fursenko 1973) total
2. Lake Saroma, Japan (Yoshida 1954) total (as *Eponides*)
3. Akkeshi Bay, Japan (Morishima and Chiji 1952) total (as *Eponides*)

Additional common species:

Trochammina inflata	1
Rosalina columbiensis	1
Trochammina japonica	2
Ammonia beccarii	2 (as *Rotalia*)
Elphidium cf. *subincertum*	2
Elphidium etigoensis	2
Eggerella advena	2 (as *Verneuilina polystroph*)

Table 13.2 *(continued)*

***Eggerella advena* association**

Salinity: 22–31‰

Temperature: 1 to 29 °C

Substrate: mud

Depth: 0–9 m

Distribution:
1. Lake Saroma, Japan (Yoshida 1954) total (as *Vernueilina polystroph*)
2. Matsushima Bay, Japan (Matoba 1970) total (as *E. scabra*)
3. Gippsland Lakes, Australia (Apthorpe 1980) total

Additional common species:

Elphidium etigoensis	1
Elphidium jenseni	1
Elphidium subgranulosum	1
Trochammina japonica	1, 2
Miliammina fusca	1, 3
Ammotium cassis	1 (as *Ammobaculites*)
Buccella frigida	1 (as *Eponides*), 2
Textularia parvula	1
Haplophragmoides canariensis	1
Ammonia beccarii	1 (as *Rotalia*), 3 (as *A. aoteanus*)
Trochammina hadai	2
Martinottiella communis	3
Reophax barwonensis	3
Ammobaculites barwonensis	3

***Haplophramoides canariensis* association**

Salinity: 1–29‰

Temperature: 9 to 31 °C

Substrate: mud, fine sand

Depth: 0–11 m

Distribution:
1. Hamana-ko, Japan (Ishiwada 1958) total
2. Labuk, Sabah (Dhillon 1968) total

Additional common species:

Ammoglobigerina globigeriniformis	1 (as *Trochammina*)
Ammonia beccarii	1 (as *Rotalia* vars.)
Ammobaculites dilatatus	2

***Miliammina fusca* association**

Salinity: 1–25‰

Temperature: 16 to 22 °C

Substrate: mud

Depth: 1 m

Distribution:
1. Busse Lagoon, Russia (Fursenko and Fursenko 1973) living, total
2. Haihe River, China (Li 1986) total
3. Gippsland, Australia (Apthorpe 1980) total

Additional common species:

Trochammina inflata	1
Ammotium cassis	1
Nonion akitaensis	2
Ammobaculites sp.	3
Ammonia beccarii	3 (as *A. aoteanus*)
Elphidium williamsoni	3 (as *E. articulatum*)
Elphidium poeyanum	3 (as *Cribroelphidium*)
Eggerella advena	3

***Trochammina inflata* association**

Salinity: 29–32‰

Temperature: 4 to 20 °C

Substrate: mud

Depth: 0–5 m

Distribution:
1. Busse Lagoon, Russia (Fursenko and Fursenko 1973) living, total
2. Hawaii – lake (Resig 1974) total

Additional common species:

Elphidium excavatum	1
Elphidium selseyense	1
Elphidium frigidum	1 (as *Cribroelphidium*)
Buccella frigida	1

***Trochammina japonica* association**

Salinity: 25–32‰

Temperature: 1 to 29 °C

Substrate: mud

Depth: 0–3 m

Distribution:
1. Lake Saroma, Japan (Yoshida 1954) total
2. Matsushima Bay, Japan (Matoba 1970) living, total

Additional common species:

Elphidium etigoensis	1
Eggerella advena	1 (as *Verneuilina polystroph*), 2 (as *E. scabra*)
Trochammina hadai	2
Ammonia beccarii	2
Elphidium subarcticum	2
Valvulineria hamanakoensis	2

E. frigidum (Table 13.1). The *Buccella frigida* association (Table 13.2) extends as far south as northern Japan. The *Miliammina fusca* association is probably controlled more by low salinity than by temperature for it is found off Russia, China and Gippsland, Australia. The *Eggerella advena* association is present in Japan and Australia. It favours moderately brackish waters and tolerates both high and low temperatures.

The most widespread fauna is the *Ammonia beccarii* association which is present from north Japan to Gippsland Lakes in Australia (Table 13.2). It shows a broad tolerance of temperature and salinity conditions. It also tolerates macrotidal conditions as in Korea and China. Off Korea the standing crops are very variable (0–500 per 10 cm^2) but diversity is consistently low (α <1–1, Chang 1983, 1984a, b, 1986, 1987; Chang and Lee 1982, 1983, 1984).

The extreme environment of the Purari River in Papua New Guinea (salinity 1‰, temperature 24–25 °C, pH 5.5–5.7) causes the standing crop to be very low (~4 per 10 cm^2, Haig and Burgin 1982). There are several studies of Australian estuaries which do not give data tables (Johnson and Albani 1973; Albani and Johnson 1976; Albani 1968, 1974, 1978).

Postmortem effects

The macrotidal Korean estuaries experience active postmortem transport of tests and the introduction of reworked fossil tests (Chang 1983, 1984a, b, 1986, 1987; Chang and Lee 1982, 1983, 1984). In the macrotidal Qiangtan Estuary, China, 85–78% of the tests are exotic and they are size sorted (75–325 μm, mean 116–142 μm) while in the mesotidal Yangtze the figures are 88–74% exotic, 75–425 μm, mean 124–134 μm (Wang and Murray 1983). By contrast the microtidal Pearl River has <9% exotic tests.

Macrotidal Port Darwin, north Australia, has extensive reworking of foraminiferal tests in the subtidal areas and this causes mixing of faunas from different environments. The estuary is marine for most of the year and this together with the mixing gives rise to very diverse dead assemblages (α 8–17 in the channel, Michie 1987).

Microtidal Gippsland Lakes have similar living and dead assemblages with little evidence of transport, but there are indications of some postmortem dissolution of calcareous tests in the more brackish parts and this leads to assemblages rich in agglutinated tests (Apthorpe 1980).

Shelf and slope

Minor and major associations are summarised in Tables 13.3 and 13.4 respectively.

Bering Sea

This marginal basin is separated from the main north Pacific by the Alaska Peninsula and Aleutian island arc. In the northeastern part the continental shelf is very wide. Communication through the Bering Strait to the arctic Chuckchi Sea is confined to near-surface water.

During the winter months, most of the area is covered by sea ice ~2 m thick and over the shelf the water is cooled from surface to bottom by convection currents. The annual temperature range is approximately –1.6 to 10 °C and shelf salinities vary from ~22 to 32.8‰ due to summer freshwater input. These waters are well oxygenated, but at depths <200 m there is cold water depleted in oxygen derived from the Pacific Ocean (0.46 ml l^{-1} O_2, 2.6–3.7 °C, 33.88–34.4‰) (Fairbridge 1966).

The total assemblages fall into three main associations: *Elphidium clavatum* (18–20 m) in inner shelf slightly brackish waters, *Eggerella advena* (22–48 m and 132–155 m) and *Cuneata arctica* (48–100 m) (Table 13.4). The *E. advena* association has accessory *Elphidium clavatum* and *C. arctica* on the inner shelf and *Trifarina angulosa*, *Epistominella exigua* and *Uvigerina juncea* on the outer shelf (data from Anderson 1963).

Saidova (1962) gave distribution maps of the calcareous taxa in the western part of the Bering Sea. Fursenko *et al.* (1979: Table 9) summarised the abundance of faunas in the northwest Bering Sea and Gulf of Anadyr, from which it can be seen that at depths of 0–50 m the principal species are *Cibicides lobatulus*, *Elphidiella arctica*, *Spiroplectammina biformis*, *Eggerella advena*, *Buccella frigida*, *Elphidium subgranulosum* and *Cassidulina subacuta*. Water temperatures are –1.7 to 7 °C and salinity 29.6–33‰.

Sea of Okhotsk

This is another marginal basin separated from the Pacific by the Kurile Islands. Depths range from 0 to 3374 m, and apart from the marginal shelves there are rises and troughs. The northern part of the sea has an essentially polar climate and the southern part summer and winter monsoons.

Table 13.3 Minor shelf and slope associations on the western margin of the Pacific Ocean (L = living, D = dead, T = total)

Species		Depth (m)	Area	Source
1. *Elphidium subarcticum*	T	14	Bering Sea	Anderson 1963
	T	100	Muroran	Ikeya 1971a
	T	0–140	Sendai	Matoba 1976c
2. *Epistominella exigua*	T	250	Bering Sea	Anderson 1963
3. *Miliammina herzensteini*	T	201–1000	Okhotsk Sea	Fursenko *et al.* 1979
4. *Globobulimina hanzawai*	T	1000	Okhotsk Sea	Fursenko *et al.* 1979
5. '*Trochammina inflata*'	T	21–50	Okhotsk Sea	Fursenko *et al.* 1979
6. *Elphidium fax*	T	0–20	Okhotsk Sea	Fursenko *et al.* 1979
7. *Cassidulina limbata*	T	0–50	NW Japan Sea	Fursenko *et al.* 1979
8. *Cassidulina grandis*	T	51–200	NW Japan Sea	Fursenko *et al.* 1979
9. *Elphidium asterineum*	T	0–50	W. Japan Sea	Fursenko *et al.* 1979
10. *Archimerismus subnodosus*	T	51–1000	W. Japan Sea	Fursenko *et al.* 1979
11. *Islandiella japonica*	T	201–1000	Central Japan Sea	Fursenko *et al.* 1979
12. *Trochammina voluta*	T	1000	SW Japan Sea	Fursenko *et al.* 1979
13. *Islandiella norcrossi*	T	80–90	E. Korea	Kim and Han 1972
14. *Cibicides refulgens*	T	30	E. Korea	Kim and Han 1972
	T	60	W. Hokkaido	Ikeya 1970
	T	0	Kushimoto	Uchio 1962b
15. *Quinqueloculina seminula*	L	60	W. Hokkaido	Ikeya 1970
	T	7	N. New Zealand	Hayward 1982
16. *Goësella flintii*	L, T	15–150	W. Hokkaido	Ikeya 1970
17. *Trochammina charlottensis*	T	150	W. Hokkaido	Ikeya 1970
18. *Cassidulina yabei*	T	190	W. Hokkaido	Ikeya 1970
19. *Islandiella islandica*	T	150	W. Hokkaido	Ikeya 1970
	T	115–598	Muroran	Ikeya 1970
20. *Adercotryma glomerata*	L	200	W. Hokkaido	Ikeya 1970
21. *Eggerelloides medius*	T	100	NW Honshu	Matoba and Honma 1986
22. *Trochammina pygmaea*	T	1700	NW Honshu	Matoba and Honma 1986
23. *Saccammina tubulata*	Y	2000–2350	NW Honshu	Matoba and Honma 1986
24. *Ammodiscus gullmarensis*	T	2650–2750	NW Honshu	Matoba and Honma 1986
25. *Textularia parvula*	T	64–73	NW Honshu	Matoba 1976a
26. *Rectobolivina raphana*	T	69–78	NW Honshu	Matoba 1976a, b
27. *Bulimina marginata*	T	77–80	NW Honshu	Matoba 1976a
28. *Bolivina pacifica*	L, T	650–875	NW Honshu	Matoba 1976a
29. *Spiroplectammina* cf. *biformis*	T	570	NW Honshu	Matoba 1976a
30. *Epistominella tamana*	L, T	80	NW Honshu	Matoba and Nakagawa 1972
31. *Nonionella stella*	L	100	NW Honshu	Matoba and Nakagawa 1972
	L, T	84–100	Muroran	Ikeya 1971a
32. *Reophax* spp.	L,T	14, 125–195	NW Honshu	Matoba and Nakagawa 1972 Matoba 1976b
33. *Rotalinoides gaimardii*	T	40	NW Honshu	Matoba 1976b
34. *Prolixoplecta exilis*	T	200	NW Honshu	Matoba 1976b
35. *Ammobaculites* sp.	T	20	NW Honshu	Matoba 1976b
36. *Lagenammina* sp.	T	30	NW Honshu	Matoba 1976b
37. *Eggerella* sp.	T	94–120	NW Honshu	Matoba 1976b
38. *Rosalina bradyi*	T	58–65	NW Honshu	Matoba 1976b
39. *Elphidium crispum*	T	8–34	NW Honshu	Matoba 1976b
	T	0	Noshiro	Uchio 1962b
	T	1	Seto	Uchio 1967
40. *Cibicides lobatulus*	T	30	NW Honshu	Matoba 1976b
	L, T	49–115	Muroran	Ikeya 1971a

Table 13.3 *(continued)*

Species		Depth (m)	Area	Source
41. *Trochammina pacifica*	T	27	NW Honshu	Uchio 1962a
42. *Trochammina hadai*	T	21–42	NW Honshu	Uchio 1962a
43. *Saccammina* sp.	T	33	NW Honshu	Uchio 1962a
44. *Nonion* cf. *pacificum*	T	50	NW Honshu	Uchio 1962a
45. *Bolivina robusta*	T	60	NW Honshu	Hasegawa 1979
	T	495–904	Banda Sea	Marle 1988
46. *Globobulimina auriculata*	T	200	NW Honshu	Hasegawa 1979
47. *Trochammina japponica*	T	500–1000	NW Honshu	Hasegawa 1979
48. *Bolivina spissa*	L,T	815–985	Muroran	Ikeya 1971a
49. *Epistominella takayanagi*	L, T	320–505	Muroran	Ikeya 1971a
50. *Hanzawaia nipponica*	L	100	Muroran	Ikeya 1971a
	T	20–50	South China	Wang *et al.* in Wang 1985
51. *Buccella frigida*	T	300	Muroran	Ikeya 1971a
52. *Nonionellina labradorica*	L, T	135–695	Muroran	Ikeya 1971a
53. *Nonionella globosa*	L	320–840	Muroran	Ikeya 1971a
54. *Nonionella* sp.	L,T	54	Muroran	Ikeya 1971a
55. *Trifarina kokozuraensis*	L, T	240	Muroran	Ikeya 1971a
	T	400	Sendai	Matoba 1976c
56. *Cassidulina complanata*	L		Miyako	Ujiié and Kusakawa 1969
57. *Elphidium miyakoensis*	D	10	Miyako	Ujiié and Kusakawa 1969
58. *Elphidium etigoensis*	D	6–78	Miyako, Yamada	Ujiié and Kusukawa 1969
59. *Hopkinsina pacifica*	L	14	Yamada	Ujiié and Kusukawa 1969
60. *Cassidulinoides parkerianus*	T	190	Sendai	Matoba 1976c
61. *Bolivina decussata*	T	250–300	Sendai	Matoba 1976c
62. *Fursenkoina apertura*	T	600–750	Sendai	Matoba 1976c
63. *Elphidium batialis*	T	1580	Sendai	Matoba 1976c
64. *Elphidium tokyoensis*	T		Tokyo	Ujiié 1962
65. *Elphidium subgranulosum*	T		Tokyo	Ujiié 1962
66. *Nonionella pulchella*	T		Tokyo	Ujiié 1962
	D	27–39	Yamada	Ujiié and Kusukawa 1969
67. *Amphistegina radiata*	T	33	Tanabe	Chiji and Lopez 1968
68. *Amphistegina madagascariensis*	T	24	Tanabe	Chiji and Lopez 1968
69. *Eratidus foliaceum*	T	7	Tanabe	Chiji and Lopez 1968
70. *Planorbulina acervalis*	T	2	Seto	Uchio 1967
71. *Cibicides haidingeri pacificus*	T	15–85	S. Korea	Polski 1959
		85–120	Taiwan Straits	Polski 1959
72. *Cassidulina carinata*	T	85–120	S. Korea	Polski 1959
73. *Elphidium magellanicum*	T	5–10	Bo Hai	Wang and Bian in Wang 1985
74. *Haynesina turberculata*	T	5–20	Bo Hai	Wang and Bian in Wang 1985
75. *Elphidium tsudai*	T	15–85	Yellow Sea	Polski 1959
76. *Ammonia ketieziensis*	T	45–90	Yellow Sea	Polski 1959, Wang *et al.* in Wang 1985
77. *Ammonia compressiuscula*	T	20–50	Yellow Sea	Wang *et al.* in Wang 1985
78. *Bigenerina taiwanica*	T	50–80	S. China	Wang *et al.* in Wang 1985
79. *Uvigerina proboscidea*	T	80–200	S. China	Wang *et al.* in Wang 1985
80. *Uvigerina peregrina*	T	150	S. China	Wang *et al.* in Wang 1985
81. *Bulimina aculaeta*	T	1564–1816	Banda Sea	Marle 1988
82. *Heterolepa mediocris*	T	78–342	Banda Sea	Marle 1988
83. *Pullenia bulloides*	T	1509–1951	Banda Sea	Marle 1988
84. *Rosalina villardeboana*	T		Hawaii	Coulbourn and Resig 1975

Table 13.3 *(continued)*

	Species		Depth (m)	Area	Source
85.	*Hauerina pacifica*	T		Hawaii	Coulbourn and Resig 1975
86.	*Quinqueloculina curta*	T		Hawaii	Coulbourn and Resig 1975
87.	*Marginopora vertebralis*	T		Hawaii	Coulbourn and Resig 1975
88.	*Cymbaloporetta bradyi*	T		Hawaii	Coulbourn and Resig 1975
89.	*Elphidium imperatrix*	T	50–60	E. Australia	Albani 1970
90.	*Zeaflorilus parri*	T	0	N. New Zealand	Hayward 1979
91.	*Elphidium novozealandicum*	T	3–28	N. New Zealand	Hayward 1982
92.	*Planoglabratella opercularis*	T	1–6	N. New Zealand	Hayward 1982
93.	*Neoconorbina pacifica*	T	17	N. New Zealand	Hayward 1982
94.	*Glabratella harmeri*	T	9	N. New Zealand	Hayward 1982
95.	*Cassidulina carinata*	T	34–41	N. New Zealand	Hayward 1982
96.	*Globocassidulina canalisuturata*	T	41	N. New Zealand	Hayward 1982
97.	*Bulimina submarginata*	T	50	N. New Zealand	Hayward *et al.* 1984

On the inner shelf, at depths down to 30–75 m, the water is warmed to 10–18 °C in the summer and cools to –1.8 to 2 °C in winter (salinity 33–34‰). The mid shelf (down to 150 m) is under the influence of cold intermediate water (–1.6 °C). Below this the outer shelf and slope are bathed in Pacific Water with temperatures of 2–2.5 °C at 750–1500 m and 1.8 °C at greater depths and salinity of 34.5‰. This 'warm' water enters via deep breaks in the Kurile island chain. The oxygen minimum (1 ml l^{-1} O_2) occurs at 750–1500 m (Fairbridge 1966).

It has been established that infaunal foraminifera live down to 25–30 cm beneath the sediment surface at water depths samples between 140 and 3356 m. The diversity is greatest in the top 5 cm of sediment and the most commonly occuring species are *Alabaminella weddellensis* and *Islandiella smechovi* (as *Discoislandiella*, Basov and Khusid 1983).

Generalised data on total assemblages (Fursenko *et al.* 1979) allow eight associations to be recognised (Tables 13.3 and 13.4). The inner shelf (to 50 m) has the *Buliminella elegantissima*, *Cuneata arctica*, *Eggerella advena*, *Trochammina inflata* and *Elphidium fax* associations. The *Islandiella kasiwazakiensis* association is present on the outer shelf and the *Miliammina herzensteini* association on the slope (200–1000 m). At greater depths is the *Globobulimina hanzawai* association.

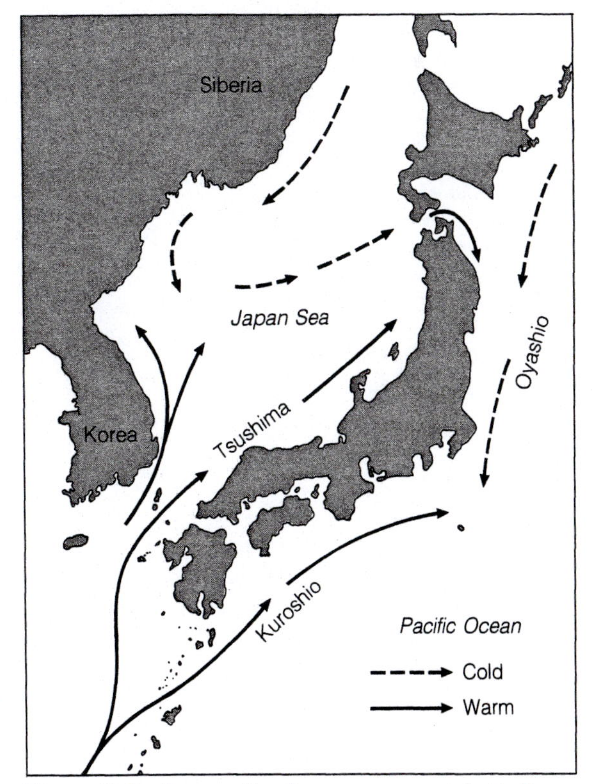

Figure 13.2 Pattern of surface-water currents in the Japan Sea (after Ujiié 1988)

Japan Sea

Unlike to Sea of Okhotsk, the Japan Sea is connected to the adjacent marginal seas and to the Pacific Ocean only by shallow channels (generally 100 m deep). Much of the Japan Sea is >3000 m deep and the greatest depth is 4049 m. The pattern of surface-water currents is shown in Fig. 13.2. Cold water bathes the Siberian shore but a warm current is present off South and North Korea. Along the Japanese coast

Table 13.4 Major shelf and slope associations on the western margin of the Pacific Ocean

***Ammonia beccarri* association**

Salinity: 31–35‰

Temperature: 5 to 27 °C

Substrate: mud, sand

Depth: 0–25 m

Distribution:
1. Tokyo, Japan (Ujiié 1962) total
2. Wakaura–Mori Harbor, Japan (Uchio 1962b) total
3. Tanake, Japan (Chiji and Lopez 1968) total
4. Yellow Sea (Wang *et al.* in Wang 1985) living, total
5. Macau (Rocha and Torquato 1967) total

Additional common species:

Elphidium tokyoensis	1
Elphidium advenum	2
Elphidium jenseni	2
Elphidium subincertum	2
Triloculina trigonula	2
Quinqueloculina laevigata	3
Ammoglobigerina globigeriniformis	3 (as *Trochammina*)
Ammonia convexidorsa	4

***Amphistegina lessonii* association**

Salinity: 34.0–34.7‰

Temperature: ~29 °C

Substrate: sand

Depth: 0–10 m

Distribution:
1. Seto: (a) (Uchio 1962b) total; (b) Uchio 1967) total
2. Banda Sea (Marle 1988) total

Additional common species:

Operculina ammonoides	2
Heterolepa dutemplei	2
Heterolepa mediocris	2

***Buliminella elegantissima* association**

Salinity: 27.7–34.3‰

Temperature: –1 to 27 °C

Substrate: sand

Depth: 0–20 m

Distribution:
1. Bering Sea (Anderson 1963) total
2. Okhotsk Sea (Fursenko *et al.* 1979) total
3. Shinano, Honshu (Uchio 1962a) total

Additional common species:

Elphidum clavatum	1, 3
Buccella frigida	1
Valvulineria sp.	1
Buccella hannai arctica	2
Ammonia beccarii	3
Textularia earlandi	3

***Cuneata arctica* association**

Salinity: 32–34‰

Temperature: –1 to 14.6 °C

Substrate: –

Depth: 0–100 m

Distribution:
1. Bering Sea (Anderson 1963) total (as *Reophax*)
2. Okhotsk Sea (Fursenko *et al.* 1979) total

Additional common species:

Eggerella advena	1
Textularia torquata	1
Elphidium clavatum	1
Buliminella elegantissima	2
Elphidium asterineus	2
Cassidulina limbata	2 (as *Cassandra*)
Islandiella umbonata	2 (as *Discoislandiella*)

Table 13.4 *(continued)*

***Eggerella advena* association**

Salinity: 31–34‰

Temperature: –1.7 to 25 °C

Substrate: mud, silt

Depth: 0–200 m

Distribution:
1. Bering Sea (Anderson 1963) total
2. Okhotsk Sea (Fursenko *et al.* 1979) total (as *E. scrippsii*)
3. W. Hokkaido (Ikeya 1970) living, total
4. Akita, Honshu (Matoba 1976b) total
5. Muroran (Ikeya 1971a) living
6. Miyako (Ujiié and Kusukawa 1969) total
7. Yellow Sea (Wang *et al.* in Wang 1985) total

Additional common species:

Elphidium clavatum	1, 5
Cuneata arctica	1 (as *Reophax*), 2
Trifarina angulosa	1 (as *Angulogerina*)
Epistominella exigua	1
Uvigerina juncea	1
Cassidulina limbata	2 (as *Cassandra*)
Islandiella umbonata	2 (as *Discoislandiella*)
Lagenammina difflugiformis	3
Spiroplectammina biformis	3
Trochammina charlottensis	3
Trochammina pacifica	3
Haplophragmoides bradyi	3
Adercotryma glomerata	3
Trochamminopsis quadriloba	3 (as *Trochammina*)
Trochammina nana	3
Textularia earlandi	4
Reophax spp.	4
Nonionella stella	5
Hopkinsina pacifica	6
Trochammina squamata	6
Buccella frigida	7
Saccammina atlantica	7 (as *Proteonina*)

***Elphidium clavatum* association**

Salinity: 31–34.5‰

Temperature: 6 to 26 °C

Substrate: muddy sand

Depth: 0–150 m

Distribution:
1. Bering Sea (Anderson 1963) total
2. Noshiro, Honshu, (Matoba 1976a) total
3. Akita, Honshu: (a) Matoba and Nakagawa 1972) total; (b) (Matoba 1976b) total
4. Shinano, Honshu (Uchio 1962a) total
5. Muroran (Ikeya 1971a) total
6. Sendai (Matoba 1976c) total

Additional common species:

Elphidium subarcticum	1, 6
Eggerella advena	1, 3b
Buliminella elegantissima	1
Buccella frigida	1, 4, 5
Cuneata arctica	1 (as *Reophax*)
Rectobolivina raphana	2,4 (as *Siphogenerina*)
Textularia earlandi	3b, 4
Saccamina sp.	4
Epistominella tamana	4
Rotalidium japonicum	4 (as *Ammonia*)
Trochammina cf. *pacifica*	4
Islandiella islandica	5 (as *Cassidulina*)
Nonionella stella	5, 6

***Globocassidulina subglobosa* association**

Salinity: 34‰

Temperature: 3 to 5 °C

Substrate: sand, mud

Depth: 120–1340 m

Distribution:
1. South Korea (Polski 1959) total (as *Cassidulina*)
2. East China Sea (Polski 1959) total (as *Cassidulina*)

Additional common species:

Elphidium advenum	1
Cassidulina carinata	1
Bolivina robusta	2
Bolivina pseudoplicata	2

Table 13.4 *(continued)*

***Haplophragmoides parkeri* association**

Salinity: 34‰

Temperature: 0.2 to 0.5 °C

Substrate: silty clay

Depth: 65–2000 m

Distribution: (all as *Thalmannammina*)
1. Nishitsugara, Honshu (Matoba and Honma 1986) total
2. Noshiro, Honshu (Matoba 1976a) total
3. Akita, Honshu (Matoba 1976b) total

Additional common species:

Trochammina pygmaea	1
Bolivina pacifica	2
Spiroplectammina cf. *biformis*	2
Reophax guttifer	3
Trochamminopsis quadriloba	3 (as *Trochammina*)

***Islandiella kasiwazakiensis* association**

Salinity: 32–34.2‰

Temperature: 0 to 8 °C

Substrate: –

Depth: 51 – >1000 m

Distribution: (all as *Planocassidulina*)
1. Okhotsk Sea (Fursenko *et al.* 1979) total
2. Central Japan Sea (Fursenko *et al.* 1979) total
3. SW Japan Sea (Fursenko *et al.* 1979) total

Additional common species:

Trochammina voluta	1, 2
Haplophragmoides parkerae	1 (as *Recurvoidella*)
Alabaminoides antarcticus	1, 3
Uvigerina akitaensis	1
Nonionellina labradorica	1
Elphidium batialis	1 (as *Criboelphidium*)
Miliammina herzensteini	2
Islandiella umbonata	2 (as *Discoislandiella*)
Islandiella japonica	3
Bolivina decussata	3

***Pararotalia nipponica* association**

Salinity: 33.1–34.9‰

Temperature: 22 to 29 °C

Subtrate: very fine sand

Depth: 0–40 m

Distribution:
1. Akita, Honshu (Matoba 1976b) total
2. Shiraki (Uchio 1962b) total (as *Ammonia*)
3. Taiwan (Huang 1971, 1983) total (as *P. ozawaia*)

Additional common species:

Murrayinella minuta	1 (as *Pararotalia*)
Reophax spp.	1
Elphidium crispum	1, 2
Lagenammina sp.	1
Cibicides lobatulus	1
Discorbis chinensis	3
Cibicidoides pseudoungerianus	3 (as *Cibicides*)
Quinqueloculina seminula	3

***Textularia earlandi* association**

Salinity: 32.7–34.2‰

Temperature: 0.5 to 26.6 °C

Substrate: silt, sand

Depth: 8–150m

Distribution:
1. West Hokkaido (Ikeya 1970) living
2. Akita, Honshu: (a) Matoba and Nakagawa 1972) living, total; (b) (Matoba 1976b) total
3. Shinano, Honshu, (Uchio 1962a) total

Additional common species:

Trochammina charlottensis	1
Haplophragmoides bradyi	1
Goësella flintii	1
Nonionella stella	2a, 3
Rectobolivina raphana	2b, 3 (as *Siphogenerina*)
Trochammina hadai	3
Trochammina cf. *pacifica*	3
Elphidium clavatum	3
Rotalidium japonicum	3 (as *Ammonia*)
Nonion cf. *pacificum*	3
Brizalina striatula	3 (as *Bolivina*)
Saccammina sp.	3
Elphidium spp.	3

Table 13.4 *(continued)*

***Uvigerina akitaensis* association**

Salinity: 32.6–34.6‰

Temperature: 9 to 11 °C

Substrate: mud

Depth: 100–840 m

Distribution:
1. East Korea (Kim and Han 1972) total
2. Akita (Matoba and Nakagawa 1972) total
3. Muroran (Ikeya 1971a) living, total

Additional common species:	
Islandiella norcrossi	1, 3 (as *Cassidulina*)
Islandiella japonica	1 (as *Globocassidulina*)
Trifarina kokozuraensis	1
Reophax spp.	2
Bolivina spissa	3
Lagena striata	3
Glandulina nipponica	3

is the warm Tsushima Current. The Surface Water extends to a depth of 200–300 m and beneath this is Japan Sea Proper Water. This is very uniform in character (0–0.5 °C, 34.0–34.1‰ and well oxygenated; Fairbridge 1966). According to Matoba (1984) and Ujiié (1988) it is only in postglacial times that the warm Tsushima Current has flowed into the Japan Sea.

West and Central Japan Sea

There is very little information in relation to the large area (Troitskaya 1969, 1973; Kim and Han 1972; Fursenko *et al.* 1979). Minor associations are listed in Table 13.3 and details of those recognised from Fursenko *et al.* (1979: Table 7) are given in Table 13.5. On the northwest shelf and that off eastern Korea, the assemblages are dominated by hyaline taxa and so too is the central deep part. In the Peter the Great Gulf agglutinated forms become common and on the southwest slope they are dominant at depths >1000 m.

Western seaboard of Japan

The warming effect of the Tsushima Current is greater in the south than in the north. At shallow depths, temperatures reach or exceed 20 °C but they decline to ~2 °C at about 200 m.

The principal foraminiferal studies are Ikeya (1970) Ishikari Bay, Hokkaido, Matoba and Honma (1986) Nishitsugaru, Matoba 1976b) Noshiro, Matoba and Nakagawa (1972) and Matoba (1976a) Akita, Uchio (1962a) Shinano, Hasegawa (1979, 1988) and Hasegawa and Takayanagi (1981) Toyama Bay and Nomura (1988) Masuda, all off Honshu. A taxonomic checklist has been provided by Takayanagi and Hasegawa (1987).

Numerous minor associations are listed in Table 13.3. With one exception, all the principal associations are from the shelf beneath the Surface Water. The calcareous dominated examples are the *Buliminella elegantissima* (0–17 m, total), *Pararotalia nipponica* (5–20 m, total), *Elphidium clavatum* (34–85 m, total) and *Uvigerina akitaensis* (100–230 m, total) associations. Agglutinated examples are the *Textularia earlandi* (8–150 m, living, total) and *Eggerella advena* (15–200 m, living, total) associations. In addition the agglutinated *Haplophragmoides parkerae* (65–2000 m, total) association spans both the shelf

Table 13.5 Associations from the western and central parts of the Japan Sea (data from Fursenko *et al.* 1979)

***Cassidulina limbata* association (as *Cassandra*)**

NW shelf, 0–50 m, 0 to 45 °C, 34‰

Additional common species:
Cassidulina grandis (as *Cassandra*)
Trifarina kokozuraensis
Uvigerina akitaensis
Eggerella advena (as *E. scrippsi*)
'*Trochammina inflata*'

***Cassidulina grandis* association (as *Cassandra*)**

NW shelf, 51–200 m, −1.4 to 3 °C, 34‰

Additional common species:
Buccella hannai arctica
Cassidulina limbata (as *Cassandra*)
Trifarina kokozuraensis
Uvigerina akitaensis

***Elphidium asterineus* association (as *Cribroelphidium*)**

Peter the Great Gulf, 0–50 m, 0 to 19 °C, 29.8–34.5‰

Additional common species:
'*Trochammina inflata*'
Lepidodeuterammina ochracea (as *Rotaliammina*)
Elphidiella recens
Buliminella elegantissima
Quinqueloculina vulgaris

***Archimerismus subnodosus* association (as *Hyperammina*)**

Peter the Great Gulf, 51–1000 m, 0.2 to 0.4 °C, 29.8–34.5‰

Additional commons species:
Elphidiella recens
Cribrostomoides scitulus
Haplophragmoides parkerae (as *Recurvoidella*)

***Islandiella japonica* association**

Central, 201–1000 m, −0.2 to 1.5 °C, 33.9–34.2‰

Additional common species:
Islandiella kasiwazakiensis (as *Planocassidulina*)
Bolivina decussata
Valvulineria sadonica
Uvigerina akitaensis

***Islandiella kasiwazakiensis* association (as *Planocassidulina*)**

Central, > 1000 m, 0.2 to 0.3 °C, 33.9–34.2‰
Southwest, 201–1000 m, 0.1 to 8.0 °C, 34.0–34.2‰

Additional common species:
Trochammina voluta
Miliammina herzensteini
Islandiella umbonata (as *Discoislandiella*)
Islandiella japonica
Alabaminoides antarcticus
Bolivina decussata

***Trochammina voluta* association**

Southwest, > 1000 m, 0.3 to 0.5 °C, 34.1–34.2‰

Additional common species:
Miliammina herzensteini
Haplophragmoides parkerae
Ammodiscus minutissimus

with Surface Water and the slope with Japan Sea Proper Water.

The standing crop size varies from 0 to >2000 per 10 cm^2. However, many of these are estimates based on much smaller faunal counts (e.g. Matoba and Nakagawa 1972; Matoba 1976a, b). There are only 14 living assemblages counts >100 : 11 Ishikari Bay, Hokkaido (Ikeya 1970) and 3 off Akita (Matoba 1976b). In Ishikari Bay the diversity is α 2–4 and the assemblages are mainly agglutinated. There is a single example from 60 m on sand with 37% porcellaneous and 63% hyaline tests. Off Akita diversity is α 6–18 and the assemblages have 0–11% agglutinated, 10–33% porcellaneous and 57–90% hyaline tests. However, it is unlikely that these few assemblages are representative. In this regard, the presence of *Elphidium clavatum* alongside warm-water species such as *Pararotalia nipponica* and *Rotalinoides gaimardii* is difficult to explain. Uchio (1962a), who studied total assemblages, suggested that *E. clavatum* might be relict from Pleistocene sediments or that perhaps it is living at the southern limit of its distribution.

The abundance of agglutinated tests in many of the assemblages and the poor preservational state of many of the calcareous tests led Matoba (1972, 1976a) to suggest that postmortem dissolution was taking place. On the shelf it is prevalent in muddy sediments even though the bottom waters are alkaline. On the slope sediment accumulation is low and the bottom waters have a pH of 7.7–7.8. Nevertheless, not all shelf areas are affected by dissolution and Nomura (1988) has pointed out that off Masuda some of the shelf agglutinated taxa have a calcareous cement whereas those of the slope do not.

Ikeya (1970) used a logarithmic (geometrical) plot to demonstrate that the dead assemblages differ from the living and he attributed this to postmortem mixing by bottom currents.

Japan–Pacific coast

As can be seen from Fig. 13.2 the south-flowing Oyashio Current brings cool water to the northern part while the Kuroshio Current brings warm water to the southern part. The Oyashio Water is nutrient-rich, has a salinity of 33.7–34.0‰ and temperature of 4 to 5 °C at 200–500 m depth. By contrast the Kuroshio Water is poor in nutrients, has a salinity of 34.5–35.0‰ and temperatures of 12–18 °C at similar depths (Fairbridge 1966). A branch of the Tsushima Current mixes warmer water with that of the Oyashio Water in the channel between Hokkaido and Honshu.

In the rocky intertidal zone of Shimoda Bay, under the influence of Kuroshio Water, Kitazato (1988b) found that the foraminifera could be divided into four groups. The phytal forms living on seaweeds, have lenticular tests and are probably suspension feeders, e.g. *Pararotalia nipponica*, *Elphidium crispum* and *E. reticulosum*. Crawling forms graze on epiphytic algae on seaweeds and they have conical tests, e.g. *Glabratella subopercularis*, *Patellina corrugata* and *Spirillina vivipara*. The attached immobile forms have a flattened umbilical side and feed on epiphytic diatoms, e.g. *Cibicides* spp., *Rosalina* spp., *Planorbulina* spp. On the muddy sediments around the base of the seaweeds there are free-living forms such as *Bolivina* spp. and *Quinqueloculina* spp.

On the cool northern shelf (off Muroran and Sendai) there are numerous minor associations (Table 13.3) and two principal associations, *Elphidium clavatum* (0–150 m) and *Eggerella advena* (28–56 m). The *Uvigerina akitaensis* association (300–840 m) is present on the slope. Ikekya (1971a, b) found that the standing crop was higher where mixed Tsushima–Oyashio waters were present than in areas under the Tsushima Current, the variation being from 10 to 1000 per 10 cm^2. Values for Miyako and Yamada bays were from 0 to 128 per 10 cm^2 (Ujiié and Kusukawa 1969). The diversity of the four assemblages with >100 living individuals is α 10–12.

On the slope off Sendai, the *Bolivina decussata* association (250–300 m) is present on muddy sand beneath Intermediate Water derived from the Oyashio. The oxygen minimum layer (100–1400 m, ~1 ml l^{-1} O_2) is characterised by an *Elphidium batialis* association (with *Bolivina spissa*) (Matoba 1976c).

From Tokyo southwards, the margin is under the influence of the warm Kuroshio Current. This is marked by the appeareance of the *Ammonia beccarii*, *Amphistegina lessonii* and *Pararotalia nipponica* inner shelf associations (Table 13.4). In Tanabe Bay the standing crop values were 2–270 per 20 g sample and the diversity was very low (α <1–2; Chiji and Lopez 1968).

Because there is little information on the living assemblages of the Pacific seaboard of Japan, it is difficult to assess the extent of postmortem effects. Ujiié and Kusukawa (1969) noted that there are major differences between the living and dead assemblages of Miyako and Yamada bays. The four living assemblages with >100 individuals are *Eggerella advena*, *Cassidulina complanata* and *Hopkinsina pacifica* associations yet all are represented by *Elphidium etigoensis* dead associations and the nominate species is very rare living. Unusually, the living assemblages are richer than the dead in agglutinated tests. It is possible that relict Pleistocene tests may be present here.

Yellow, East and South China seas

In the Yellow Sea the water is turbid due to the great influx of river-borne detritus from the Chinese mainland. Surface salinities are slightly brackish (30–31‰ in Bo Hai and 31–33‰ in Huang Hai) and become more brackish close to the major rivers. Much of the area is mesotidal but along the coast of South Korea it is macrotidal. Bo Hai experiences a wide variation in temperature from freezing with sea ice in the winter to ~20 °C and in the Huang Hai to ~28 °C in summer. During spring and summer the surface waters are separated from the cooler bottom waters (temperature 6 –8 °C) by a thermocline at ~30 m (Fairbridge 1966).

The warm Kuroshio Current flows east of Taiwan but a secondary branch passes west of the island and rejoins the main stream to the north. It flows through the East China Sea and a branch extends into the Yellow Sea. The characteristics of the South China Sea are controlled by the monsoonal climate. During the summer salinities in the top 100 m are lowered due to heavy runoff but during the winter they rise to ~34.5‰.

A reconnaissance study of the whole area was carried out by Polski (1959). Early Chinese studies (Wang *et al.* 1978, 1979; Wang 1980) were translated and updated in Wang (1985) and additional data are given in Wang *et al.* (1988).

The *Ammonia beccarii* association is widely present on the inner shelf of the Yellow and East China seas and it extends into the estuaries. On the middle to outer shelf there is an *A. compressiuscula* association (with *A. ketienziensis* and *Bolivina robusta*) probably from waters below the thermocline. At depths <70 m in the northern Yellow Sea is the *Eggerella advena* association rich in agglutinated forms.

Taiwan Straits has an essentially warm-water fauna (Huang 1971, 1983), but with the exception of the *Pararotalia nipponica* association it is difficult to recognise others from the data given.

Waller (1960) divided the famous of the South China Sea Shelf into three but gave no quantitative data. Three different assemblages were recognised by Wang *et al.* (in Wang 1985, see Table 13.3). Larger foraminifera are common around islands in association with coral reefs. Off Hainan living *Calcarina calcarinoides* are present on calcareous algae but most of the records are of dead tests from sediment. Li and Wang (in Wang 1985) list 12 species from Hainan of which *C. hainanensis*, *C. spengleri* and *Operculina ammonoides* are dominant. Twenty-nine species are present in the Xisha Islands, with dominant *C. calcarinoides* and *C. hispida*, while in the Zhongsha Islands there are 26 species with dominant *C. hispidula* and *Baculogypsinoides spinosus*. Larger foraminifera commonly form >80% of the total foraminiferal assemblages. The taxonomy of the island fauna is described by Cheng and Zheng (1978) and Zheng (1979, 1980).

A total of 280 species of agglutinated foraminifera have been recorded from the East China Sea where they comprise from <10 to >50 % of the total benthic assemblages. Zheng and Fu (1988) distinguished five assemblages:

1. Inner shelf, characterised by *Arenoparrella asiatica*;
2. Middle shelf, *Textularia foliacea* with local abundance of *Pseudoclavulina gracilis* and *A. asiatica*;
3. Middle to outer shelf, *Bigenerina nodosaria* commonly with *Spiroplectinella wrightii* (as *Spirorutilis*);
4. Outer shelf to slope, dominated by *Spiroplectinella* spp.;
5. Slope to trough, highest diversity with common *Martinottiella minuta*, *M. communis* and *Eggerella bradyi*.

The Philippines margin of the South China Sea has been studied by Graham and Militante (1959) and Glenn *et al.* (1981). The latter recorded living distributions on Apo Reef off Mindoro and found the major habitats to be sediments, algal mat and algal turf. Standing crops were highest in sediment and sediment –algal substrates of the outer reef slope and flat.

Coarse sands and gravels have attached forms such as *Glabratella*, *Neoconorbina*, *Spirillina*, *Discorbis*, *Cymbaloporetta* and juvenile hyaline tests. Sediments coated with an algal mat have similar diverse faunas with miliolids. Algal turf growing on upright dead branching coral in the intertidal zone support low attached forms and miliolids. Although substrate plays an important role, many genera are present on more than one substrate type. For example, *Amphistegina*, *Rosalina*, *Calcarina*, *Acervulina*, *Elphidium*, *Peneroplis*, *Vertebralina*, *Quinqueloculina*, *Sorites* and *Textularia* are found on both algal mats and algal turf.

Graham and Militante (1959) described 248 benthic

species but gave no quantitative distributional data. More than 87% of the taxa are calcareous with ~30% being porcellaneous.

Because of the generally high rates of sediment accumulation the living and total assemblages in the Yellow and East China seas are very similar. Polski (1959) noted that inner shelf forms are generally well preserved, and Waller (1960) pointed out that some reworked fossil forms are also well preserved. Some tests have been glauconitised (Waller 1960) or replaced with phosphorite (Polski 1959.)

Malay Archipelago

The major taxonomic study is that of Millett (1898–1904). Hofker (1968) gave taxonomic notes on the fauna of Jakarta Bay (Java) and in 1978 he listed 462 and described 53 species from off Indonesia. He stressed that temperature is a more important ecological control than depth *per se* and listed the average temperatures favoured by certain deeper-water species.

Benthic foraminifera are important components of the carbonate sediments on the inner shelf to the southeast of Borneo (Boichard *et al.* 1985). The most widely distributed larger form is *Amphistegina* which occurs in all subenvironments and is adapted to withstand currents and abrasion. *Heterostegina* and *Spiroclypeus* occur in well-oxygenated waters at depths of 30–40 m, whereas *Calcarina* occurs between 2 and 30 m. *Operculina* is found in depressions at 40–50 m with blackened bioclasts which may indicate lower oxygen levels. *Alveolinella* favours lower slope depths of 30–40 m in the protection of reefs or other barriers. Miliolids are ubiquitous in these shallow waters. No information was given on the environments of life of these forms.

Postmortem dissolution of calcareous tests is obvious at 1600 m and strongly developed off Indonesia, at 3200 m in the Sulawesi Sea and 4200 m in the Molucca Sea (Hofker 1978). Below these depths the assemblages are predominantly or wholly agglutinated.

Banda Sea

At depths <100 m is the *Amphistegina lessonii* association, and from 78 to 324 m the *Heterolepa mediocris* association (Table 13.3) beneath the warm surface waters (temperature up to 29 °C, salinity 34.0–34.7‰). The Banda or Indonesian Intermediate Water has a *Bolivina robusta* association (495–904 m) while the Indonesian Deep Water has the *Pullenia bulloides* (1509–1951 m) and *Bulimina aculeata* (1564–1816 m) associations (Marle 1988, >125 μm fraction). At depths below 2500 m dissolution of calcareous tests leaves an agglutinated assemblage (Basov 1981).

The shallow carbonate shelf sediments from southeast of Borneo contain discoloured foraminifera. Those that are yellow-orange are coated in cryptocrystalline limonite and are found in water depths of <47 m. The blackened forms appear to be coated with a chlorite clay mineral and they occur at depths >35 m. This mineral may form in reduced conditions some centimetres below the sediment surface (Boichard *et al.* 1985).

East Australia

A little ecological information was given in the mainly taxonomic study by Collins (1958) of the Great Barrier Reef. He noted that in the coarse surf zone sediments only large *Baculogypsina* and *Marginopora* are present, the latter being up to 2.2 cm in diameter and sufficiently abundant to be made into necklaces for tourists! In deeper, current-scoured, offshore waters are large tests of *Operculina*, *Heterostegina* and *Operculinella*. *Marginopora vertebralis* reproduces asexually when 2 or more years old and at test volumes of ~300 mm^2 at depths of 8–10 m. Greatest reproductive activity takes place in the spring. At greater depths growth is slower and a larger size is reached before reproduction takes place because the environment is not optimal (Ross 1972, 1974).

Non-quantitative résumés of the shelf faunas off central and southern Queensland have been given by Palmieri (1976a, b). Albani (1970) recorded an *Elphidium imperatrix* association from 50 to 60 m off Port Hacking. Additional common species include *Triloculina trigonula* and *Quinqueloculina lamarckiana*.

At Lizard Island in the northern Great Barrier Reef of Australia, strong currents and periodic storms and hurricanes lead to long-distance transport (km) of soritids, although on the windward reef flat transport is less as heavier tests are trapped in depressions (Baccaert 1976). However, around Heron Island in the southern part, Jell *et al.* (1965) observed that the larger foraminifera live in protected environments of the reef flats and the dead distributions are closely similar.

Nevertheless, mobile sands under the influence of strong waves and currents contain some abraded tests although they lack small forms. Some of these sands contain brown and black stained tests especially of *Marginopora* and *Alveolinella*. The colour is due to iron and manganese and the staining takes place under reducing conditions beneath the sediment surface. Erosion and bioturbation return these grains to the surface (Maiklem 1967). Relict tests are present from 30 to 200 m on the shelf off central Queensland (Palmieri 1976b).

New Zealand

There is a marked temperature grandient from north to south in the surface waters from ~20 to 13 °C in summer (February) to 15 to 8 °C in winter (August, Sverdrup *et al.* 1942). Thus the region spans the warm to cool temperate zones.

Mainly taxonomic studies have been carried out by Vella (1957), Hedley *et al.* (1965, 1967) and Eade (1967) and many taxa appear to be endemic. The few ecological studies are of the inner shelf off North Island (Hayward 1979, 1981, 1982; Hayward *et al.* 1984). Four major associations are present: *Cibicides marlboroughensis*, *Elphidium charlottensis*, *Glabratella zealandica* and *Trochulina dimidiata* (Table 13.6). All are developed on sandy substrates in areas of moderate energy. No details of temperature or salinity are known. In addition, several minor associations have been recorded (Table 13.3). Some of the total assemblages are very diverse (α 6.5–34 around the Cavalli Islands). Hayward (1982) suggested that the higher values could be due either to postmortem mixing or the situation of the area at the junction of subtropical and temperate faunas. The former seems to be the principal cause.

The single study concerning South Island is that of Kustanowich (1964) on Milford Sound, a silled fjord. The bottom waters are normal marine and well oxygenated. In the entrance the total assemblages are dominated by *Evolvocassidulina orientalis* (as *Cassidulinoides*) and *Saidovina karreriana* (as *Loxostomum*) or *Hoeglundina elegans* with *Saidovina karreriana* and *Pacinonion novozealandicum* (as *Astrononion*) at depths of 60 m. The deep basin at 140–150 m is characterised by *Bulimina denudata* and *B. marginata*.

Nonionellina flemingi from 60 to 80 m water depth shows a cline from small adults in the warmer northern waters to large adults in the cooler southern waters – this has been attributed to delayed maturation and reproduction under cooler conditions (Lewis and Jenkins 1969).

Planktonic : benthic ratio

Data are limited in coverage. The abundance of planktonic tests on the Japan Sea shelf off Japan is generally low (Matoba and Nakagawa 1972, Maratoba 1976a) and may be affected by dissolution. In the Korean Strait values rises to >60% due to the influence of the Tsushima Current (Kim and Han 1972).

Off the south China coast there is a seaward increase in the abundance of planktonic tests but they are relatively infrequent in the Yellow Sea (Polski 1959). The pattern in Taiwan Strait is somewhat complex but nevertheless shows the path of the Taiwan Current (Huang 1971, 1983).

In the East China Sea planktonic tests are abnormally abundant off the Yangtze and Qiantang rivers due to the local current pattern. In this area the size of the planktonic tests decreases from >1000 μm on the slope to 300–800 μm in mid shelf to <300 μm on the inner shelf (Wang *et al.* in Wang 1985).

Tropical island faunas

The west Pacific is a region of island chains which provide isolated shallow-water habits in an otherwise vast area of deep water. Biogenic carbonate sediments, in which larger foraminifera are an important component, predominate on shallow shelves and in lagoons isolated from the open ocean by reefal barriers. By far the majority of samples are of total assemblages from beaches, and in most cases the foraminifera did not live there but in the adjacent subtidal region.

Eleven larger foraminiferal associations are listed in Table 13.7. The most widespread is the *Amphistegina lessonii* association recorded over a broad depth range (0–38 m). The *Baculogypsina sphaerulata* and *Calcarina spengleri* associations are restricted to the western part while the *Marginopora vertebralis* association is recorded only in the eastern part. The other associations are either rare or inadequately sampled.

The distribution of individual taxa clearly shows that sampling, especially from subtidal areas, is inadequate. New Caledonia has 23 species, perhaps partly because it has been intensively sampled (>800 samples, Debenay 1985a). *Amphistegina* is probably

Table 13.6 Associations endemic to the New Zealand shelf

***Cibicides marlboroughensis* association**

Salinity: –

Temperature: –

Substrate: fine to coarse sand

Depth: 1–87 m

Distribution:
1. Cavalli Islands (Hayward 1982)
2. West North Island (Hedley *et al.* 1965)

Additional common species:

Elphidium novozealandicum	1
Glabratella zealandica	1 (as *Pileolina*)
Quinqueloculina seminula	1
Elphidium charlottensis	1

***Elphidium charlottensis* association**

Salinity: –

Temperature: –

Substrate: sand, shell gravel

Depth: 0–6 m

Distribution:
1. Cavalli Islands (Hayward 1982) total
2. Bay of Islands (Hayward 1981) total

Additional common species:

Elphidium oceanicum	1
Cibicides marlboroughensis	1
Glabratella zealandica	2 (as *Pileolina*)
Quinqueloculina seminula	2

***Glabratella zealandica* association**

Salinity: –

Temperature: –

Substrate: fine to coarse sand, gravel

Depth: 0–31 m

Distribution:
1. Cavalli Islands (Hayward 1982) total (as *Pileolina*)
2. Bay of Islands (Hayward 1981) total (as *Pileolina*)
3. Chicken Islands (Hayward *et al.* 1984) total (as *Pileolina*)

Additional common species:

Elphidium charlottensis	1, 2, 3
Trochulina dimidiata	1 (as *Discorbis*)
Cibicides marlboroughensis	1
Quinqueloculina seminula	1, 3
Quinqueloculina triangularis	3

***Trochulina dimidiata* association**

Salinity: –

Temperature: –

Substrate: shelly coarse sand

Depth: 5–7 m

Distribution:
1. Cavalli Islands (Hayward 1982) total (as *Discorbis*)
2. Chicken Islands (Hayward *et al.* 1984) total (as *Discorbis*)

Additional common species:

Quinqueloculina seminula	1, 2
Glabratella zealandica	2 (as *Pileolina*)
Elphidium charlottensis	2

Table 13.7 Principal association of tropical islands

***Amphistegina lessonii* association**

Depth: 0–38 m, beach and lagoon

Distribution:

1. Palau: (a) (Lessard 1980) total; (b) (Hallock 1984) living
2. Agrihan, Mariana Islands (Lessard 1980) total
3. Caroline Islands (Lessard 1980) total
4. Kapingamarangi Atoll: (a) (McKee *et al.* 1959) total; (b) (Lessard 1980) total
5. Oahu, Hawaii (Hallock 1984) living
6. Johnston Island (Lessard 1980) total
7. Solomon Islands (Hughes 1985) total
8. New Caledonia (Debenay 1985a) total
9. Tuamoto: (a) (Lessard 1980) total; (b) (Sournia 1976) living

Additional common species:

Calcarina spengleri	1a (as *Tinoporus*)
Baculogypsina sphaerulata	2, 3
Marginopora vertebralis	3, 6
Cymbaloporetta bradyi	6
Miliolinella circularis	6
Amphistegina papillosa	8
Operculina gaimardii	8
Amphistegina quoyi	9b

***Amphistegina madagascariensis* association**

Depth: 0–45 m, beach, reef flat, lagoon

Distribution:

1. Kapingamarangi Atoll (McKee *et al.* 1959) total
2. Marshall Islands (Cushman *et al.* 1954) total
3. Gilbert Islands (Todd 1961) total
4. Hawaiian Islands (Coulbourn and Resig 1975) total
5. Solomon Islands (Hughes 1977) total

Additional common species:

Marginopora vertebralis	1, 2, 3, 4
Heterostegina suborbicularis	2, 3
Calcarina hispida	2
Miniacina miniacea	2
Homotrema rubra	2
Spirolina arietina	3
Quinqueloculina laevigata	4
Hauerina pacifica	4
Elphidium macellum	5
Ammonia papillosa	5
Ammonia parkinsoniana	5

***Amphistegina radiata* association**

Depth: 0–80 m, beach, lagoon

Distribution:

1. Ryukyu Islands (Matsumaru and Matsuo 1976) total
2. Chichi–Jima (Matsumaru and Matsuo 1976) total
3. New Caledonia (Debenay 1985a) total

Additional common species:

Operculina gaimardii	3
Amphistegina quoyi	3
Amphistegina lessonii	3

***Baculogypsina sphaerulata* association**

Depth: intertidal

Distribution:

1. Ryukyu Islands: (a) (Matsumaru and Matsuo 1976) total; (b) (Sakai and Nishihira 1981) living; (c) (Lessard 1980) total
2. Guam, Mariana Islands; (a) (Matsumaru and Matsuo 1976) total; (b) (Lessard 1980) total
3. Caroline Islands (Lessard 1980) total

Additional common species:

Calcarina delicata	1a
Calcarina spengleri	1a, c, 2b, 3 (as *Tinoporus*)
Amphistegina radiata	1a
Calcarina defrancii	1b
Calcarina calcar	1b, c (as *Tinoporus*)
Calcarina hispida	1c (as *Tinoporus*)
Miliolinella circularis	2b
Amphistegina lessonii	3
Marginopora vertebralis	3

***Calcarina calcar* association**

Depth: 0–3 m

Distribution:

1. Palau (Hallock 1984) living
2. New Caledonia (Lessard 1980) total (as *Tinoporus*)

Additional common species:

Calcarina spengleri	1
Baculogypsina sphaerulata	2

Table 13.7 *(continued)*

***Calcarina delicata* association**

Depth: intertidal

Distribution:
1. Ryukyu Islands (Matsumaru and Matsuo 1976) total

Additional common species:

Amphistegina radiata	1
Baculogypsina sphaerulata	1
Calcarina spengleri	1

***Calcarina hispida* association**

Depth: 0–26 m

Distribution:
1. Tobi Island (Lessard 1980) total (as *Tinoporus*)
2. Solomon Islands (Hughes 1977) Total

Additional common species:

Calcarina spengleri	2
Amphistegina radiata	2

***Calcarina spengleri* association**

Depth: 0–16 m, beach, reef flat

Distribution:
1. Ryukyu Islands (Lessard 1980) total (as *Tinoporus*)
2. Palau (Lessard 1980) total (as *Tinoporus*)
3. Yap (Lessard 1980) total (as *Tinoporus*)
4. Caroline Islands (Lessard 1980) total (as *Tinoporus*)
5. Marshall Islands; (a) (Cushman *et al.* 1954) total; (b) (Lessard 1980) total (as *Tinoporus*)
6. Gilbert Islands (Todd 1961) total
7. Solomon Islands (Hughes 1977) total

Additional common species:

Baculogypsina sphaerulata	1, 4, 6, 7
Amphistegina lessonii	2, 4, 5b
Peneroplis pertusus	2
Marginopora vertebralis	5a, b, 6
Homotrema rubra	5a
Miniacina miniacea	5a
Carpenteria proteiformis	5a
Miliolinella circularis	5b
Amphistegina madagascariensis	6, 7
Cymbaloporetta bradyi	6
Cymbaloporetta squamosa	6
Calcarina hispida	7

***Marginopora vertebralis* association**

Depth: 0–2 m, reef flat

Distribution:
1. Hawaiian Islands (Coulbourn and Resig 1975) total
2. Johnston Islands (Lessard 1980) total
3. New Caledonia (Debenay 1985a) total
4. Fiji (Smith 1968) living

Additional common species:

Amphistegina madagascariensis	1
Hauerina pacifica	1
Cymbaloporetta bradyi	2

***Operculina ammonoides* association**

Depth: 30–48 m

Distribution:
1. Solomon Islands (Hughes 1977) total

Additional common species:
Operculina complanata

***Operculina gaimardii* association**

Depth: 2–28 m

Distribution:
1. New Caledonia (Debenay 1985a) total

Additional common species:

Nummulites cumingii	1 (as *Operculinella*)
Heterostegina operculinoides	1

the most abundant genus.The taxonomy of its species has not yet been stabilised (see Larsen 1976; Todd 1976; Debenay 1985b). According to Todd, *A. madagascariensis* characterises shallow water, *A. lessonii* is lagoonal at 37–64 m and *A. radiata* is an outer shelf–upper slope form (based on total assemblages). However, Larsen considers *A. madagascariensis* to be a junior synonym of *A. lessonii*. Todd noted that *A. madagascariensis* can hang beneath the water surface by means of its pseudopodia and suggested that this might be a means of dispersal.

Baculogypsina sphaerulata and *Calcarina* spp. have broadly similar distributions being confined to the western part but, as noted by Todd (1960) and Matsumaru and Matsuo (1976), the northern limits of the two taxa do not coincide (see Fig. 13.3). Lessard (1980) considered that they could not extend their geographic range because of unfavourable current directions and perhaps because they have a short zygotic stage which is inadequate to enable them to be transported over great distances. In *Baculogypsina* the asexually produced young are formed in brood chambers and once they are released they immediately settle on the substrate, usually attaching themselves to algae. It is not known whether they reproduce sexually (Sakai and Nishihira 1981).

Marginopora vertebralis is present throughout the area and is often very abundant both in the intertidal zone and in reefal environments. It lives attached to a great variety of substrates including plants and coarse carbonate bioclasts encrusted with algae. It can withstand exposure to the air and to rainfall (Smith 1968; Haig 1988). Its distribution on seagrass was studied in a Papuan lagoon over a 6-week period (Severin 1987) and it was concluded that spatial variability is greater than temporal variability even though the area was subject to periodic subaerial exposure. In this lagoon the distribution of living *Alveolinella quoyi* is confined to algal-covered coral rubble and around the base of living coral heads, in sheltered areas at depths of 3–12 m. The standing crop averaged 40 per square metre with 20–40 individuals per patch. Dead tests show a greater depth range, from 5 to >50 m, and they are dominant bioclasts in the carbonate sediment (Lipps and Sevein 1985).

In Scilly Atoll Lagoon (Society Islands) the living

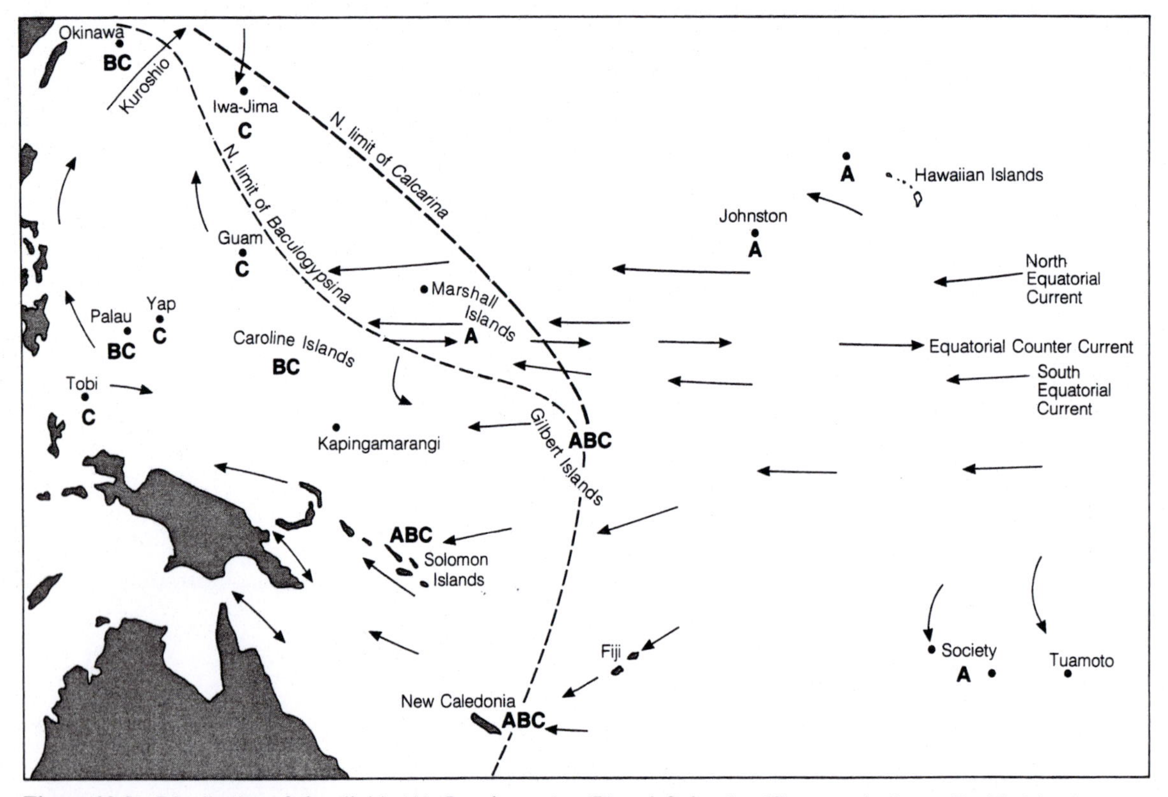

Figure 13.3 Distribution of alveolinids (A), *Baculogypsina* (B) and *Calcarina* (C) on tropical west Pacific islands. Surface currents after Lessard (1980) and Sverdrup *et al.* (1942)

assemblages associated with the alga *Microdictyon* at 18–52 m are dominated by species of *Textularia* with less common but diverse miliolids (Salvat and Vénec-Peyré 1981). Both here and on Moorea (French Polynesia) at depths of <8 m the living assemblages are dominated by attached trochamminids which bioerode pits in carbonate sand grains (Vénec-Peyré 1985a, b, 1988 a, b). This is thought to be the source of detrital grains for test construction.

On the whole, the smaller foraminifera of these tropical islands have received little attention. The distinctive genus *Bolivinella* is widely distributed although not in great abundance and has not yet been recorded living. It is an inner shelf form mainly confined to depths of <60 m (Hayward and Brazier 1980).

Postmortem effects

Many of the larger foraminifera samples are from beaches where postmortem concentration of tests has undoubtedly taken place. Detailed studies of a New Caledonia lagoon (Debenay 1985a, 1986, 1987a, b, 1988a, b, c) have shown that in channels with powerful tidal currents the bottom may be swept clean of sediment, but in quieter channels and in dunes sheltered behind the barrier reef, foraminifera are major contributors to the bioclastic sediments. They are transported according to the local hydrodynamic conditions. Nevertheless, the sediments preserve a fairly reliable record of the penetration of ocean water into the lagoon.

However, in the lagoon of Scilly Atoll Vénec-Peyré and Salvat (1981) recorded 24 living species but 100 dead. They concluded that many tests were transported into the lagoon from outside (none was large).

Coulbourn and Resig (1975) investigated the use of foraminifera as sediment tracers in a bay on Hawaii. Six associations were recognised, the most important being the *Rosalina villardeboana* and *Hauerina pacifica* associations in the channel fine sands and the *Amphistegina madagascariensis* association on the reef flat (Table 13.3). The latter species undergoes some offshore transport where wave action is strong. Overall, only a quarter of the species showed evidence of transport.

Summary

With the limited data available the main conclusions which can be drawn concern only the lagoonal and shallow shelf assemblages, for very little is known of those of the slope.

The *Buccella frigida* association is present as far south as the lagoons of northern Japan while the warmer-water *Ammonia beccarii* association reaches its northern limit here. On the inner shelf the cold-water *Cuneata arctica* association extends from the Bering Sea to the Sea of Okhotsk, while the *Elphidium clavatum* and *Eggerella advena* associations range as far south as the Japan Sea and northern Japan on the Pacific seaboard. These correlate with the south-flowing cold surface-water current.

The western boundary current (Kuroshio) brings warm-water inner-shelf faunas particularly to the Pacific margin of Honshu. On the inner shelf, the *Ammonia beccarii* association extends only as far north as Tokyo, whereas in lagoons (which no doubt achieve higher summer temperatures than the adjacent shelf) it is present from the northern tip of Japan. Larger foraminifera are present from the Ryukyu Islands in the north to the Brisbane area in Australia. These limits approximate to the 25 °C surface-water summer isotherm and the 18 °C winter isotherm. Larger foraminifera contain endosymbionts and they are adapted to life in clear oligotrophic waters.

New Zealand spans the southern limit of the subtropical zone in its northern part and the temperate zone for the rest. Although many taxa there are cosmopolitan, the majority of the dominant forms are endemic.

CHAPTER 14

Eastern margin of the Pacific Ocean

Due to its plate tectonic setting, much of the coastline of the eastern margin of the Pacific Ocean is rugged. Climatically it includes northern, central and southern wet zones with dry zones in between (see Fig. 14.1A). The principal surface-water currents are eastern boundary currents which originate in high latitudes. These are broad and sluggish, they transport cool water into low latitudes and they are commonly associated with upwelling: California Current (Northern Hemisphere), Peru Current (Southern Hemisphere). The generalised oceanography of the North American margin is shown in Fig. 14.1. The California Current develops as one branch of the Subarctic Current, the other flowing around the Gulf of Alaska. Between the California Current and the coast there is the northward-flowing Davidson Countercurrent. Upwelling from shallow depths takes place off southern California in summer and somewhat earlier off Baja California. Upwelling off Peru takes place mainly during the winter (Fairbridge 1966). From the Gulf of California to Ecuador, the surface waters are essentially tropical, but strong winter upwelling takes place in the Gulf of Panama and reduces surface temperatures (Crouch and Poag 1987). The Gulf of California shows minor salinity variations, but the annual temperature range of the surface waters in the northern part is considerable (26 –30 °C, Fairbridge 1966).

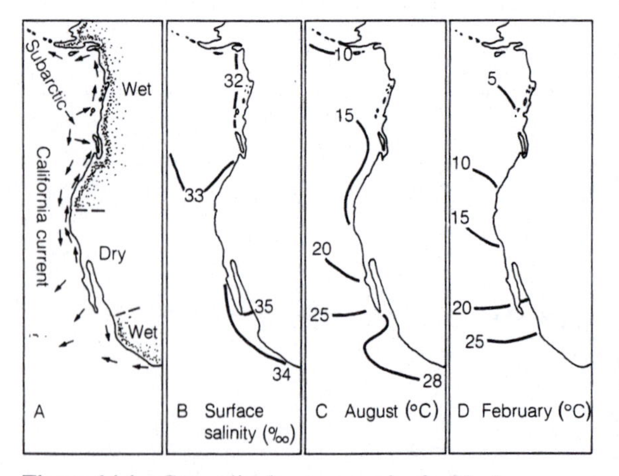

Figure 14.1 Generalised oceanography for North America. (A) surface currents and climatic zones; (B) surface salinities (‰); (C) surface temperatures °C), August; (D) surface temperatures (°C), February (based on Sverdrup *et al.* 1942)

Off the North American coast at depths of ~200 to 1200 m and from 28° N to the equator is a body of water with oxygen values of <0.25 ml l^{-1} and a similar situation prevails off Peru (Sverdrup *et al.* 1942). Oxygen-deficient waters are also present in silled basins on the California borderland and in the Gulf of California. In these areas laminated sediments are accumulating.

The majority of foraminiferal studies have been carried out on the North American margin. A bibliography of North American foraminifera has been given by Culver (1980). Many of the ecological papers give taxonomic notes and illustrations but in addition the following are useful: Galloway and Wissler (1927), Todd and Low (1967) and McCulloch (1977). Bandy (1963a) has reviewed the dominant marginal marine faunas from southern California and the Gulf of California. Regional distribution studies include those of Culver and Buzas (1985, 1986, 1987) for Alaska to Central America, Crouch and Poag (1987) for Central America, Boltovskoy (1976) for South America and Ingle and Keller (1980) for California to South America.

All the places referred to in this chapter are shown on Fig. 14.2.

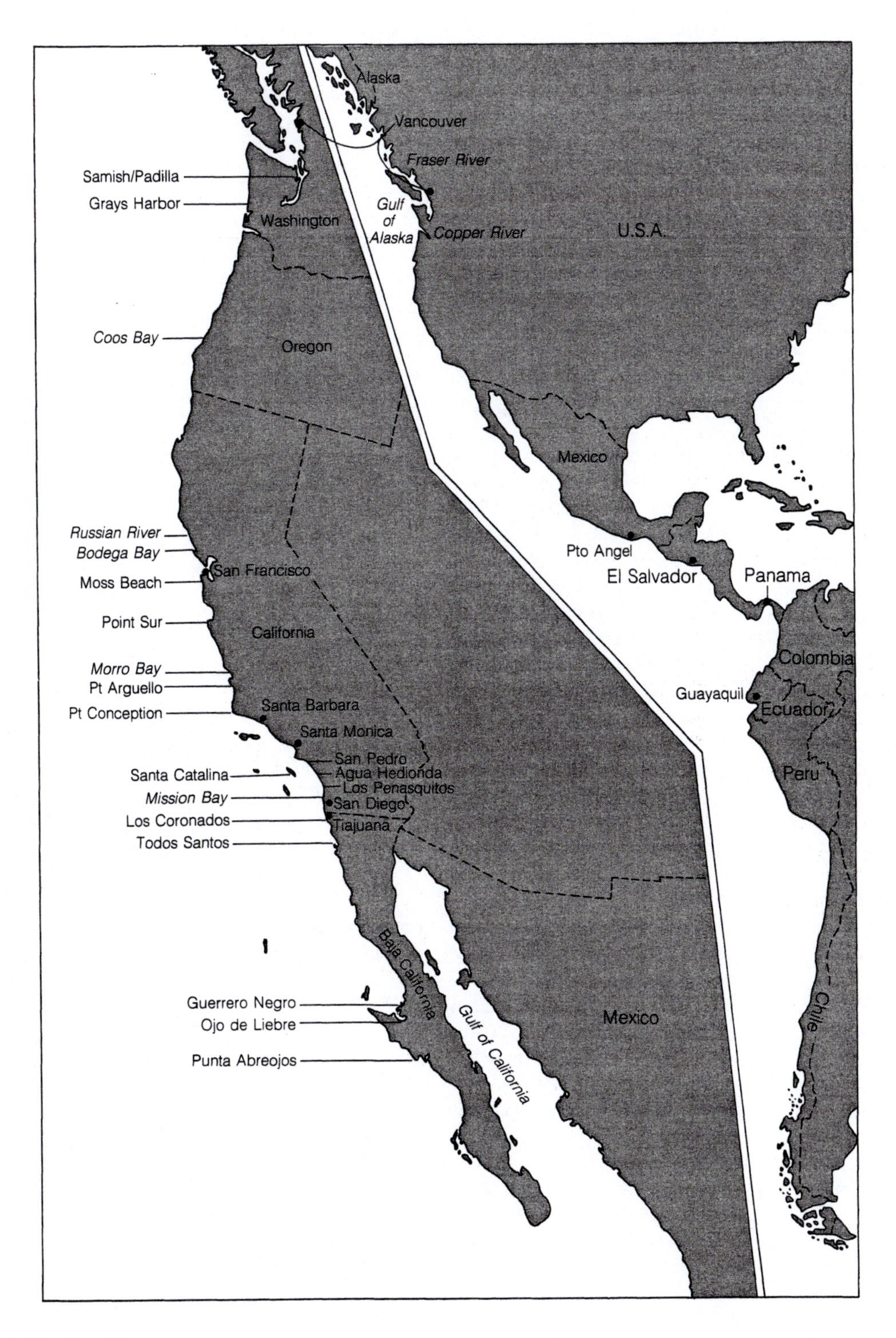

Figure 14.2 Localities

Marshes

Examples have been studied from the Gulf of Alaska (Copper River 60° N) to Baja California (Ojo de Liebre Lagoon 28° N). These range from brackish through normal marine to hypersaline. Three principal associations are widely distributed (Table 14.1, Fig. 14.3). The *Trochammina inflata* association is present in the most extreme northerly area (Copper River) where the marsh lies adjacent to a glacier and is itself covered in snow during the winter (Phleger 1967). Together with the *Jadammina macrescens* association, it is characteristic of the high marsh with *Salicornia*. Low marshes are mainly occupied by the *Miliammina fusca* association. The *Protoschista findens* association is present from Coos Bay, Oregon, to Tiajuana, California.

There are only two minor associations in the northern brackish areas but ten in the normal marine and hypersaline southern areas (Table 14.2). This reflects an overall increase in diversity from north to south although this also varies according to the salinity conditions. The brackish marshes have diversities of α <1 to 2, normal marine marshes α <1.5 and hypersaline examples α <1 to 5. There is a great range in standing crop with high values in the northern brackish examples (51–8600 per 10 cm^2 in Coos Bay, Phleger 1967), and in Mission Bay (2–4400, generally >1000 per 10 cm^2 Phleger 1967) and and lower values in southern brackish marshes (0–567 per 10 cm^2, San Diego area, Scott 1976b, 0–108 per 10 cm^2, Estero de Punta Banda, Todos Santos Bay, Walton 1955) and hypersaline examples (0–800, Guerrero Negro, Phleger 1965b). All the brackish marshes are almost exclusively composed of agglutinated taxa whereas the normal marine and hypersaline examples have abundant porcellaneous and hyaline tests.

An intensive study of the normal marine Mission Bay marsh revealed that the area is more variable than might be expected. The salinity ranged from 29 to 51‰, mean 39‰, temperature 5–33 °C, mean 18 °C and pH 6.6–8.4, mean 7.5 (Phleger and Bradshaw 1966). Part of the salt content of the water is from salt discharged from plant metabolism.

In a mangrove swamp by Guayaquil, with salinity 11.54–12.80‰, temperature 24–29 °C and pH 6.5–7.0, Boltovskoy and Vidarte (1977) recorded dominant *Arenoparrella mexicana, Miliammina fusca* and *Trochammina inflata* together with common *Ammoastuta inepta, Siphotrochammina lobata, Warrenita palustris* (as *Sulcophax*), *Tiphotrocha comprimata* and *Ammotium salsum*. Some calcareous taxa with protoplasm were also found in this acidic environment.

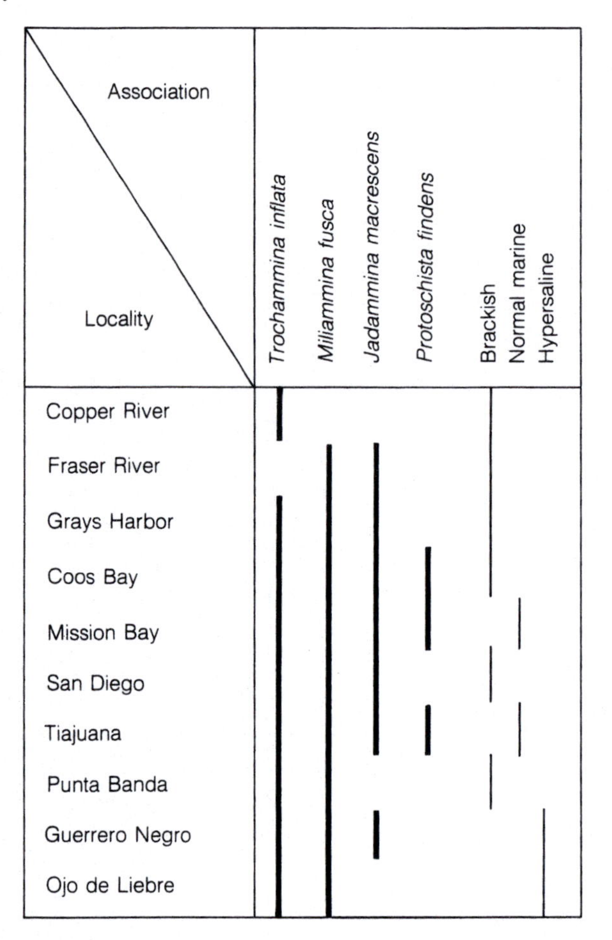

Figure 14.3 Distribution of marsh associations along the Pacific seaboard of North America

The geographic distribution of marsh species (Fig. 14.4) shows that *Trochammina inflata, Miliammina fusca, Jadammina macrescens* and *Ammotium salsum* are cosmopolitan. Phleger (1967) suggested that marsh species could be grouped into biogeographic assemblages like those of molluscs. However, Scott (1976b) showed that some agglutinated species are more widely distributed than Phleger reported. Therefore he concluded that the biogeographic assemblages are not as distinct as previously thought. He further considered that *Trochammina inflata* var. of Phleger and present from Copper River to Fraser River were *J. macrescens*. Figure 14.4 shows major faunal changes between 49 and 47° N. The mangrove faunas from Guayaquil resemble that of the California assemblage.

Table 14.1 Principal association from tidal marshes from the Pacific seaboard of North America

***Jadammina macrescens* association**

Salinity: 2–56‰

Temperature: 13 to 20 °C

Substrate: –

Depth: intertidal marsh

Distribution:

1. Fraser River, British Columbia (Phleger 1967) living (as *Trochammina* and *J. polystoma*)
2. Gray's Harbor, Washington (Phleger 1967) living (as *Trochammina* and *J. polystoma*)
3. Coos Bay, Oregon (Phleger 1967) living (as *Trochammina* and *J. polystoma*)
4. Mission Bay, California: (a) (Phleger 1967) living (as *J. polystoma*); (b) (Scott 1976a) living, total (as *J. polystoma*)
5. San Diego (Scott 1976a) total (as *T. inflata* var. *macrescens* and *J. polystoma*)
6. Tiajuana Slough, California (Scott 1976a) living, total (as *J. polystoma*)
7. Guerrero Negro, Baja California (Phleger 1965b) living (as *J. polystoma*)

Additional common species:

Species	Sites
Haplophragmoides subinvolutum	1, 2, 3
Miliammina fusca	1, 2, 3, 4a, b, 7
Trochammina inflata	2, 3, 4a, b, 5, 7
Haplophragmoides sp.	2
Miliolinella elongata	4a
Quinqueloculina laevigata	4a
Discorinopsis aguayoi	4a
Elphidium translucens	4b (as *Cribroelphidium*)
Spirillina sp.	5
Quinqueloculina seminula	6
Triloculina sp.	6
Elphidium spp.	7
Ammonia beccarii	7

***Miliammina fusca* association**

Salinity: –

Temperature: –

Substrate: –

Depth: intertidal

Distribution:

1. Fraser River, British Columbia (Phleger 1967) living
2. Gray's Harbor, Washington (Phleger 1967) living
3. Coos Bay, Oregon (Phleger 1967) living
4. Mission Bay, California (a) (Phleger 1967) living, (b) (Scott 1976a) living, total
5. Tiajuana Slough, California (Scott 1976a) living, total
6. Estero de Punta Banda, Baja California (Walton 1955) living, dead
7. Guerrero Negro, Baja California (Phleger 1965b) living
8. Ojo de Liebre, Baja California (Phleger 1967) living, total

Additional common species:

Species	Sites
Ammobaculites sp.	1
Ammotium cf. *salsum*	1, 2
Jadammina macrescens	1, 2, 4a (as *J. polystoma* and *T. macrescens*), 4b, 6, 7, (as *J. polystoma*)
Pseudoclavulina sp.	1
Haplophragmoides subinvolutum	2
Ammobaculites exiguus	2,3
Trochammina inflata	2, 4a, b, 5, 8
Protoschista findens	3, 4b, 5
Elphidium sandiegoensis	4a
Ammonia beccarii	4b
Elphidium translucens	4b (as *Cribroelphidium*)
Elphidium sp.	6,7
Saccammina sp.	6 (as *Proteonina*)
Glabratella sp.	7
Miliolidae	8

Table 14.1 *(continued)*

***Protoschista findens* association**

Salinity: –

Temperature: –

Substrate: –

Depth: intertidal

Distribution:
1. Coos Bay, Oregon (Phleger 1967) living
2. Mission Bay, California (Scott 1976a) total
3. Tiajuana Slough, California (Scott 1976a) total

Additional common species:

Ammobaculites exiguus	1
Ammotium cf. *salsum*	1
Miliammina fusca	1, 2, 3
Ammonia beccarii	2
Elphidium translucens	2 (as *Cribroelphidium*)
Trochammina inflata	3
Rosalina columbiensis	3

***Trochammina inflata* association**

Salinity: –

Temperature: –

Substrate: –

Depth: intertidal

Distribution:
1. Copper River, Alaska (Phleger 1967) living
2. Gray's Harbor, Washington (Phleger 1967) living
3. Coos Bay, Oregon (Phleger 1967) living
4. Mission Bay, California (a) (Phleger 1967) living, (b) (Scott 1976a) total
5. San Diego, California (Scott 1976a) living, total
6. Tiajuana Slough, California (Scott 1976a) living, total
7. Estero de Punta Banda, Baja California (Walton 1955) dead
8. Guerrero Negro, Baja California (Phleger 1965b) living
9. Ojo de Liebre, Baja California (Phleger 1967) living, total

Additional common species:

Jadammina macrescens	1, 2, 3, 4a (as *J. polystoma* and *Trochammina*), 4b, 6, 7, 8 (as *J. polystoma*)
Miliammina fusca	3, 4a, b
Discorinopsis aguayoi	4a, b
Protoschista findens	4b
Quinqueloculina spp.	4b
Glabratella ornatissima	4b
Polysaccammina ipohalina	5
Arenoparrella mexicana	7
Quinqueloculina oblonga	8 (as *Miliolinella)*
Elphidium spp.	8
Glomospira sp.	8
Miliolidae	9

Table 14.2 Minor associations on tidal marshes from the Pacific seaboard of North America (L = living, D = dead, T = total)

Association		Area	Source
1. *Haplophragmoides subinvolutum*	L	Gray's Harbor	Phleger 1967
2. *Elphidium lene*	L	Gray's Harbor	Phleger 1967
3. *Miliolinella elongata*	L	Mission Bay	Phleger 1967
4. *Reophax* sp.	L	Mission Bay	Phleger 1967
	L	Guerrero Negro	Phleger 1965b
5. *Elphidium translucens*	T	Mission Bay	Scott 1976a
	T	Tiajuana Slough	Scott 1976a
6. *Glabratella ornatissima*	T	Mission Bay	Scott 1976a
7. *Spirillina* sp.	T	San Diego	Scott 1976b
8. *Ammonia beccarii*	T	Tiajuana Slough	Scott 1976a
9. *Quinqueloculina* spp.	L	Guerrero Negro	Phleger 1965b
10. *Elphidium* spp.	L	Guerrero Negro	Phleger 1965b
11. *Glomospira* sp.	L	Guerrero Negro	Phleger 1965b
12. *Rosalina columbiensis*	L	Guerrero Negro	Phleger 1965b
13. Miliolidae	L, T	Ojo de Liebre	Phleger 1967

Table 14.3 Minor associations of benthic foraminifera from lagoons on the Pacific seaboard of North America (L = living, D = dead, T = total)

Association		Depth (m)	Area	Source
1. *Buccella frigida*	T	0	Samish–Padilla	Scott 1974
2. *Elphidiella hannai*	T	0	Samish–Padilla	Scott 1974
3. *Miliammina fusca*	T	0	Samish–Padilla	Scott 1974
4. *Trochammina pacifica*	L, T	0	Samish–Padilla	Jones and Ross 1979, Scott 1974
5. *Elphidium frigidum*	T	0–3	Russian River	Erskian and Lipps 1977
6. *Miliolinella* sp.	T	1	Agua Hedionda	Scott *et al.* 1976
7. *Elphidium translucens*	T	3–6	San Diego	Scott *et al.* 1976
8. *Brizalina vaughani*	L, T	4–16	San Diego	Scott *et al.* 1976
9. *Bulimina marginata*	T	12	San Diego	Scott *et al.* 1976
10. *Elphidium sandiegoensis*	T	5	San Diego	Scott *et al.* 1976
11. *Triloculina* sp.	L, D	?	Punta Banda	Walton 1955
12. *Glabratella* sp.	L	?	Guerrero Negro	Phleger and Ewing 1962, Phleger 1965b

Postmortem changes

Transport of tests on to marshes does not seem to be a significant process in any of the areas studied. However, where calcareous forms are present in the living assemblages, postmortem dissolution may take place. For instance, Bradshaw (1968) noted that in Mission Bay, pH <7.6 was present for half the time and he inferred that $CaCO_3$ was deposited for half the time and being dissolved for half the time. Cores studied by Scott (1976a) had exclusively agglutinated assemblages in the top 2.47 m and 0.7 m in holes 1 and 2 even though there is a notable calcareous component in the surface sediments. He considered that dissolution took place soon after death.

Lagoons

Minor associations are listed in Table 14.3 and major associations in Table 14.4.

The only cool brackish examples studied are the

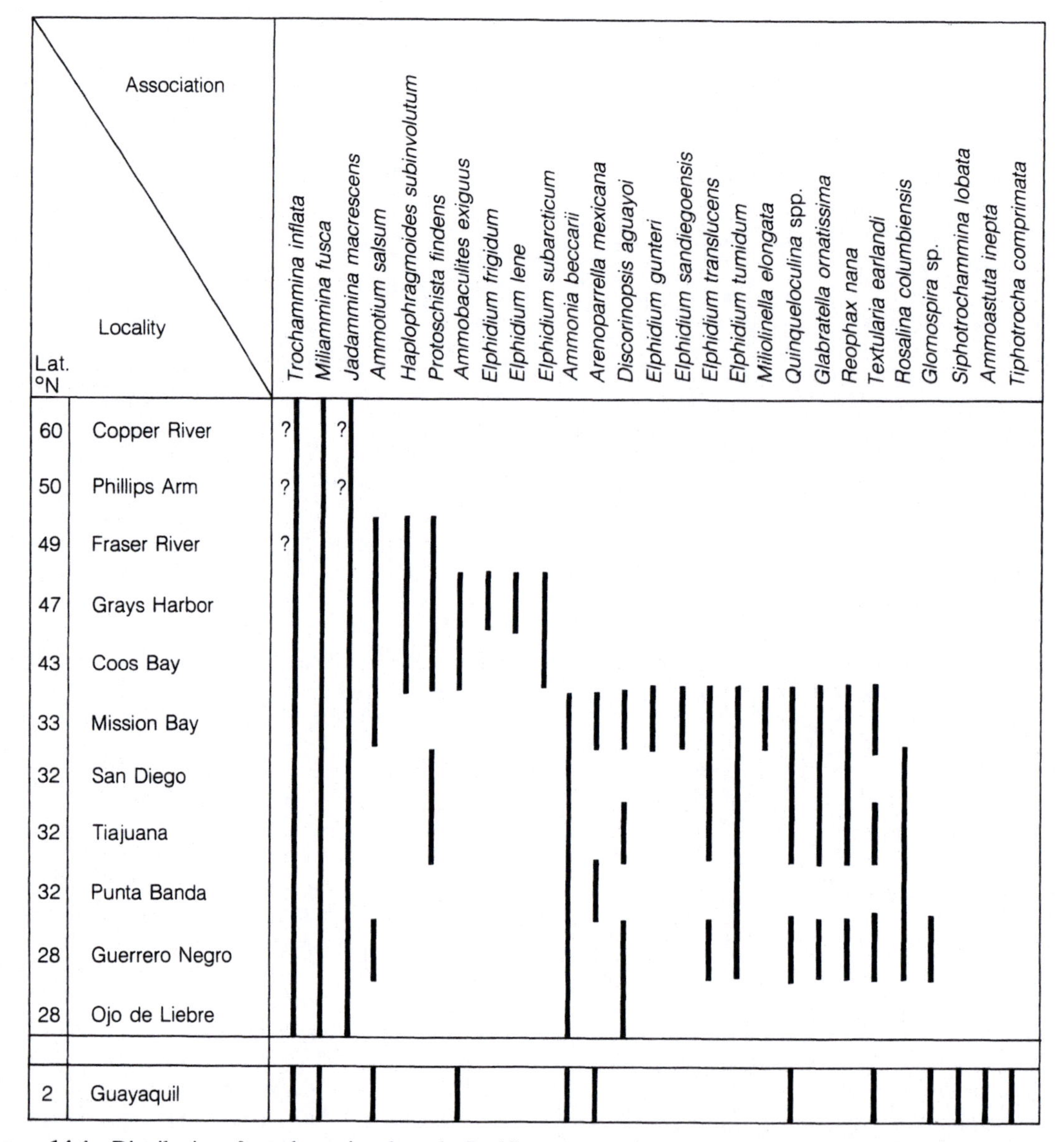

Figure 14.4 Distribution of marsh species along the Pacific seaboard of North America (based on Phleger 1965b, 1967 and Scott 1976a, b); ? indicates uncertainty of identification

Samish–Padilla bays of Washington (Scott 1976b). Salinities rarely fall below 26‰ and temperatures range from 5 to 20 °C. Four associations are recognised: *Miliammina fusca, Trochammina pacifica, Buccella frigida* and *Elphidiella hannai* (Table 14.3). Of these, the *T. pacifica* association is best developed and living assemblages are dominated by this from late autumn to early spring (Jones and Ross 1979). During the summer, calcareous taxa become more important and standing crop and diversity reach their peak values. *Ammonia beccarii* reaches its northernmost recorded occurrence here (Scott 1976b).

Russian River in northern California has salinities of 23–34‰, temperatures of 11–14 °C and is microtidal. The total assemblages are dominated by the *Elphidium frigidum* association (as *Cribrononion* of Erskian and Lipps 1977). Brackish San Francisco Bay has an *Ammonia beccarii* association (Arnal *et al.* 1980).

The principal associations are all from lagoons in

Table 14.4 Principal associations of benthic foraminifera from lagoons on the Pacific seaboard of North America

***Ammonia beccarii* association**

Salinity: 7–68‰

Temperature: 8 to 27 °C

Substrate: fine sand

Depth: <1–10 m

Distribution:
1. San Francisco, California (Arnal *et al.* 1980) total
2. Los Penasquitos, California (Scott *et al.* 1976) total
3. San Diego, California (Scott *et al.* 1976) total
4. Ojo de Liebre, Baja California (Phleger and Ewing 1962) living (as *Streblus*)
5. Gulf of California (Bandy 1961) total (as *Streblus tepidus*)

Additional common species:

Elphidiella hannai	1
Elphidium incertum	1
Elphidium translucens	2 (as *Cribroelphidium*)
Elphidium tumidum	2 (as *Cellanthus*)
Rosalina columbiensis	2
Bulimina marginata	3
Reophax nana	3
Textularia earlandi	3
Quinqueloculina spp.	4, 5
Elphidium spp.	4

***Buliminella elegantissima* association**

Salinity: 31–35‰

Temperature: –

Substrate: fine sand and silt

Depth: 2–16 m

Distribution:
1. Agua Hedionda, California (Scott *et al.* 1976) living, total
2. San Diego, California (Scott *et al.* 1976) living, total

Additional common species:

Elphidium translucens	1 (as *Cribroelphidium*)
Quinqueloculina seminula	1
Rosalina columbiensis	1, 2
Ammonia beccarii	1
Brizalina vaughani	1, 2
Bulimina marginata	1, 2
Textularia earlandi	1

***Cibicides fletcheri* association**

Salinity: 34–37‰

Temperature: 16 to 27 °C

Substrate: fine sand

Depth: 0–25 m

Distribution:
1. Guerrero Negro, Baja California (Phleger and Ewing 1962) total
2. Ojo de Liebre, Baja California (Phleger and Ewing 1962) total

Additional common species:

miliolids	
Trochulina lomaensis	1, 2 (as *'Rotalia lomaensis-versiformis'*)
Quinqueloculina spp.	2

***Elphidium tumidum* association**

Salinity: 32–68‰

Temperature: 17 to 25 °C

Substrate: fine sand and silt

Depth: <2 m

Distribution:
1. Agua Hedionda, California (Scott *et al.* 1976) total (as *Cellanthus*)
2. Los Penasquitos, California (Scott *et al.* 1976) total (as *Cellanthus*)
3. Estero de Punta Bunda, Baja California (Walton 1955) living, dead

Additional common species:

Trochammina inflata	1, 3
Ammonia beccarii	1, 2
Elphidium translucens	1 (as *Cribroelphidium*)
Trochulina sp.	3 (as *Lamellodiscorbis*)
Saccammina sp.	3 (as *Proteonina*)
Reophax sp.	3

Table 14.4 *(continued)*

***Trochulina lomaensis* association**

Salinity: 34–37‰

Temperature: 15 to 25 °C

Substrate: fine sand

Depth: 0–25 m

Distribution:

1. Guerrero Nergro, Baja California (Phleger and Ewing 1962) living, total (as '*Rotalia lomaensis-versifornis*')
2. Ojo de Liebre, Baja California (Phleger and Ewing 1962) total (as '*Rotalia lomaensis-versiformis*')

Additional common species:

Rosalina columbiensis	1
Trochammina kellettae	1
miliolids	1
Cibicides fletcheri	1, 2
Dyocibicides biserialis	2

***Quinqueloculina* spp. association**

Salinity: 34–68‰

Temperature: 17 to 27 °C

Substrate: fine sand

Depth: 0–25 m

Distribution:

1. Los Penasquitos, California (Scott *et al.* 1976) total
2. Guerrero Negro, Baja California (a) (Phleger and Ewing 1962) living, (b) (Phleger 1965b) living
3. Ojo de Liebre, Baja California (Phleger and Ewing 1962) living, total

Additional common species:

Ammonia beccarii	1, 3 (as *Streblus*)
Rosalina columbiensis	1, 2b
Buliminella elegantissima	2a
Elphidium spp.	2a, b
Trochulina lomaensis	2a, 3 (as '*Rotalia lomaensis-versiformis*')
Spirillina vivipara	2b
Trochammina spp.	2b
Glabratella sp.	2b
Brizalina striatella	2b, 3 (as *Bolivina*)
Cornuspira involvens	3

***Reophax nana* association**

Salinity: 31–35‰

Temperature: –

Substrate: fine sand

Depth: 2–12 m

Distribution:

1. Agua Hedionda, California (Scott *et al.* 1976) total
2. San Diego, California (Scott *et al.* 1976) living, total

Additional common species:

Textularia earlandi	1, 2
Buliminella elegantissima	1, 2
Bulimina marginata	1
Brizalina vaughani	1, 2
Ammonia beccarii	1, 2
Elphidium translucens	2 (as *Cribroelphidium*)
Elphidium sandiegoensis	2
Brizalina acutula	2

***Rosalina columbiensis* association**

Salinity: 31–68, mainly 31–35‰

Temperature: 18 to 24 °C

Substrate: fine sand

Depth: 0–25 m

Distribution:

1. Agua Hedionda, California (Scott *et al.* 1976) living, total
2. Los Penasquitos, California (Scott *et al.* 1976) total
3. San Diego California
4. Ojo de Liebre, Baja California (Phleger and Ewing 1962) living

Additional common species:

Elphidium translucens	1, 2, 3, (as *Cribroelphidium*)
Buliminella elegantissima	1,3
Trochammina inflata	2
Ammonia beccarii	2
Quinqueloculina seminula	2
Quinqueloculina laevigata	3
Dyocibicides biserialis	3
Gavelinopsis campanulata	3
Textularia earlandi	3
Trochulina lomaensis	4 (as '*Rotalia lomaensis-versiformis*')

Table 14.4 *(continued)*

***Textularia earlandi* association**

Salinity: 31–35‰

Temperature: –

Substrate: fine sand

Depth: 2–13 m

Distribution:

1. Agua Hedionda, California (Scott *et al.* 1976) total
2. San Diego, California (Scott *et al.* 1976) living, total

Additional common species:

Reophax nana	1, 2
Buliminella elegantissima	2
Bulimina marginata	2
Ammonia beccarii	2

southern California and Baja California. Here the waters are either essentially normal marine (32–37‰), e.g. *Buliminella elegantissima, Cibicides fletcheri, Trochulina lomaensis, Reophax nana* and *Textularia earlandi* associations or normal marine to hypersaline (31–68‰), e.g. *Ammonia beccarii, Elphidium tumidum, Quinqueloculina* spp. and *Rosalina columbiensis* associations (Table 14.4). The standing crop values of the extreme Los Penasquitos Lagoon are much lower than those of the normal marine and hypersaline examples (Table 14.5). Organic production in Ojo de Liebre lagoon is high with average fixation of 47.2 mg C m^{-3} day^{-1} (Phleger and Ewing 1962). Diversities range overall from α 2.5. In terms of wall structure, the normal marine living assemblages from Agua Hedionda and San Diego lagoons extend the field reported by Murray (1973) (see Fig. 14.5).

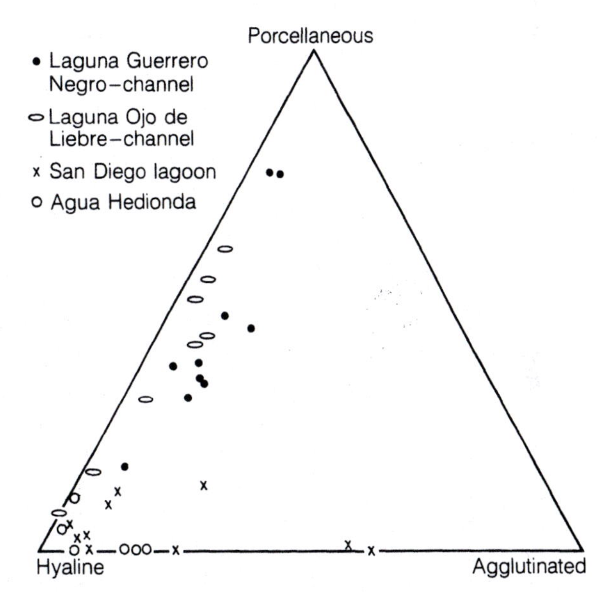

Figure 14.5 Ternary plot of wall structure of normal marine lagoons (data from Phleger and Ewing 1962, Scott *et al.* 1976)

Postmortem changes

In none of the lagoons discussed does transport appear to be a major postmortem modifying influence although Scott *et al.* (1976) note that *Rosalina columbiensis* is carried into California lagoons by tidal currents. Dissolution below the top 1 cm of sediment in Samish Bay leads to the loss of calcareous forms (Jones and Ross 1979). Boreholes in Agua Hedionda

Table 14.5 Diversity and standing crop data, southern California and Baja California lagoons (data from Phleger and Ewing 1962, Phleger 1965b, Scott *et al.* 1976)

Lagoon type	Salinity (‰)	α diversity	Standing crop /10 cm²
Brackish–hypersaline	15–68		
Los Penasquitos	Mainly 40–60	—	1–85
Normal marine			
Agua Hedionda	32–34	3–35	1–238
San Diego	31–35	2.5–8	1–184
Guerrero Negro	35	4–8	60–778
Ojo de Liebre (outer)	34–35	4–11	100–1652, Av. 100–200
Hypersaline			
Ojo de Liebre (inner)	42–47	4–6	7–257, Av. 200–250

and San Diego lagoons showed that subrecent assemblages are similar to the present living ones (Scott *et al.* 1976).

Rocky shores

The high-energy rock platform of Moss Beach, California, yielded 60 species but only 5 make up more than 75% of the total assemblage, *Rosalina columbiensis, Glabratella ornatissima, Protelphidium* sp., *Cibicides lobatulus*, and *Saccammina* aff. *S. alba. Rosalina columbiensis* lives attached to hydroids and bushy bryozoans where it is accompanied by *C. lobatulus*. Rock surfaces and surf-grass holdfasts are dominated by *G. ornatissima*. Most living individuals were found close to the low-water mark or in rock pools. This minimises exposure to drying and to great variations in temperature. Steinker (1976) considers that to overcome the problem of disturbance by waves, asexual reproduction is carried out in a protective cyst.

Rosalina columbiensis and *Glabratella ornatissima* were also found to be common in rocky tide pools along the Oregon coast (Detling 1958). However, *R, columbiensis* was not recorded by Cooper (1961) who studied tide-pool samples from Oregon to California. He found the principal living species to be *Cibicides fletcheri, Elphidiella hannai, Elphidium microgranulosum, G. ornatissima* (as *Discorbis*), *G. pyramidalis* and miliolids. The diversity ranges from α 4 to 13. Tide-pool samples are dominated by hyaline taxa, but up to 30% of tests are porcellaneous and up to 20% agglutinated although the latter commonly forms <5%.

Beaches

Two beach samples from the Gulf of Alaska were dominated by *Elphidium clavatum* (Bergen and O'Neil 1979).

Beaches from Oregon to San Francisco are characterised by living *Trochammina kellettae, Buccella tenerrima* and *Glabratella ornatissima* (as *Discorbis*) while those in southern California are dominated by *Cibicides fletcheri, Trochulina lomaensis* (as *Rotorbinella*) and *Trochammina kellettae* (Cooper 1961). Diversity ranges from α 3 to 7. All assemblages are dominated by hyaline tests and lack porcellaneous forms. In Santa Monica Bay, the dominant living forms are *C. fletcheri, D. monicana, Elphidium translucens, Trochulina lomaensis* (as *Rotorbinella*) and, locally, miliolids (Reiter 1959). From September to November, the diversity and standing crop were high but these decreased from December to March and then started to rise again in April. These changes are partly due to storms and partly to temperature changes limiting reproduction.

It has been shown that *Glabratella ornatissima* is well adapted to the turbulent nearshore environment (Erskian and Lipps 1987). In winter, when wave activity is at its peak, a sparse population of plastogamous pairs and adult agamonts is present in intertidal and subtidal sediments which are mobile. During the spring juvenile agamonts are released and settle on algae to seek protection. The food supply increases due to upwelling and the adult agamonts reproduce. The standing crop increases to reach a peak in the summer. By the end of the summer, newly formed plastogamous pairs and adult agamonts have recolonised the sediment with which they are dispersed by the winter waves.

At Guerrero Negro, Baja California, the total assemblages of the beaches are dominated by *Cibicides fletcheri* with accessory *Trochulina lomaensis* (as '*Rotalia lomaensis-versiformis*' of Phleger 1965b). The adjacent barrier has these same two species almost equally dominant. In the Gulf of California a beach assemblage dominated by *Quinqueloculina* spp. with *C. fletcheri* and *Elphidium crispum* was recorded by Bandy (1961) in waters of 11–33 °C.

Postmortem changes

Assemblages from rocky shorelines are not preserved *in situ*. The dead tests are either transported elsewhere or destroyed in the high-energy conditions. For the same reasons, beach assemblages also stand little chance of preservation without modification. In Santa Monica Bay, the dead tests are redistributed by waves and currents and there is an inverse relationship between the median diameter of the sand and the number of foraminiferal tests (Reiter 1959).

Shelf and slope

The shelf is narrow throughout the length of the Americas from the Gulf of Alaska to the southern tip of Chile. Because of its plate tectonic setting, the margin off southern California is faulted to give rise to a series of silled basins, the California borderland, between the shelf and the ocean proper.

A large number of minor associations are listed in Table 14.6. These are represented at few localities and often by few samples. Some may be artificially caused by inconsistent taxonomic usage, but the main reason for the large number is the great latitudinal span of the eastern Pacific seaboard.

The principal associations are listed in Table 14.7 and their geographic distribution is summarised in Fig. 14.6. It should be noted that the distribution of individual species is broader than that of the association to which they give their name. Scott *et al.* (1976) consider that *Bulimina denudata* is a synonym of *B. marginata* and this should be borne in mind although *B. denudata* has been used here following the majority of authors.

From Fig. 14.6 it can be seen that, on the basis of

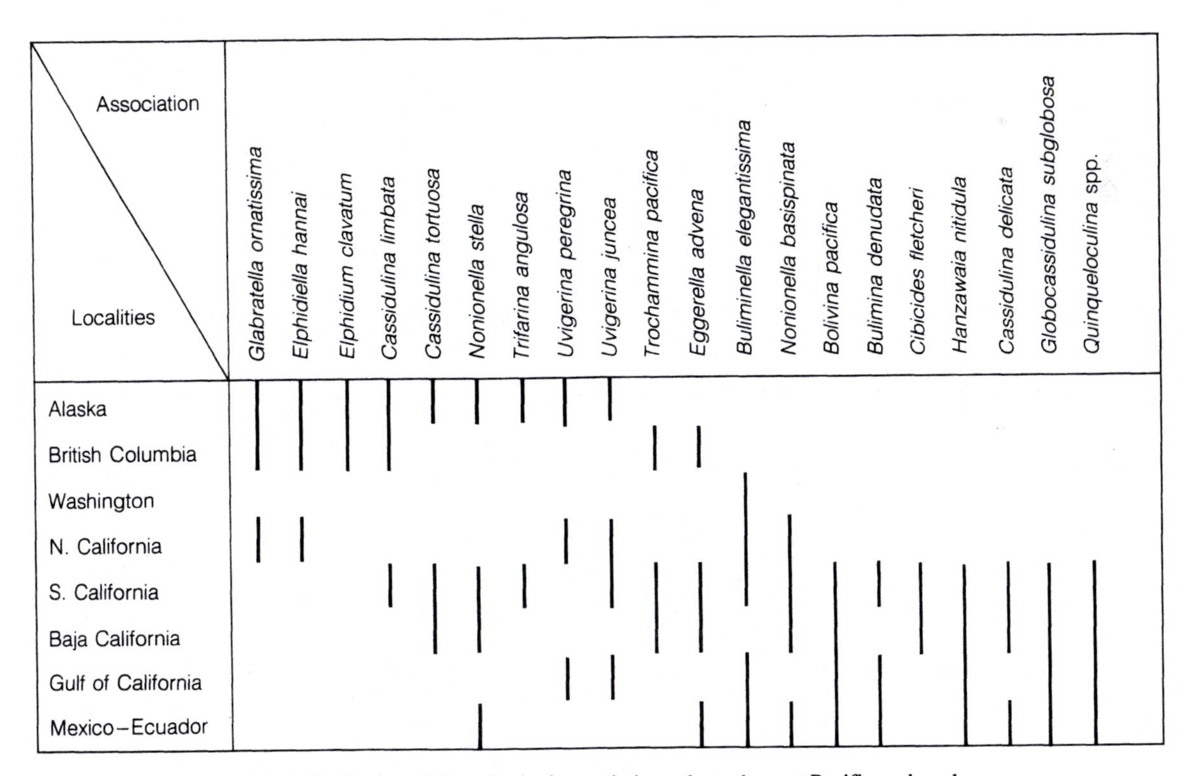

Figure 14.6 Geographical distribution of the principal associations along the east Pacific seaboard

Table 14.6 Minor shelf and slope associations from the Pacific seaboard of North and South America (L = living, D = dead, T = total)

Association		Depth (m)	Area	Source
1. *Elphidium bartletti*	T	18	Gulf of Alaska	Echols and Armentrout 1980
2. *Elphidium oregonense*	T	35	Gulf of Alaska	Echols and Armentrout 1980
3. *Buccella frigida*	T	35–46	Gulf of Alaska	Echols and Armentrout 1980
4. *Uvigerina* cf. *senticosa*	T	1372	Gulf of Alaska	Echols and Armentrout 1980
	T	2438–2743	Gulf of California	Bandy 1961
	T	2835	Peru	Boltovskoy and Totah 1987
5. *Nonionella pulchella*	T	66	Gulf of Alaska	Bergen and O'Neil 1979
6. *Islandiella norcrossi*	T	90–242	Gulf of Alaska	Bergen and O'Neil 1979, Echols and Armentrout 1980
7. *Recurvoides turbinatus*	T	119–146	Gulf of Alaska	Echols and Armentrout 1980
	T	250	British Columbia	Cockbain 1963
	T	3942	Chile	Bandy and Rodolfo 1964
8. *Cassidulina californica*	T	92–300	Gulf of Alaska	Bergen and O'Neil 1979, Echols and Armentrout 1980
	T	122	San Francisco	Bandy 1953a
9. *Reophax scorpiurus*	T	300–302	Gulf of Alaska	Echols and Armentrout 1980
	L	40	Todos Santos	Walton 1955
	T	5609–6250	Peru	Bandy and Rodolfo 1964
10. *Cibicides lobatulus*	T	146	Gulf of Alaska	Bergen and O'Neil 1979
11. *Epistominella naraensis*	T	80–243	Gulf of Alaska	Echols and Armentrout 1980
12. *Epistominella pacifica*	T	165–900	Gulf of Alaska	Bergen and O'Neil 1979, Echols and Armentrout 1980
13. *Nonionellina labradorica*	T	180	Gulf of Alaska	Bergen and O'Neil 1979, Echols and Armentrout 1980
14. *Buliminella tenuata*	T	564	Gulf of Alaska	Bergen and O'Neil 1979
	T	823	Mexico	Bandy and Arnal 1957
15. *Ehrenbergina compressa*	T	595	Gulf of Alaska	Bergen and O'Neil 1979
16. *Trifarina fluens*	T	650	Gulf of Alaska	Bergen and O'Neil 1979
17. *Cassidulina depressa*	T	216	Gulf of Alaska	Echols and Armentrout 1980
	T	46–503	San Diego	Uchio 1960b
18. *Cassidulina cushmani*	T	994–1696	Gulf of Alaska	Bergen and O'Neil 1979
	D	885–1600	El Salvador	Smith 1964
	T	1223	Peru	Resig 1981
19. *Cassidulina lomitensis*	T	1174	Gulf of Alaska	Bergen and O'Neil 1979
20. *Eponides subtener*	T	1244	Gulf of Alaska	Bergen and O'Neil 1979
21. *Karreriella parkerae*	T	2037	Gulf of Alaska	Bergen and O'Neil 1979
22. *Fursenkoina complanata*	T	2144	Gulf of Alaska	Bergen and O'Neil 1979
23. *Hyperammina elongata*	T	2281–2516	Gulf of Alaska	Bergen and O'Neil 1979
24. *Buccella* spp.	T	16–76	British Columbia	Cockbain 1963
25. *Ammotium planissimum*	T	28	British Columbia	Cockbain 1963
26. *Adercotryma glomerata*	T	123	British Columbia	Cockbain 1963
27. *Elphidium lene*	L	22–40	Bodega Bay, Calif.	Lankford and Phleger 1973
28. *Buccella tenerrima*	L	31	Morro Bay, Calif.	Lankford and Phleger 1973
29. *Saccammina atlantica*	T	19	Santa Monica	Zalesny 1959
	D	27	Todos Santos	Walton 1955

Table 14.6 *(continued)*

Association		Depth (m)	Area	Source
30. *Discorbis monicana*	T	23	Santa Monica	Zalesny 1959
	D	7–27	Santa Catalina	McGlasson 1959
31. *Gaudryina arenaria*	T	60	Santa Monica	Zalesny 1959
32. *Bolivina acuminata*	T	60–80	Santa Monica	Zalesny 1959
	L, D	64–141	Santa Catalina	McGlasson 1959
	L	146	San Diego	Uchio 1960b
	T	123	Costa Rica	Bandy and Arnal 1957
33. *Bolivina argentea monicana*	T	274–576	Santa Monica	Zalesny 1959
34. *Bolivinita minuta*	T	402–494	Santa Monica	Zalesny 1959
	D	355–452	San Diego	Uchio 1960b
	T	640	Costa Rica	Bandy and Arnal 1957
35. *Brizalina acutula*	L	6–88	Santa Catalina	Lankford and Phleger 1973, McGlasson 1959
	L	113	Mexico	Phleger 1965c
	T	110	Panama	Bandy and Arnal 1957
36. *Cancris sagra*	L, D	55–161	Santa Catalina	McGlasson 1959
37. *Bolivina quadrata*	L, D	38	Santa Catalina	McGlasson 1959
38. *Cassidulina minuta*	L, D	24–183	Santa Catalina	McGlasson 1959
	T	2134–2438	Gulf of California	Bandy 1961
	L	27–46	Mexico	Phleger 1965c
39 *Elphidium rugulosum*	L,D	2–7	Santa Catalina	McGlasson 1959
40. *Elphidium translucens*	D	7–66	Santa Catalina	McGlasson 1959
	L	29	Todos Santos	Walton 1955
	L	16	Baja California	Lankford and Phleger 1973
41. *Textularia schencki*	D	29	Santa Catalina	McGlasson 1959
	L	18	Gulf of California	Phleger 1964a
	T	0–35	Panama	Golik and Phleger 1977
42. *Brizalina argentea*	T	372–531	Santa Barbara	Harman 1964
	L, T	503–897	Todos Santos	Walton 1955
43. *Suggrunda eckisi*	T	346–588	Santa Barbara	Harman 1964
44. *Bolivina seminuda*	T	576	Santa Barbara	Harman 1964
	T	457	Nicaragua	Phleger 1965c
	T	121–224	Peru	Resig 1981
45. *Nouria polymorphinoides*	L	39	San Diego	Lankford and Phleger 1973
46. *Reophax gracilis*	L	69–287	San Diego	Uchio 1960b
47. *Trochammina rhumbleri*	L	164	San Diego	Uchio 1960b
48. *Reophax micaceous*	L	355	San Diego	Uchio 1960b
49. *Nonionella globosa*	L	380	San Diego	Uchio 1960b
50. *Uvigerina curticosta*	L	408	San Diego	Uchio 1960b
51. *Bolivina subargentea*	L	452	San Diego	Uchio 1960b
51. *Trochulina turbinata*	L	37	San Diego	Uchio 1960b
52. *Goësella flintii*	L	77–428	San Diego	Uchio 1960b
	L	71	Todos Santos	Walton 1955
53. *Buccella angulata*	L, D	80–84	San Diego	Uchio 1960b
54. *Rosalina columbiensis*	L,D	35	San Diego	Uchio 1960b
	L	6–20	Baja California	Lankford and Phleger 1973
55. *Placopsilina bradyi*	L	168	San Diego	Uchio 1960b

Table 14.6 *(continued)*

	Association		Depth (m)	Area	Source
56.	*Neoconorbina terquemi*	L	9	San Diego	Uchio 1960b
57.	*Fursenkoina apertura*	L	293–630	San Diego	Uchio 1960b
58.	*Epistominella sandiegoensis*	T	512	San Diego	Uchio 1960b
59.	*Cassidulina subcarinata*	T	521	San Diego	Uchio 1960b
60.	*Brizalina vaughani*	L	30–31	San Diego–Todos Santos	Lankford and Phleger 1973
		L	80–84	Gulf of California	Phleger 1964a, 1965c
		T	35	Panama	Bandy and Arnal 1957
61.	*Buccella parkerae*	L	13	Todos Santos	Lankford and Phleger 1973
62.	*Globobulimina* spp.	L, D	201–420	Todos Santos	Walton 1955
63.	*Bolivina spissa*	L	640	Todos Santos	Walton 1955
		T	970	Mexico	Bandy and Arnal 1957
		T	962	Chile	Ingle *et al.* 1980
64.	*Alveolophragmium columbiense*	L	37	Todos Santos	Walton 1955
65.	*Chilostomella ovoidea*	L	686	Todos Santos	Walton 1955
66.	*Reophax curtus*	L, D	51	Todos Santos	Walton 1955
67.	*Ammonia* spp.	D	62–198	Todos Santos	Walton 1955
68.	*Discorbis* spp.	D	64	Todos Santos	Walton 1955
69.	*Elphidium tumidum*	D	49	Todos Santos	Walton 1955
70.	*Reophax nana*	L	16	Baja California	Lankford and Phleger 1973
		L	11	Gulf of California	Phleger 1964a
71.	*Trochammina kellettae*	L	9	Baja California	Lankford and Phleger 1973
		L	18	Gulf of California	Phleger 1964a
72.	*Epistominella bradyana*	T	72–366	Gulf of California	Bandy 1961
		T	520–1354	Costa Rica	Bandy and Arnal 1957
73.	*Bolivina plicata*	T	610–914	Gulf of California	Bandy 1961
74.	*Epistominella smithi*	T	914–1219	Gulf of California	Bandy 1961
		D	1700	El Salvador	Smith 1964
		L	796	Peru	Bandy and Rodolfo 1964
75.	*Textularia conica*	T	2–38	Gulf of California	Brenner 1962
76.	*Elphidium gunteri*	T	9–30	Gulf of California	Brenner 1962
77.	*Trochulina avalonensis*	T	5	Gulf of California	Brenner 1962
78.	*Peneroplis pertusus*	T	1–8	Gulf of California	Brenner 1962
79.	*Ammonia beccarii*	L, T	6–27	Gulf of California	Brenner 1962, Phleger 1964
		D	20–37	El Salvador	Smith 1964
		T	2–25	Panama	Bandy and Arnal 1957, Golik and Phleger 1977
80.	*Hanzawaia concentrica*	T	5–22	Gulf of California	Brenner 1962
81.	*Trifarina occidentalis*	T	35	Gulf of California	Brenner 1962
82.	*Triloculina inflata*	T	5	Gulf of California	Brenner 1962
83.	*Tretomphalus bulloides*	T	9	Gulf of California	Brenner 1962
84.	*Elphidium incertum*	T	3–26	Gulf of California	Brenner 1962, Phleger 1964a
85.	*Elphidium crispum*	T	5	Gulf of California	Brenner 1962
86.	*Eponides babsae*	T	11	Gulf of California	Brenner 1962
87.	*Planulina exorna*	T	74	Gulf of California	Brenner 1962
88.	*Buccella hannai*	T	4–18	Gulf of California	Brenner 1962

Table 14.6 *(continued)*

Association		Depth (m)	Area	Source
89. *Cancris auricula*	L	31	Gulf of California	Phleger 1964a
90. *Fursenkoina pontoni*	L	40	Gulf of California	Phleger 1964a
91. *Trifarina jamaicensis*	L	62	Gulf of California	Phleger 1964a
92. *Bulimina marginata*	L	75	Gulf of California	Phleger 1964a
93. *Cancris panamensis*	L	91	Gulf of California	Phleger 1964a
	L	82	El Salvador	Smith 1964
	T	98–104	Panama	Bandy and Arnal 1957
94. *Planulina ornata*	L	58	Mexico	Phleger 1965c
95. *Fursenkoina sandiegoensis*	L	50–77	Mexico	Phleger 1965c
	T	25–1600	Panama	Golik and Phleger 1977
96. *Bulimina clava*	T	1144	Mexico	Bandy and Arnal 1957
97. *Uvigerina proboscidea*	T	1354–1601	Mexico	Bandy and Arnal 1957
	T	1912	Panama	Bandy and Arnal 1957
98. *Brizalina striatula*	D	80–82	El Salvador	Smith 1964
99. *Textularia vola*	T	9	Nicaragua	Bandy and Arnal 1957
100. *Uvigerina incilis*	D	144	El Salvador	Smith 1964
	T	50–59	Costa Rica	Bandy and Arnal 1957
101. *Epistominella obesa*	D	435–450	El Salvador	Smith 1964
102. *Bolivina subadvena*	D	800	El Salvador	Smith 1964
103. *Chilostomella cushmani*	L	885	El Salvador	Smith 1964
104. *Cibicides mckannai*	T	95	Panama	Bandy and Arnal 1957
105. *Uvigerina excellens*	T	1026	Panama	Bandy and Arnal 1957
106. *Lagenammina difflugiformis*	T	2600–3278	Panama	Golik and Phleger 1977
107. *Haplophragmoides bradyi*	T	2489	Colombia	Bandy and Rodolfo 1964
108. *Eilohedra levicula*	L	2489	Colombia	Bandy and Rodolfo 1964
	L	4606–5929	Serena, Chile	Bandy and Rodolfo 1964
	T	1948	Chile	Ingle *et al.* 1980
109. *Ammodiscus tenuis*	T	1171–1180	Ecuador	Bandy and Rodolfo 1964
110. *Pseudonodosiella nodulosa*	T	6006	Peru	Bandy and Rodolfo 1964
111. *Rhabdammina abyssorum*	T	3404–3495	Peru	Bandy and Rodolfo 1964
	T	3149–3257	Chile	Bandy and Rodolfo 1964
112. *Hormosina globulifera*	T	4634–5314	Peru	Bandy and Rodolfo 1964
	T	3541	Chile	Bandy and Rodolfo 1964
113. *Cassidulina neocarinata*	L	1932	Chile	Bandy and Rodolfo 1964
114. *Epistominella exigua*	L, T	2634–6011	Chile	Bandy and Rodolfo 1964, Ingle *et al.* 1980
115. *Pseudoparrella subperuviana*	T	365–454	Peru	Resig 1981
116. *Trifarina carinata*	T	632	Peru	Resig 1981
117. *Bolivina costata*	T	82–2165	Peru	Resig 1981
118. *Alabaminella weddellensis*	T	3530	Peru	Boltovskoy and Totah 1987
119. *Valvulineria inflata*	T	179–5797	Chile	Bandy and Rodolfo 1964
120. *Hoeglundina elegans*	T	1932–3142	Chile	Bandy and Rodolfo 1964
121. *Bolivina rankini*	T	135–200	Chile	Ingle *et al.* 1980
122. *Uvigerina auberiana*	T	800–1326	Chile	Ingle *et al.* 1980
123. *Uvigerina peregrina dirupta*	T	428–2568	Chile	Ingle *et al.* 1980
124. *Spiroplectammina biformis*	T	4500	Chile	Ingle *et al.* 1980

Table 14.7 Principal shelf and slope association from the Pacific seaboard of North and South America

***Bolivina pacifica* association**

Salinity: 33.2–33.5‰

Temperature: 8 to 11.6 °C

Substrate: silt

Depth: 37–503 m

Distribution:
1. Santa Catalina (McGlasson 1959) living
2. San Diego (Uchio 1960b) living
3. Todos Santos (Walton 1955) living, total
4. Gulf of California (Phleger 1964a) living
5. Mexico (Phleger 1965c) living
6. Peru (Phleger and Soutar 1973) living

Additional common species:

Bolivina acuminata	1, 2
Cancris auricula	1 (as *C. sagra*), 2
Globobulimina auriculata	1 (as *Bulimina*)
Nonionella basiloba	1
Bulimina denudata	1, 2
Brizalina acutula	1 (as *Bolivina*)
Nonionella basispinata	1
Hanzawaia nitidula	1
Globobulimina pacifica	2
Nonionella stella	2, 3
Bolivina spissa	2
Goësella flintii	2
Textularia sandiegoensis	2
Reophax gracilis	2
Arenoparrella oceanica	2
Fursenkoina apertura	2
Globobulimina spp.	3
Cancris panamensis	4
Epistominella cf. *sandiegoensis*	5
Bolivina subargentea	5
Suggrunda eckisi	5

***Bulimina denudata* association**

Salinity: mainly 33.5–35‰

Temperature: 9 to 25 °C

Substrate: sand, mud

Depth: 10–200 m

Distribution:
1. Santa Monica, California (Zalesny 1959) total
2. Gulf of California: (a) (Brenner 1962) total; (b) (Phleger 1964a) living
3. Panama: (a) (Bandy and Arnal 1957) total; (b) (Golik and Phleger 1977) total

Additional common species:

Bolivina acuminata	1
Globocassidulina subglobosa	1 (as *Cassidulina*)
Uvigerina juncea	1
Nonionella stella	1, 2b
Buccella frigida	1
Gaudryina arenaria	1
Cassidulina tortuosa	1
Cassidulina limbata	1
Nonionella basispinata	1, 3b
Globobulimina auriculata	1 (as *Bulimina*)
Hanzawaia concentrica	2a (as *Cibicides*)
Ammonia beccarii	2a (as *Streblus*)
Buccella hannai	2a (as *Eponides*)
Reophax gracilis	2b
Brizalina acutula	3a (as *Bolivina*)
Cassidulina minuta	3a
Uvigerina incilis	3a
Fursenkoina sandiegoensis	3b
Brizalina vaughani	3b (as *Bolivina*)

***Buliminella elegantissima* association**

Salinity: 33.2–34.3‰

Temperature: 8 to 20 °C

Substrate: –

Depth: mainly 4–40 m, (2) 73–302 m

Distribution:
1. Washington – Baja California (Lankford and Phleger 1973) Living
2. Santa Monica (Zalesny 1959) total
3. San Pedro (Bandy *et al.* 1964c) total
4. San Diego (Uchio 1960b) living, total
5. Gulf of California (Phleger 1964a) living
6. Mexico (Phleger 1965c) living

Additional common species:

Miliolids	1
Triloculina inflata	1
Nonionella basispinata	1 (as *Florilus*), 2, 4, 5
Buccella parkerae	1
Brizalina vaughani	1, 3 (as *Bolivina*)
Bulimina denudata	2
Nonionella stella	2
Suggrunda eckisi	2
Bolivina argentea monicana	2
Elphidium spp.	4, 5
Cassidulina depressa	4
Bolivina pacifica	5

Table 14.7 *(continued)*

***Cassidulina delicata* association**

Salinity: 33.5–35‰

Temperature: 5 to 9.3 °C

Substrate: silt

Depth: 478–897 m

Distribution:
1. Santa Monica (Zalesny 1959) total
2. San Diego (Uchio 1960b) living, total
3. Todos Santos (Walton 1955) dead
4. Nicaragua (Bandy and Arnal 1957) total

Additional common species:	
Chilostomella ovoidea	1
Epistominella smithi	1, 2
Bolivinita minuta	1
Bolivina spissa	1
Valvulineria araucana	1
Buliminella tenuata	1
Cassidulina minuta	1
Cassidulina subcarinata	2
Cassidulina laevigata	3

***Cassidulina limbata* association**

Salinity: 33.5–34‰

Temperature: 5 to 13 °C

Substrate: mud, sand, pebbles

Depth: 38–306 m

Distribution:
1. Gulf of Alaska: (a) (Bergen and O'Neil 1979) total; (b) (Echols and Armentrout 1980) total
2. British Columbia (Cockbain 1963) total
3. Santa Catalina (McGlasson 1959) dead
4. San Diego (Uchio 1960b) living, total

Additional common species:	
Cibicides fletcheri	1a, 2
Cibicides lobatulus	1a, 2
Eponides repandus	1a
Cassidulina californica	1a, 2
Cassidulina tortuosa	1a, 3, 4
Islandiella norcrossi	1b (as *Cassidulina*)
Elphidium clavatum	1b
Elphidiella nitida	2
Buccella spp.	2
Globocassidulina subglobosa	3, 4 (as *Cassidulina*)
Globobulimina pacifica	4
Bolivina pacifica	4
Cassidulina depressa	4
Trifarina angulosa	4 (as *Angulogerina*)

***Cassidulina tortuosa* association**

Salinity: 33.2–34.5‰

Temperature: 5 to 13 °C

Substrate: sand

Depth: 37–190 m

Distribution:
1. Gulf of Alaska (Bergen and O'Neil 1979) total
2. Santa Monica (Zalesny 1959) total
3. Santa Catalina (McGlasson 1959) dead
4. San Diego (Uchio 1960b) living, total
5. Todos Santos (Walton 1955) dead

Additional common species:	
Cassidulina limbata	1, 3, 4
Cibicides lobatulus	1
Eponides repandus	1
Glabratella ornatissima	1 (as *Trichohyalus*)
Bolivina pacifica	2
Bolivina acuminata	2
Bulimina denudata	2
Globocassidulina subglobosa	3, 4, 5, 6 (as *Cassidulina*)
Cibicides fletcheri	3, 4, 5
Trifarina angulosa	4 (as *Angulogerina*)
Nonionella stella	4
Rosalina campanulata	4

***Cibicides fletcheri* association**

Salinity: 33.4‰

Temperature: 13 to 20 °C

Substrate: fine sand

Depth: 2–124 m

Distribution:
1. Santa Catalina (McGlasson 1959) dead
2. San Diego (Uchio 1960b) total
3. Todos Santos (Walton 1955) living, dead

Additional common species:	
Trochulina turbinata	1 (as *Rotorbinella*), 2 (as *Rosalina*)
Textularia schencki	1
Cibicides gallowayi	1
Miliolids	1, 3
Trochulina versiformis	1 (as *Rotorbinella*)
Rosalina campanulata	2
Elphidium tumidum	3
Globocassidulina subglobosa	3 (as *Cassidulina*)
Cassidulina tortuosa	3
Bolivina acuminata	3

Table 14.7 *(continued)*

Ammonia spp.	3 (as *Rotalia*)
Discorbis spp.	3
Buccella frigida	3
Elphidium translucens	3

Eggerella advena **association**

Salinity: 30–35‰

Temperature: 8 to 25 °C

Substrate: sand

Depth: 24–202 m

Distribution:
1. Gulf of Alaska (Echols and Armentrout 1980) total
2. British Columbia (Cockbain 1963) total
3. San Diego (Uchio 1960b) total
4. Todos Santos (Walton 1955) dead
5. Mexico (Phleger 1965c) living

Additional common species:

Reophax curtus	1
Buccella frigida	1
Elphidium clavatum	1
Trochammina pacifica	2, 4
Cassidulina californica	2
Cassidulina limbata	2
Spiroplectammina biformis	2
Trochammina discorbis	2
Adercotryma glomerata	2
Recurvoides turbinatus	2
Nonionella stella	3
Cassidulina depressa	3
Elphidium spp.	3
Alveolophragmium columbiense	4 (as *Labrospira*)
Saccammina atlantica	4 (as *Proteonina*)
Hanzawaia nitidula	5
Trochammina charlottensis	5

Elphidiella hannai **association**

Salinity: 25–34‰

Temperature: 7 to 15 °C

Substrate: mud, sand

Depth: 1–206 m

Distribution:
1. Gulf of Alaska: (a) (Bergen and O'Neil 1979) total (as *E. nitida*); (b) (Echols and Armentrout 1980) total
2. British Columbia (Cockbain 1963) total (as *E. nitida*)
3. Russian River, Calif. (Erskian and Lipps 1977) total
4. San Francisco (Bandy 1953a) total

Additional common species:

Elphidium clavatum	1a, b, 4 (as *E. articulatum*)
Glabratella ornatissima	1b (as *Trichohyalus*)
Buccella frigida	1b
Eggerella advena	1b
Reophax curtus	1b
Buccella spp.	2
Cassidulina limbata	2
Buccella tenerrima	3
Buliminella elegantissima	3, 4
Elphidium translucens	3 (as *Cribroelphidium*)
Gaudryina arenaria	4

Elphidium clavatum **association**

Salinity: 32.5–34‰

Temperature: 4 to 13 °C

Substrate: mud, sand

Depth: 22–190 m

Distribution:
1. Gulf of Alaska: (a) (Bergen and O'Neil 1979) total; (b) (Echols and Armentrout 1980) total
2. British Columbia (Cockbain 1963) total

Additional common species:

Buccella frigida	1a, b
Nonionella pulchella	1a
Cribrostomoides jeffreysii	1a
Epistominella vitrea	1a
Nonionella labradorica	1a (as *Florilus*), 1b
Epistominella pacifica	1a
Islandiella norcrossi	1b (as *Cassidulina*)
Cassidulina depressa	1b
Elphidiella hannai	1b
Quinqueloculina sp.	1b
Recurvoides turbinatus	1b
Cassiduina limbata	1b
Glabratella ornatissima	1b (as *Trichohyalus*)
Cassidulina cf. *teretis*	1b
Epistominella naraensis	1b
Eggerella advena	1b, 2
Cibicides lobatulus	1b
Nonionella stella	1b
Trochammina pacifica	2
Spiroplectammina biformis	2

Table 14.7 *(continued)*

***Glabratella ornatissima* association**

Salinity: 33.8–34‰

Temperature: 5 to 19 °C

Substrate: sand

Depth: 20–50 m

Distribution:

1. Gulf of Alaska: (a) (Bergen and O'Neil 1979) total (as *Trichohyalus*); (b) (Echols and Armentrout 1980) total (as *Trichohyalus*)
2. British Columbia (Cockbain 1963) total (as *Discorbis*)
3. Russian River, Calif. (Quinterno and Gardner 1987) total

Additional common species:

Elphidium clavatum	1b
Elphidiella hannai	1b, 3
Cassidulina limbata	2, 3
Rosalina columbiensis	3
Cibicides fletcheri	3
Elphidium sp.	3

***Globocassidulina subglobosa* association**

Salinity: 33–34‰

Temperature: 8 to 20 °C

Substrate: mud, sand

Depth: 37–428 m, (5) 1171 m

Distribution:

1. Santa Catalina (McGlasson 1959) living, dead (as *Cassidulina*)
2. San Diego (Uchio 1960b) living, total (as *Cassidulina*)
3. Todos Santos (Walton 1955) living, dead
4. Gulf of California (Brenner 1962) total (as *Cassidulina*)
5. Ecuador (Bandy and Rodolfo 1964) total (as *Cassidulina*)

Additional common species:

Cassidulina minuta	1
Gaudryina atlantica pacifica	1
Trifarina angulosa	1, 2 (as *Angulogerina*)
Cassidulina limbata	1, 2, 3
Cassidulina tortuosa	1, 2
Bolivina pseudoplicata	1
Cibicides fletcheri	1, 2, 3
Göesella flintii	2
Globobulimina pacifica	2
Nonionella stella	2, 3
Cassidulina depressa	2
Epistominella sandiegoensis	2
Bolivinita minuta	2 (as *Bolivina*)
Buccella frigida	3
Ammonia spp.	3 (as *Rotalia*)
Bolivina acuminata	3
Bolivina pacifica	3
Globobulimina spp.	3
Textularia conica	4
Discorbis floridensis	4
Cibicides floridanus	4
Uvigerina auberiana	5

***Hanzawaia nitidula* association**

Salinity: 33.5–35‰

Temperature: 10 to 30 °C

Substrate: sand

Depth: 9–200 m

Distribution:

1. Santa Catalina (McGlasson 1959) living
2. Gulf of California (Phleger 1964a) living
3. Nicaragua to Panama (Bandy and Arnal 1957) total
4. Panama (Golik and Phleger 1977) total

Additional common species:

Trifarian angulosa	1 (as *Angulogerina*)
Bolivina pacifica	1
Bolivina tongi	1
Cancris sagra	1
Globocassidulina subglobosa	1 (as *Cassidulina*)
Portatrochammina simplissima	2 (as *Trochammina pacifica*)
Nonionella basispinata	2
Textularia schencki	2
Nouria polymorphinoides	2
Quinqueloculina lamarckiana	3
Trochulina versiformis	3 (as *Rotorbinella*)
Bulimina denudata	3
Fursenkoina schreibersiana	3 (as *Virgulina*)
Brizalina vaughani	3 (as *Bolivina*)
Nonionella atlantica	3
Planulina ornata	3
Cibicides fletcheri	3
Brizalina acutula	3 (as *Bolivina*)
Bolivina costata	3
Epistominella bradyana	3
Cancris panamensis	4
Epistominella sandiegoensis	4
Cibicides floridanus	4
Uvigerina juncea	4
Planulina exorna	4
Brizalina striatula	4 (as *Bolivina*)

Table 14.7 *(continued)*

***Nonionella basispinata* association**

Salinity: 33.2–35‰

Temperature: 8 to 20 °C

Substrate: silt, sand

Depth: 10–90 m

Distribution:
1. Russian River, Calif. (Quinterno and Gardner 1987) total
2. Santa Catalina (McGlasson 1959) living
3. San Diego (Uchio 1960b) living
4. Todos Santos (Walton 1955) living
5. Baja California: (a) (Lankford and Phleger 1973) living (as *Florilus*); (b) (Phleger 1964a) living
6. Mexico (Phleger 1965c) living

Additional common species:

Elphidium excavatum	1
Nonionellina labradorica	1 (as *Florilus*)
Bolivina pacifica	2
Bolivina quadrata	2
Nonionella stella	3
Alveolophragmium columbiense	3
Buliminella elegantissima	3, 6
Saccammina atlantica	4 (as *Proteonina*)
Ammoscalaria pseudospiralis	5
Nonionella basispinata	4, 5, 6
Trochammina charlottensis	4
Bolivina pacifica	4, 5, 7
Reophax gracilis	4
Trochammina kellettae	4
Bolivina acuminata	4, 5
Ammotium planissimum	4, 5
Globocassidulina subglobosa	4 (as *Cassidulina*)
Cassidulina depressa	4
Elphidium spp.	4
Epistominella smithi	4
Miliolids	5
Eggerella advena	5
Saccammina atlantica	5 (as *Proteonina*)
Alveolophragmium columbiense	5 (as *Labrospira*)
Recurvoides spp.	5
Reophax curtus	5
Reophax scorpiurus	5
Globobulimina spp.	5
Uvigerina peregrina	5
Hanzawaia nitidula	6
Cancris panamensis	7
Bulimina marginata	7
Ammoscalaria pseudospiralis	7
Brizalina acutula	7 (as *Bolivina*)
Fursenkoina sandiegoensis	7
Bolivina seminuda	7

***Nonionella stella* association**

Salinity: 33.5–35‰

Temperature: 9 to 25 °C

Substrate: sand

Depth: 15–176 m

Distribution:
1. Gulf of Alaska (Echols and Armentrout 1980) total
2. Santa Barbara – San Diego (Lankford and Phleger 1973) living
3. Santa Monica (Zalesny 1959) total
4. San Diego (Uchio 1960b) living, total
5. Todos Santos (Walton 1955) living
6. Gulf of California (Phleger 1964a) living
7. Mexico (Phleger 1965c) living

Additional common species:

Nonionellina labradorica	1
Brizalina vaughani	2
Goësella flintii	2, 4, 5
Bulimina denudata	3, 5
Buliminella elegantissima	3, 5, 7
Fissurina lucida	3

***Quinqueloculina* spp. association**

Salinity: 33.4–35‰

Temperature: 11 to 30 °C

Substrate: sand

Depth: 0–47 m

Distribution:
1. Santa Catalina (McGlasson 1959) dead
2. Los Coronados (Lankford and Phleger 1973) living
3. Todos Santos (Walton 1955) living
4. Gulf of California: (a) (Bandy 1961) total; (b) (Brenner 1962) Total; (c) (Phleger 1964a) living (as Miliolidae)
5. Costa Rica (Bandy and Arnal 1957) total

Additional common species:

Discorbis monicana	1
Elphidium rugulosum	1
Buliminella elegantissima	3
Eggerella advena	3
Nonionella stella	3
Peneroplis pertusus	4b
Textularia conica	4b

Table 14.7 *(continued)*

Eponides babsae	4b
Amphisorus hemprichii	4b
Bolivina paula	4b
Articulina lineata	4b
Elphidium incertum	4b
Triloculina fichteliana	4b
Elphidium gunteri	4b
Hanzawaia concentrica	4b (as *Cibicides*)
Pyrgo denticulata	4b
Planorbulina acervalis	4b
Nodobaculariella atlantica	4b
Elphidium crispum	4b
Reussella pacifica	4c
Rosalina campanulata	4c (as *Rotorbinella*)
Hanzawaia nitidula	5

***Trifarina angulosa* association**

Salinity: 33.5–34.5‰

Temperature: 4 to 11.6 °C

Substrate: sand

Depth: 62–550 m

Distribution:
1. Gulf of Alaska (Echols and Armentrout 1980) total
2. Santa Monica (Zalesny 1959) total (as *Angulogerina*)
3. Santa Catalina (McGlasson 1959) dead (as *Angulogerina*)
4. San Diego (Uchio 1960b) total (*Angulogerina*)
5. Chile (Ingle *et al.* 1980) total

Additional common species:

Uvigerina peregrina	1, 4, 5
Cassidulina californica	1
Islandiella norcrossi	1 (as *Cassidulina*)
Bolivina decussata	1
Cassidulina tortuosa	2, 3, 4
Cassidulina limbata	2, 3, 4 (as *Cassidulina*)
Bolivina perrini	2
Nonionella stella	2
Cibicides fletcheri	2, 4
Epistominella smithi	4
Cassidulina depressa	4
Bolivina interjuncta	5

***Trochammina pacifica* association**

Salinity: 33.4–34.5‰

Temperature: 7 to 20 °C

Substrate: sand

Depth: 17–162 m

Distribution:
1. British Columbia (Cockbain 1963) total
2. Santa Monica (Zalesny 1959) total
3. Todos Santos (Walton 1955) dead
4. Baja California (Lankford and Phleger 1973) living

Additional common species:

Eggerella advena	1, 3
Spiroplectammina biformis	1
Buliminella elegantissima	2
Bulimina denudata	2
Nonionella stella	2
Goësella flintii	2
Ammonia spp.	3 (as *Rotalia*)
Reophax scorpiurus	3
Saccammina atlantica	3 (as *Proteonina*)
Aveolophragmium columbiense	3 (as *Labrospira*)
Nonionella basispinata	4 (as *Florilus*)

***Uvigerina juncea* association**

Salinity: 32–35‰

Temperature: 4 to 19 °C

Substrate: mud, fine sand

Depth: 65–450 m

Distribution:
1. Gulf of Alaska (Echols and Armentrout 1980) total
2. Russian River, Calif. (Quinterno and Gardner 1987) total
3. Santa Monica (Zalesny 1959) total
4. Panama (Golik and Phleger 1977) total

Additional common species:

Cassidulina californica	1
Islandiella norcrossi	1 (as *Cassidulina*)
Globobulimina spp.	2
Nonionella basispinata	2
Bulimina denudata	3
Trifarina angulosa	3 (as *Angulogerina*), 3
Globobulimina auriculata	3 (as *Bulimina*)
Bolivina acuminata	3
Cassidulina limbata	3
Epistominella sandiegoensis	4
Hanzawaia nitidula	4
Cassidulina minuta	4
Cancris panamensis	4

Table 14.7 *(continued)*

***Uvigerina peregrina* association**

Salinity: 34–34.5‰

Temperature: 2 to 13 °C

Substrate: silt

Depth: 142–2134 m

Distribution:
1. Gulf of Alaska: (a) (Bergen and O'Neil 1979) total; (b) (Echols and Armentrout 1980) total
2. Russian River, Calif. (Quinterno and Gardner 1987) total
3. San Francisco (Bandy 1953a) total (as *U. hollicki*)
4. Point Arguello (Bandy 1953a) total
5. Gulf of California (Bandy 1961) total
6. Peru (Resig 1981) total
7. Chile (Ingle *et al.* 1980) total

Additional common species:

Eilohedra levicula	1a (as *Eponides*)
Epistominella pacifica	1a, 2
Cassidulina californica	1b
Trifarina angulosa	1b, 3 (as *Angulogerina*)
Bulimina striata mexicana	1b, 7
Bolivina spissa	1b
Nonionella stella	3
Cassidulina delicata	4
Epistominella smithi	5, 6
Fursenkoina spinosa	5
Valvulineria araucana	5
Cassidulina cushmani	6
Bolivina interjuncta	7
Epistominella exigua	7
Valvulineria inflata	7

principal association distribution, the major faunal change is between northern and southern California. Culver and Buzas (1985) recognised a major change off British Columbia (52–55° N) another at Point Conception (34° N, Culver and Buzas 1986) and another at Puerto Angel (16° N, Culver and Buzas 1987). Further discussion of the biogeography will be given on p. 258.

Gulf of Alaska

A large number of associations based on total assemblages are listed in Table 14.6 and 14.7. Those most abundantly represented (based on Bergen and O'Neil 1979 and Echols and Armentrout 1980) are the *Elphidium clavatum* (22–178 m), *Cassidulina tortuosa* (57–190 m), *C. californica* (58–300 m), *Epistominella naraensis* (80–243 m), *Islandiella norcrossi* (80–242 m) and *C. cushmani* (994 –1696 m) associations. A seasonal thermocline develops during the summer with surface temperatures up to 15 °C, but at greater depths temperatures are in the range 4–7 °C. Coarse substrates extend to depths of ~60 m, and deeper than this silty clays and clayey silts are present.

Echols and Armentrout (1980) stained their samples and gave comments on the distribution of living individuals. They noted that *Cassidulina californica, C. limbata* and *Cibicides lobatulus* make a distinctive fauna which characterises a bank at the outer edge of the inner shelf. They considered it to be a modern assemblage present on rock outcrops or on relict sediments. Bergen and O'Neil (1979) termed this their 'gold' assemblage. Quinterno *et al.* (1980) considered that it might be relict as some of the foraminifera are glauconitised.

British Columbia

Total assemblages from the Juan de Fuca and Georgia straits fall into three principal associations: *Elphidiella hannai* (34–206 m), *Cassidulina limbata* (40–192 m) and *Eggerella advena* (89–175 m). The *E. advena* association is present in Georgia Strait in waters which are slightly brackish (salinity 30–31.5‰). Most of the additional common species are agglutinated (Table 14.7) and calcareous tests are infrequent. The *C. limbata* association dominates Juan de Fuca Strait where salinities are normal (33.8‰). The *Elphidiella hannai* association (as *E. nitida* of Cockbain 1963) occupies an intermediate area under the influence of normal marine water. All the additional common species in these two associations are calcareous hyaline.

On the open continental shelf, Bornhold and Giresse (1985) have reported contemporary glauconite infillings in *Cassidulina californica, C. tortuosa, C. teretis, Cibicides fletcheri* and *C. lobatulus* in muddy sands and sandy muds at depths of 100–500 m.

Northern California

In the area off Russian River the *Elphidiella hannai* and *Glabratella ornatissima* associations are present on the inner shelf (Erskian and Lipps 1977; Quinterno and Gardner 1987). Mid-shelf sands, at 50–90 m, have a *Nonionella basispinata* association, mid-shelf to upper bathyal (90–450 m) have the *Uvigerina juncea* association on silt and at 600–840 m there is a *U peregrina* association. An oxygen minimum zone (with <0.4 ml l^{-1} O_2) is present between 500 and 1000 m.

Southern California

This area has been more intensively studied than any other part of the Pacific seaboard of North America and perhaps for this reason a greater number of associations has been recognised. The principal ones are shown together with generalised oceanographic data in Fig. 14.7. The surface waters are variable in salinity and temperature and the base of the thermocline varies from ~18 m off Santa Monica (Zalesny 1959) to ~75 m off San Diego (Uchio 1960b). Associations which occur only in the shallow waters above the thermocline are the *Quinqueloculina* spp., *Cibicides fletcheri, Discorbis monicana, Trochammina pacifica, Nonionella basispinata* and *Eggerella advena* associations, all on sandy substrates. Several associations have their upper depth limit at 37 m (*E. advena, Cassidulina tortuosa, Globocassidulina subglobosa* and *Bolivina pacifica*). The *Trifarina angulosa* and *C. limbata* associations have their upper depth limit close to the base of the thermocline. The *C. delicata* association (549–842 m, not plotted on Fig. 14.7) occurs in low-oxygen waters. These distribution patterns are based on living, dead and total assemblage data. Not all associations have documented living assemblages and for those that do have, the depth ranges are not always the same as the total.

The living assemblages off San Diego were sampled on six separate occasions by Uchio (1960b). The standing crop values showed little variation from season to season and were generally high on fine sands, silts and clayey silts. Average values were ~540 to 630 per 10 cm^2 but maximum values exceed 1000 per 10 cm^2. Similar values were recorded by Phleger (1964b). Coarse sands have lower standing crops averaging 146–149 per 10 cm^2. On the inner shelf at depths of 0–24 m, diversity values are low (α 2–4) but elsewhere they are moderate to high (α 5–19). Murray (1973) showed that in general, α <4 indicated depths <24 m, α <12 depths <82 m and α >16 depths of more than 825 m. In terms of wall structure, all living assemblages from deeper than 24 m are dominantly hyaline (40–95%) with some agglutinated (5–60%) and porcellaneous (0–12%) tests.

Uchio was much concerned with the depth distribution of the foraminifera. He recognised seven living depth assemblages but only four total ones and, of course, the boundaries differed between living and total zones. Furthermore, the total assemblages between 40 and 510 m commonly have >20% *Cassidulina* spp. whereas the living rarely reach 10%. Uchio considered many of these dead tests to be reworked from Pleistocene foraminiferal sands. Many of Uchio's samples contain <1% living tests and in some cases more than 50% of the dead tests are possibly of Pleistocene age. McGlasson (1959) also noted major differences between living and dead assemblages from the shelf off Santa Catalina Island.

Stable isotope data show that *Bolivina, Cassidulina, Globobulimina* and *Uvigerina* have test calcite in approximate ^{18}O equilibrium with the water. Furthermore, the $\delta^{13}C$ values of *Cassidulina* spp. are almost identical to those of the ambient dissolved inorganic carbon, whereas *Hoeglundina elegans* is enriched and *Brizalina argentea, U. curticosta, U. peregrina* and *G. pacifica* are depleted. The latter is infaunal but the others are believed to be epifaunal (Grossman 1984a).

Apart from the oxygen-deficient basins discussed below, coastal upwelling brings an oxygen-minimum zone into contact with the sea floor. Off Point Sur, in central California, the boundaries of the oxygen minimum zone (as defined by 0.5 ml l^{-1} O_2) occur at 500–525 m and 1000–1025 m. A minimum value of 0.27 ml l^{-1} O_2 occurs at depths of 700–750 m and beneath this total assemblages comprise nearly 80% *Brizalina argentea* and *Bolivina spissa. Cassidulina* spp. are most abundant along the upper edge of the oxygen minimum zone and *Globobulimina* spp. along the lower edge. Mullins *et al.* (1985) suggest that nutrient recycling takes place with abundant denitrifying bacteria at <0.3 ml l^{-1} O_2 in the centre and

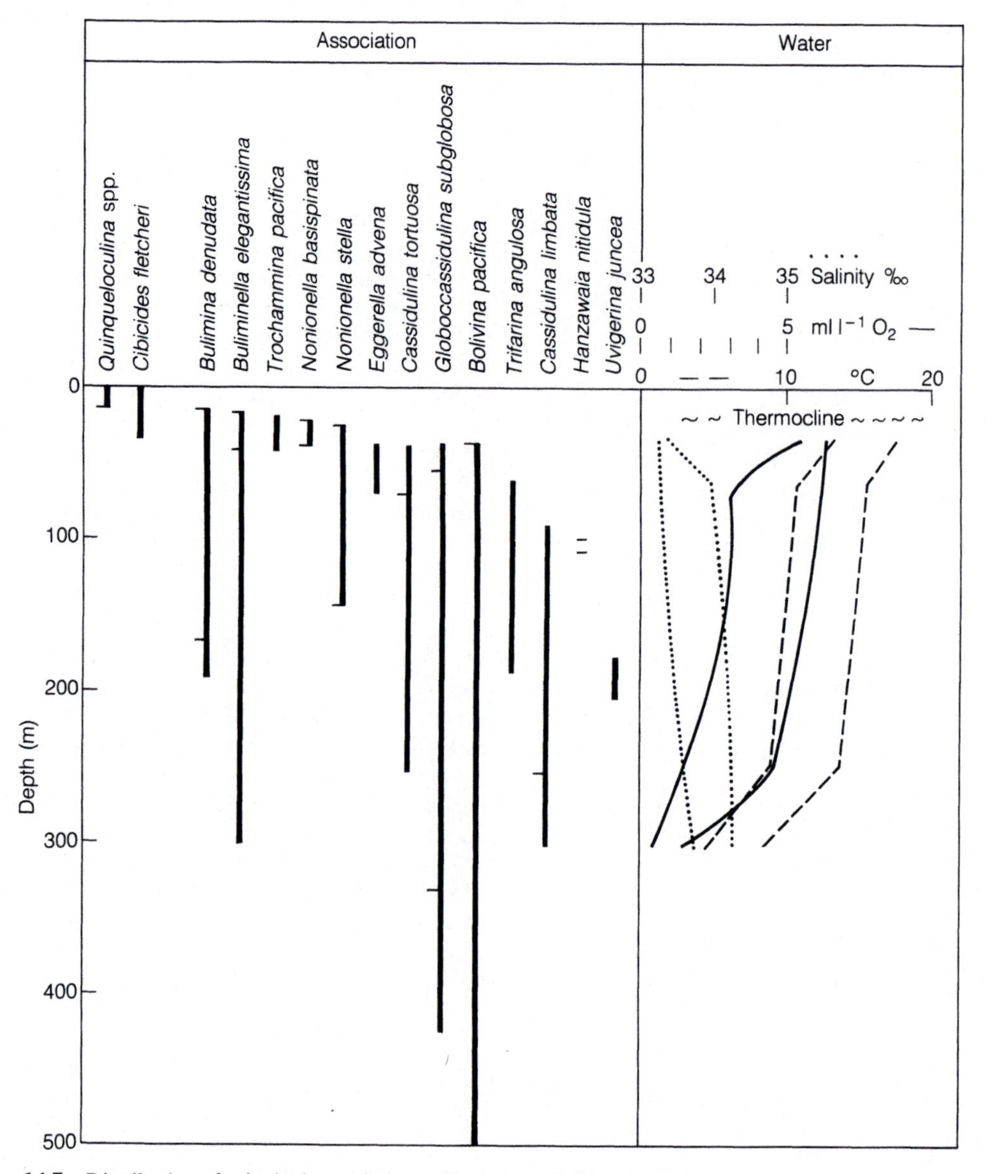

Figure 14.7 Distribution of principal associations off southern California. Ticks on side of bars show limits of living occurrences (oceanographic data based on Bandy 1953a)

nitrifying bacteria at 0.5 ml l^{-1} O_2 at the edges of the oxygen minimum zone.

Postmortem changes

The dynamic relationships between living and dead assemblages from this area have been discussed by Douglas *et al.* (1980). There is probably seasonal variation in the production of tests on the shelf due to temperature and food supply controls on reproduction. In the slope and basin environments the rank order of the top five to six species varies little from season to season suggesting that reproduction is not seasonally controlled. However, in the Santa Monica and San Pedro basins, *Cassidulina delicata, C. bradshawi, Fursenkoina bramletti, F. apertura* and *Alliatina primitiva* are present living only in the winter while *Uvigerina curticosta, Nonionella fragilis* and *Bolivinita minuta* are present only in the summer. Nevertheless, these changes are not thought to produce major differences between the living and dead assemblages.

Postmortem selective preservation, mixing and transport are very important. Douglas *et al.* point to widespread destruction of agglutinated tests and selective dissolution/preservation of calcareous tests. Excellent preservation occurs in the laminated sediments deposited where bottom waters have <0.2 ml l^{-1} dissolved oxygen because sulphate-reducing bacteria increase the alkalinity of the pore waters. Similarly, the oxygen level controls the presence or absence of deposit-feeding macrobenthos which also contribute to test destruction.

Douglas *et al.* concluded that, on the shelf, reworked Pleistocene foraminifera are abundant, biological predation causes significant selective destruction of tests and living and dead assemblages are therefore dissimilar. On the offshore banks and terraces, the living and dead assemblages are similar although thick-walled forms are selectively preserved while delicate forms are lost (fields IIIA, IVA). Downslope transport modifies dead assemblages on slopes and may lead to their introduction on to basin floors. Bandy (1964c) considered that displaced shelf tests formed ~3% of typical basin floor assemblages, but ~25% of submarine fan assemblages and up to 78% of submarine fan sand assemblages!

Pollution

Major outfalls transport sewage into shallow waters along the coast of southern California. Studies of the ecological effects include Resig (1960), Watkins (1961), Bandy *et al.* (1964a, b, 1965a, b).

In each case, beneath the suspended sewage field there are few or no living foraminifera ('dead zone'), but away from this, either as an aureole or as a downcurrent field, the standing crop increases to twice or more than that of the adjacent unpolluted shelf. The abundance of individual species may increase as much as 5000 times. These changes are undoubtedly the result of an increase in organic material which provides a source of food. In several cases there is a dramatic increase in agglutinated taxa such as *Eggerella advena* and *Trochammina pacifica*, but in other cases hyaline *Buliminella elegantissima* in favoured. Some species which would normally be present do not tolerate this pollution, e.g. *Nonionella basispinata* and *N. stella*. Therefore, the polluted zone has reduced diversity. The typical shelf sediments are olive grey but under the influence of sewage they become black and anoxic beneath the surface. Dead assemblages are commonly enriched in agglutinated tests even where the living assemblages are predominantly hyaline, and this change is attributed to postmortem dissolution of calcareous tests.

Basins

From southern California to northern Baja California is the continental borderland, a region of basins separated by narrow banks lying between the narrow shelf and the ocean basin proper. The banks act as sills between basins. The sill depths are <750 m in the northern nearshore examples and 1000–1100 m in the offshore examples. The inshore basins are rapidly infilling with sediment but the offshore basins are starved of sediment. At depths >300 m the water has a salinity of 34.2‰, temperature <10 °C, is oxygen-deficient (<2 ml l^{-1}), nutrient-rich and it flows from south to north. Within each basin the water is more or less uniform below sill depth. Upwelling from a depth of <100 m supports high surface productivity over the inshore basins.

Studies of the foraminifera have been made by Resig (1958), Zalesny (1959), Bandy (1963b), Harman (1964), Phleger and Soutar (1973), Douglas and Heitman (1979) and Douglas (1981), but few of these papers give numerical details of assemblages so associations cannot be defined.

The standing crop of the inshore basins ranges from 100 to 1200 per 10 cm^2 compared with values of 0–650 per 10 cm^2 above the sill (Phleger and Soutar 1973; Douglas 1981). Although the oxygen concentration is low in the basin bottom water, it is sufficient for the needs of the foraminifera and they consume about 3.5% of that available. Oxygen is therefore not a limiting factor (Phleger and Soutar 1973). However, low oxygen does exclude macrobenthic predators (the sediments are laminated) so that the foraminifera are not cropped and this accounts for the higher standing crops in the basins (Douglas 1981). The abundance of dead tests shows a comparable trend. In the offshore basins, standing crop is lower than above the sill depth (0–700 compared with 80–800 per 10 cm^2, Douglas and Heitman 1979) perhaps due to the lower fertility of the overlying waters.

Living assemblages from inshore basin slopes are dominated by *Epistominella smithi* and *Uvigerina curticosta* while those of offshore basins have *Eilohedra levicula* (as *Eponides*), *Gyroidina io* and *Hoeglundina elegans* in addition. Basin floor assemblages comprise common *Fursenkoina apertura*, *Cassidulinoides cornuta* and *Buliminella tenuata*, with *F. bramlettei*, *F. seminuda* and *Textularia earlandi* in inshore basins and *Loxostomum pseudobeyrichi*, *Chilostomella ovoidea* and *Bolivina pacifica* in offshore basins. The latter species is opportunistic and commonly occurs in areas of sediment disturbance (Douglas and Heitman 1979). It was also noted that *Epistominella smithi* and *B. pacifica* live at shallower depths during the summer when oxygen levels are depressed.

The diversity of living assemblages is $H(S)$ 1.5–2.5 and 2.0–2.7 for inshore and offshore basins respectively (Douglas 1981). The pattern of diversity values for the dead assemblages reflects the downslope transport of material by slides and debris flows in this tectonically unstable area. Both living and dead assemblages are dominated by hyaline tests. Agglutinated forms are rare and porcellaneous forms absent.

Within the bolivinids there are different morphotypes in different environments. Lutze (1964b) described how *Brizalina argentea* (as *Bolivina*) showed a cline from broad, keeled tests with spines and six to eight pairs of chambers with ribs, living in well-oxygenated slope waters, to slender, non-keeled, non-spinose tests either lacking ribs or having short ribs on two to three pairs of chambers, living in low-oxygen waters. Similar trends were noted by Harman (1964) and Douglas (1981) in bolivinids in general. The latter observed that small, prolate forms such as *B. vaughani* live in well-oxygenated shelf waters (4–6 ml $l^{-1}O_2$), and large flat forms such as *B. spissa* and *B. argentea* are characteristic of low-oxygen waters (<2 ml l^{-1} O_2).

Baja California

Most of the available data are from Todos Santos Bay (Walton 1955; Kaesler 1966). Walton took samples on a seasonal basis. The most abundant living assemblages are of the *Nonionella stella* association (18–176 m) yet no dead assemblages are of this kind. All other living associations are represented by few samples. There is no single dominant dead association but the most common are the *Cibicides fletcheri* (11–124 m), *Trochammina pacifica* (22–37 m) and *Globocassidulina subglobosa* (35–198 m) associations. As with southern California there are major differences between the living and dead assemblages.

The standing crop ranged from 0 to 923 per 10 cm^2 but only a quarter of the samples had >100 per 10 cm^2. The diversity ranges from α 1.5–9.5. All assemblages are mixtures of hyaline and agglutinated tests with <10% porcellaneous forms (Murray 1973).

Off Punta Abrejos, the shelf has very low-oxygen values (0.1 ml l^{-1} O_2, 75–100 m) during July and August. The standing crop in August 1961 was 450–4750 per 10 cm^2 over a depth range of 71–183 m and the dominant taxa were species of *Bolivina*, *Bulimina* and *Uvigerina* (Phleger and Soutar 1973).

Gulf of California

The main studies are those of Bandy (1961), Brenner (1962), Phleger (1964a, 1965c), Streeter (1972 – a reinterpretation of Phleger's 1964a data) and Schrader *et al.* (1983). Although seven principal associations are represented (Fig. 14.6) only the *Quinqueloculina* spp., *Nonionella stella* and *Hanzawaia nitidula* associations are represented by more than a few samples. Among the many minor associations only the *Peneroplis pertusus* association is common. The poor representation of each of the many associations recognised may be due to the relatively small number of samples studied from this large and ecologically diverse area.

Some of the standing crops are very large (492 per 10 cm^2 at 13 m, 702 per 10 cm^2 at 732 m in the southern gulf, 372–632 per 10 cm^2 at 18–40 m in the northern gulf, Phleger 1964a) and may reflect local upwelling or the influence of input of nutrients from nearby rivers. The average for depths of 900–3166 m was 26 per 10 cm^2. Additional data in Phleger (1965c) show a range of 3–2948 per 10 cm^2 over a depth range of 7–594 m. Diversity of the living assemblages is α 1.5–12 for the depth range 0–180 m and α 3.5–6.5 for depths >180 m (Murray 1973). From 0 to 33 m the living assemblages have mixtures of porcellaneous (0–50%), agglutinated (0–60%) and hyaline (18–100%) tests. Those from deeper than 33 m have 49–100% hyaline and 0–51% agglutinated tests.

In the southern gulf, tropical species are present. Brenner (1962) listed *Amphisorus hemprichii*, *Articulina lineata*, *Sphaerogypsina globula* (as *Gypsina*) and *Peneroplis pertusus*. Crouch and Poag (1987) considered the Gulf of California to be the Sonoran Subprovince of the tropical Panamanian Province.

Two of Streeter's (1972) factor assemblage five samples have <100 living individuals and both are dominated by '*Bolivina* other spp.'. They come from 97 and 732 m and may be from low-oxygen waters. Schrader *et al.* (1983) took cores from different parts of the gulf specifically to investigate the low-oxygen environments. They found that the assemblages have low diversity and are dominated by a few species that favour low-oxygen conditions, e.g. *Bolivina seminuda, Buliminella tenuata* and *Bolivina subadvena*. They considered that few tests had been transported in although Bandy (1961) believed that 70–100% of basin faunas were allochthonous. A study of living assemblages is necessary to resolve this difference of opinion. Schrader *et al.* also consider that postmortem dissolution is active at, or near, the sediment–water interface under anaerobic conditions, although they observed that preservation of benthic tests was most excellent in the top 5 cm of their cores.

Southern Mexico to Ecuador

The faunas of this area are rather poorly known. Sources include Bandy and Arnal (1957, Mexico to Panama), Smith (1963, 1964, El Salvador), Guzman *et al.* (1987, Costa Rica), Golik and Phleger (1977, Gulf of Panama), Bandy and Rodolfo (1964, Colombia, Ecuador), Boltovskoy and Gulancañay (1975, Ecuador), Miro and Gulancañay (1974, Ecuador), Boltovskoy (1976, biogeography) and Crouch and Poag (1987, biogeography).

Six of the principal associations are represented by single samples: *Quinquloculina* spp. (4 m), *Buliminella elegantissima* (9 m), *Nonionella basispinata* (11 m), *Eggerella advena* (55 m), *Cassidulina delicata* (778 m) and *Globocassidulina subglobosa* (1171 m). Four are better represented: *Hanzawaia nitidula* (9–200 m), *Bulimina denudata* (25–80 m), *Uvigerina juncea* (65–200 m) and *Bolivina pacifica* (201–238 m) (see Table 14.7 and Fig. 14.6). There are many minor associations (Table 14.6).

Substrate type appears to be a significant control on the distribution of the *Quinqueloculina* spp. (coarse shelly sands) and the *Bulimina denudata* (mud) associations (Golik and Phleger 1977). *Buliminella elegantissima* is most abundant on sands having a median diameter of >0.5 mm (Smith 1964).

Off El Salvador, the standing crop on the shelf varied from 14 to 243 per 10 cm^2 with only 3 values above 100, while on the slope, from 435 to 1700 m, the range was 2–100 per 10 cm^2. Two samples from 3100 and 3200 m yielded only 7 and 8 per 10 cm^2 (Smith 1964). A similar pattern was recorded in the Gulf of Panama with a range of 0–600 per 10 cm^2 and a mean of 68 per 10 cm^2 on the shelf, and 0–150, means 20 per 10 cm^2 on the slope (Golik and Phleger 1977). The standing crop averaged 27.5 per 10 cm^2 where the salinity was <34‰ and 116.1 per 10 cm^2 where it was >35‰. Off El Salvador oxygen minimum waters (0.3 ml l^{-1} O_2) extend from ~140 to 1500 m while in the Gulf of Panama the range is ~300 to ~500 m. A site on a coral reef at 5 m off Costa Rica had a standing crop of 40 per 10 cm^2 (Guzman *et al.* 1987).

Detailed biometric and distributional studies of the bolivinids off El Salvador showed that shelf species are small (mean length ~250 μm) whereas slope species are large (mean length ~500 μm). Individuals reach a maximum size and abundance at 800–900 m the position of the oxygen minimum and nitrogen maximum. *Bolivina subadvena* has a thick wall and large pores (may be infaunal), and *Loxostomum pseudobeyrichii* (as *Bolivina*) has a thin wall with fine pores (may be epifaunal) (Smith 1963).

Stable isotopic analyses show that *Hanzawaia concentrica* has δ^{18} O–0.45 to 0.79‰ while *Bolivina subadvena, Uvigerina peregrina* and *U. excellens* show enrichment ($\delta^{18}O$ + 2.43 to + 2.69‰) (Smith and Emiliani 1968).

Peru–Chile

The continental shelf is narrow and the steep slope leads to the adjacent trench. Bandy and Rodolfo (1964), Ingle *et al.* (1980) Resig (1981) and Boltovskoy and Totah (1987) have provided data on total assemblages, most of which are minor associations (Table 14.6). A single box core from a depth of 180 m off Peru had a standing crop of 800 per 10 cm^2 and *Bolivina* cf. *pacifica* formed >95% of the living assemblage. This is an area of very low oxygen (Phleger and Soutar 1973). The shelf at 90–133 m off northern Chile is also an area of very low oxygen (0.1 ml l^{-1}) and yielded living *Bolivina punctata, B. costata* and a few small individuals of other taxa (Boltovskoy 1972). General comments on the distribution of individual species off central Chile have been given by Boltovskoy and Theyer (1970). More detailed studies by Ingle *et al.* (1980) have shown a clear correlation between faunas and water masses: *Uvigerina peregrina* association (142 m) from the oxygenated waters of the Peru–Chile Current, *Bolivina* aff. *rankini* association (with *B. interjuncta* and *Trifarina fluens*, 135–200 m) from the oxygen-deficient, poleward-flowing countercurrent, *T. angulosa* (274 m), *U. peregrina dirupta* (428 m) and *U. auberiana* (800–860 m) associations from oxygen-rich Antarctic Intermediate Water and *Bolivina spissa* association (962 m, with *Cibicides mckannai* and *U. auberiana*) from deep oxygen minimum water. Below this, oxygenated Deep Water has the *Eilohedra levicula* (1948 m) and *Epistominella exigua* (2634 m) associations. The CCD lies at 4000 m and a single assemblage at 4500 m was entirely agglutinated: *Spiroplectammina biformis* association with *Glomospira gordialis, Eratidus foliaceum* (as *Ammobaculites*) and *Ammoglobigerina globigeriniformis* (as *Trochammina*).

Different size classes of *Uvigerina peregrina* from off Peru have different $\delta^{18}O$ values with the difference between the 210–297 and 420–500 μm size classes increasing with water depth from ~0.3‰ at 365 m to ~1.4‰ at 810 m. *Uvigerena peregrina* and *Globocassidulina subglobosa* from the oxygen-minimum zone show enrichment of up to 1.5‰ in $\delta^{13}C$ relative to deeper-water specimens from more oxygenated waters. This is against the expected trend and Dunbar and Wefer (1984) suggest that it reflects decreased growth rates.

On the Peru shelf, an area noted for upwelling, hyaline benthic tests are replaced by carbonate-fluorapatite (Glenn and Arthur 1988). This does not take place at the sediment surface but below in the zone of bacterial sulphate reduction.

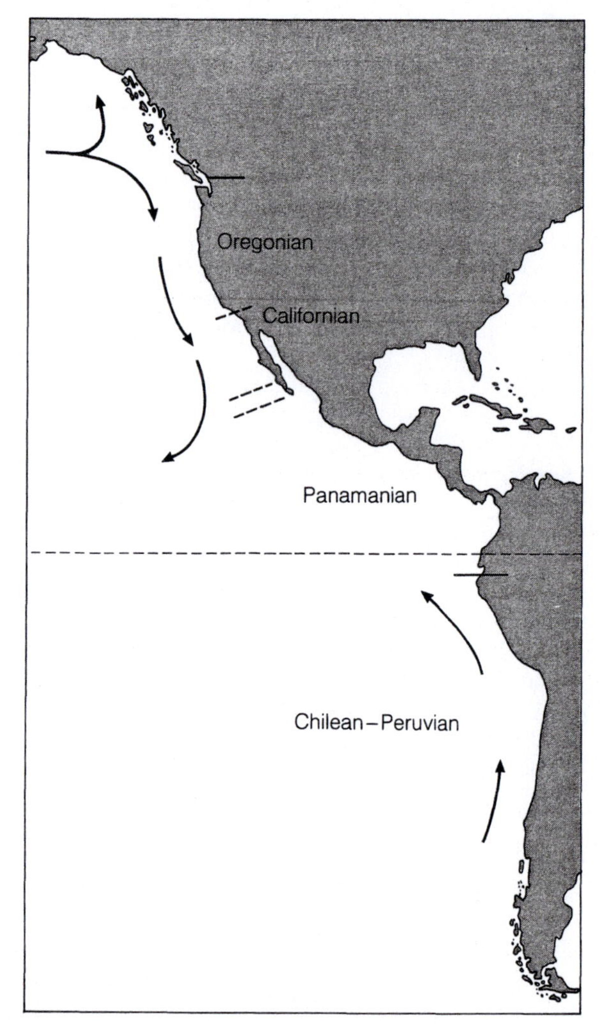

Figure 14.8 Major biogeographic divisions along the eastern Pacific seaboard (see text for sources)

Biogeography

The shallow-water faunas show distinct changes between high and low latitudes, and although it is possible to recognise 'provinces' their boundaries are often diffuse. Boltovskoy and Wright (1976) identifed a North American Pacific cold-water fauna extending south to Vancouver (48° N), a temperate-water fauna from there to the southern tip of Baja California, a warm-water Panamanian Province extending south to Ecuador (3° S) and a Chilean–Peruvian temperate-water fauna along the entire length of South America.

Within the northern temperate-water province, it has already been shown that the marsh faunas show distinct differences between the Oregonian and Californian subprovinces (Fig. 14.4). Shelf assemblages show a major change at Point Conception which is broadly in agreement with the marsh faunas. Crouch and Poag (1987) had a broad transition zone off southern Baja California as the northern limit of the Panamanian Province. Within the later they distinguished a Sonoran Subprovince in the Gulf of California from the main Isthmian Subprovince. The southern boundary of the Panamanian subprovince is abrupt and is controlled by the westward diversion of the cool Peru –Chile Current. Similarly its northern boundary is controlled by the diversion of the Californian Current. Thus these eastern boundary currents are the principal influences on biogeography (Fig. 14.8).

Summary and conclusions

Although a large number of associations have been recognised, only 20 are widely distributed (Fig. 14.6, Table 14.7). Their occurrence is controlled by a complex of parameters that includes substrate type and water mass (Ingle and Keller 1980). A special feature of this seaboard is the presence of bottom waters with very low oxygen contents. These support specialised faunas which are epifaunal, have generally large tests, with a high surface to volume ratio and are exclusively calcareous.

Most studies have concentrated on depth distribution patterns based mainly on total assemblages. Ingle and Keller (1980) have reviewed this topic and shown that certain depth boundaries correspond to environmental changes. Thus, at 40–50 m there is the lower limit of wave disturbance of the sea floor, the base of the photic zone is ~100 m, the average shelf–slope break at 130 m, the oxygen-minimum layer extends from between 200–500 m and 800–1200 m, the base of the permanent thermocline and top of the Deep Water is between 1500 and 2000 m and the CCD causes an increase in agglutinated tests at depths >3000–3500 m. However, they caution that most of these boundaries are subject to variation due to major changes in ocean circulation. Apart from this, because the margin is tectonically active, downslope transport of material takes place by a variety of mechanisms and on a variety of scales. Large-scale movements have been described (Field and Edwards 1980) and no doubt some of the extended depth ranges based on total or dead assemblages are due to this cause.

There are still major gaps in the knowledge of foraminifera from the east Pacific margin, especially off the northern USA and from Baja California southwards.

CHAPTER 15

Pacific Ocean

Introduction

This is the largest and deepest of the oceans. All the deep water originates off Antarctica and flows northwards at depths >2500 m. In the South Pacific the bottom-water temperatures are 0.5–2.5 °C and salinity 34.70–34.74‰. It takes roughly 1000 years for the water to reach the North Pacific and during this time oxygen is reduced and phosphate levels increased. The AABW flows as a contour current and passes from one basin to the next through fracture zones in the ocean ridges. The CCD lies at 4200–4500 m over much of the ocean but it is depressed to 5000 m beneath the equatorial zone of high productivity and becomes shallower (<3000 m) close to the continental margins. Consequently carbonate sediments are restricted to the East Pacific Rise and Lord Howe Rise–New Zealand Platform. Elsewhere the sediments are carbonate-poor or non-carbonate abyssal clays (Kennett 1982).

Taxonomic studies of deep-sea Pacific Ocean foraminifera include Saidova (1970, 1975) and Smith (1973).

The localities discussed in this chapter are shown in Fig. 15.1.

General features

Following early studies of parts of the ocean (Saidova 1961a, b, c, 1963, 1970) Saidova (1976) compiled maps showing the distribution and abundance of agglutinated and calcareous tests (see Fig. 15.1). From this it can be seen that the ridges and rises are not only the sites of occurrence of calcareous tests but they are also the areas of greatest abundance of agglutinated ones. The deep basinal abyssal plains are below the CCD and they are also beneath the surface central water masses. These are very poor in nutrients and therefore have low plankton productivity. This in turn means that little phytodetritus reaches the abyssal ocean floor so that standing crops are also low.

Bernstein and Hessler (1978) took box cores from such an oligotrophic abyssal plain in the North Pacific. The total assemblages larger than 297 μm showed randomly distributed patchiness on a scale of <100 cm^2 up to several kilometres. The faunas were considered to be diverse and of high density in comparison with metazoans from such areas, but it must be remembered that probably the majority of the foraminifera were dead and had accumulated over hundreds or even thousands of years. Indeed Bernstein and Meador (1979) recorded standing crops of 1–3 per 10 cm^2 and demonstrated that the patchiness of living forms may be on a scale of <100 cm^2 and some species may be randomly distributed. However, some patches may persist for more than one generation. A more detailed study in the Panama Basin based on two samples collected at 3912 m from a submersible yielded standing crops of 12 per 10 cm^2 and diversity of α 3.5–4. *Dendronina arborescens* (as *Dendrophyra*) was the most abundant species living in the top 2 cm of sediment which is flocculent. However, it was virtually absent, both living and dead, at 5 cm below the sediment surface. This species is believed to be an epifaunal immobile passive suspension feeder. Kaminski *et al.* (1988) consider that the dead tests are destroyed below the sediment surface within 9 months. *Nodulina dentaliniformis* (as *Reophax*) was common living at 2–15 cm (deepest sample) below the surface. It is believed to be infaunal and to generate fine burrows. A recolonisation experiment showed that *N. dentaliniformis* and *Reophax excentricus* were opportunistic whereas *Dendronina arborescens* was not.

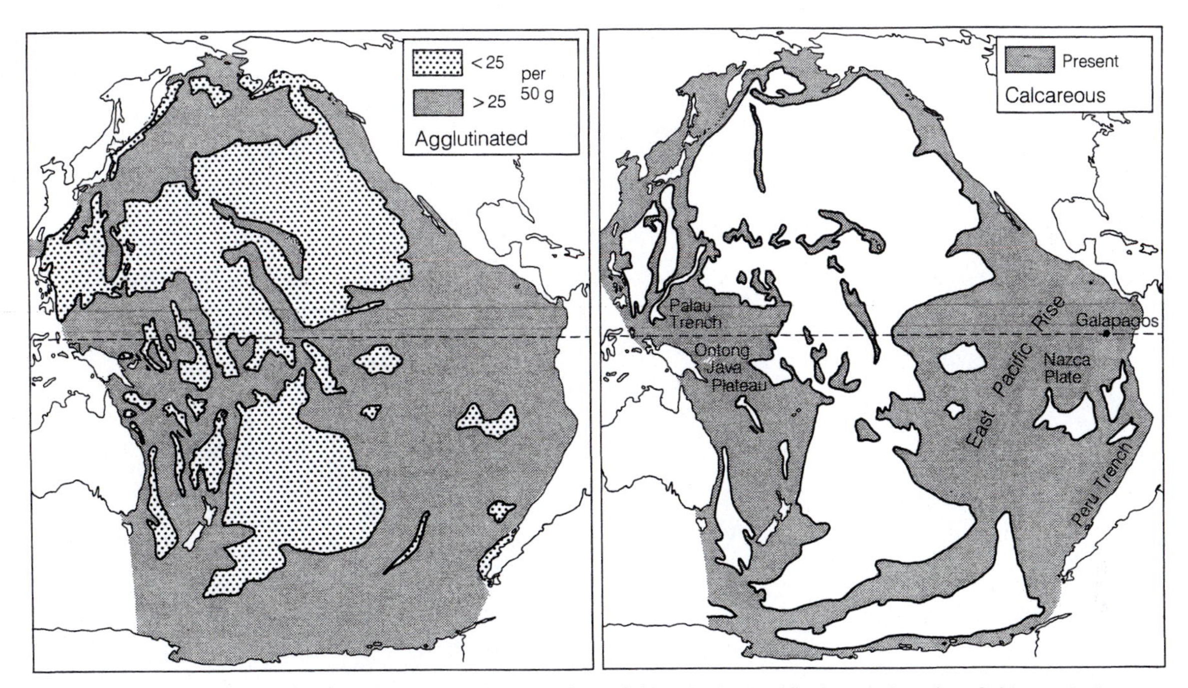

Figure 15.1 Distribution of agglutinated and calcareous foraminifera in the Pacific Ocean (based on Saidova 1976)

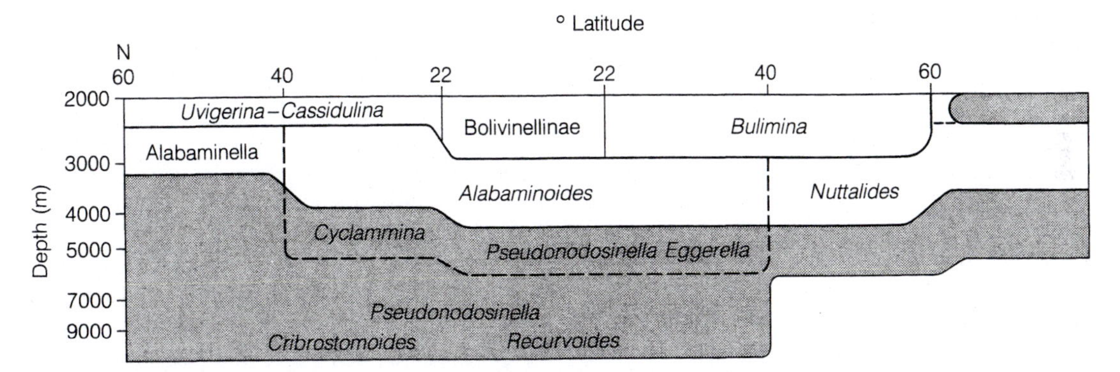

Figure 15.2 Generalised distribution of deep-sea foraminifera in a N–S transect of the Pacific Ocean (based on Saidova 1976). Dotted – agglutinated, unshaded – calcareous assemblages

The generalised distribution of genera of foraminifera is a north–south transect (Fig. 15.2) and clearly shows the effects of the CCD on their composition.

Associations

Only a few papers give numerical data on assemblages (Burke 1981; Resig 1981; Nienstedt and Arnold 1988; Hughes 1988). From this the minor and principal associations listed in Tables 15.1 and 15.2 have been defined but these give only a glimpse of the Pacific deep-sea faunas.

The *Siphouvigerina interrupta* association (1611–3057 m) occurs on the Ontong Java Plateau beneath Deep Oxygen Minimum Water and Deep Water. The *Epistominella exigua* association (1758–3990 m) is present in Pacific Deep Water. Its lower boundary is close to the lysocline which lies at 3500 m off the Ontong Java Plateau (Burke 1981) the

Tables 15.1 Minor associations from the Pacific Ocean (T=total)

	Association		Depth	Area	Source
1.	*Hyperammina friabilis*	T	4430	North	Smith 1973
2.	*Adercotryma glomerata*	T	4650	North	Smith 1973
3.	*Cribrostomoides subglobosum*	T	5000	North	Smith 1973
4.	*Ammoglobigerina globigeriniformis*	T	5500	North	Smith 1973
5.	*Glomospira gordialis*	T	5610	North	Smith 1973
6.	*Pullenia bulloides*	T	2342–3451	Ontong Java Plateau	Burke 1981
		T	3117	Nazca Plate	Resig 1981
7.	*Fijiella simplex*	T	3002	Ontong Java Plateau	Burke 1981
8.	*Astrononion stelligerum*	T	3179	Ontong Java Plateau	Burke 1981
9.	*Bulimina translucens*	T	3410–3748	Nazca Plate	Resig 1981
10.	*Bolivina costata*	T	4246	Nazca Plate	Resig 1981
11.	*Cassidulina depressa*	T	3198–3455	Nazca Plate	Resig 1981
12.	*Repmanina charoides*	T	4246	Nazca Plate	Resig 1981
13.	*Nodulina dentaliniformis*	T	3968–6309	Peru Trench	Resig 1981
14.	*Placopsilinella aurantiaca*	T	6391–6505	Peru Trench	Resig 1981
15.	*Recurvoides contortus*	T	4852	Nazca Plate	Resig 1981
16.	*Rhabdammina abyssorum*	T	5791	Peru Trench	Resig 1981
		T	3009	East Pacific Rise	Neinstedt and Arnold 1988

Globocassidulina subglobosa association (2211–2876 m) is mainly associated with Deep Oxygen Minimum water (O_2 – 3.94 ml l^{-1}). The *Eponides turgidus* association is developed on organic-rich sediments beneath the westward deflection of the Peru Current off north Peru and the region of equatorial upwelling (Resig 1981). The *Nuttallides umboniferus* association is mainly found between the lysocline and the CCD in AABW (Resig 1981). Beneath the CCD only agglutinated assemblages occur.

Other observations

The Pacific Ocean is the major area of formation of manganese nodules and they cover approximately 70% of the floor in the northern part. Various types of encrusting foraminifera are associated with the nodules (Riemann 1983; Mullineaux 1987, 1988). Most are thought to be suspension feeders from their tube and mat-like morphologies. Riemann has suggested that foraminiferal stercomata (waste products) contain iron and manganese which contribute to the formation of the nodules; this view was supported by Mullineaux (1988) although, as he pointed out, proof is still lacking.

On the East Pacific Rise, the Galapagos hydrothermal mounds cause local temperature anomalies in both the water and the surface sediment. Nienstedt and Arnold (1988) recorded temperature elevations of 2 °C in the top 15 cm of sediment and found that *Uvigerina auberiana* has an increased length : breadth ratio (3.70) compared with normal areas (2.54–2.82). They speculated that this morphological variation is temperature-related. The hydrothermal mound total assemblages are more diverse (*H*(*S*) 3.03–3.55 compared with 1.06–2.88 for other seamounts), and have *U. auberiana*, *Chilostomella oolina*, *Sphaeroidina bulloides*, *Oridorsalis tener* (as *Eponides*) total assemblages compared with small *Epistominella exigua* – *Nuttallides umboniferus* from other seamounts at similar depths. *Uvigerina auberiana* is selectively preyed upon probably by a naticid gastropod which, having found its prey, systematically bores into each chamber to feed on the contents (Arnold *et al.* 1985). Presumably the food gained is greater than the energy expended in boring.

Postmortem changes

The effects of dissolution on calcareous tests have already been noted. However, the agglutinated forms also suffer differential postmortem destruction as many taxa have poorly cemented walls which are prone to disaggregate. Kaminski *et al.* (1988) propose the following sequence according to their preservation potential from good to poor: large *Rhizammina* sp., *Saccammina* sp., *Ammoglobigerina globigeriniformis*, *Reophanus ovicula*, *Cribrostomoides subglobosum*,

Table 15.2 Principal associations from the Pacific Ocean

***Alveolophragmium nitidum* association**

Salinity: Pacific Bottom Water

Temperature: –

Substrate: –

Depth: 6560–7230 m

Distribution:
1. N. Pacific (Smith 1973) total

Additional common species:

Ammobaculites filaformis	1
Ammoglobigerina globigeriniformis	1 (as *Trochammina*)
Cribrostomoides scitulus	1 (as *Alveolophragmium*)
Hyperammina friabilis	1

***Epistominella exigua* association**

Salinity: –

Temperature: 1.55 to 1.65 °C

Substrate: Carbonate ooze

Depth: 1758–3990 m

Distribution:
1. Nazca Plate (Resig 1981) total
2. East Pacific Rise (Nienstedt and Arnold 1988) total

Additional common species:

Eponides turgidus	1 (as *Gyroidina*)
Nuttallides umboniferus	1

***Eponides turgidus* association**

Salinity: –

Temperature: –

Substrate: Organic – rich ooze

Depth: 2734–4622 m

Distribution:
1. Nazca Plate (Resig 1981) total (as *Gyroidina*)

Additional common species:

Nuttallides unboniferus	1
Epistominella exigua	1
Globocassidulina subglobosa	1 (as *Cassidulina*)
Bulimina translucens	1

***Globocassidulina subglobosa* association**

Salinity: 34.58–34.74‰

Temperature: 1.4 to 3.9 °C

Substrate: –

Depth: 2211–2876 m

Distribution:
1. Ontong Java Plateau (Burke 1981) total (as *Cassidulina*)
2. Nazca Plate (Resig 1981) total (as *Cassidulina*)

Additional common species:

Siphotextularia sp.	1
Siphouvigerina interrupta	1
Pullenia bulloides	1
Epistominella exigua	2
Nuttallides umboniferus	2

***Nuttallides umboniferus* association**

Salinity: 34.7‰

Temperature: 1.5 to 1.7 °C

Substrate: Calcareous ooze

Depth: 2850–4649 m

Distribution:
1. Ontong Java Plateau (Burke 1981) total
2. Nazca Plate (Resig 1981) total
3. East Pacific Rise (Nienstedt and Arnold 1988) total

Additional common species:

Epistominella exigua	1, 2, 3
Pullenia bulloides	1
Globocassidulina subglobosa	2 (as *Cassidulina*), 3
Eponides turgidus	2 (as *Gyroidina*)

***Siphouvigerina interrupta* association**

Salinity: 34.58–34.74‰

Temperature: 1.4 to 3.9 °C

Substrate: –

Depth: 1611–3057 m

Distribution:
1. Ontong Java Plateau (Burke 1981) total

Additional common species:

Epistominella exigua	1
Globocassidulina subglobosa	1 (as *Cassidulina*)

Recurvoides spp., *Hormosina globulifera*, *Reophax excentricus*, *Dendronina arborescens*, *Nodulina dentaliniformis*, *Pelosina* sp.

Since most of the fauna living in the sediment is infaunal this has consequences for sediment mixing. From a study of box cores taken in carbonate sediments, Swinbanks and Shirayama (1984) concluded that bioturbation is slow because of the low density of living organisms and their probably low activity. They consider that it takes around 1000 years to turn over a layer 10 cm thick. Since the sedimentation rate is unlikely to exceed 2 cm per 1000 years this means that the faunas seen in the top few centimetres of sediment may span several thousand years.

The Pacific Ocean is bordered by trenches and zones of earthquake activity. Displacement of shallower-water sediments downslope must be a common occurrence in these areas. One impressive example is the presence of carbonate turbidite layers at 8053 m in the Palau Trench. The associated foraminifera include *Amphistegina* spp., *Calcarina spengleri* and *Calcarina hispida* showing that a shallow reefal environment was the source (Yamamoto *et al.* 1988).

Summary

The foraminiferal faunas of the Pacific are the least known of those of the deep sea. In contrast to the Indian and Atlantic oceans, which are principally areas of carbonate sedimentation with calcareous assemblages, the Pacific has a shallower CCD and greater depth and hence mainly agglutinated assemblages at depths greater than around 4000 m.

CHAPTER 16

Southern Ocean

The Southern Ocean lies between Antarctica and the subtropical convergence (~40 to 50° S). To the north it continues into the southern parts of the Atlantic, Indian and Pacific oceans. Around Antarctica the shelves are deep (300–500 m) and particularly wide in the Weddell Sea and Ross Sea. In these two areas the Antarctic Ice Shelf floats on the sea over large areas and extends to a depth of >400 m below the surface. Indeed icebergs and sea-ice plough furrows in the sea floor at depths down to 500 m. They gouge tracks a few metres deep and tens of metres wide and produce diamictites over 50% of the shelf (Barnes and Lien 1988).

From an ecological point of view, the single most important oceanographic process is deep-water upwelling (Foldvik and Gammelsrød 1988). During the winter low-pressure areas develop around Antarctica. Clockwise winds around the lows produce a westward-flowing coastal current termed the East Wind Drift, while to the north of the lows the winds are westerlies and the reluctant eastward circulation is known as the West Wind Drift. In the Southern Hemisphere wind-induced Ekman transport is directed to the left of the wind stress. Hence, the East Wind Drift has southerly Ekman transport and the West Wind Drift northerly Ekman transport. Between them is the Antarctic Divergence of surface waters which permits upwelling of deeper water.

The Weddell Sea is the site of formation of very cold, dense AABW which cascades northwards into the deepest parts of the ocean. At intermediate depths, south-flowing, high-salinity warm Atlantic Deep Water mixes with surface and deep Antarctic waters to give rise to the CDW which flows from west to east. The relationship between the three basic water masses is shown in Fig. 16.1. During the summer <10% of the Southern Ocean is ice-covered but during

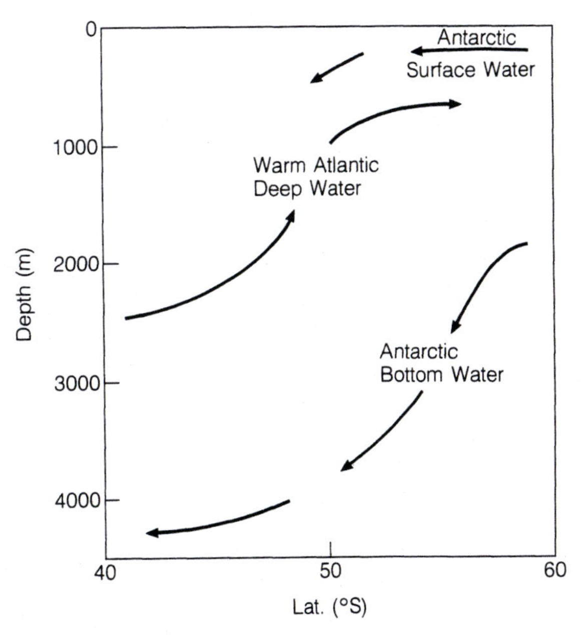

Figure 16.1 Generalised circulation in the Southern Ocean (based on Deacon 1963)

the winter this rises to ~60%. The pack ice is only 1–2 m thick.

The term 'cryogenic sediments' has been applied to marine deposits that form in association with ice (Kellogg and Kellogg 1988). There has long been discussion as to whether there is benthic life beneath the floating ice shelf because phytoplankton, the primary producers, must be absent. There are few data but Kellogg and Kellogg note that no infauna has been recorded from beneath the Ross Ice Shelf. However, currents supply food sufficient to support an abundant and diverse fauna within 100 km of the calving margin of the McMurdo Ice Shelf. Lipps *et al.* (1977) reported an agglutinated assemblage from mud at a depth of 68.5 m beneath a proglacial lake with a 2.5–4.5 m thick ice cover throughout the year, but this is clearly not representative of conditions beneath the thick ice shelf.

Parts of the shelf receive a very low terrigenous input and, because of high productivity, sediments rich in organic carbon and opaline silica from diatom frustules are deposited, e.g. McMurdo Sound (Dunbar *et al.* 1989). This sediment is resuspended, especially during the winter, to give a near-bottom nepheloid layer. This favours organisms that feed from suspension and indeed deposit-feeders are rare.

The principal taxonomic studies are those of Heron-Allen and Earland (1932), Earland (1934a, b, 1936), Parr (1950), McKnight (1962), Pflum (1966), Kennett (1967), Echols (1971) and Nomura (1984). Non-quantitative studies and those considering possible depth zonations are those of Saidova (1961a), Bandy and Echols (1964), Theyer (1971) and Lena (1975). The agglutinated component of the dominating calcareous assemblages of the Kerguelen Plateau has been discussed by Lindenberg and Auras (1984).

The localities mentioned in the chapter are shown on Fig. 16.2.

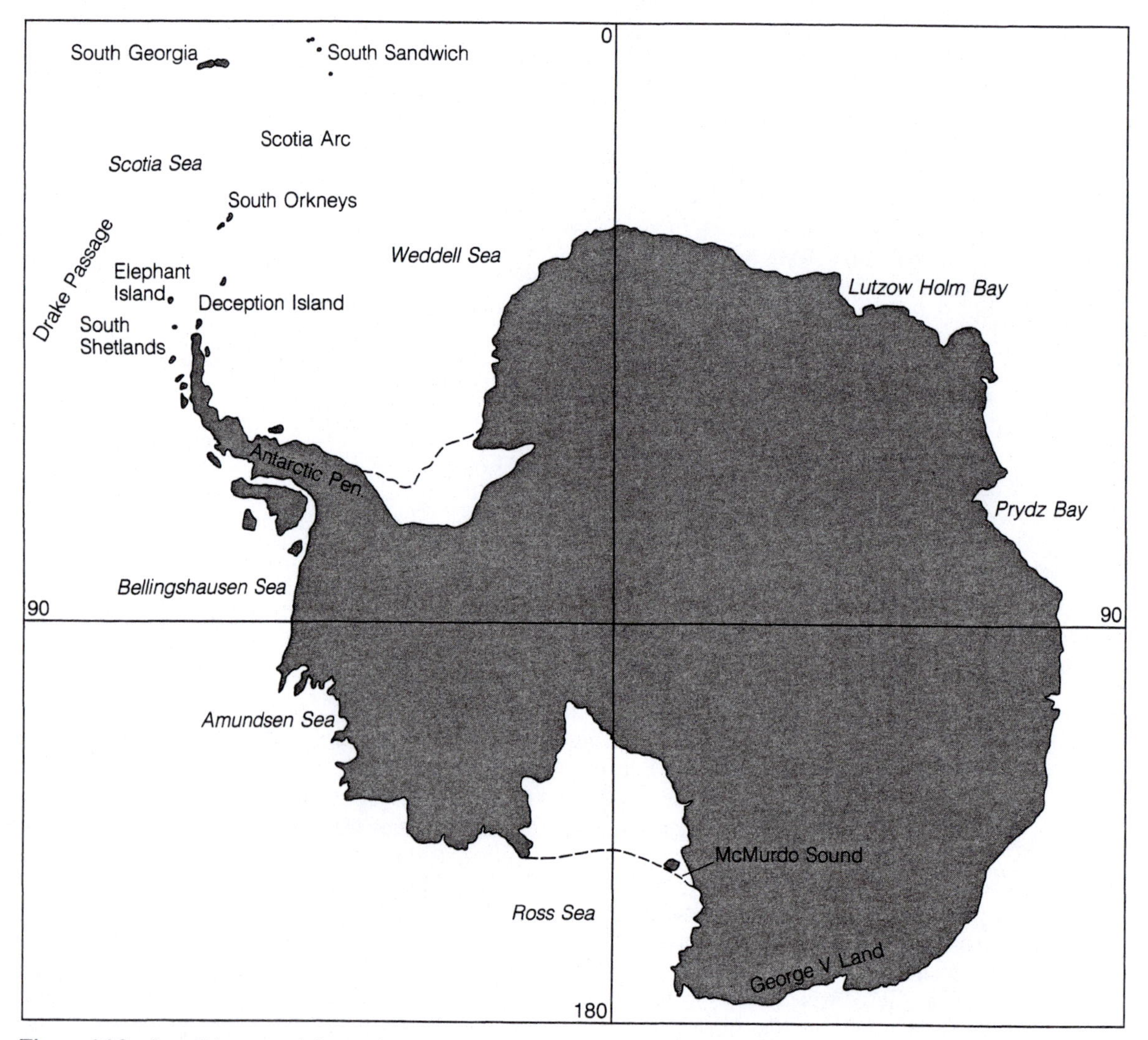

Figure 16.2 Localities around Antarctica

Nearshore

Intertidal environments are subject to extreme climatic conditions including very low temperatures, severe wave attack and ice scour. On the Antarctic Peninsula, very shallow (~1 m) sand and gravel areas protected from wave attack support living *Cibicides refulgens* and *Rosalina globularis* (Delaca and Lipps 1976). Submarine parts of cliffs have dense growths of macroalgae and sponges. The latter support assemblages of *Astrononion stelligerum*, *Cassidulinoides parkerianus*, *Cibicides refulgens*, *Cribrostomoides jeffreysii*, *Cassidulina crassa* (as *Globocassidulina*), *Pullenia subcarinata*, *Pyrgo elongata*, *Nodulina dentaliniformis* (as *Reophax*), *Rosalina globularis*, *Tolypammina vagans*, *Portatrochammina malovensis*, *Lepidodeuterammina ochracea* (as *Trochammina*) and *Turritellella shoneana* (Temnikov 1976). Forms found abundantly with tunicates include *Cibicides refulgens*, *R. globularis*, *Tolypammina vagans* and *Turritellella shoneana*.

Two sites at depths of 30 and 33 m were sampled monthly by scuba divers for 1 year. Biometrical studies of *Rosalina globularis* revealed three periods when the greatest spiral diameter dropped – November, March and June. There were interpreted as periods of reproduction. The summer period of reproduction (November) is related to a diatom bloom and is characterised by low juvenile mortality, a rapid growth rate and tests having smaller prolocular diameters. Winter reproduction takes places in March and June, has high juvenile mortality, slow growth rates and the tests have larger prolocular diameters. The summer form has an epiphytic habit which leads to a flattened test, while the winter form is mobile in a low-nutrient sedimentary environment and the test is more inflated (Showers 1980).

Experiments on feeding showed that *Rosalina globularis* in normal seawater feeds over a temperature range of 0–12 °C while *Cassidulina crassa* (as *Globocassidulina*) does so from 0 to 11 °C. Experiments on the effects of salinity showed that for 50% feeding at 0 °C, *R. globularis* was more euryhaline than other shallow-water Antarctic taxa: (salinity 4–51‰), *C. crassa* 8–51‰; *Lepidodeuterammina ochracea* (as *Trochammina*) and *Cibicides refulgens* 17–51‰; *Cribrostomoides jeffreysii* 17–35‰ (Showers *et al.* 1977).

Shallow (30–40 m) sediment substrates from Arthurs Harbor, Anvers Island showed different faunal dominance according to relative exposure to wave action (Stockton 1973). The common species are *Hippocrepinella hirudinea*, *Portatrochammina malovensis* (as *Trochammina*), *Psammosphaera fusca*, *Nodulina dentaliniformis* (as *Reophax*) and *Cassidulina crassa*. In sheltered water at 7–38 m on King George Island the dominant forms are *Portatrochammina malovensis* and *Bolivinellina pseudopunctata* (as *Bolivina*) Li and Zhang (1986).

In these shallow-water environments, foraminifera are ingested by members of both the hard and soft substrate communities. On hard substrate, suspension-feeding terebellid polychaetes and holothurians consume many foraminifera and they form up to ~25% of the gut contents. Generalised herbivores such as *Nacella concinna* (limpet) and *Sterechinus neumayeri* (echinoid) consume large quantities of foraminifera, but these form only a minor component of their food. This non-selective feeding may be a major source of density-independent mortality among foraminifera. Fewer foraminifera per consumer are ingested on soft substrates, but they may nevertheless be an important trophic resource for mud-dwelling meiofauna. The most abundant foraminifera (*Portatrochammina malovensis* and *Lepidodeuterammina ochracea*) are taken by selective predators but without harm to the test. A bivalve (*Philine alata*) crushes the test during digestion. Other deposit-feeding bivalves, asteroids and ophiuroids take foraminifera. Soft-sediment standing crops are reduced in size by this trophic activity (Brand and Lipps 1982).

Shelf to deep sea

Nothing is known of benthic assemblages from those areas of the shelf covered by ice shelves several hundred metres thick. Beyond this area total assemblages have been studied from piston core tops or dredge trawls. Discussion of these results is organised here from the Weddell Sea around Antarctica to the Scotia Sea.

Minor associations are listed in Table 16.1 and principal associations in Table 16.2. Little is known of the habit or feeding strategies of these species.

The depth distribution of the principal associations is shown in Fig. 16.3.

Weddell Sea

Two variants of Antarctic Surface Water are present: Saline Shelf Water (34.51–34.84‰, –2.2 to –1.5 °C) on the ice-covered southwestern continental shelf

Table 16.1 Minor benthic associations from the shelf to deep sea in the Southern Ocean (L=living, T=total)

	Species		Depth (m)	Area	Source
1.	*Rhabdammina linearis*	T	459–1054	Weddell Sea	Anderson 1975a
2.	*Portatrochammina wiesneri*	T	512	Weddell Sea	Anderson 1975a
3.	*Oridorsalis tener*	T	?	Weddell Sea	Echols 1971
		T	3989–4099	Drake Passage	Herb 1971
4.	*Nonionella iridea*	T	2105	Scotia Sea	Echols 1971
		T	1500–2924	South Georgia	Echols 1971
5.	*Cribrostomoides sphaeriloculus*	T	3854	Scotia Sea	Echols 1971
6.	*Eggerella parkerae*	T	2485–2937	Scotia Sea	Echols 1971
7.	*Martinottiella antarctica*	T	2800–3468	Scotia Sea	Echols 1971
8.	*Reophax* sp.	T	3084	Scotia Sea	Echols 1971
			550	Amundsen Sea	Kellogg and Kellogg 1987
9.	*Textularia wiesneri*	T	2722	Scotia Sea	Echols 1971
10.	*Repmanina charoides*	T	3993–4644	Rim, South Sandwich Trench	Echols 1971
11.	*Bulimina aculeata*	T	636–1032	South Georgia	Echols 1971
12.	*Cassidulinoides parkerianus*	T	531	South Georgia	Echols 1971
13.	*Pseudobolivina antarctica*	T	2189–2627	South Orkney	Echols 1971
		T	558–1244	South Sandwich	Echols 1971
14.	*Fursenkoina earlandi*	L	851–1279	South Orkney	Echols 1971
		L	128–796	McMurdo Sound	Ward *et al.* 1987
15.	*Psammosphaera fusca*	T	2159	South Orkney	Echols 1971
16.	*Haplophragmoides quadratus*	L	2621	South Orkney	Echols 1971
17.	*Ammoflintina argentea*	T	805	South Sandwich	Echols 1971
18.	*Adercotryma glomerata*	T	1841	South Sandwich	Echols 1971
19.	*Haplophragmoides parkerae*	T	1244	South Sandwich	Echols 1971
20.	*Bulimina aculeata*	T	870–1900	Lutzow Holm Bay	Uchio 1960a
21.	*Globocassidulina* spp.	T	124–483	George V Land	Milam and Anderson 1981
22.	*Rosalina* spp.	T	95	George V Land	Milam and Anderson 1981
23.	*Uvigerina bassensis*	T	284–507	George V Land	Milam and Anderson 1981
24.	*Reophax subdentaliniformis*	L	254–755	McMurdo Sound	Ward *et al.* 1987
25.	*Reophax pilulifer*	L	660–856	McMurdo Sound	Ward *et al.* 1987
		T	223–1120	Drake Passage	Herb 1971
26.	*Pseudobolivina antarctica*	L, D	266–854	McMurdo Sound	Ward *et al.* 1987
		T	640	E. Ross Sea	Pflum 1966
27.	*Deuterammina glabra*	D	266–550	McMurdo Sound	Ward *et al.* 1987
28.	*Cassidulina laevigata*	T	1670	Ross Sea	McKnight 1962
		T	1670	Queen Maud Land	McKnight 1962
29.	*Pseudobulimina chapmani*	T	365	Ross Sea	McKnight 1962
30.	*Trochammina gaboensis*	T	269	Ross Sea	Kennett 1968
31.	*Globocassidulina* sp.	T	357	Ross Sea	Kennett 1968
32.	*Globocassidulina biora*	T	530	Ross Sea	Kennett 1968
33.	*Bathysiphon discreta*	T	410–1290	Ross Sea	Kennett 1968, Pflum 1966
34.	*Reophax subfusiformis*	T	470	Ross Sea	Kennett 1968
35.	*Cassidulinoides porrectus*	T	90–399	Ross Sea	Kennett 1968
36.	*Ammoglobigerina globigeriniformis*	T	584	Amundsen Sea	Kellogg and Kellogg 1987
37.	*Portatrochammina malovensis*	T	0–150	Deception Island	Finger and Lipps 1981
38.	*Nonionella bradii*	T	0–150	Deception Island	Finger and Lipps 1981
39.	*Fursenkoina fusiformis*	T	0–150	Deception Island	Finger and Lipps 1981
40.	*Textularia pseudoturris*	T	878	Drake Passage	Herb 1971
41.	*Ammolagena clavata*	T	3312–3975	Drake Passage	Herb 1971
42.	*Discanomalina vermiculata*	T	247–293	Drake Passage	Herb 1971

Table 16.1 *(continued)*

	Species		Depth (m)	Area	Source
43.	*Cibicides fletcheri*	T	42–4136	Drake Passage	Herb 1971
44.	*Psammosphaera fusca*	T	3074–4086	Drake Passage	Herb 1971
45.	*Rupertina stabilis*	T	2782–2827	Drake Passage	Herb 1971
46.	*Cibicides corpulentus*	T	1806–2013	Drake Passage	Herb 1971
47.	*Hormosina normani*	T	4758	Drake Passage	Herb 1971
48.	*Cribrostomoides crassimargo*	T	3715–4090	Drake Passage	Herb 1971
49.	*Ammobaculites agglutinans*	T	1437	Drake Passage	Herb 1971
50.	*Recurvoides contortus*	T	681–1409	Drake Passage	Herb 1971
51.	*Fontbotia wuellerstorfi*	T	~200	Drake Passage	Herb 1971
52.	*Hoeglundina elegans*	T	~500	Drake Passage	Herb 1971

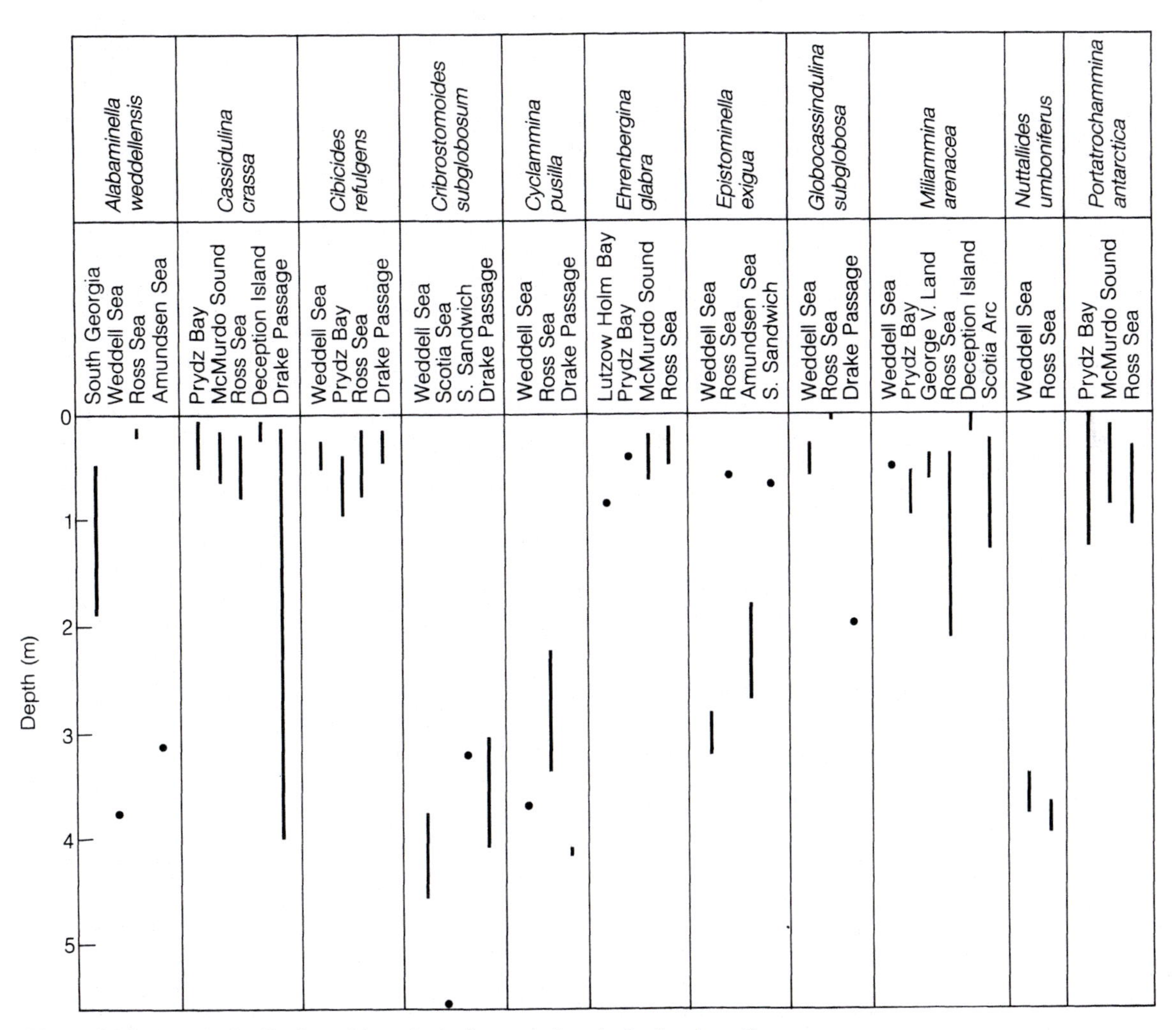

Figure 16.3 Depth distribution of the principal associations in the Southern Ocean

Table 16.2 Principal benthic associations from the shelf and deep sea of the Southern Ocean

***Alabaminella weddellensis* association**

Salinity: 34.29–34.17‰

Temperature: –1.9 to 1.0 °C

Substrate: ?

Depth: 164–3770 m

Distribution:

1. South Georgia (Echols 1971) living, total (as *Eilohedra*)
2. Weddell Sea (Echols 1971) total (as *Eilohedra)*
3. Lutzow Holm Bay (Uchio 1960a) total (as *Eponides*)
4. Ross Sea: (a) (McKnight 1962) total; (b) (Pflum 1966) total (both as *Eponides*)
5. Amundsen Sea (Pflum 1966) total (as *Eponides*)

Additional common species:

Pullenia subsphaerica	1
Haplophragmoides quadratus	1
Nonionella iridea	1
Epistominella exigua	2, 3, 4b, 5
Nuttallides umboniferus	2
Trifarina angulosa	3
Trifarina earlandi	4a (as *Angulogerina*), 3b (as A. *earlandi*)
Cassidulina crassa	4a
Ehrenbergina glabra	4a
Astrononion sp.	4a
Sigmoilina umbonata	4a
Globocassidulina subglobosa	4b (*Cassidulina*)
Uvigerina asperula	5
Rosalina globularis	5b
Psammophax consociata	5b
Cibicides lobatulus	6, 8
Cibicides spp.	6
Reophax spp.	6
Bulimina marginata	6
Ehrenbergina pacifica	8

***Cassidulina crassa* association**

Salinity: 33.96–35.19‰

Temperature: –1.9 to 0.6 °C

Substrate: fine sand

Depth: 50–4008 m

Distribution:

1. Prydz Bay (Quilty 1985) total (as *Globocassidulina*)
2. McMurdo Sound (Ward *et al.* 1987) living, dead (as *Globocassidulina*)
3. Ross Sea: (a) (McKnight 1962) total; (b) (Kennett 1968) total (as *Islandiella porrecta*)
4. Deception Island (Finger and Lipps 1981) total (as *Globocassidulina*)
5. Drake Passage (Herb 1971) total

Additional common species:

Trifarina earlandi	1, 2, 3a, (as *Angulogerina*), 3b
Cibicides refulgens	1, 3b
Ehrenbergina glabra	2, 3a, b
Portatrochammina antarctica	2, 3b (as *Trochammina*)
Alabaminella weddellensis	3a (as *Eponides*)
Oridorsalis tener	3a (as *Eponides*)
Trochamminopsis conica	3a (as *Trochammina*)
Nonionella bradii	3a
Fursenkoina davisi	3a (as *Virgulina*)
Cribrostomoides jeffreysii	3b
Trochammina gaboensis	3b
Rosalina globularis	3b, 4
Epistominella exigua	3b
Globocassidulina sp.	3b
Lepidodeuterammina ochracea	4 (as *Rotaliammina*)
Psammosphaera fusca	5
Saccammina spherica	5
Gyroidina neosoldanii	5
Ammolagena clavata	5
Pyrgo murrhina	5
Martinottiella communis	5
Discanomalina vermiculata	5
Cibicides fletcheri	5

Table 16.2 *(continued)*

***Cibicides refulgens* association**

Salinity: <34.51‰

Temperature: –1.9 to 1.5 °C

Substrate: coarse sand and gravel

Depth: 136–950 m

Distribution:
1. Weddell Sea (Anderson 1975a) total
2. Prydz Bay (Quilty 1985) total
3. Ross Sea: (a) (Kennett 1968) total; (b) (Osterman and Kellogg 1979) total
4. Drake Passage (Herb 1971) total

Additional common species:

Ehrenbergina glabra	1, 2
Globocassidulina subglobosa	1 (*Cassidulina*)
Cassidulina crassa	1, 2 (as *Globocassidulina*)
Trochammina gaboensis	3a
Rosalina globularis	3a
Trifarina earlandi	3a
Globocassidulina sp.	3a
Portatrochammina antarctica	3a (as *Trochammina*)
Cibicides lobatulus	3b
Rosalina villardeboana	4

***Cribrostomoides subglobosum* association**

Salinity: AABW/ Weddell Sea Bottom Water

Temperature: –

Substrate: –

Depth: 3022–5587 m

Distribution:
1. Weddell Sea (Echols 1971) total
2. Scotia Sea (Echols 1971) total
3. Sandwich (Echols 1971) total
4. Drake Passage (Herb 1971) total

Additional common species:

Adercotryma glomerata	1
Tritaxis inhaerens	3
Cyclammina pusilla	4
Ammobaculites agglutinans	4
Hormosina normani	4
Psammosphaera fusca	4
Hyperammina elongata	4
Cyclammina orbicularis	4
Cribrostomoides crassimargo	4

***Cyclammina pusilla* association**

Salinity: Weddell Sea Bottom Water

Temperature: –

Substrate: mud

Depth: 2195–4176 m

Distribution:
1. Weddell Sea (Anderson 1975a) total
2. Ross Sea: (a) (Kennett 1968) total; (b) (Pflum 1966) total
3. Drake Passage (Herb 1971) total

Additional common species:

Cribrostomoides subglobosum	1, 2a, 3
Haplophragmoides rotulatum	2a
Hyperammina elongata	3

***Ehrenbergina glabra* association**

Salinity: 33.96–35.19‰

Temperature: –1.9 to 0.6 °C

Substrate: fine to coarse sand

Depth: 100–830 m

Distribution:
1. Lutzow Holm Bay (Uchio 1960a) total
2. Prydz Bay (Quilty 1985) total
3. McMurdo Sound (Ward *et al.* 1987) living, dead
4. Ross Sea: (a) (McKnight 1962) total; (b) (Osterman and Kellogg 1979) total; (c) (Webb 1988) dead

Additional common species:

Cibicides refulgens	1, 2
Miliolinella subrotunda	2
Reophax subdentaliniformis	3
Trifarina earlandi	3, 4c
Cassidulina crassa	3, 4 (as *Globocassidulina*)

Table 16.2 *(continued)*

***Epistominella exigua* association**

Salinity: 34.2–34.71‰

Temperature: –1.9 to 1.0 °C

Substrate: mud to pebbles

Depth: 658–3199 m

Distribution:
1. Weddell Sea (Echols 1971) total
2. S. Sandwich (Echols 1971) living, total
3. Ross Sea (Kennett 1968) total
4. Amundsen Sea (Pflum 1966) total

Additional common species:

Alabaminella weddellensis	1 (as *Eilohedra*), 4 (as *Eponides*)
Islandiella quadrata	1
Nuttallides umboniferus	1
Fursenkoina earlandi	2
Textularia earlandi	2
Trifarina earlandi	3
Cibicides refulgens	3
Globocassidulina subglobosa	4 (as *Cassidulina*)
Bolivina pseudoplicata	4
Trifarina angulosa	4 (as *Angulogerina*)

***Globocassidulina subglobosa* association**

Salinity: <34.51‰

Temperature: –1.9 to –1.5 °C

Substrate: ?

Depth: 26–1971 m

Distribution:
1. Weddell Sea (Anderson 1975a) total (as *Cassidulina*)
2. Ross Sea (Osterman and Kellogg 1979) total
3. Drake Passage (Herb 1971) total (as *Cassidulina*)

Additional common species:

Cassidulinoides crassa	1
Ehrenbergina glabra	1
Cibicides refulgens	1
Cassidulinoides porrectus	2
Psammosphaera fusca	3

***Miliammina arenacea* association**

Salinity: 34.45–35.19‰

Temperature: –1.9 to 0.6 °C

Substrate: muddy sand

Depth: 0–2100 m

Distribution:
1. Weddell Sea (Anderson 1975a) total
2. S. Georgia (Echols 1971) total
3. S. Orkney (Echols 1971) total
4. S. Sandwich (Echols 1971) total
5. Prydz Bay (Quilty 1985) total (as *Miliammina*)
6. George V Land (Milam and Anderson 1981) total (as *M. earlandi*)
7. Ross Sea: (a) (McKnight 1962) total; (b) (Kennett 1968) total; (c) (Osterman and Kellogg 1979) total
8. Deception Island (Finger and Lipps 1981) total

Additional common species:

Portatrochammina wiesneri	1
Saccorhiza ramosa	1
Portatrochammina antarctica	2, 7b (as *Trochammina*)
Epistominella exigua	3
Fursenkoina earlandi	3, 4
Globocassidulina subglobosa	3 (as *Cassidulina*)
Pullenia simplex	3
Haplophragmoides canariensis	4, 6
Miliammina lata	4, 5
Conotrochammina bullata	4
Portatrochammina eltaninae	4
Adercotryma glomerata	6
Pseudobolivina antarctica	6, 7b (as *Textularia*)
Uvigerina bassensis	6
Alabaminella weddellensis	7a
Reophax subfusiformis	7b
Trochammina inconspicua	7b
Bathysiphon discreta	7b
Reophanus ovicula	7b (as *Hormosina*)
Trochammina rossensis	7c
Lagenammina difflugiformis	7c
Portatrochammina malovensis	8 (as *Trochammina*)
Nonionella bradii	8
Fursenkoina fusiformis	8 (as *Stainforthia*)

Table 16.2 *(continued)*

***Nuttalides umboniferus* association**

Salinity: >34.60‰, AABW

Temperature: −1.0 to −0.4 °C

Substrate: ?

Depth: 3374–3932 m

Distribution:
1. Weddell Sea: (a) (Anderson 1975a) total; (b) (Echols 1971) total
2. Ross Sea (Osterman and Kellogg 1979) total (as *Osangularia*)

Additional common species:

Cyclammina pusilla	1a
Epistominella exigua	1a

***Portatrochammina antarctica* association**

Salinity: 34.45–35.19‰

Temperature: −1.9 to 0.6 °C

Substrate: mud, fine sand

Depth: 9–1275 m

Distribution:
1. Prydz Bay (Quilty 1985) total (as *Trochammina*)
2. McMurdo Sound (Ward *et al.* 1987) living, dead
3. Ross Sea: (a) (McKnight 1962) total (as *Trochammina*); (b) (Kennett 1968) total (as *Trochammina*); (c) (Webb 1988) living

Additional common species:

Cribrostomoides jeffreysii	2, 3b
Deuterammina glabra	2 (as *Trochammina*)
Trifarina earlandi	2
Pseudobolivina antarctica	2, 3a, b (as *Textularia*)
Miliammina arenacea	2, 3a, b
Reophax subdentaliniformis	2, 3b
Portatrochammina eltaninae	2
Cassidulina crassa	2 (as *Globocassidulina*)
Trochamminopsis conica	3a (as *Trochammina*)
Recurvoides contortus	3a
Miliammina lata	3a
Textularia earlandi	3b
Trochammina inconspicuans	3b
Trochammina gaboensis	3b
Reophax subfusiformis	3b

***Portatrochammina eltaninae* association**

Salinity: 34.7‰ AABW

Temperature: –

Substrate: –

Depth: 1058–6167 m

Distribution:
1. Scotia Sea (Echols 1971) total
2. S. Sandwich Trench (Echols 1971) total
3. S. Orkney (Echols 1971) total
4. S. Sandwich (Echols 1971) total

Additional common species:

Adercotryma glomerata	1
Cribrostomoides sphaeriloculus	1, 2
Cribrostomoides subglobosum	1
Conotrochammina kennetti	2
Reophax sp.	2, 3
Trochammina multiloculata	3
Textularia wiesneri	3
Conotrochammina bullata	3
Cribrostromoides arenacea	3
Alabaminella weddellensis	3 (as *Eilohedra*)
Pseudobolivina antarctica	4

Table 16.2 *(continued)*

***Trifarina angulosa/earlandi* association**

Salinity: 33.96–35.19‰

Temperature: −1.9 to 0.6 °C

Substrate: muddy sand

Depth: 110–1919 m

Distribution:

1. S. Georgia (Echols 1971) total (as *Angulogerina earlandi*)
2. Weddell Sea (Anderson 1975a) total (as *T. angulosa*)
3. Lutzow Holm Bay (Uchio 1960a) total (as *Angulogerina angulosa*)
4. McMurdo Sound (Ward *et al.* 1987) living, dead (as *T. earlandi*)
5. Ross Sea: (a) (McKnight 1962) total (as *Angulogerina earlandi*); (b) (Kennett 1968) total (as *T. earlandi*); (c) (Osterman and Kellogg 1979) total (as *T. earlandi*); (d) (Pflum 1966) total (as *Angulogerina angulosa*)
6. Amundsen Sea (Kellogg and Kellogg 1987) total (as *T. earlandi*)
7. Bellingshausen Sea (Pflum 1966) total (as *Angulogerina angulosa*)
8. Drake Passage (Herb 1971) total (as *Angulogerina angulosa*)

Additional common species:

Cassidulinoides parkerianus	1
Fursenkoina fusiformis	1
Globocassidulina subglobosa	2, 3 (as *Cassidulina*)
Cassidulina crassa	2, 5b (as *Islandiella porrecta*), 5d
Ehrenbergina glabra	2, 4, 5b
Cibicides refulgens	2, 5b, 6
Alabaminella weddellensis	3, 7 (as *Eponides*)
Epistominella exigua	3
Portatrochammina antarctica	4

and Fresh Shelf Water (<34.51‰, −1.89 to −1.5 °C) on the eastern continental shelf down to a depth of 550–700 m. Warm Atlantic Deep Water (34.45–34.69‰, −0.36 to 0.20 °C) flows along the southwest slope and then mixes with Saline Shelf Water to form AABW (Anderson 1975a).

In general, assemblages from beneath Fresh Shelf Water are dominantly (>90%) calcareous but some from 235 to 713 m are thought to have suffered partial dissolution. Those from beneath Saline Shelf Water are 75–100% agglutinated. Slope assemblages from the area where AABW is forming are calcareous in part (e.g. *Nuttallides umboniferus*) at depths of 2050–3800 m, but the deepest abyssal plain areas are dominated by *Cyclammina pusilla* and *Cribrostomoides subglobosum* (Anderson 1975).

From data in Echols (1971) and Anderson (1975a) nine principal associations are recognised (Table 16.2, Fig. 16.4). Fresh Shelf Water – *Cibicides refulgens* (235–507 m), *Globocassidulina subglobosa* (260–567 m) and *Trifarina angulosa/earlandi* (348–713 m) associations; Saline Shelf Water – *Miliammina arenacea* association (490 m); AABW – *Epistominella exigua* (2796–3199 m), *Nuttallides umboniferus* (3374–3768 m), *Alabaminella weddellensis* (3770 m), *Cyclammina pusilla* (3694 m) and *Cribrostomoides subglobosum* (3737 m) associations. Thus, the relationship between the preservation of calcareous assemblages or their partial or total dissolution is dependent primarily on bottom water mass rather than on water depth. The Saline Shelf Water from beneath the ice shelf is rich in CO_2 and therefore corrosive to $CaCO_3$ (Anderson 1975b).

Anderson (1975a) recorded 160 species in this area which he described as '. . . probably the world's most severe glacial-marine environment'.

Prydz Bay

Five principal associations are present: *Cassidulina crassa* (73–525 m), *Cibicides refulgens* (345–950 m) and *Ehrenbergina glabra* (400 m) associations, all dominated by calcareous forms; *Portatrochammina antarctica* (9–1275 m) and *Miliammina arenacea* (505–950 m) associations which are dominantly agglutinated. The CCD lies at 1264–1500 m and does not seem to be the primary control on assemblage composition since calcareous and agglutinated assemblages span similar depth ranges in adjacent areas (Quilty 1985).

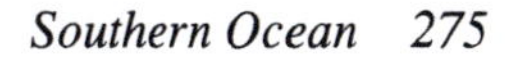

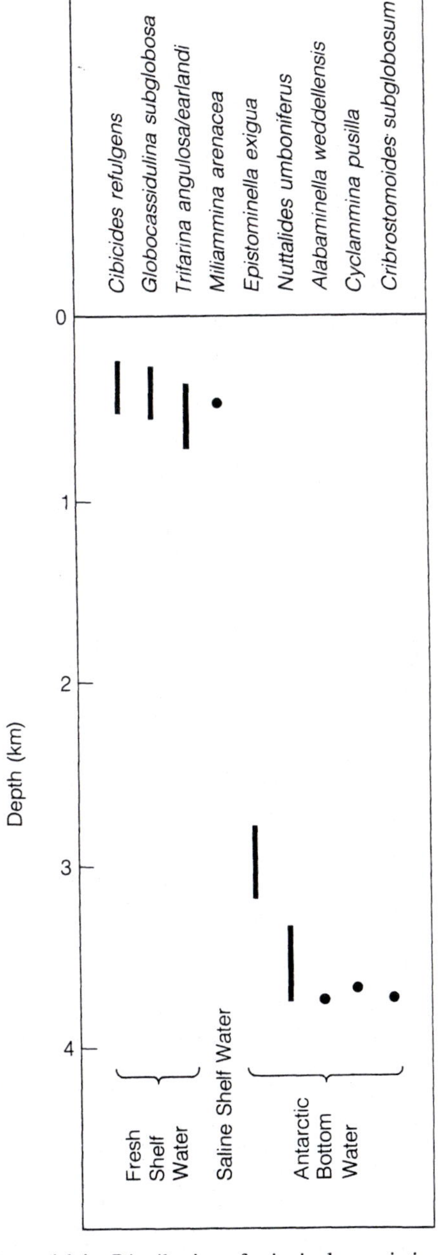

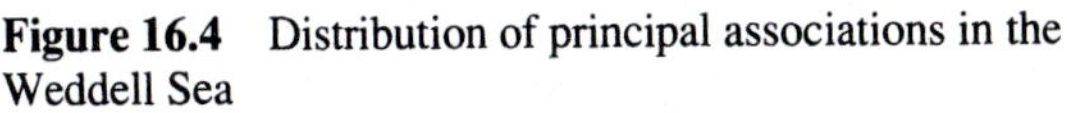

Figure 16.4 Distribution of principal associations in the Weddell Sea

McMurdo Sound and Ross Sea

McMurdo Sound (maximum depth 900 m) is covered in ice for more than 9 months per year and the bottom waters have very constant salinity and temperature conditions (33.96–34.99‰, –1.95 to –1.14 °C; part of the Antarctic Surface Water). The sediment is mainly fine sand and mud (Ward *et al.* 1987). Sediments from water depths <7 m may be infiltrated and covered with anchor ice (aggregations of ice platelets). Small basins with a sill depth of ~18 m may become anoxic in the late summer (January) if the organic input is high (Bernhard 1987).

At depths of <26 m, the summer standing crop values ranged from 41 to 238 per 10 cm^2 with the lowest values from anchor ice at 4 m and the highest from boulders at 26 m. A seasonally anoxic basin showed a change from 55 per 10 cm^2 in early summer to 189 per 10 cm^2 when productivity was high and anoxia developed. Diversity was low ($H(S)$ 0.67–1.10). Three species were common: *Cassidulinoides porrectus*, *Globocassidulina* cf. *G. biora* and *Portatrochammina antarctica* (Bernhard 1987).

The most abundant macrofaunal organism at 3–40 m is the pecten *Adamussium colbecki* and the surface of its shells forms > 90% of the hard substrate in Explorers Cove, McMurdo Sound. Four species attached to these shells were abundant: *Cibicides refulgens*, *Rosalina globularis*, *Lepidodeuterammina ochracea* (as *Trochammina*) and *Portatrochammina malovensis* (as *Trochammina*). *Cibicides refulgens* averaged 39 per 10 cm^2. *Rosalina globularis* and *C. refulgens* formed agglutinated tubes around their pseudopodia, commonly one test diameter in length. It is thought that these tubes aided feeding on suspended material by greatly enlarging the effective feeding area. Altogether 30 species were found in the sediment but only 7 on the pectens (Mullineaux and DeLaca 1984).

The sea floor beneath permanent ice is oligotrophic. The sea floor at 597 m depth below ice 420 m thick showed no evidence of infauna although foraminiferal tests were present in the sediment (Lipps *et al.* 1979). In shallow water with ice 4–5 m thick at a depth of ~40 m and bathed by nutrient-free waters from beneath the Ross Ice Shelf, DeLaca *et al.* (1980) found a large arborescent agglutinated form, *Notodendrodes antarctikos*. This feeds on material in suspension and it is believed that swimming by *Adamussium colbecki* stirs up bottom material including benthic diatoms, bacteria and flocculent organic matter which serves as food for the foraminifer. *Notodendrodes antarctikos* also utilises dissolved nutrients. When food and nutrients are not available its metabolism becomes very slow. Lipps and Delaca (1980) suggest that this large form may live to a great age while waiting for conditions to become suitable for reproduction.

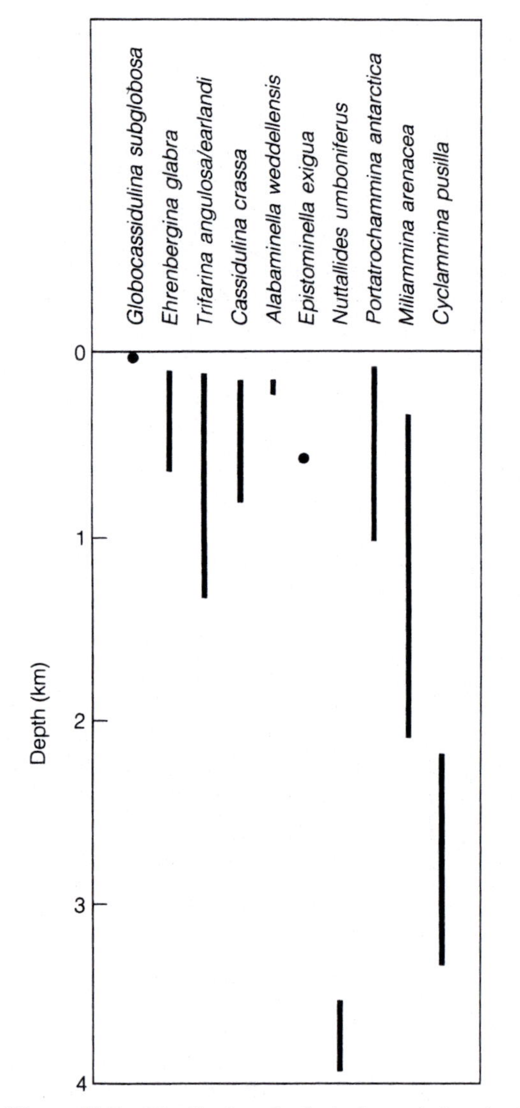

Figure 16.5 Distribution of principal associations in the Ross Sea

From studies of total assemblages by McKnight (1962), Pflum (1966), Kennett (1968), Osterman and Kellogg (1979) and Ward *et al.* (1987) 11 principal associations can be recognised (Fig. 16.5). The calcareous ones are the *Globocassidulina subglobosa* (26 m), *Ehrenbergina glabra* (110–620 m), *Trifarina angulosa/earlandi* (110–1335 m), *Cibicides refulgens* (136–790 m), *Cassidulina crassa* (139–800 m), *Alabaminella weddellensis* (164–210 m), *Epistominella exigua* (583) and *Nuttallides umboniferus* (3549 –3932 m) associations. The agglutinated examples are the *Portatrochammina antarctica* (79–1015 m), *Miliammina arenaca* (320–2100 m) and *Cyclammina pusilla* (2195–3370 m) associations. Kennett found that all the calcareous assemblages were restricted to depths <550 m and the agglutinated ones to depths >430 m, but this is clearly not true when a larger number of samples is considered. As in the Weddell Sea, water from beneath the ice shelf is CO_2 rich and $CaCO_3$ corrosive (Osterman and Kellogg 1979). However, this relationship is not true of the George V Shelf where dissolution-relict agglutinated assemblages are believed to be the product of organic-rich siliceous muds and oozes (Milam and Anderson 1981).

Amundsen and Bellingshausen seas

Only seven samples were studied by Pflum (1966) and all yielded calcareous assemblages of the *Trifarina angulosa/earlandi* (265–463 m), *Epistominella exigua* (1765–2685 m) and *Alabaminella weddellensis* (3150 m) associations. Pflum attributed the rarity of agglutinated tests to postmortem destruction but there is no evidence to confirm this. Indeed, of the five samples with >100 individuals described by Kellogg and Kellogg (1987) two were agglutinated: *Reophax* spp., and *Ammoglobigerina globigeriniformis* (as *Trochammina*).

Drake Passage and Scotia Sea

This area is under the influence of CDW which flows from west to east. The foraminiferal data are from Herb (1971, Drake Passage) and Echols (1971, Scotia Sea). To the south, the principal associations are calcareous on the shelf (*Cibicides refulgens* association, 156–436 m) but agglutinated in deeper water (*Cribrostomoides subglobosum*, 3022–3186 m, and *Cyclammina pusilla*, 4077–4176 m). In addition, there are three minor calcareous associations at depths of 200–500 m, and seven minor agglutinated associations from depths ≥~650 m (see Table 16.1). Thus, the deeper waters are corrosive with respect to $CaCO_3$ (Fig. 16.6).

To the north, calcareous assemblages extend to greater depth with the *Cassidulina crassa* (104–4008 m), *Trifarina angulosa/earlandi* (1814–1919 m) and *Globocassidulina subglobosa* (1953–1917 m) associations. The *Cribrostomoides subglobosum* association is present from 3138 to 4099 m.

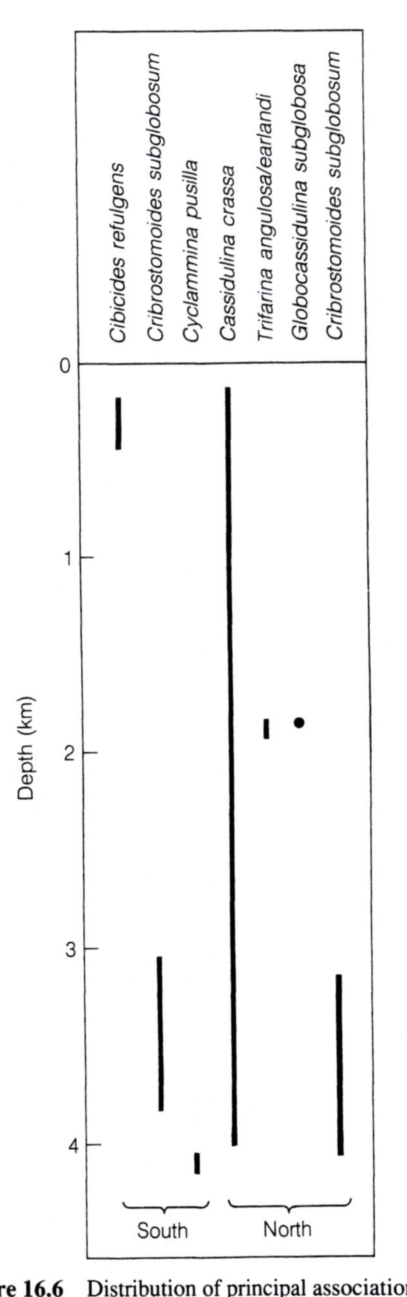

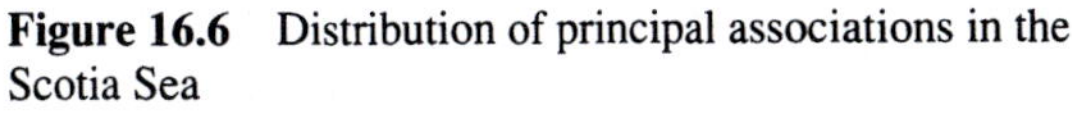

Figure 16.6 Distribution of principal associations in the Scotia Sea

Around the Scotia Arc (South Georgia, South Sandwich and South Orkney) the distribution of water masses is complex and poorly understood. The most widely distributed assemblages fall into the *Miliammina arenacea* (201–1199 m, 0–150 m around Deception Island) and *Portatrochammina eltaninae* (1058– 6167 m) associations. Two calcareous associations are present off South Georgia (*Trifarina angulosa/earlandi*, 329 m, and *Alabaminella weddellensis*, 1900 m), while off South Sandwich the *Epistominella exigua* association (658 m) and *Cribrostomoides subglobosum* association (3206 m) are found.

Echols noted that on the South Sandwich Ridge, standing crop values were 144–188 per 10 cm^2 at <1000 m and 56–180 per 10 cm^2 at >1000 m while on the slope south of South Georgia they were only 33–66 per 10 cm^2. The highest value, 400 per 10 cm^2, was from the centre of the Scotia Sea.

Recolonisation following volcanic disturbance

Deception Island, off the Antarctic Peninsula, is an area of active volcanicity, with a drowned caldera connected to the open sea by a channel. Material collected in 1927 (Earland 1934a) lacked calcareous taxa and was a *Miliammina arenacea* association. This may be normal for conditions during a long period of volcanic inactivity. Following eruptions during 1969 and 1970, Finger and Lipps (1981) studied total assemblages from collections in 1972, 1973, 1974, 1975 and 1976. Only one of the 1972 samples yielded a rich fauna, but those from the 1974–76 collections were fruitful and showed that recolonisation of the area had been considerable.

Only summaries of the data were given by Finger and Lipps, but it appears that in the caldera, the succession of associations was as follows: 1972 *Miliammina arenacea*, 1973 *Nonionella bradii*, 1974 *Fursenkoina fusiformis* (as *Stainforthia*), 1975 and 1976 *M. arenacea*. On the outer slopes, the *Cassidulina crassa* association was present from 1973 until 1975 (given as *Globocassidulina crassa* plexus). The diversity values for the total assemblages from the caldera were α 1.3–1.6 compared with α 2.1–8.7 for the outer slope. The caldera area is more variable than that of the slope as it receives volcanic debris and fluids, seasonal runoff and seasonal phytoplankton blooms, and is both muddy and turbid. Finger and Lipps note that the recolonisation of the area is patchy and, by comparison with faunal lists from earlier studies, conclude that although the fauna as a whole has changed little during the past 50 years, there is local adjustment to a new state of equilibrium following the 1969–70 volcanic eruption.

Summary

Although the area of the Southern Ocean is enormous, because it has an essentially latitudinal circulation system with the same processes influencing the development of the water masses throughout, the distribution of the principal associations is remarkably constant. Calcareous assemblages are present generally at depths < 800 m (*Cassidulina crassa*, *Cibicides refulgens*, *Ehrenbergina glabra* and *Trifarina angulosa/earlandi* associations), and less commonly at intermediate and greater depths (e.g. *Epistominella exigua* and *Nuttallides umboniferus* associations, Fig. 16.3). Agglutinated assemblages are present throughout although most commonly at depths <1000 m. Although living and dead assemblages have not been distinguished, the agglutinated-rich assemblages are known to be associated with corrosive bottom waters in which calcareous taxa were either unable to live or if they did their tests were subsequently dissolved.

Postmortem changes

From the studies of Mullineaux and DeLaca (1984) it is known that large numbers of tests of *Cibicides* and *Rosalina* which live attached to pecten shells are contributed to the sediment on death. At depths of <500 m, mixing or disturbance of bottom sediments through ice scour is a major process although as all the areas sampled are beyond the ice shelf, such action can only be from icebergs. Uchio (1960a) notes that some sediments exposed at the surface may be premodern and if this is the case relating them to modern water masses will be misleading.

Planktonic : benthic ratio

Because of the fertility of Southern Ocean waters, large numbers of planktonic tests are contributed to the sediments but their presence or absence on the sea floor depends on the local dissolution potential of the bottom waters.

CHAPTER 17

Arctic Ocean

Introduction

The Arctic Ocean has been described as a 'polar Mediterranean' region for it is an enclosed sea, partly ice-covered, and surrounded by vast polar and subpolar deserts (Rey 1982). The waters are stratified and three main layers are recognised: Surface Water (0–200 m) with variable salinity and temperature, Atlantic Water (200 to ~900 m) with salinity of 34.9‰ and temperatures of 0.5 °C at ~400 to 600 m and close to 0 °C from 600 to 900 m, and Arctic Bottom Water at depths >900 m with salinity 34.95‰ and temperature about –0.5 °C.

The Surface Water has a permanent ice layer in its upper part over a large area of the Arctic Ocean. The variable salinities are due to the large input of fresh water from rivers. The Atlantic Water enters the Arctic Ocean via the Fram Strait between Greenland and Spitsbergen. The Arctic Bottom Water is very stable in its properties and is prevented from circulating freely below 1400 m by the Lomonosov Ridge.

Results from Baffin Bay are included here but those from the Barents Sea are included in Chapter 9. The localities discussed are shown in Fig. 17.1.

Taxonomic reviews have been undertaken by Loeblich and Tappan (1953), Green (1960) and Lagoe (1977) and reviews of distributions by Todd and Low (1980) and Lagoe (1980). Most ecological studies have been based on total assemblages and relatively few samples have been analysed from this vast area.

Shelf to deep sea

The continental shelf is of variable width (maximum 800 km off Siberia) and depth. Minor associations are listed in Table 17.1. Typical shelf and deep-sea species (Table 17.2) are infaunal or epifaunal and probably all are detritivores. Ten major associations based on total assemblages can be recognised (Table 17.3).

Shelf areas overlain by seasonally ice-free waters have *Elphidium clavatum* (7–60 m), *Eggerella advena* (7–155 m) or *Cuneata arctica* (48–100 m) associations. Additional common species include *Buccella frigida*, *Haynesina orbiculare*, *Islandiella islandica* and *I. norcrossi*. All other associations are from beneath areas of permanent ice cover and they extend from the shelf to 3709 m, the greatest depth sampled (Fig. 17.2). None of these show any clear correlation with the named water masses except perhaps the *Stetsonia horvathi* association which has its upper limit at ~900 m. Substrate differences may account for the distributions of associations from similar depths, but if this is so there are no reliable data to demonstrate it. All the sediments appear to be fine-grained with high proportions of mud and silt, especially at depths >200 m. Lagoe (1977) believed that (unspecified) biotic factors may be important distributional controls.

Agglutinated and calcareous tests are present at depths down to ~1000 m, but deeper than this the faunas are almost exclusively calcareous and dominantly hyaline. There is no evidence that the deep bottom waters are undersaturated with respect to calcium carbonate. The diversity of the total assemblages from all depths is low, with $H(S)$ generally 1.4–2.3 and only a few values in the range 0.6–1.4 (Lagoe 1976).

In the Bering and Chukchi seas, which are seasonally ice-free, the *Eggerella advena* and *Cuneata arctica* associations commonly have >70% agglutinated tests. These are rare in the East Siberian and Laptev seas (Todd and Low 1966) and the Kara Sea (Todd and Low 1980). However, the Kara Sea is said to be characterised by low-diversity agglutinated

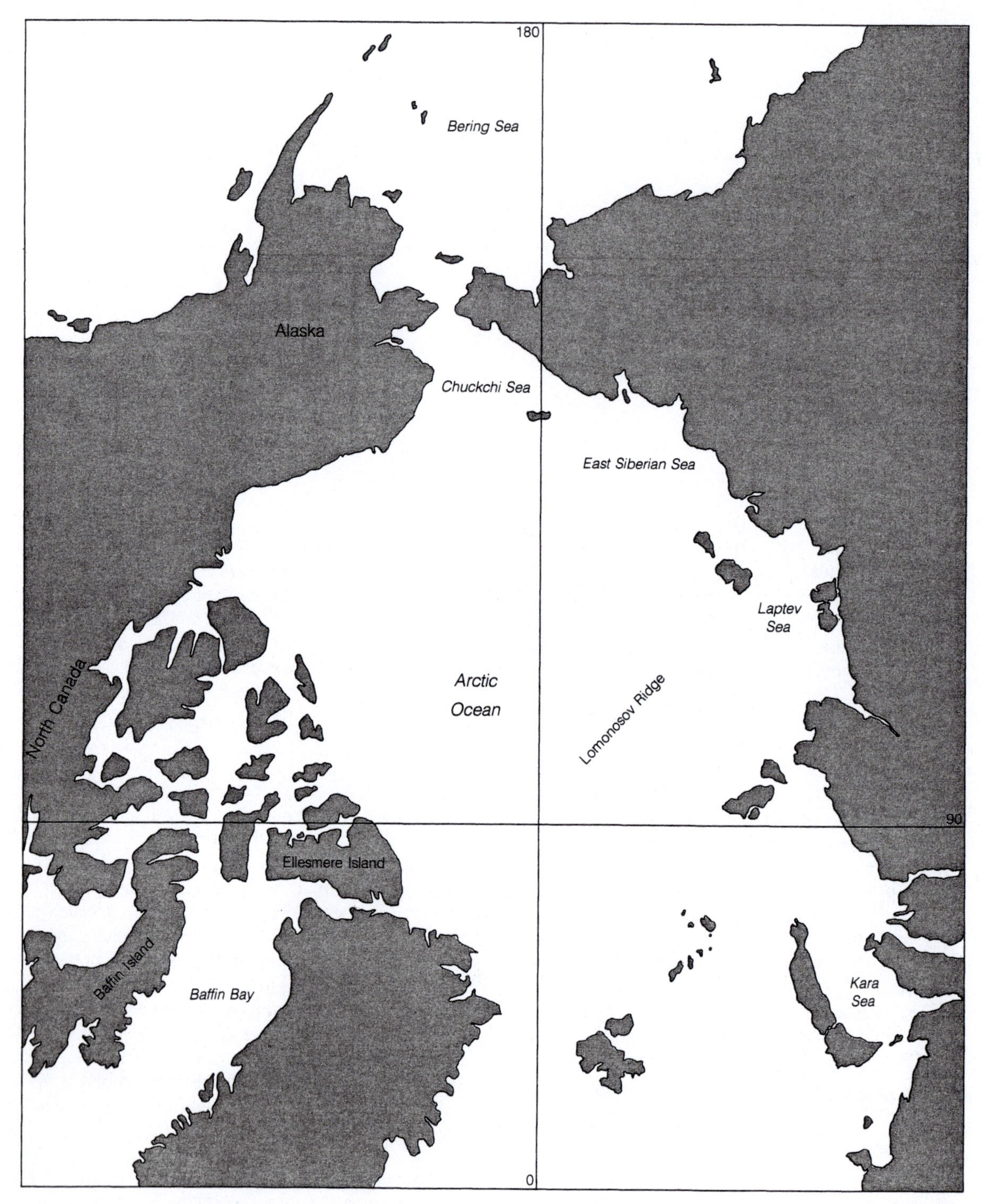

Figure 17.1 Localities mentioned in the text

Table 17.1 Minor benthic assocations from the Arctic Ocean (T = total assemblages)

Association		Depth (m)	Area	Source
1. *Elphidium subarcticum*	T	14	Bering Sea	Anderson 1963
	T	55	Chuckchi Sea	Cooper 1964
2. *Epistominella exigua*	T	250	Bering Sea	Anderson 1963
3. *Trochammina squamata*	T	42	Chukchi Sea	Cooper 1964
4. *Eoeponidella pulchella*	T	15	Chukchi Sea	Cooper 1964
5. *Glabratella wrightii*	T	52	Chukchi Sea	Cooper 1964
6. *Haynesina orbiculare*	T	17	Chukchi Sea	Cooper 1964
	T	1	Alaska	Lagoe 1979a
7. *Buccella frigida*	T	14–52	Chukchi Sea	Cooper 1964
8. *Textularia earlandi*	T	161–800	Baffin Island	Schafer and Cole 1986
9. *Adercotryma glomerata*	T	31–592	N. Canada	Phleger 1952b
10. *Cibicides lobatulus*	T	124	N. Canada	Phleger 1952b
11. *Islandiella norcrossi*	T	38–155	N. Canada	Phleger 1952b
12. *Islandiella islandica*	T	38	N. Canada	Phleger 1952b
13. *Fursenkoina fusiformis*	T	126	N. Canada	Phleger 1952b
14. *Spiroplectammina biformis*	T	43	N. Canada	Vilks 1969
	T	90–670	Baffin Island	Schafer and Cole 1986
15. *Cribrostomoides crassimargo*	T	70–159	N. Canada	Vilks 1969
16. *Trochamminella bullata*	T	200	N. Canada	Vilks 1969
17. *Buliminella elegantissima*	T	1288	Central	Lagoe 1977
18. *Valvulineria arctica*	T	1971	Central	Lagoe 1977
19. *Elphidium excavatum alba*	T	0	Alaska	Haman 1973

Table 17.2 Typical shelf to deep-sea species

Species	Mode of life	Feeding strategy
Eggerall advena	Infaunal, free	Detritivore
Elphidium clavatum	Infaunal, free	Detritivore
Cuneata arctica	Infaunal, free	Detritivore
Textularia torquata	?	?
Trochammina nana	?	?
Epistominella arctica	? Epifaunal	?
Eponides tumidulus horvathi	?	?
Cassidulina teretis	Infaunal, free	? Detritivore
Oridorsalis tener	Epifaunal	?
Stetsonia horvathi	?	?

assemblages whereas the East Siberian Sea is dominated by *Elphidium*, *Buccella* and *Islandiella*.

Discussion

In reviews of Arctic Foraminifera, Lagoe (1979b, 1980) stated that water-mass properties play a dominant role in the foraminiferal distribution patterns. However, this appears to be an over-simplification and Fig. 17.2 clearly shows that the associations defined here are not water-mass dependent.

Macrofaunal biomass is low (40 times lower than that off Antarctica). The food supply to the deeper parts is principally organic matter in faecal pellets from zooplankton together with rare carcasses of larger animals. Since the primary productivity of the surface waters is low the benthic food supply must also be low, hence the low biomass (Marshall in Rey 1982). Furthermore, the primary productivity is compressed into a very short summer season so the Arctic benthos must have slow growth, deferred maturity and greater longevity. Organisms need to be *K*-strategists, able to store enough food to enable them to survive the long unfavourable season. Dunbar (in Rey 1982) considers that this is the reason why diversity is so low. In copepods, the lifetime is extended to 2 years and then reproduction takes place. However, a two-phase cycle is present with each group reproducing in alternate years but reproductively isolated, i.e. the year 1 group is immature and the year 2 group reproduces. This may also be true of the benthos including the foraminifera.

Table 17.3 Shelf–deep sea associations from the Arctic Ocean

Cuneata arctica **association**

Salinity: 32–34‰

Temperature: –1.8 to 4.0 °C

Substrate: muddy sand

Depth: 48–100 m

Distribution:
1. Bering Sea (Anderson 1963) total (as *Reophax*)
2. Chukchi Sea (Cooper 1964) total (as *Reophax*)
3. Baffin Island (Schafer and Cole 1986) total (as *Reophax*)

Additional common species:

Eggerella advena	1, 2
Textularia torquata	1
Spiroplectammina biformis	1, 3
Buccella frigida	2

Eggerella advena **association**

Salinity: 31–34‰

Temperature: –1.8 to 9.0 °C

Substrate: muddy sand

Depth: 7–155 m

Distribution:
1. Bering Sea (Anderson 1963) total
2. Chukchi Sea (Cooper 1964) total
3. N. Canada (Phleger 1952b) total

Additional common species:

Elphidium clavatum	1, 2
Cuneata arctica	1, 2 (as *Reophax*)
Uvigerina juncea	1
Epistominella exigua	1
Trifarina angulosa	1 (as *Angulogerina*)
Elphidium bartletti	2
Haynesina orbiculare	2 (as *Elphidium*)
Textularia torquata	2
Eoeponidella pulchella	2 (as *Asterellina*)
Portatrochammina lobata	2 (as *Trochammina*)
Islandiella islandica	2 (as *Cassidulina*)
Islandiella norcrossi	3 (as *Cassidulina*)
Spiroplectammina biformis	3

Elphidium clavatum **association**

Salinity: 14–34‰

Temperature: –2 to 10 °C

Substrate: muddy sand

Depth: 1–60 m

Distribution:
1. Bering Sea (Anderson 1963) total
2. Chukchi Sea (Cooper 1964) total
3. N. Canada (Phleger 1952) total
4. Alaska (Lagoe 1979a) total

Additional common species:

Eggerella advena	1, 2
Elphidium subarcticum	1
Buliminella elegantissima	1
Buccella frigida	1, 2
Cuneata arctica	1 (as *Reophax*)
Elphidium bartletti	2, 3 (as *Cribroelphidium*)
Glabratella wrightii	2 (as *Rosalina*)
Haynesina orbiculare	2 (as *Elphidium*), 4 (as *Protelphidium*)
Eoeponidella pulchella	2 (as *Asterellina*)
Elphidiella arctica	2
Islandiella islandica	3 (as *Cassidulina*)
Islandiella norcrossi	3 (as *Cassidulina*)
Cibicides lobatulus	3
Fursenkoina fusiformis	3 (as *Virgulina*)

Table 17.3 *(continued)*

***Epistominella arctica* association**

Salinity: 34.95‰

Temperature: –0.5 to 0.2 °C

Substrate: silty sandy clay

Depth: 1181–2810 m

Distribution:
1. Ellesmere Island (Green 1960) total
2. Central (Lagoe 1977) total

Additional common species:

Species	Distribution
Eponides tumidulus horvathi	1 (as *Valvulineria*)
Oridorsalis tener	1, 2 (as *Eponides*)
Stetsonia horvathi	1, 2
Epistominella sp.	2
Buliminella elegantissima	2
Valvulineria arctica	2
Cassidulina teretis	2

***Cassidulina teretis* association**

Salinity: 34.5–34.9‰

Temperature: –0.5 to 0.4 °C

Substrate: mud

Depth: 193–1696 m

Distribution:
1. N. Canada (Vilks 1969) total (as *Islandiella*)
2. Ellesmere Island (Green 1960) total

Additional common species:

Species	Distribution
Cribrostomoides subglobosum	1
Nummoloculina sp.	2
Valvulineria arctica	2
Stetsonia horvathi	2
Epistominella naraensis	2
Oridorsalis tener	2 (as *Eponides*)

***Eponides tumidulus horvathi* association**

Salinity: 34.95‰

Temperature: –0.5 to 0.4 °C

Substrate: silty sandy clay

Depth: 1537–2424 m

Distribution:
1. Ellesmere Island (Green 1960) total
2. Central (Lagoe 1977) total

Additional common species:

Species	Distribution
Oridorsalis tener	1, 2 (as *Eponides*)
Stetsonia horvathi	1, 2
Cassidulina teretis	1
Discorbis sp.	1
Fontbotia wuellerstorfi	2 (as *Planulina*)

***Oridorsalis tener* association**

Salinity: 34.95‰

Temperature: –0.5 to 0.0 °C

Substrate: sandy clay

Depth: 1142–2760 m

Distribution:
1. Ellesmere Island (Green 1960) total (as *Eponides*)
2. Central (Lagoe 1977) total (as *Eponides*)

Additional common species:

Species	Distribution
Eponides tumidulus horvathi	1 (as *Valvulineria*), 2
Quinqueloculina ackneriana	1
Stetsonia horvathi	1, 2
Triloculina trihedra	1
Cassidulina teretis	1
Islandiella islandica	1 (as *Cassidulina*)
Epistominella arctica	1, 2

Table 17.3 *(continued)*

***Stetsonia horvathi* association**

Salinity: 34.95‰

Temperature: –0.5 to 0.0 °C

Substrate: silty sandy clay

Depth: 878–3709 m

Distribution:
1. Ellesmere Island (Green 1960) total
2. Central (Lagoe 1977) total

Additional common species:

Cassidulina teretis	1
Oridorsalis tener	1, 2 (As *Eponides*)
Valvulineria arctica	1
Epistominella arctica	2
Brizalina arctica	2 (as *Bolivina*)
Buliminella elegantissima	2
Eponides tumidulus horvathi	2
Epistominella sp.	2

***Textularia torquata* association**

Salinity: 31.5‰

Temperature: –1.5 °C

Substrate: mud

Depth: 31–457 m

Distribution:
1. N. Canada: (a) (Phleger 1952b) total; (b) (Vilks 1969) total

Additional common species:

Spiroplectammina biformis	1a, b
Eggerella advena	1a
Trochammina nana	1a, b
Adercotryma glomerata	1a (as *Haplophragmium*)

***Trochammina nana* association**

Salinity: 28–34.9‰

Temperature: –1.0 to 0.0 °C

Substrate: mud

Depth: 17–629 m

Distribution:
1. N. Canada: (a) (Phleger 1952b) total; (b) (Vilks 1969) total
2. Alaska–Canada (Lagoe 1979b) total
3. Baffin Island (Schafer and Cole 1986) total

Additional common species:

Saccammina atlantica	1a (as *Proteonina*), 1b, 2
Cribrostomoides crassimargo	1a (as *Labrospira*), 1b, 2
Adercotryma glomerata	1a (as *Haplophragmium*)
Trochamminella bullata	1a (as *Trochammina*), 1b, 2
Spiroplectammina biformis	1a, b, 2
Cribrostomoides jeffreysii	1b, 2
Atlantiella atlantica	1b, 2 (as *Trochamminella*)
Textularia torquata	1b, 2
Recurvoides turbinatus	1b, 2
Saccammina spherica	1b, 2
Cassidulina teretis	1b
Textularia earlandi	3

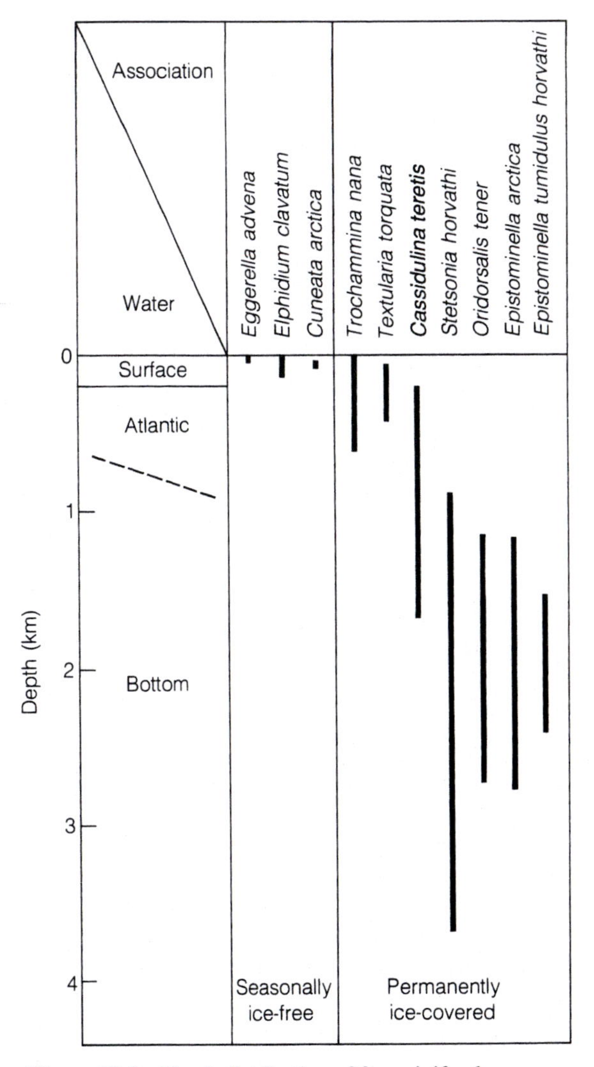

Figure 17.2 Depth distribution of foraminiferal associations

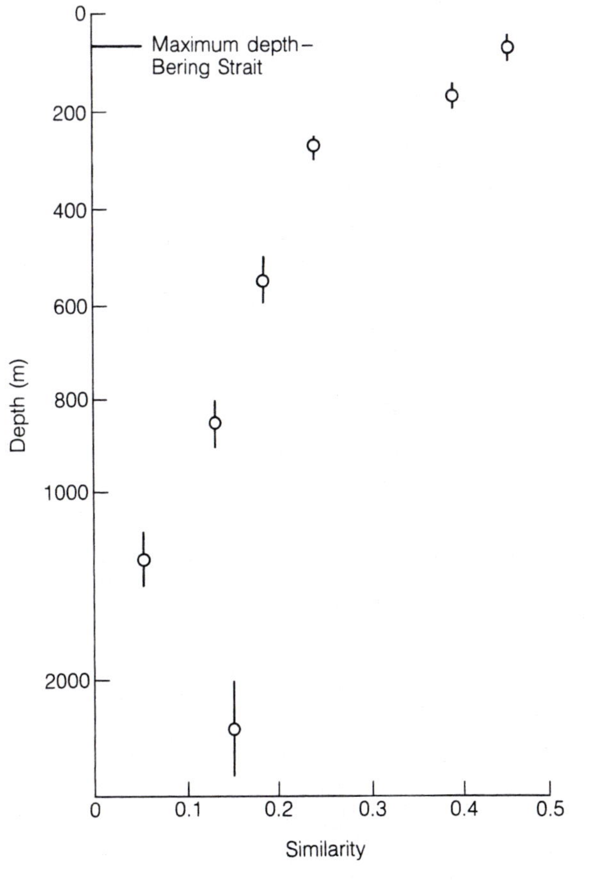

Figure 17.3 Similarity of assemblages between the Arctic Ocean and the eastern Gulf of Alaska (from Lagoe 1980). Similarity = $2w/(a + b)$, where w is the number of species common to the two areas and $(a + b)$ the total number of species in the two areas. Total similarity yields a value of 1.0 and total dissimilarity 0.0

The low diversity of Arctic assemblages causes the absence of taxa common in other deep-sea areas, e.g. *Cyclammina*, *Martinottiella*, *Eilohedra*, *Pullenia*, *Uvigerina* (Lagoe 1980). The Bering Strait has acted as an effective barrier to the migration of deeper-water species (Fig. 17.3) and the isolation of the Arctic Ocean is at least as old as late Miocene.

Fontbotia wuellerstorfi (as *Planulina*) shows a –0.8 to –0.9‰ depletion in $\delta^{18}O$ and –1.2 to –0.9 ‰ $\delta^{13}C$ relative to calcite deposited in isotopic equilibrium with the ambient seawater. This is attributed to the incorporation of metabolic CO_2 and the test is very close to the $\delta^{13}C$ of ΣCO_2 of the bottom water (Aksu and Vilks 1988).

Postmortem changes

Lack of data on living and dead assemblages precludes an analysis of postmortem changes. Phleger (1952b) noted that in the Canadian Arctic island area shallow-water foraminifera are transported by grounded ice from nearshore to deeper-water areas. However, this process does not appear to be significant in the Arctic Ocean proper.

Planktonic : benthic ratio

An almost monospecific assemblage of sinistral *Neogloboquadrina pachyderma* lives in Arctic Ocean surface waters. Shelf seas have few planktonic tests, but from 200 to 500 m the values increase to >90% planktonic and >500 m to generally >85‰ planktonic (Green 1960; Lagoe 1980). Because of the carbonate saturation of the deep water, no dissolution takes place.

CHAPTER 18

Summary of modern distribution patterns and characteristics of assemblages

Details of the distributions of associations of species in different environments throughout the world have already been described in Chapters 6–17. The more important abiotic ecological parameters include water temperature, salinity, energy and turbidity, and the nature of the substrate. Although one factor may appear to be most significant, the total range of a species or an association is invariably controlled by a plexus of factors. Biotic parameters include the availability of food and the effects of predation. Neither is limiting but both influence the abundance of individuals. To a certain extent foraminifera have evolved non-competitive feeding strategies which are most clearly seen in shallow-water environments where herbivores, detritivores, omnivores and passive suspension feeders may coexist. Exclusion of macrobenthic predators under experimental conditions or in dysaerobic/anoxic environments leads to higher standing crop values. In response to oligotrophic conditions where nutrients are in very low supply, certain foraminifera have evolved an endosymbiotic relationship with algae which enables them to achieve both moderate densities and large size.

Foraminifera have high infant mortality and for those adults which survive to reproduce this process terminates their life. The life cycle of such forms ranges from days to several years. The longest life cycles are believed to be those of tropical larger foraminifera and shallow cold-water forms, but nothing is known of the length of life of deep-sea forms. There are seasonal (and longer period) fluctuations in the abundance of individual species, and seasonal replacement of the dominant form is documented both from obviously seasonal-influenced shallow-water environments and from the deep sea where there is seasonal input of phytodetritus (Gooday 1988).

There are commonly marked differences between living and dead assemblages from the same area. The causes include differing rates of productivity between species (but this is not usually a major effect) and various postmortem processes including transport (loss or gain) and destruction of tests (disintegration of agglutinated and dissolution of calcareous forms). These processes are discussed in detail in Chapters 4 and 5.

Résumé of modern faunas

The general attributes of modern faunas are given in Table 18.1.

Diversity

In terms of diversity, the lowest values are typical of the most stressed environments (Fig. 18.1). Values of α <5 generally indicate brackish or hypersaline marginal marine environments but may also indicate normal marine environments with a high dominance of a single species. Where α >7 normal marine shelf to slope or hypersaline shelf are indicated. From the limited data on $H(S)$, values <0.6 indicate brackish waters while values >2.1 indicate normal marine environments. The diversity of dead and fossil assemblages may be increased or diminished by the postmortem processes described in Chapter 5.

Wall structure

Although there is some overlap of environments certain features stand out (Fig. 18.2). The porcellaneous component exceeds 20% only in normal marine and

Table 18.1 Summary of the occurrence of genera >10 ‰ (living, dead or total) in modern environments. Diversity and wall structure values are based on living assemblages

Brackish marsh

$\alpha < 1$ to 3, range < 1 to 5, $H(S)$ 0–1.8
High agglutinated content

Agglutinated:	Calcareous:
Ammotium	*Ammonia**
Arenoparrella	*Elphidium** (unkeeled)
Haplophragmoides	*Haynesina**
Jadammina	
Miliammina	* = *rarely preserved*
Tiphotrocha	*dead*
Trochammina	

Marine marsh

$\alpha < 1$ to 2
Up to 50‰ porcellaneous

Agglutinated:	Calcareous:
Ammotium	simple miliolids
Arenoparrella	
Miliammina	

Hypersaline marsh

$\alpha < 1$ to 7
Up to 100‰ porcellaneous

Agglutinated:	Calcareous:
Ammotium	simple miliolids
Arenoparrella	
Miliammina	

Brackish mangal

$\alpha < 1$ to 3
High agglutinated content

Agglutinated:
Ammotium
Arenoparrella
Haplophragmoides
Miliammina
Siphotrochammina
Tiphotrocha
Trochammina

Brackish lagoon

$\alpha < 1$ to 5
No porcellaneous component except in near-marine examples

Agglutinated:	Calcareous:
Ammobaculites	*Ammonia*
Ammotium	*Buccella*
Eggerella	*Elphidiella*
Eggerelloides	*Elphidium* (unkeeled)
Miliammina	*Haynesina*
Trochammina	

Marine lagoon

α 3 to 12
High porcellaneous content

	Calcareous:
	Ammonia
	Archaias
	Nonion
	Peneroplis
	Quinqueloculina
	Triloculina

Hypersaline lagoon

$\alpha < 1$ to 6, rarely 7
High porcellaneous content

	Calcareous:
	Ammonia
	Peneroplis
	Quinqueloculina
	Triloculina

Inner shelf (0 – 100 m)

α 3 to 19, $H(S)$ 0.6–2.75
Occasionally > 20‰ porcellaneous content

Agglutinated:	Calcareous:
Adercotryma	*Ammonia*
Aveolophragmium	*Amphistegina*
Ammobaculites	*Archaias*
Ammoglobigerina	*Asterigerina*
Ammoscalaria	*Asterigerinata*
Bigenerina	*Asterorotalia*
Cribrostomoides	*Astrononion*
Cuneata	*Baculogypsina*
Eggerella	*Bolivina*
Eggerelloides	*Bolivinellina*
Eratidus	*Brizalina*
Gaudryina	*Buccella*
Göesella	*Bulimina*
Lagenammina	*Buliminella*
Lepidodeuterammina	*Calcarina*
Martinottiella	*Cancris*
Nouria	*Cassidella*
Portatrochammina	*Cassidulina*
Recurvoides	*Cibicides*
Reophax	*Cibicidoides*
Saccammina	*Discorbis*
Spiroplectammina	*Elphidiella*
Textularia	*Elphidium* (keeled and unkeeled)
Trochammina	*Eoeponidella*
	Epistominella
	Fursenkoina

Table 18.1 *(continued)*

Gavelinopsis	**Outer shelf (100 – 200 m)**	
Glabratella	α 5 to 19, $H(S)$ 0.6–2.75	
Globocassidulina	Agglutinated:	Calcareous:
Hanzawaia	*Adercotryma*	*Bolivina*
Haynesina	*Archimerismus*	*Brizalina*
Heterolepa	*Cribrostomoides*	*Buccella*
Heterostegina	*Eggerella*	*Bulimina*
Homotrema	*Goësella*	*Cancris*
Marginopora	*Haplophragmoides*	*Cassidulina*
Miniacina	*Portatrochammina*	*Cassidulinoides*
Murrayinella	*Recurvoides*	*Cibicides*
Neoconorbina	*Reophax*	*Cibicidoides*
Neoponides	*Saccamina*	*Ehrenbergina*
Nonion	*Spiroplectammina*	*Elphidiella*
Nonionella	*Textularia*	*Epistominella*
Nonionellina		*Eponides*
Nonionoides		*Fursenkoina*
Operculina		*Gavelinopsis*
Pararotalia		*Globobulimina*
Peneroplis		*Globocassidulina*
Planorbulina		*Hyalinea*
Planulina		*Islandiella*
Quinqueloculina		*Nonionella*
Rectobolivina		*Nonionellina*
Rectuvigerina		*Rectobolivina*
Reussella		*Trifarina*
Rosalina		*Uvigerina*
Rotalinoides		
Suggrunda	**Upper slope (200–2000 m)** (= **upper bathyal**)	
Tretomphaloides	α 1 to 22. $H(S)$ 0.75–4.1	
Triloculina	Agglutinated:	Calcareous:
Trochulina	*Adercotryma*	*Alabaminella*
Uvigerina	*Ammobaculites*	*Alabaminoides*
Valvulineria	*Ammoflintina*	*Bolivinellina*
	Ammoglobigerina	*Bolivinita*
	Archimerismus	*Brizalina*
	Bathysiphon	*Bulimina*
	Cribrostomoides	*Buliminella*
	Deuterammina	*Cassidulina*
	Glomospira	*Chilostomella*
	Goësella	*Cibicides*
	Haplophragmoides	*Cibicidoides*
	Karreriella	*Discanomalina*
	Miliammina	*Ehrenbergina*
	Portatrochammina	*Eilohedra*
	Psammosphaera	*Epistomaroides*
	Recurvoides	*Epistominella*
	Reophax	*Fontbotia*
	Repmanina	*Fursenkoina*
	Textularia	*Globobulimina*
	Trochammina	*Globocassidulina*
	Trocahmminella	*Gyroidina*
	Trochamminopsis	*Hoeglundina*

Table 18.1 *(continued)*

	Islandiella
	Melonis
	Nummoloculina
	Nuttallides
	Oridorsalis
	Osangularia
	Pseudoparrella
	Pullenia
	Rectuvigerina
	Reussella
	Robertinoides
	Sphaeroidina
	Stetsonia
	Suggrunda
	Trifarina
	Trochulina
	Uvigerina
	Valvulineria
Lower slope (2000–4000 m) (= lower bathyal)	
α 1 to 22, $H(S)$ 0.75–4.1	
Agglutinated:	Calcareous:
Adercotryma	*Alabaminella*
Ammolagena	*Bulimina*
Cribrostomoides	*Cassidulina*
Cyclammina	*Cibicidoides*
Eggerella	*Ehrenbergina*
Haplophragmoides	*Eilohedra*
Hyperammina	*Epistominella*
Lagenammina	*Eponides*
Martinottiella	*Fontbotia*
Nodulina	*Globocassidulina*
Psammosphaera	*Hoeglundina*
Pseudobolivina	*Laticarinina*
Reophax	*Melonis*
Repmanina	*Nuttallides*
Rhabdammina	*Oridorsalis*
Sigmoilopsis	*Pullenia*
Textularia	*Rupertina*
Abyssal (> 4000 m)	
Agglutinated:	Calcareous:
Cyclammina	*Alabaminella*
Hormosina	*Cibicidoides*
	Epistominella
	Eponides
	Fontbotia
	Globocassidulina
	Nuttallides
	Pullenia

hypersaline lagoons and marshes and it is normally <20% in shelf seas. Postmortem influences may change the proportions to a greater or lesser extent.

Genera

The occurrence of genera with an abundance of >10% in living, dead or total assemblages is summarised in Table 18.1. The overall distribution of many genera may be more widespread but only in low abundance. More detailed ecological requirements of selected genera are listed in Appendix B and line drawings of some representatives are given in Fig. 18.3.

Biogeographic boundaries

Because the modern oceans have an essentially layered water-mass structure with the densest, i.e. coldest and most saline, waters in the deepest parts and the warmer and/or least saline waters in their upper parts, faunas on the slope and in the ocean basins generally show distribution patterns related to those of the water masses. However, on the shelf and especially in the shallow parts, temperature is a principal control on biogeography. Other factors such as geographic barriers (land or broad expanses of deep water) may also be important.

As a generalisation faunal boundaries can be considered as of three types (Fig. 18.4). Type A represents a fauna showing varying degrees of stenothermy, for example, tropical larger foraminifera. Type B shows the abrupt change from one fauna to another. There may be overlap between species ranges, but commonly there is a high diversity zone marking the region of transition caused by the lateral fluctuation of the boundary. Type C shows a progressive transition with no clearly defined boundary. An example of type B is the Gulf Stream–Virginian coastal water transition off Cape Hatteras (Murray 1969) while the European seaboard is an example of type C.

The distribution of larger foraminifera in the Atlantic and Pacific oceans shows the effects of extensive deep-water barriers to migration between the favoured side (west) and the unfavoured side (east). In addition, plate tectonic processes have led to the development of shallow water on island arcs on the west side of these two oceans and this has also contributed to the observed distribution patterns.

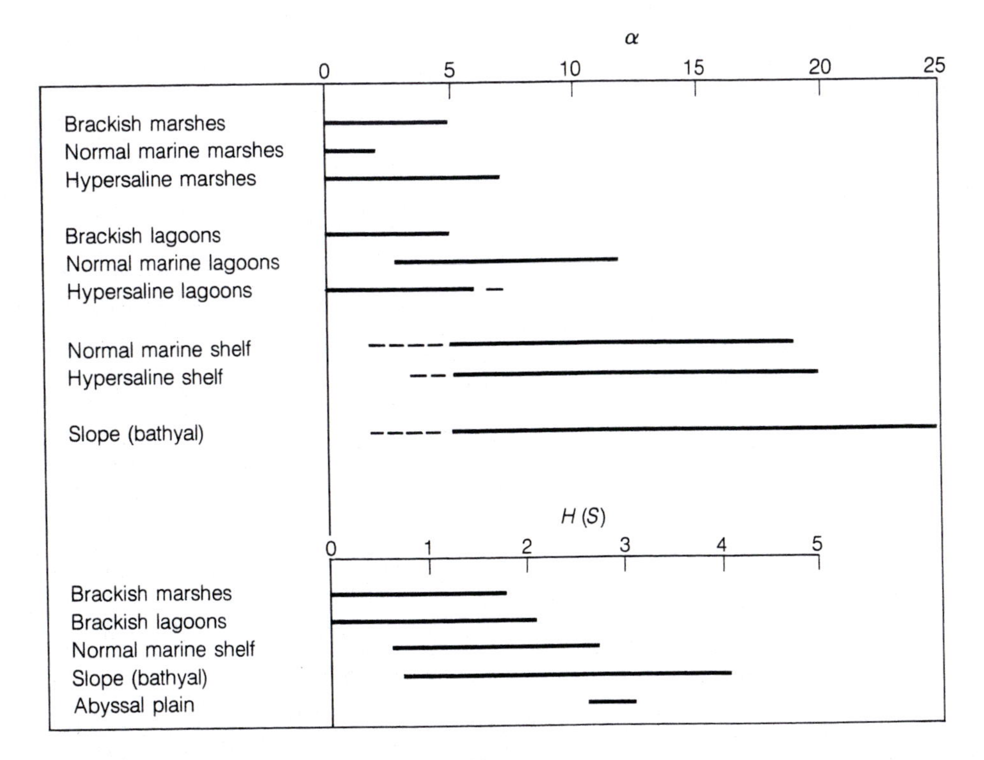

Figure 18.1 Summary of diversity data for living assemblages

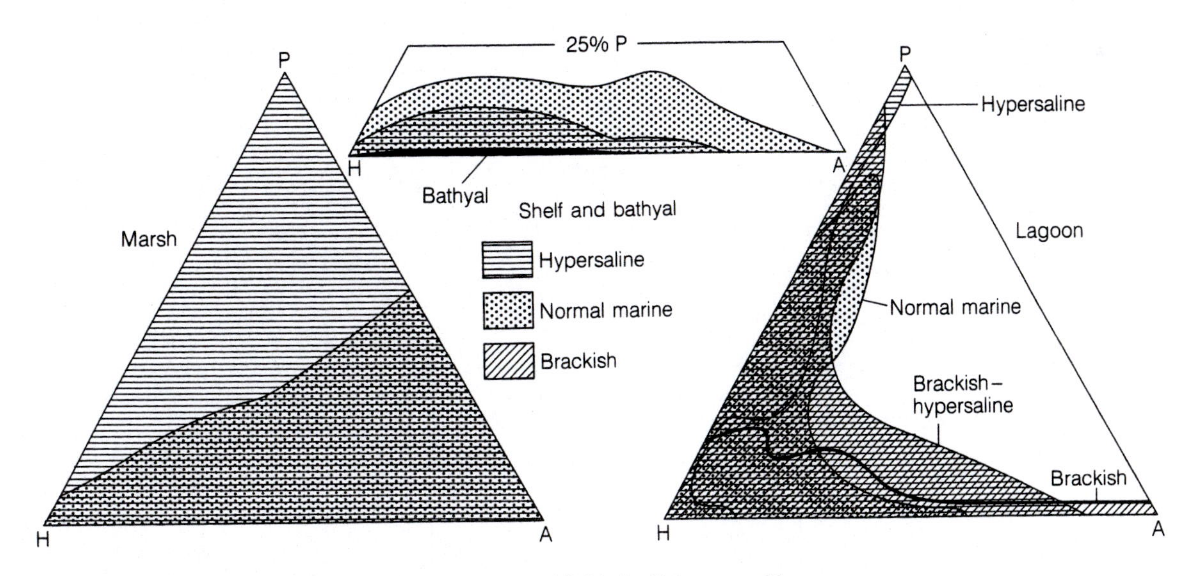

Figure 18.2 Summary of wall structure environmental fields for living assemblages

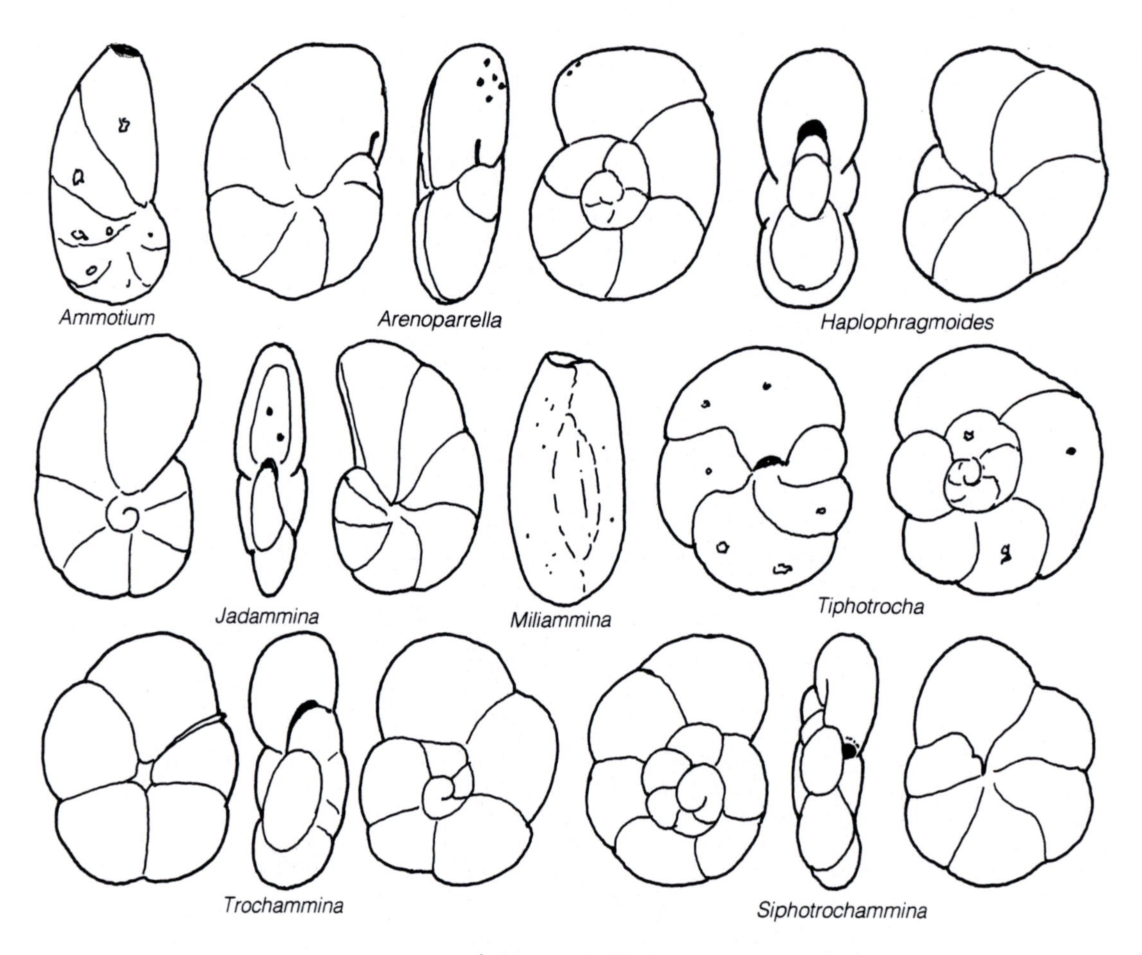

Figure 18.3 Line drawings of some genera from different environments (based on the author's photographs except for the following which are after the named authors: *Haplophragmoides*, Haynes 1981; *Arenoparrella, Tiphotrocha, Siphotrochammina, Elphidiella, Lagenammina, Portatrochammina, Recurvoides, Spiroplectammina, Nonionoides*, Loeblich and Tappan 1988; *Ammotium*, Lutze 1965; *Archaias, Asterorotalia, Tretomphaloides*, Brady 1884

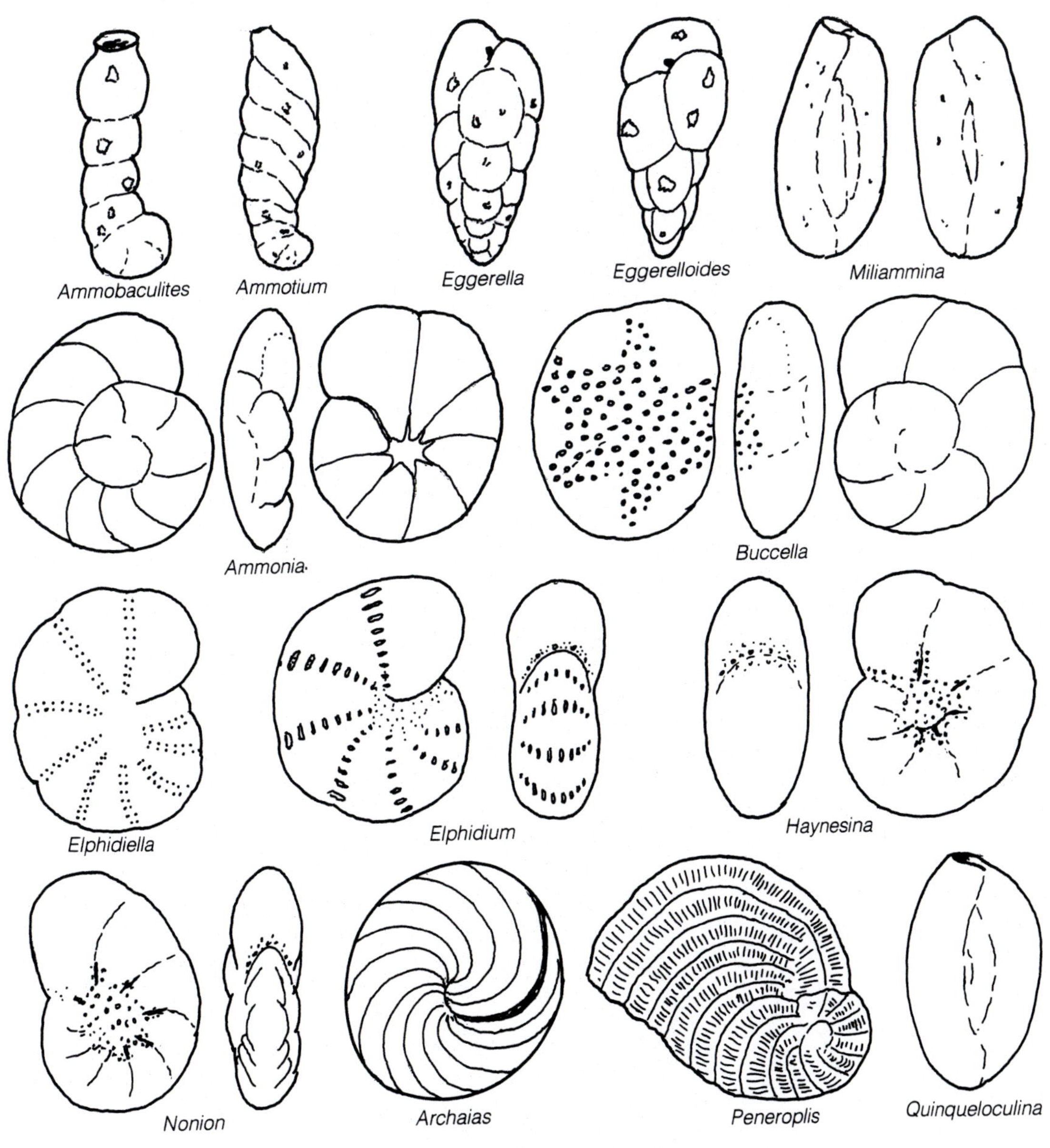

Lagoons and innermost shelf

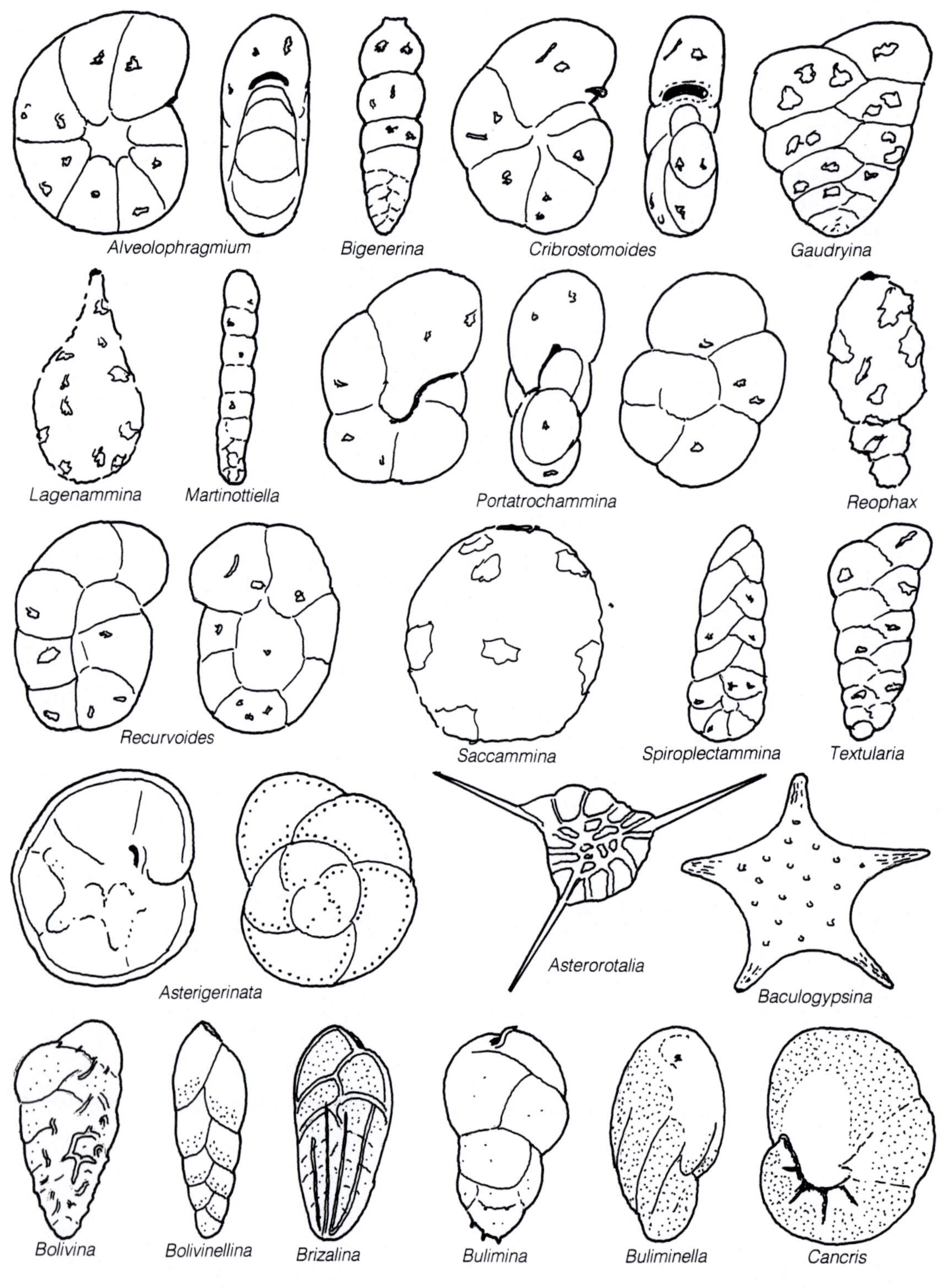
Alveolophragmium
Bigenerina
Cribrostomoides
Gaudryina
Lagenammina
Martinottiella
Portatrochammina
Reophax
Recurvoides
Saccammina
Spiroplectammina
Textularia
Asterigerinata
Asterorotalia
Baculogypsina
Bolivina
Bolivinellina
Brizalina
Bulimina
Buliminella
Cancris

Shelf

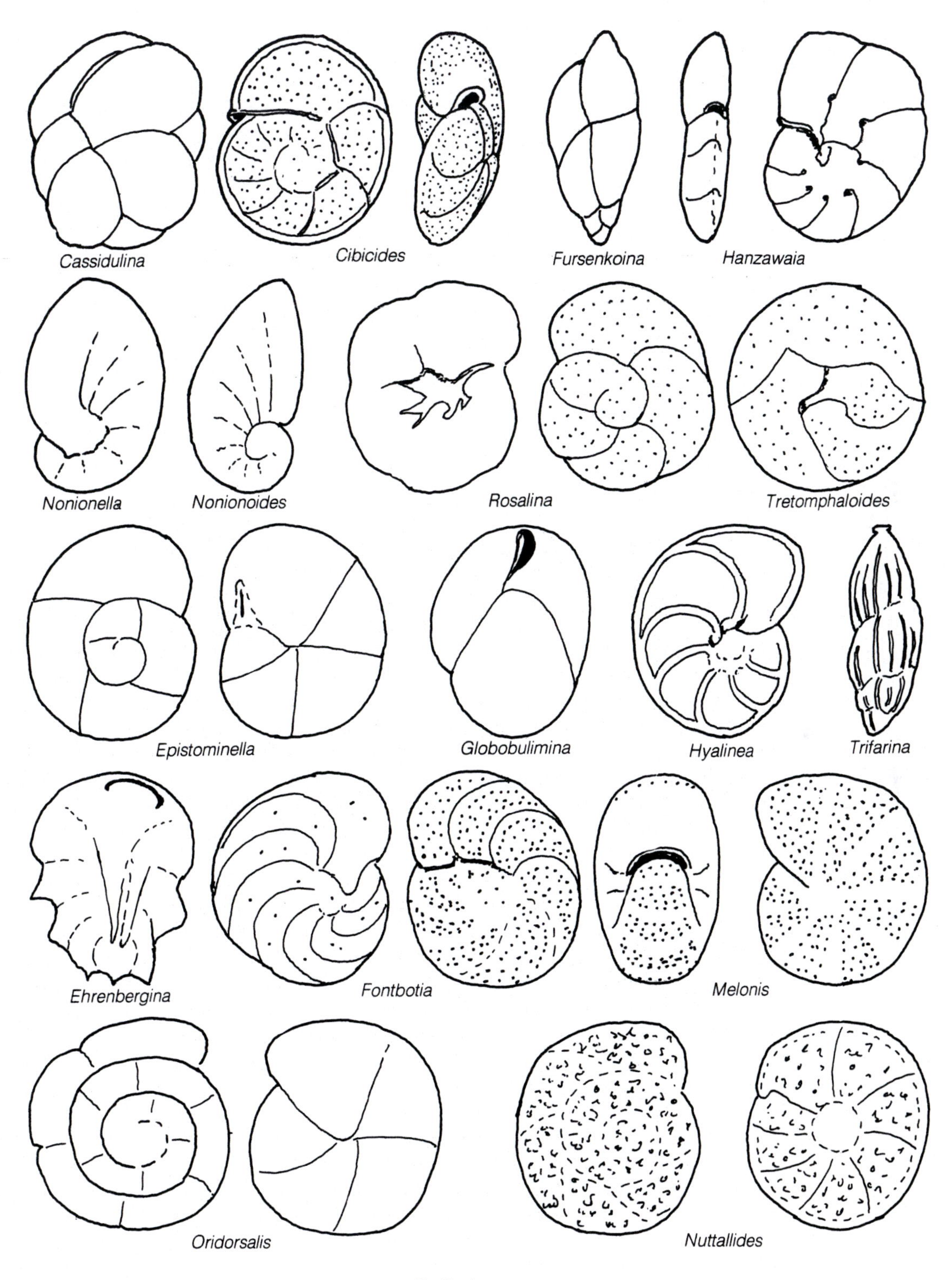

Shelf – deep sea

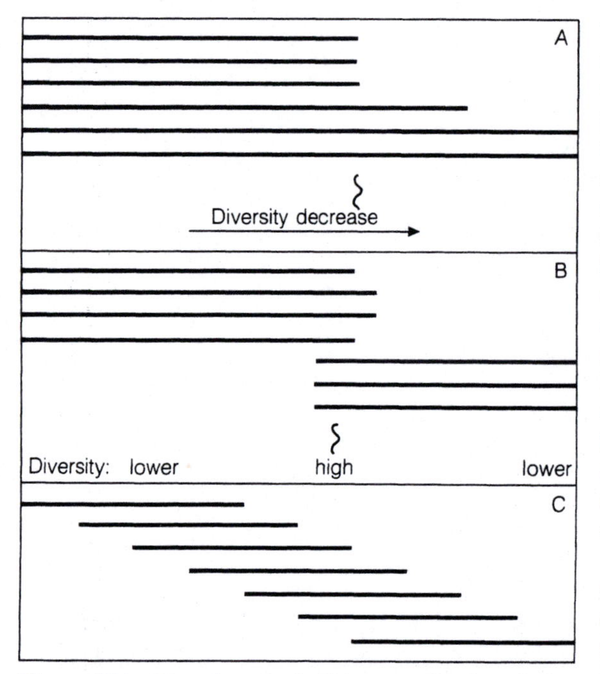

Figure 18.4 Three hypothetical biogeographic boundaries

Depth distribution

Traditionally, geologists have had an almost obsessive interest in determining the depth of past seas and this influence can be seen in the approach to many ecological studies. However, depth is probably not a limiting factor although other more important parameters are closely related to it, e.g. temperature, salinity, density, light penetration, etc. As shown in previous chapters, individual associations and the species comprising them have well-defined local depth distributions controlled by differences of water mass and/or substrate. For any given association the depth range changes from one area to another. This confirms that depth *per se* is not a controlling factor. Nevertheless, individual genera do have distinctive broad depth ranges and these are summarised in Appendix B. In combination with other assemblage attributes discussed above, it is therefore possible to recognise broad depth divisions such as inner shelf, upper slope, etc. but less easy to determine that a particular assemblage represents, say, 160–170 m depth. Haake *et al.* (1982) and Culver (1988) have described methods of calculating a depth estimate for recent and subrecent samples.

On theoretical grounds the abundance of a given species should be highest where conditions are at their optimum, although the depth range might extend both shallower and deeper than this level. In Fig. 18.5A the effects of varying amounts of downslope transport on an initial living distribution have been calculated assuming that production leads to one adult dead test from each parent. The result is that the peak abundance of the dead assemblage is both reduced in height and moved downslope, and furthermore the total depth range is greatly extended. In Fig. 18.5B the effect on a succession of species is shown diagramatically. These results suggest that caution should be exercised in attempting to define precise palaeodepths.

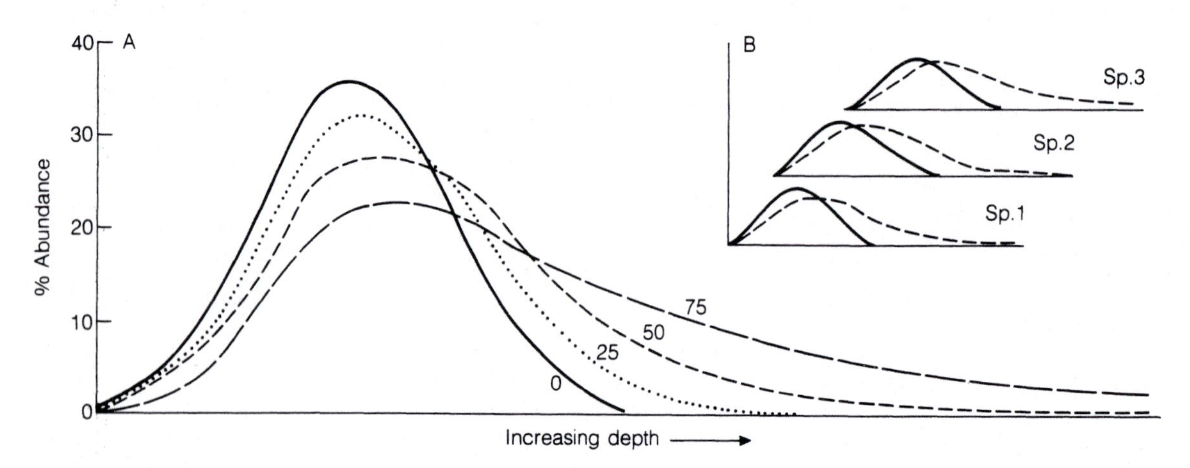

Figure 18.5 (A) The effects of varying percentages of postmortem downslope transport on a species; (B) diagrammatic effect on a succession of species

Comments on certain faunas

Agglutinated faunas

The interest in this group has been sufficient to lead to international workshops the first of which was held in 1981. Notwithstanding this, the ecological significance of these faunas remains controversial.

There are two basic wall types with either solely organic or organic plus secreted calcareous cement. Calcareous cemented agglutinated forms are present on continental shelves such as the English Channel and Arabian Gulf and to a minor extent in the bathyal zone (e.g. *Eggerella bradyi*, Weston 1984). Organic-cemented forms are known from virtually all modern environments from supratidal marshes to the deepest trenches.

In many cases agglutinated foraminifera are present together with calcareous taxa. However, some fossil assemblages consist solely of organic-cemented agglutinated tests (often termed 'flysch-type' assemblages). As pointed out by Scott *et al.* (1983), these may be primary or they may result from the alteration of assemblages originally containing some calcareous tests.

Modern environments having bottom waters corrosive with respect to $CaCO_3$ include tidal marshes, some lagoons or estuaries, parts of epicontinental seas (e.g. the Baltic) and those parts of the ocean deeper than the local CCD. All these areas support living assemblages of organic-cemented agglutinated foraminifera. However, dead assemblages of these forms are more widely distributed, especially in muddy sediments with corrosive pore waters in which calcareous tests are destroyed. Such circumstances must also have existed in the past. In addition, it is known that the level of the CCD has varied through time (see Kennett 1982). From a comparison of the modern agglutinated component of the Newfoundland slope with the Maastrichtian to Palaeogene flysch-type assemblages, Scott *et al.* (1983) concluded that they show a high degree of similarity. It follows that the fossil examples must have shown relative enrichment in their agglutinated component by postmortem dissolution. It seems most likely that agglutinated assemblages in the fossil record are primarily the product of diagenetic processes (notably dissolution of calcareous tests to leave a residue dominated by agglutinated forms) rather than characteristic of any particular environment (or depth of water).

Diverse modern agglutinated assemblages have not been reported from dysaerobic waters so it is unlikely that fossil examples are indicative of these conditions.

Anoxic–dysaerobic assemblages

Areas of intense upwelling with associated high primary productivity lead to the accumulation of fine-grained sediments with an organic carbon content 10–20 times greater than normal. Much of the organic material is in the form of faecal pellets. Where the organic carbon content exceeds 4–5% the sediments have an unusually high water content with a high undisturbed shear strength but high sensitivity (Keller 1982). This may allow epifaunal foraminifera to remain on the sediment surface.

Commonly such deposits occur beneath waters depleted in oxygen to a greater or lesser extent. From macrofaunal studies it is known that the diversity is reduced where the oxygen level falls below 1.0 ml l^{-1} but the effects of the meiofauna are less obvious. Rhoads and Morse (1971) classified such environments as anaerobic, dysaerobic and aerobic with 0–0.1, 0.1–1.0, and >1.0 ml l^{-1} O_2 respectively. However, as pointed out by Wignall and Myers (1988), this ignores short-term fluctuations in oxygen levels and they suggested that the term 'dysaerobic' should be used only in a broad sense to denote oxygen-restricted bottom waters. Under anaerobic (anoxic) conditions exclusion of the infauna and seasonal variation in sediment/organic input leads to laminated sediments. Fluctuating anaerobic/dysaerobic conditions lead to alternations of laminated and unlaminated sediments.

In oxygen-deficient basins off California where laminated sediments are accumulating, the standing crop of the foraminifera is higher than that of the adjacent shelf due to the exclusion of macrobenthic predators (see Ch. 2). In nearshore basins with an oxygen level of 0.1–0.9 ml l^{-1} the principal genera are *Bolivina*, *Brizalina*, *Loxostomum*, *Epistominella*, *Uvigerina*, *Fursenkoina*, *Cassidulinoides*, *Buliminella*, and *Textularia*, while the offshore basins with 0.25–1.1 ml l^{-1} O_2 have in addition *Cassidulina*, *Eponides* and *Hoeglundina* (Douglas and Heitman 1979). *Bolivina* is the principal form off Chile (Boltovskoy 1972) and Peru (Phleger and Soutar 1973).

Using a morphotype approach Bernhard (1986) determined from a small number of modern assemblages that elongate flattened (i.e. *Bolivina* shaped) and tapered (i.e. *Fursenkoina*/*Buliminella* shaped) tests are typical of anoxic and dysaerobic environments. Flattened planispiral and lenticular forms are also present. These results are in accord with those presented above. Flattened morphologies could reduce sinking into the soft sediment (if the test lays flat on the surface) or increase the surface area for oxygen exchange (if the test is held upright).

Mitochondria beneath pores might enhance oxygen uptake (Leutenegger and Hansen 1979) so coarsely perforated tests would be an adaptation to low-oxygen conditions. However, using the experimental results (of Bradshaw 1961) on oxygen consumption Phleger and Soutar (1973) concluded that an oxygen level of 0.1 ml l^{-1} was adequate for foraminifera living in these environments. Nevertheless, a test size <250 μm is common and the walls are typically thin, unornamented and highly perforate. Bernhard suggests that such forms should predominate during a prolonged anoxic episode. Alternatively, during a short-lived anoxic period these forms would survive whereas the larger forms would not.

Although some agglutinated taxa (*Textularia*, *Reophax*) are present in oxygen-deficient environments, they are not generally abundant. The absence of wall pores may limit their uptake of oxygen. In many cases the pore waters in anoxic sediments are alkaline due to the activities of sulphate-reducing bacteria and therefore preservation of calcareous tests is good (Douglas and Heitman 1979), but in other cases corrosive waters may cause dissolution (Schrader *et al.* 1983) and therefore postmortem enrichment of agglutinated tests. Under totally anoxic conditions (i.e. no oxygen) benthic foraminifera are absent.

The occurrence of pyrite infilling foraminiferal tests is not an indicator of anoxic bottom waters but rather the presence of microenvironments in an otherwise oxic environment, as shown by the work of Love and Murray (1963). Nevertheless, larger amounts of pyrite may be formed in dysaerobic and anoxic than in more oxic environments.

Larger foraminifera

These are characterised by the presence of endosymbionts and by a large test volume. Growth rates far exceed those of forms lacking symbionts (Ross 1974). Larger foraminifera and other zooxanthelate organisms such as hermatypic corals are highly adapted to oligotrophic (nutrient-deficient) conditions. It has been cogently argued that eutrophication (an increase in nutrients) is detrimental to such organisms and may in the past have led to widespread extinction in reef communities (Hallock and Schlager 1986; Hallock 1988a). In particular, eutrophication stimulates the growth of plankton, thus increasing the turbidity of the water. The euphotic zone (depth to which 1% of midday surface irradiation penetrates) extends to 50–100 m in the Caribbean and to 100–150 m in the Indo-Pacific. Light and water energy are considered to be the most important factors controlling the distribution of larger foraminifera (Hottinger 1980, 1983). In the Gulf of Aqaba the lower limit of larger foraminifera at 130 m represents the 0.5% level of surface light intensity. In shallower waters, if the light intensity becomes too high then the symbionts may move to the shaded side of the test in *Amphisorus*, *Marginopora* or *Sorites* (Leutenegger 1977) or the individual may burrow beneath the sand as in the case of *Alveolinella* (Hottinger 1983). Depth and light relationships in the Gulf of Aqaba are shown in Fig. 12.2.

The role of water energy is summarised in Table 18.2. Certain faunas are missing from some areas. Thus the high-energy fauna and several deep eutrophic species are missing from Hawaii, but the backreef and shallow slope assemblages resembles that of the western Pacific. Nevertheless, there are great similarities between the faunas of open backreefs and shallow slopes, and Hallock (1988b) describes these as 'morphologic and ecologic generalists among the algal symbiont-bearing foraminifera'.

The imperforate porcellaneous wall structure is a barrier to gas exchange. Those forms with algal symbionts have structural features such as pits which may enhance gas exchange and light penetration (Hansen and Dalberg 1979). Those forms having chlorophycean symbionts (Fig. 2.2) have higher nutritional requirements and need higher light levels than others. In the extremely oligotrophic waters of the Pacific the only eutrophic areas are in bays and lagoons of high islands. Therefore only one eutrophic species (*Laevipeneroplis proteus*) is present. In the oligotrophic waters *Marginopora* and *Sorites* (dinophycean symbionts) and *Alveolinella* (diatom symbionts) flourish. By contrast the more eutrophic Carribbean supports several chlorophycean-bearing species (*Peneroplis* spp., *Cyclorbiculina compressa*, *Archaias angulatus*) (Hallock 1988b).

Larger rotaliine foraminifera with their hyaline perforate wall structure are adapted to light gradients and are intolerant of eutrophication. Therefore they have achieved higher diversity and have extended into deeper waters in the Indo-Pacific, whereas in the Caribbean they are poorly developed in the more eu-

Table 18.2 Geographic distribution of some larger foraminifera in relation to habitat (modified from Hallock 1988b). *A.* = *Archaias*, *Al.* = *Alveolinella*, *Am.* = *Amphistegina*, *Ap.* = *Amphisorus*, *B.* = *Baculogypsina*, *Ba.* = *Baculogypsinoides*, *Bo.* = *Borelis*, *Br.* = *Brookina*, *C.* = *Calcarina*, *Cy.* = *Cycloclypeus*, *H.* = *Heterostegina*, *He.* = *Heterocyclina*, *M.* = *Marginopora*, *N.* = *Nummulites*, *O.* = *Operculina*, *S.* = *Sorites*

Habitat	Depth (m)	Caribbean	Red Sea	W. Pacific	Hawaii
High-energy reef margin	<10			*B. sphaerulata* *Ba. spinosus* *C. splengleri* *M. vertebralis*	
Moderate to low-energy back reef	<10	*A. angulatus* *Am. lobifera*	*C. calcar* *Am. lobifera*	*C. calcar* *Am. lobifera*	*Am. lobifera*
Open back reef to shallow	<40	*Ap.* sp. *Bo. pulchra* *S.* spp. *Am. gibbosa* *H. antillarum* *Br.* spp.	*Ap. hemprichii* *Bo. schlumbergeri* *S.* spp. *Am. lessonii*	*Ap. hemprichii* *Bo.* spp. *S.* spp. *Am. lessonii* *C. hispida* *Al. quoyi*	*Ap. hemprechii* *Bo.* spp. *S.* spp. *Am. lessonii* *H. apogama*
Deep euphotic	>30	*Am. radiata*	*H. depressa* *Am. bicirculata* *Am. papillosa* *O.* spp. *N.* sp	*Am. radiata* *H. depressa* *Am. bicirculata* *Am. papillosa* *O.* spp. *N.* sp.	*H. depressa* *Am. bicirculata* *O. ammonoides*
Deepest euphotic	>60		*He. tuberculata*	*Cy. carpenteri* *H. operculinoides*	

trophic conditions (Hallock 1988b). Maximum test size is realised under extreme oligotrophic conditions. Shallow-water forms are biconvex and the test becomes flatter and has the greatest surface area to volume ratio at the depth limit of ~130 m.

It has long been recognised that larger symbiont-bearing foraminifera are indicative of warm waters. Murray (1973) noted that all occurrences are encompassed by the 25 °C surface-water isotherm for the southern and northern summers. Temperatures in excess of 20–22 °C are probably necessary for reproduction. Larsen (1976) considered that all occurrences of *Amphistegina* were delimited by the 14 °C surface-water winter isotherm. Laboratory experiments have shown that *A. lobifera* (as *A. radiata*) and *A. lessonii* (as *A. madagascariensis*) ceased to move below 12 and 16 °C respectively (Zmiri *et al.* 1974).

As shown by Murray (1987b) there is a marked diversity gradient from the tropics to the fringes of distribution, but there is also a decrease in diversity from west to east in the Pacific Ocean (compare Hawaii and the west Pacific in Table 18.2) and from shallow to deeper water in any given area.

Apart from the role of light, energy, oligotrophy, temperature and salinity in determining geographic distribution patterns, past plate tectonic activity has played an important role. Adams (1983) considers that plate tectonic processes during the Cenozoic created and maintained the three main faunal provinces (Central American, Mediterranean and Indo-west Pacific). Some features of modern distributions are summarised in Fig. 18.6. Provincialism is a characteristic feature with each taxonomic group showing one or more taxa with a more restricted distribution than the rest: *Alveolinella* compared with *Borelis*, *Cycloclypeus* and *Nummulites* compared with *Operculina* and *Heterostegina*. Even within the ubiquitous *Amphistegina* group *A. gibbosa* is in the Atlantic, *A.*

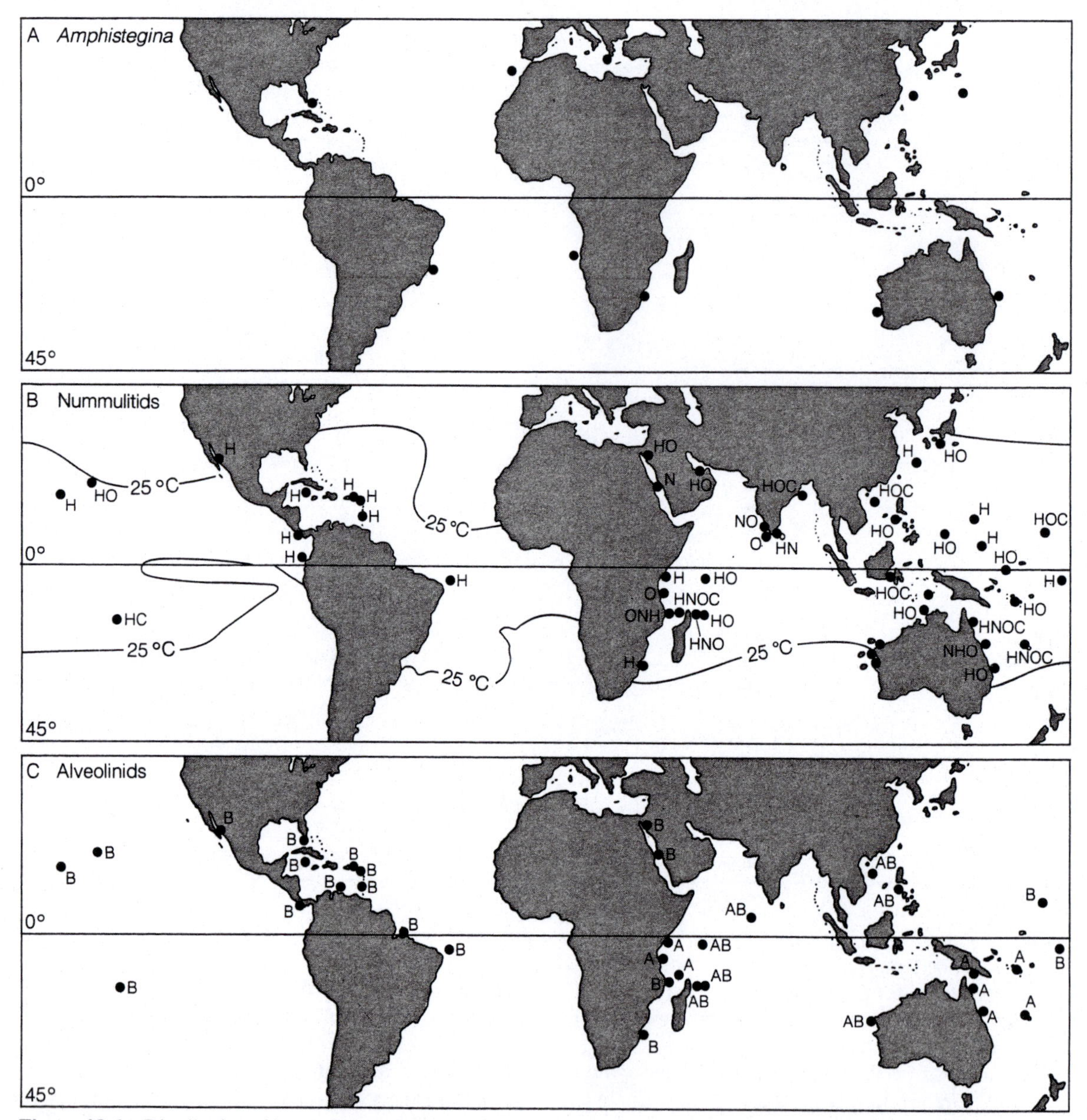

Figure 18.6 Distribution of larger foraminifera based on the literature. (A) *Amphistegina*, northern and southern limits; (B) nummulitids (C = *Calcarina*, H = *Heterostegina*, N = *Nummulites*, O = *Operculina*) and surface-water isotherms for summer in both hemispheres (after Sverdrup *et al.* 1942); (C) alveolinids (A = *Alveolinella*, B = *Borelis*)

lobifera the Mediterranean, and *A. lessonii*, *A. bicirculata*, *A. papillosa* and *A. radiata* in the Indo-Pacific (Morariu and Hottinger 1988).

Much remains to be learned of the niches occupied by the various larger foraminifera and of the post-mortem processes which lead to the transport and concentration of their tests. Hallock and Glenn (1986) provided a idealised model of modern distributions and discussed how fossil examples compare with this (see Fig. 18.7).

Deep sea

Since hydrostatic pressure increases by 1 atm for each 10 m depth, organisms living at 4000 m experience a

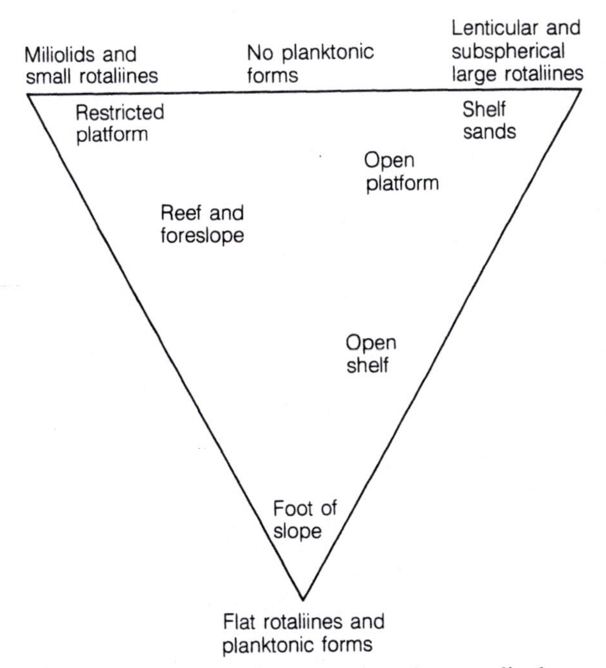

Figure 18.7 Ternary diagram to show the generalised distribution of carbonate environments in terms of benthic morphotype and the presence or absence of planktonic tests (after Hallock and Glenn 1986)

pressure of 401 atm. Slight compression (4% at 1100 atm) causes the concentration of salts in the water to be higher than that at the surface. Furthermore, increased pressure depresses the freezing-point of water (–1.8 °C at the surface, –9 °C at 1000 atm; data from Macdonald 1975). Organisms have physiological adaptations to this high-pressure environment.

Life in the deep sea is dependent on primary producton in the sunlit surface waters and the rain of organic detritus and faecal pellets to the sea bed. The standing crop and biomass of the macrofauna decline exponentially with depth (Rex 1981) yet the metabolic rate (and hence food consumption) of deep-sea organisms is around one-tenth of that of shallow-water species (Macdonald 1975).

Biologists have been surprised at the high diversity of deep-sea faunas and theories to account for it range from the stability–time hypothesis of Sanders (1968) to the biological disturbance theory of Dayton and Hessler (1972). In the former, physical stability of the environment is thought to promote diversification, while in the latter biological disturbance such as cropping is though to be responsible. Both theories consider some species to be immigrants. In the absence of long-term studies the explanations remain controversial. However, as far as benthic foraminifera are concerned, the evolution of deep-sea species is slow (many modern species go back to the Miocene and some to the Palaeogene). In the absence of seasonality in the environment (except perhaps the input of phytodetritus, Gooday 1988, Tyler 1988), with low densities and high diversity, asexual reproduction may be the norm and sexual reproduction rather rare. Under such circumstances evolution would be slow.

Notwithstanding the high diversity only a small number of associations have been recognised. Three characterised by epifaunal taxa are widely distributed (Table 18.3). The *Epistominella exigua* and *Fontbotia wuellerstorfi* associations are also present in the Norwegian Sea and Gulf of Mexico but are absent from the Mediterranean. The *Nuttallides umboniferus* association occurs at somewhat shallower depths in the Weddell Sea (mainly 3374–3800 m) and McMurdo Sound (3549–3932 m) than it does in the ocean basins (generally >4000 m). In the Pacific it occurs principally between the lysocline and the CCD.

Three associations characterised by infaunal taxa, namely the *Globocassidulina subglobosa*, *Melonis barleeanum* and *Uvigerina peregrina* associations, are

Table 18.3 Distribution of epifaunal associations and deep-sea water masses in the oceans

Association	Atlantic	Indian	Pacific	Southern (areas of occurence)
Epistominella exigua	NEADW	–	Pacific Deep Water	Weddell Sea McMurdo Sound Amundsen/Bellingshausen
Fontbotia wuellerstorfi	NADW	NADW	–	–
Nuttallides umboniferus	AABW	AABW/IBW	AABW	Weddell Sea McMurdo Sound

also widely distributed in areas having sediments with elevated organic carbon contents. Unlike the epifaunal associations, they are less clearly related to water masses. At depths greater than the local CCD, exclusively agglutinated assemblages are present.

Oceanic sedimentation rates are low and this is of significance in connection with infaunal taxa. Corliss (1985) found living individuals down to a depth of 15 cm. Taking a reasonable sedimentation rate of 2.5 cm per 10^3 years this means that at 10 cm below the surface the modern infauna is being added to the dead epifauna of 4000 years ago. This has consequences for the interpretation of both faunal change and the iostopic record of fossil deposits.

Planktonic : benthic ratio

The association of planktonic and benthic foraminifera in sediment reflects two separate and independent sources of material. The benthic forms may have lived and died in the depositional area and may or may not have been subject to loss or gain through transport. By contrast, planktonic forms live in the water column and their accumulation in bottom sediments is *always* the result of sedimentation, that is, transport and deposition. On a continental slope not subject to significant current activity all sizes of planktonic tests may be sedimented beneath the water column in which they lived. In other cases planktonic foraminifera are transported by water currents into unfavourable environments, for example, on to a continental shelf. In this case they die and the empty tests settle to the sea floor according to their relative rate of settling (which is size dependent) and the amount of lateral transport. This is important because the bottom sediment in such an area may contain an indigenous unsorted benthic assemblage of wide size range together with an allochthonous planktonic assemblage which is very well size sorted.

From an analysis of this topic Murray (1976b) concluded that the following generalisations can be made:

1. Upper continental slope: wide range of sizes with juveniles and adults, maximum diversity for the latitudinal position, high planktonic:benthic ratio (>70 : < 30);
2. Outer shelf (open sea): wide range of sizes with juveniles and adults of some species, juveniles only of others, somewhat lower diversity than the adjacent slope, planktonic:benthic ratio 40–70 : 60–30;
3. Middle shelf (open sea): high proportion of juveniles with only a few species represented by adults, low diversity, planktonic:benthic ratio 10–60 : 90–40;
4. Inner shelf (open sea): almost exclusively juveniles, very low diversity, planktonic:benthic ratio <20 : >80

Very wide shelves and enclosed epicontinental seas have few (juvenile) planktonic forms.

Tau value

Gibson (1988) noted that whereas the percentage of planktonic tests generally shows a progressive increase across the shelf and down the slope, the number of species of benthic foraminifera may level off on the outer shelf and fluctuate on the slope. He introduced the value tau (number of benthic species × percentage of planktonic tests) as a depth indicator. Using data from the Gulf of Mexico he found the following: tau <100 at depths <40 m, increase to 1000 from 40 to 1000 m deep, increase to 10 000 at depths of 1000–2000 m. Each geographic area may have its own absolute values but relative depth changes can be determined by this method. Of courses, on broad shelves and epicontinental seas low values of tau might be associated with depths considerably greater than 40 m. For example the Ekofisk area of the North Sea (depth 70 m) yields tau values of 0–4 while the Forties area (depth 100–130 m) close to the ocean has values of 532–1288 (data from Murray 1985).

CHAPTER 19

Palaeoecology

To a large extent palaeoecological studies are based on the assumption that 'the present is the key to the past'. In this context it is necessary to consider two questions: How representative are modern environments? Are the ecological relationships seen in modern faunas applicable to the fossil record?

How representative are modern environments?

Several aspects need to be considered starting with the most recent and working back through time.

Man's influence

For several thousand years man has caused environmental change. Early clearance of land for cultivation led to increased soil erosion which aided the silting up of estuaries and increased the supply of nutrients. Since the Industrial Revolution there has been a major change in the input of nutrients from fertilisers and sewage and this is having a profound effect on the ecosystem. Walsh (1988) has documented the consequences of this eutrophication. For instance in the North Sea there has been an increase in biomass of various components of the phytoplankton. From simulation models involving oxygen, nitrogen and carbon budgets it is suggested that these increases in algal biomass lead to anoxia, denitrification and increased production of hydrogen sulphide. Such events may just be man-induced analogues of natural processes but others, such as the introduction of toxins and heavy metals, are not. Pollution of the sea is an increasing problem and for this reason comments have been made on it in Chapters 6–17.

Postglacial sea-level rise

Following the last glacial event sea-level has risen by ~125 m in the past 10 000 years. Former land areas have been flooded to form broad continental shelves. Rivers formerly adjusted to a lower sea-level have adjusted their profiles by silting up their valleys. In many areas little sediment now reaches the shelves and so they are regions having a veneer of relict sediment formed during the initial transgression and not yet buried because of the low sedimentation rate. In geological terms the shelf and coastal environments are very young. This may help to explain why the faunas of discontinuous environments, such as estuaries, each separated from the next by normal marine inner shelf, are similar to one another. There has been insufficient time for evolutionary changes to occur even though the faunas are isolated one from the other. Of course, there is the alternative that brackish taxa are represented by stenohaline ecophenotypes on the inner shelf.

Origin of oceanic bottom waters

The basic principles governing physical oceanography can be assumed to be constant but the intensity of activity must have varied through time. Important controls are the shapes and depths of the ocean basins, which are the result of plate tectonic processes, and climate. The principles of palaeoceanography and the history of the ocean basins are concisely summarised by Kennett (1982). Major differences between ancient and modern oceans include the development of widespread anoxia at certain times in the Mesozoic, and a less pronounced thermohaline structure prior to the late Eocene with bottom waters possibly forming in low latitudes through evaporative increase in density. In modern oceans bottom waters form in high latitudes (especially around Antarctica and in the North Atlantic) due to cooling which leads to sinking of the denser water. The first introduction of cold water into the bottom of the oceans was a late Eocene–early Oligocene event (development of the psychrosphere) associated with climatic cooling. The intensity of formation of cold bottom water and of the rate of circulation in the deep ocean is controlled by the climate, being greatest during colder periods.

Are the ecological relationships seen in modern faunas applicable to the fossil record?

The validity of this assumption depends on the timescale being largely true for the Neogene, less so for the Palaeogene and progressively less so for earlier times. This is for two reasons. First, many of the major taxonomic group either did not exist or were less important in pre-Cenozoic time. Second, some taxa appear to change their ecological requirements although in reality they may just have undergone niche expansion or niche contraction. Thus Van der Zwaan (1983) suggests that competition limits the realised niche; where there are numerous competitors a species will inhabit a more restricted niche than where there are few competitors. Probably for this reason *Planulina*, *Siphonina* and *Uvigerina* colonised the deep sea from a shelf origin (Van Morkhoven *et al.* 1986). Differences of niche size are clearly evident in the Lagenina and in *Nummulites* which occupied large ranges in the Jurassic and Eocene respectively.

The general attributes of assemblages, such as trends in diversity, wall structure and role of post-mortem influences, should be applicable from the late Palaeozoic onwards. However, as discussed below, the isotopic signature of seawater is controlled both by temperature and the volume of water locked up as ice and consequently the isotopic record varies through time.

It may therefore be concluded that modern environmental data can be applied to the interpretation of palaeoecology as long as these differences are borne in mind. The detail and accuracy of environmental reconstruction will inevitably decrease with progressively older fossil examples.

Methods of interpretation

Many authors make direct comparison between their fossil material and a selection of papers describing the ecology of modern environments. This generally produces adequate results as can be seen from several of the case-studies described below.

For Neogene assemblages containing a high proportion of extant species, direct comparison can be made with the modern data base using multivariate methods (cluster analysis, principal coordinate analysis, etc).

As the major part of this book is a compilation of data on modern environments, emphasis is placed here on methods that make use of this.

Procedure

Unlithified sediments can be processed to give whole specimens. Lithified sediment may be studied in thin section, acetate peels (for limestone), or as polished and lightly etched rock surfaces using back-scattered imagery in a scanning electron microscope. Assemblage counts should be made to determine diversity, percentage occurrence of each benthic species, size distribution and nature of preservation, and planktonic: benthic ratio.

Comparison should then be made with Fig. 5.8 to determine the quality of preservation. The most reliable assemblages are those of fields I and II, then I-A and II-A. It may be possible to obtain palaeoecological information from assemblages falling in fields I-C, II-C, III, III-A and III-C. Those from all other fields have undergone too much alteration to make a palaeoecological analysis worth while or meaningful (although recognition of severe bed-load transport will itself be of environmental value).

Method 1 (for Cenozoic assemblages)

Inspect Figs 18.1 and 18.2 and Table 18.1 to determine which environments are possibilities. Look up details of the abundant genera in Appendix B. Compare the planktonic : benthic ratio, maximum size of the planktonic test and tau value with the data in Chapter 18. If the assemblage falls in field II use the proportion of tests $< 200\ \mu m$ as a guide to tidal influence (p. 48).

For Neogene assemblages determine the association based on the dominant species and compare it with its modern analogue.

Method 2 (for late Mesozoic assemblages)

Inspect Figs 18.1 and 18.2 and Table 18.1 to determine which environments are possibilities. Compare the planktonic : benthic ratio, maximum size of planktonic test and tau value with the data in Chapter 18. There may be some limited help from the details of genera listed in Appendix B.

Palaeoecological interpretations

It is beyond the scope of this book to give a résumé of the palaeoecology and Phanerozoic benthic foraminiferal assemblages for the entire world. The main objective is to define principles and to give examples of their application. To this end a selection of case-studies are presented to show the variety of assemblages in time and space. Apart from the interpretations made by the original authors, additional amplification of detail is given where appropriate.

For some fossil assemblages it may be possible to derive ecological data from independent sources (see case-studies 7 and 8 below) and this may then be applied to other fossil assemblages. However, in the absence of such reliable data it is unwise to generate circular arguments by using other authors' interpretations unless they can be substantiated.

Palaeoecological studies are usually of stratigraphic sections and commonly the aim is to determine a sequence or trend of environmental changes (case-studies 1, 2, 5 below). In addition, some attempt to build up a geographic pattern of environmental distributions during short time intervals (case-study 9 below). Both types of study contribute to a fuller understanding of biostratigraphy and evolution by determining whether or not environmental change is the cause of the biotic events.

Case-study 1. Pleistocene interglacial, northwest Germany

The Ludingworth Borehole near Cuxhaven yielded marine interglacial sandy silts between 14.5 and 19 m depth. Glacigenic sediments are present above and below. Knudsen (1988) divided these silts into three foraminiferal assemblage zones from base to top.

Assemblage zone L1

α 3, hyaline, < 60% *Elphidium albiumbilicatulum*; tests showing some dissolution; much plant debris. Field I-C?

This was interpreted as intertidal and representing the start of a marine transgression.

The only modern *Elphidium albiumbilicatulum* association from Europe is from a depth of 10 m in Oslo Fjord (Table 9.3), but since dissolution has affected preservation of the fossil example caution must be exercised in direct comparisons.

Assemblage zone L2

α 1.5–2.5, hyaline, dominated mainly by *Haynesina germanica* (as *Nonion*) with additional *Buccella frigida*, *Elphidium albiumbilicatulum* and *E. williamsoni*; 2.25 m thick.

Interpreted as temperate with a water depth 10–20 m.

This is essentially a *Haynesina germanica* association which is known from marshes (Table 9.2) and lagoons (Table 9.5). These have salinities of 0–35‰ (often with a large diurnal variation), a temperature range of 0–32 °C and depth from intertidal to 20 m.

Assemblage zone L3

α 1–2, hyaline, dominated by *Ammonia batavus* with *Elphidium albiumbicatulum*, *Haynesina germanica* and *E. gunteri*; 2 m thick.

This was interpreted as intertidal with reduced salinities.

This is an *Ammonia beccarii* association from either a marsh (Table 9.2) or lagoon (Table 9.5). The absence of true marsh species such as *Jadammina macrescens* or *Trochammina inflata* and the presence of *Elphidium gunteri* favours a lagoon. The *A. beccarii* association is best developed at salinities >10‰ and temperatures of 15–20 °C at depths of <10 m.

The North Sea adjacent to Cuxhaven is mesotidal and there is every reason to suppose that this was the case during the time of deposition of these Eemian interglacial deposits. Knudsen has chosen to emphasise changes in water depth during a presumed transgression, from 0 to 10–20 m back to 0. However, the deposits are only 4.5 m thick and they have not experienced great post-depositional compaction. A much smaller variation in depth, beyond that due to the tidal cycle, together with a change from tidally very variable brackish (*Haynesina germanica*) to less variable and less brackish (*Ammonia beccarii*) conditions may be a more realistic interpretation.

Case-study 2. Early Pliocene, Mexico

Lithologically monotonous siltstones from around salt domes contain diverse assemblages of foraminifera. Kohl (1985) constructed a summary palaeobathymetric curve using the >194 μm fraction. This shows a progressive shallowing from ~600 m at the base to ~40 m at the top.

The assemblages are exceptionally diverse with α10–45 and most values ≥18. Because most of the taxa are extant, direct comparison can be made with the modern associations (Ch. 7). From the top down the following can be recognised:

Hanzawaia concentrica association (samples 29, 21, Kohl depths 40, 95 m), salinity 34–36‰, temperature 17–31 °C, 20–105 m.
Uvigerina peregrina parvula association (as *U. parvula*, samples 63, 66, 39, 40, Kohl depths 130, 160, 180, 190 m), salinity 35–36‰, temperature 14–20 °C, 77–155 m except for one exceptional range to 960 m).
Brizalina subaenariensis mexicana association (as *B. subaenariensis*, samples 43, 45, 49, Kohl depths 180–200 m), salinity 35–36‰, temperature 5–18 °C, 105–274 m except for one value of 1828 m.
Uvigerina peregrina association (sample 46, Kohl depth 250 m), salinity 35–36‰, temperature 5–15 °C, 35–1417 m.
Brizalina subaenariensis mexicana association (sample 52, Kohl depth 250 m) as above.
Planulina exorna association (sample 4, Kohl depth 250 m), salinity 35–36‰, temperature 18–31 °C, 31–42 m.
Cibicidoides floridanus association (as *Cibicides* samples 58, 72, Kohl depths 250, 400 m), salinity 36‰, temperature 16–22 °C, 67–122 m.

Three samples do not match any modern association: *Bolivina imporcata* association (samples 54 and 35), and *Uvigerina* sp. association (sample 79). Two others are not dominated by any one species.

From this analysis it can be seen that the progressive deepening from sample 29 (top) to 52 is confirmed, but in addition it is matched by a progressive decrease in water temperature. However, near the bottom of the sequence conditions were perhaps shallower (*Cibicidoides floridanus* association) and sample 4 (*Planulina exorna* association) represents a displaced assemblage as recognised by Kohl.

Case-study 3. Miocene–Pliocene, California

The southern Californian borderland is very active tectonically due to its plate tectonic setting. Bathyal Miocene and Pliocene basinal sediments were uplifted above sea-level during the Pleistocene. It has long been known that the sediments were deposited rapidly by gravity mechanisms. It was also established by Bandy (1953b) that a great deal of subsidence took place during deposition. He produced the first subsidence/palaeobathymetric curve, the latter based on foraminiferal studies.

Ingle (1967) examined several sections in detail. He used a reference data base of the depth distribution of dominant living species for the interpretation of the fossil assemblages. He found that in a single turbidite unit the lower-graded sandy unit contained 50% inner shelf, 14% outer shelf, 31% upper bathyal and 2% middle and lower bathyal species, whereas the top muds contained 70% upper and 30% middle bathyal species. Assuming that the middle bathyal species are indigenous, 70% of the fauna in the muds and 95% of those in the graded interval are displaced downslope. No details were given of the size range or quality of preservation (presumably field II) although it was noted that reworked Miocene specimens had 'filled tests, abraded surfaces and a patina of hydrous iron oxide'. However, Brunner and Ledbetter (1987) found that in Pleistocene turbidites the tests were small (generally < 150 μm) and well sorted. The shape varied from platy in silts to more compact and elongate in muds.

Taking one of the studied sections, namely the Pliocene Repetto Hills section, displaced foraminifera average 67% and reach a maximum of 87% in some samples. On the basis of the inferred indigenous component, Ingle determined the water depth to have been 2000–2600 m. The most abundant species overall is *Pseudoparrella subperuviana* (as *Epistominella*) and off Peru this association extends from 365 to 454 m (Table 14.6). If downslope transport had not been recognised the depth would have been grossly underestimated.

Case-study 4. Neogene of the northeastern Atlantic Ocean

The modern faunas from this area fall into four assemblages (defined by varimax factor analysis) and which show a clear correlation with water masses (Table 19.1). Samples from the Miocene and Pliocene of Deep Sea Drilling Project (DSDP) sites >2000 m deep have essentially the same faunas which are readily divided into the same four factor assemblages. From this Murray (1988) was able to demonstrate that an AABW influence has been present to the south of the Azores Ridge from the mid-Miocene to the present, whereas to the north of the ridge it was present only up to the late Pliocene. A spectacular increase in importance is shown by NEADW from the late Pliocene to the present.

Essentially the same results are obtained making direct comparisons with the principal associations defined in Chapter 9.

Table 19.1 Recent assemblages defined by *Q*-mode varimax factor analysis from depths > 2000 m in the NE Atlantic Ocean (from Murray *et al.* 1986)

Water mass	Factor	Principal species
North Atlantic Deep Water (NADW)	1	*Fontbotia wuellerstorfi* *Globocassidulina subglobosa* *Cibicidoides kullenbergi* *Oridorsalis umbonatus*
Antarctic Bottom Water (AABW)	2	*Nuttallides umboniferus*
North East Atlantic Deep Water (NEADW)	3	*Epistominella exigua*
Mediterranean Water (MW)	4	*Cassidulina obtusa* *Globocassidulina subglobosa*

Case-study 5. Late Eocene, northern Italy

The succession in the Possagno area consists of ~600 m of mudrocks capped by algal–coral limestone with larger foraminifera. The sequence is regarded as the product of a rapidly filled sedimentary basin. Grünig and Herb (1980) recognised five ecologic zones showing a progressive upward shallowing.

Ecologic zone 1

α 5.5–11, hyaline; dominant genera *Cibicidoides, Lenticulina, Planulina, Rhabdammina, Uvigerina, Nuttallides*; planktonic component 75–90%.

Primarily on the basis of the high abundance of planktonic tests this was interpreted as representing a water depth of ~1000 m.

With the exception of *Nuttallides* (bathyal–abyssal) the other genera occur from the shelf to the bathyal zone. Von Morkhoven *et al.* (1986) give the upper depth limit of *N. truempyi* as 500–700 m. Tau values are 1800–3330, suggesting depths of somewhat less than 1000 m.

Ecologic zone 2

α 5–9, hyaline; dominant genera *Bolivina, Lenticulina, Rhabdammina, Aragonia, Uvigerina, Heterolepa, Stilostomella*; planktonic component 70–90% in lower, 30–65% in upper part.

On the basis of the reduced proportion of planktonic forms and the occurrence of hispid *Uvigerina* (said to have an upper depth limit of 600 m, Frerichs 1970) Grünig and Herb interpreted the water depth as 600–1000 m.

The extant genera are all found in shelf to bathyal environments, but Van Morkhoven *et al.* (1986) give the upper limit of *Aragonia* as 1000–1500 m.

Ecologic zone 3

α 5–15, mainly hyaline but up to 30% agglutinated; dominant genera *Uvigerina, Trochammina, Tritaxia, Heterolepa, Trifarina, Bolivina, Globocassidulina, Lenticulina, Stilostomella, Bulimina, Gavelinella, Eponides*; planktonic forms 10–50%.

From the high diversity, change from hispid to ribbed *Uvigerina* and reduced planktonic abundance this was interpreted as representing 600 m at the base and 150 m at the top.

Ecologic zone 4

α 9–12, mainly hyaline but up to 30% porcellaneous; dominant genera *Gyroidinoides, Heterolepa, Bolivina, Nonion, Nummulites, Operculina, Eponides*; planktonic <10%.

This zone is 160 m thick but larger foraminifera are

present only in the top 45 m where they form up to 30% of the total. Grünig and Herb considered the depth to be >120–150 m at the base and 30–40 m at the top.

Ecologic zone 5

α 3.5–6.5, mainly hyaline with up to 20% porcellaneous, dominant genera *Quinqueloculina*, *Eponides*, *Heterolepa*, *Asterigerina*, *Pararotalia*, *Nummulites*, *Operculina*, *Discocyclina*; planktonic 0–10% (excluding reworked tests).

Grünig and Herb considered that water depth shallowed from 30–40 m at the base to 0 m at the top. The tau values of 0–250 are consistent with such shallow depths. The transition to the overlying limestones was interpreted as due to the reduced supply of terrigenous material.

These interpretations are consistent with the data on modern foraminifera presented in this book. However, in addition it can be concluded that although salinities probably remained normal, temperatures were lower in the deeper waters of ecologic zones 1–3 than in zones 4 and 5. The latter must have had temperatures in excess of 22 °C for at least part of the year. The upward reduction in sediment supply must have been matched by increased transparency of the water (which Grünig and Herb considered to be marked between zones 3 and 4) and probably a decrease in fertility, both favouring the development of the larger foraminifera.

Case-study 6. Eocene of the Paris Basin, France

From detailed studies Murray and Wright (1974) concluded that four groups of assemblages are present, each representing a different environment.

Assemblage 1

α 5–17, wall structure mainly hyaline, dominant genera *Cibicides*, *Nonion*, *Anomalinoides*, *Eponides*, *Textularia*, *Elphidium*, *Pararotalia* and polymorphinids; cross-bedded and laminated sands.

The diversity suggests normal marine lagoon, shelf or slope, or hypersaline shelf. The wall structure data are not diagnostically helpful. The genera are a mix of infaunal and epifaunal types, with herbivore, detritivore and passive suspension feeders. All have a normal marine shelf environment in common but *Elphidium* and *Pararotalia* indicate inner shelf. Murray and Wright suggested a temperature of 16 °C based on the abundance of *Cibicides* but modern *Pararotalia* is confined to warmer waters. In summary, these assemblages are interpreted as normal marine, warm, inner shelf.

Assemblage 2

α 5–13, rarely up to 24, porcellaneous and/or hyaline, dominant genera *Orbitolites*, *Alveolina* (as *Fasciolites*), peneroplids, *Discorbis*, *Cibicides*, *Nonion*, *Cancris*, *Valvulina* and polymorphinids; bioclastic sands.

The diversity suggests normal marine lagoon, shelf or slope, or hypersaline shelf. The wall structure data narrow this down to normal marine or hypersaline lagoon. The modern analogue of *Orbitolites* (i.e. *Marginopora*) is endosymbiotic, epiphytic, normal marine clear waters, at depths of 0–45 m in tropical lagoons or inner shelf. The modern analogues of *Alveolina* (i.e. *Alveolinella* and *Borelis*) are also endosymbiotic on algal-coated substrates of tropical lagoons or inner shelf. Summer temperatures > 22 °C are necessary for reproduction. The other genera are all indicative of normal marine inner shelf. Murray and Wright concluded that the environment was shallow (0–35 m), salinity 36–40‰, temperature > 22 °C in summer, and with a seagrass/seaweed flora. A palaeogeographic reconstruction of the middle Eocene (Calcaire Grossier) shows an extensive epicontinental shelf/lagoon.

Assemblage 3

α1–3, wall structure mainly porcellaneous, dominant genera *Quinqueloculina*, *Triloculina*, *Spirolina*; sand with cerithiid gastropods.

The diversity indicates a marginal marine environment such as a marsh or lagoon and the dominance of porcellaneous walls confirms this for hypersaline examples. From Table 18.1 it can be seen that peneroplids are not present on marshes and modern *Spirolina* is epiphytic in hypersaline (37–50‰), warm (18–26 °C) shallow lagoons.

Assemblage 4

α <2, wall structure hyaline, dominant genera *Rosal-*

ina and/or *Protelphidium* and *Elphidium* (unkeeled); associated with charophytes, smooth-valved ostracods and *Hydrobia*.

The diversity indicates very abnormal salinities. The absence of porcellaneous walls suggests brackish conditions. The modern analogue of *Protelphidium* is *Haynesina*, a brackish genus. It therefore seems likely that these represent a brackish lagoon although modern *Rosalina* are confined to marine environments.

Case-study 7. Bathyal Palaeocene, central Atlantic

Independent geological and geophysical evidence suggests that site 366 underwent little or no subsidence during the Cenozoic. Its present water depth is 2860 m and the late Palaeocene was cored between 740 and 805 m below the sea floor. Saint-Marc (1986) analysed the >160 μm fraction from 16 samples. From his Table 2 it can be seen that *Nuttallides truempyi* is generally most abundant and that other species ≥ 10% include *Gyroidinoides* spp., *Cibicidoides* cf. *C. alleni* and *Gavelinella beccariiformis*. Thus, this is a Palaeocene bathyal assemblage from ~2000 m water depth. According to Van Morkhoven *et al.* (1986) *N. truempyi* and *G. beccariiformis* have their upper depth limits at 500–700 m while that of *C. alleni* is at 25–100 m.

Case-study 8. Late Cretaceous, eastern USA

This passive margin has undergone subsidence without complex structural deformation. Olsson and Nyong (1984) argued that during Campanian and Maastrichtian times the palaeoslope was uniform in the New Jersey–Delaware area because the structure contours are parallel and uniformly spaced. Hence distance down dip would provide a constraint on the determination of palaeodepth.

Following log transformation of species abundances to reduce the effects of faunal dominance, five biofacies were recognised using *R*-mode cluster analysis. These were interpreted as depth-related, and together with data on benthic diversity (α index), proportion of planktonic tests, and morphological trends, used as a basis for a palaeodepth model. Adjacent biofacies overlap in faunal composition and water depth.

Biofacies 1

α 4–6, *Gavelinella pinguis*, *G. nelsoni*, *Lenticulina pseudosecans*, *Globulina gibba* with *Citharina multicostata*, *C. suturalis*, *Cibicides harperi* and *Nonionella* sp.; low planktonic abundance, mainly juveniles.

Interpreted as inner shelf, 10–50 m.

Biofacies 2

α 5–10, *Clavulina trilatera*, *Praebulimina carseyae*, *P. aspera*, *Hoeglundina supracretacea*, *Gaudryina rugosa*, *G. depressus* and *Pullenia americana*; 8–25% planktonic.

Interpreted as mid shelf, 50–100 m.

Biofacies 3

α 8–13, *Bolivina incrassata gigantea*, *Coryphostoma plaitum*, *Gyroidinoides cretacea*, *G. nitidus*, *Loxostomum eleyi* and *Bolivinopsis rosula*; 30–70% planktonic.

Interpreted as outer shelf, 100–200 m.

Biofacies 4

α 6–9, *Dorothia oxyconica*, *Bathysiphon vitta*, *Loxostomum eleyi*, *Osangularia cordieriana* and *Pseudogaudryinella capitosa*; no information on planktonic abundance.

Interpreted as upper slope (200–400 m) due to the reduced abundance of praebuliminids and gavelinellids, the absence of shelf forms such as *Gyroidinoides* and the incoming of deep-water agglutinated forms.

Biofacies 5

α 6–8, *Pullenia cretacea*, *Stensioeina exculpta*, *Osangularia cordierana*, *Gavelinella ammonoides* with *Cibicides* sp.; *Verneuilina* sp.; *Heterostomella americana* and *Gyroidinoides globosa*; planktonic average 90%.

Interpreted as representing 400–800 m water depth.

Comparison of these interpretations with the data in this book shows the following:

1. The trends in diversity are consistent with a deepening from biofacies 1 to 3.

2. The trends in size and abundance of planktonic tests show a deepening from biofacies 1 to 5.
3. All the biofacies appear to represent normal marine salinities.
4. Comparisons based on genera are not very fruitful because many have no living representatives.

Thus, the general trend of increasing depth shown by the five biofacies is confirmed but the absolute depths cannot be. Nevertheless, as this model is well constrained by the structural simplicity of the area it represents a well-argued palaeoecological interpretation.

Nyong and Olsson (1984) using data from COST (Continental Offshore Stratigraphic Test) and DSDP sites extended their model into the abyssal zone. As with modern forms, many of Cretaceous species extended over very great depth ranges.

Case-study 9. Middle Jurassic, England

The lower Bajacian is developed as a series of limestone from which whole foraminifera have been extracted (Morris 1982). With two exceptions all the assemblages are dominated by *Lenticulina muensteri* and only six other species occur in >10% abundance. Individuals are size sorted with respect to lithology, being small in the marls and ~1 mm in coarse oosparites. Apart from winnowing there is mixing of algal and sediment-dwelling foraminifera.

This is a good example of the effects of postmortem transport on Mesozoic fossil assemblages.

Case-study 10. Jurassic, North Sea

The Toarcian Drake Member of the Dunlin Formation consists of laminated and silty clays with ammonites, belemnites and glauconite, and is considered to have been deposited in pro-deltaic environments. The foraminiferal assemblages are almost entirely agglutinated and five genera are dominant: *Trochammina* 5–100%, *Verneuilinoides* 0.1–43%, *Lagenammina* 0.1–60%, *Reophax* 0.1–20% and *Ammobaculites* 0.1–33%. Nagy *et al.* (1984) consider that salinity was near normal because of the belemnites and that high turbidity and rapid sedimentation were the principal controls on the agglutinated assemblages. They noted that the few calcareous taxa present show no signs of dissolution. Following a comparison with modern deltaic assemblages they concluded that perhaps Jurassic nodosariids were not adapted to deltaic environments and they pointed out that all calcareous genera reported from modern deltas originated after the Jurassic.

If it is assumed that the agglutinated assemblages are original, what can be learned of the environment? *Trochammina* is very euryhaline but *Lagenammina* (30–35%) and *Reophax* (marine) suggest near-normal salinities (cf. Appendix B). In terms of temperature, the only common denominator is temperature, i.e. around 10–20 °C. All are found from shallow depths to bathyal. Thus, they could represent a pro-delta environment but it is difficult to be certain from the limited data.

Stable isotopes and palaeoceanography

The stable isotope record is invaluable in the study of palaeoceanography. In the early phase of development of ideas, oxygen isotopes were thought to be indicators of palaeotemperature. Epstein *et al.* (1953) defined the relationship as

$$T = 16.5 - 4.3\,(\delta - A) + 0.14\,(\delta - A)^2,$$

where T is the temperature (°C), $\delta = \delta^{18}O$ of the sample and A the isotopic composition of seawater. To solve the equation for T, A must be known. The relative importance of T and A is still much discussed (Berger 1981). It is now generally believed that the isotopic composition of seawater has changed over time and that the most likely cause is the 'ice effect' (Imbrie *et al.* 1973; Shackleton and Opdyke 1973). The light isotope is preferentially stored in ice so during glacial periods the oceans become enriched in ^{18}O. The magnitude of this effect is estimated to be $\delta^{18}O$ 0.9–1.4‰ (Olaussen 1965; Shackleton 1967, 1986).

Changes in ice volume cause globally synchronous changes in seawater $\delta^{18}O$. Therefore, a downcore record of $\delta^{18}O$ is a measure of ice volume and the pattern of change is a potentially valuable stratigraphic tool (Shackleton 1967). However, as already noted in Chapter 3, isotopic fractionation is caused by vital effects and different species show varying degrees of disequilibrium. In order to interpret the $\delta^{18}O$ record of benthic foraminifera, two assumptions must be made: that the disequilibrium factor for a given species has not changed through time, and that the adjustment factors used to standardize species to a common scale (Table 19.2) are correct. Zahn *et al.* (1986) have ques-

Table 19.2 Adjustment factors (from Shackleton and Hall, 1984)

Species	^{18}O	^{13}C
Uvigerina peregrina	0.0	0.9
Fontbotia wuellerstorfi	0.64	0.0
Globocassidulina subglobosa	–0.1	0.5
Hoeglundina elegans	–0.4	–1.3
Melonis pompilioides	0.3	0.6
Melonis barleeanum	0.3	1.0
Nuttallides umboniferus	0.0	1.0
Sphaeroidina bulloides	–0.1	–0.1

tioned these assumptions for the *Uvigerina peregrina* group.

A further complication is that in comparison with present-day values, those of the North Atlantic glacial events deviate more than those of Pacific glacial events. This suggests a cooling in the glacial deep North Atlantic (Shackleton 1986), thus adding a temperature effect. Such problems will have to be resolved before a 'standard' benthic oxygen isotope record can be established. Shackleton (1986) believes that detailed records of species other than *Uvigerina* spp. should be obtained from the Pacific. This would 'ensure that there are no time-dependent or water-mass-dependent departures from isotopic equilibrium in any species. When this has been done, it should be possible to create a standard deep Pacific oxygen isotope record with better than 1 ka resolution and better than 0.05 per mil (‰) accuracy.'

It follows that if the temporal changes in the oxygen isotopic composition of the oceans is due to the ice effect, then there should be a correlation between the oxygen isotope value and glacio-eustatic sea-level. A figure of $\delta^{18}O$ 0.1‰ per 10 m of sea-level change (Shackleton and Opdyke 1973) has been confirmed by independent evidence from Barbados (Fairbanks and Matthews 1978). Nevertheless, the estimation of former sea-level is sometimes complicated by temperature effects. For example, in the Pacific the record in isotope stage 5 is thought to include a $\delta^{18}O$ 0.4‰ component equal to a cooling of 1.5 °C. Thus, some interglacials may have had a similar amount of land ice to today, yet colder deep waters (Shackleton 1986).

In a recent review of the deep-sea carbon isotope record Berger and Vincent (1986) pointed out that 'it is a many-factor problem, usually much more complex than the oxygen-isotope record. A multitude of factors cannot be constrained sufficiently by considering one or two stratigraphic signals only. Multi-species, perhaps multi-variant records are necessary, as well as background information on carbonate and organic carbon accumulation and preservation.' The interpretation of the record from benthic foraminifera must be set in this context.

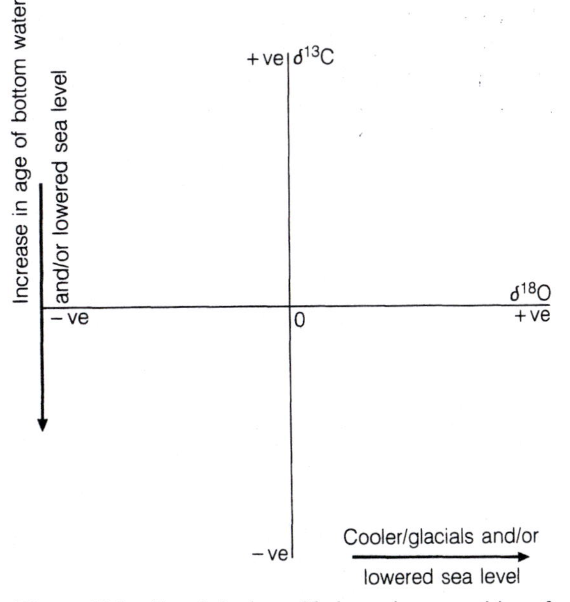

Figure 19.1 Trends in the stable isotopic composition of benthic foraminifera in relation to sea-level, temperature or age of bottom water

The first detailed record of $\delta^{13}C$ of benthic foraminifera was from the subantarctic Pacific (Shackleton and Kennett 1975) and showed that, throughout the Cenozoic, the benthic values were lower than those of the planktonic foraminifera. As already discussed, this is due to surface-water productivity leading to relative depletion in ^{13}C at depth. But apart from this vertical fractionation, there is also an ocean circulation influence leading to basin fractionation. Superimposed on this, the record showed long-term fluctuations which resemble those of sea-level curves. There is ^{13}C enrichment of the oceans during periods of high sea-level. This is attributed to the role of the external reservoir of carbon.

In summary, if the disequilibrium factors of a given benthic species are constant through geological time, trends may be cautiously interpreted as a Fig. 19.1.

Conclusions

The case-studies discussed above give an indication of the varying amounts of detail which it is possible to achieve in palaeoecological interpretations. Ideally, such studies should be carried out as part of an integrated programme with sedimentological and geochemical (especially stable isotope) analysis as is normally the case in deep-sea studies. The recognition of environmental change has much to add to our understanding of basin subsidence, eustatic changes of sea-level, past water masses, palaeoclimatology and hence to evolution (causes of local appearances and extinctions) and to biostratigraphy.

APPENDIX A

Methods

Sampling

It is impractical to carry out a census of the living foraminifera in, for example, an entire estuary or lagoon. Therefore, it is necessary to take samples of manageable size in the hope that they will be representative of the region being studied. All sampling methods are inaccurate to some degree and all ecologists have limited time and money which restricts the number of samples taken. The reviews by Holme and McIntyre (1984) and Andrew and Mapstone (1987) give details of the background theory to sampling. However, in many cases studies of foraminifera are carried out alongside others with different objectives, and the opportunity to take samples has to be taken as it arises. Only rarely is it possible to plan a sampling programme solely for benthic foraminiferal ecology.

Sampling modern foraminifera

High-quality results can be obtained only from samples that are collected and handled with care. The objective of the study may be a qualitative survey of the relative abundance of the species present, a quantitative assessment of the number per unit area for each species or both.

Types of sampler

A comprehensive review of the equipment available is given by Holme and McIntyre (1984). For foraminiferal studies a sampler should ideally take a representative sample of known area, and the sample should not suffer loss through washing on its passage from the sea floor to the deck of the ship.

Many authors have used the small gravity corer designed by Phleger (1951). This takes a core with a cross-sectional area of 10 cm^2. The top 1 cm is normally sliced off giving a sample of 10 cm^2 volume. Corers are generally suitable only for sampling cohesive sediments such as muds or muddy sands. They are less satisfactory for clean sands and gravels. Also, because they free fall into the sediment, they are preceded by a pressure wave which may wash some of the surface layer just prior to impact. A loss of a 1 mm strip around the edge of the 3.6 cm diameter Phleger corer would reduce the area sampled from 10 to 9 cm^2.

The author uses a sampler which is lowered gently on to the sea floor; this is activated by the cable going slack causing the sample bucket to scrape off the top 1 cm of sediment over an area of 100 cm^2, thus giving a sample of 100 cm^3 volume. The bucket is sealed against a closure plate so that no loss through washing takes place (Murray and Murray 1987).

A third type of sampler suitable for quantitative studies is a multiple corer mounted on a frame. The core tubes are slowly pushed into the sediment after the sampler has come to rest on the seabed. The tops and bottoms are sealed by valves which prevent loss through washing (Holme and McIntyre 1984: 221). As with other corers this works best on muddy substrates.

For qualitative studies of shallow-water environments various types of grabs may be used. However, these have imperfect closure so that loss of sample by washing is a common problem. The Shipek grab is particularly prone to such loss of the surface sediment layer.

In the intertidal zone it is convenient to use a 1 cm high ring of plastic which can be pressed into the sediment, then a plastic plate is slid underneath to isolate the sample. More refined instruments to achieve this have been described by Hoskin (1974) and Chang (1987).

Some studies make use of samples collected by scuba divers. A comparative analysis of samples collected by divers and those collected with a Smith–McIntyre corer has shown no significant differences (Vénec-Peyré *et al.* 1981).

Area of sample

Douglas (1979), using methods described by Dennison and Hay (1967), determined that a sample of 90 cm^2 is necessary to be reasonably certain of collecting a species with an adult diameter of 500 μm and a density on the sea floor (standing crop) of 0.01%. The standard sample of 10 cm^2 is too small to give reliable results for rarer species and for regions of low standing crop.

Preservation

Formalin: dissolve one part of buffered formalin in nine parts water to give a 4% solution. Walker *et al.* (1974) advised adding calcium chloride to achieve neutrality.

Place the sample in a jar of twice the sample volume, top up with filtered seawater to a level four-fifths of the jar volume, add neutralised formalin in an amount equal to 10% of the water in the jar, seal the jar and shake to mix the contents (Boltovskoy and Wright 1976).

Ethanol: place the sample in a jar, add an equal volume of ethanol and shake to mix (author's method).

Staining methods

Rose Bengal has been widely used to differentiate living from dead foraminifera since the technique was first described by Walton in 1952. Protoplasm is stained bright red whereas test walls and organic linings are either unstained or take on a light pink coloration. The stain is not vital, that is, it does not detect life and indeed most samples are fixed in a preservative before the staining is carried out. In general, it is assumed that tests containing protoplasm were either alive or only recently dead at the time of collection, although Boltovskoy and Lena (1970b) have demonstrated that protoplasm may survive for weeks or months. Of course, it is necessary to check that the red colour is not caused by clusters of bacteria or other organisms using the test as a refuge.

The technique works best for smaller hyaline foraminifera, although even for these the protoplasm of only the last few chambers may be stained. Thick hyaline tests, many agglutinated tests and thick porcellaneous tests obscure the stained contents when dry. This problem can be overcome by wetting such individuals with a moistened brush or by examining the sample wet if such forms are abundant. In some cases, tests may be broken open to examine the contents. The problem with some agglutinated forms is that the protoplasm may occupy as little as 10% of the test volume (Altenbach 1987).

The use of rose Bengal has been criticised by various authors and some of their comments have been summarised by Walker *et al.* (1974). Most of the difficulties arise form the fact that dead and even broken tests take up some colour (Reiter 1959; Green 1960; Boltovskoy 1963) but, as stated above, this is rarely as intense as that of protoplasm. However, there is some variation in response between species. Martin and Steinker (1973) observed that although the dead tests of some species did not take up rose Bengal stain, both empty and protoplasm-filled tests of the agglutinated species, *Valvulina oviedoiana* were stained to the same degree. They also noted that some miliolid tests are strongly stained on the outer surface and such forms should not be considered as 'living'. Finally, some species are believed to stain green rather than red (Lutze 1964a).

A variety of stains were tested in controlled experiments by Walker *et al.* (1974). The best results were obtained using heated acetylated Sudan black B. This stained the last six or seven chambers dark blue-black in 96.8% of the specimens tested (compared with 70.3% using rose Bengal), and furthermore, no dead tests were stained. However, this technique is time-consuming and the authors recommended using heated saturated Sudan black B for bulk sediment samples although the stain penetrates only the last two to seven chambers. Nevertheless, it is difficult to recognise staining in agglutinated species but easy to do so if rose Bengal is used (Bernhard 1988).

A different approach is to carry out an ATP assay. Adenosine-5'-triphosphate (ATP) is present in all living organisms but on death it immediately degrades. Bernhard (1988) determined that in Antarctic sediments, living foraminifera (established by ATP assay) averaged 30% of the total assemblage. Subsamples which were heat killed and then stained gave living values of 18.5% using Sudan black B and 47.2% using rose Bengal. Thus, the former underestimates the living content and the latter overestimates it when previously dead protoplasm-containing tests are present.

Some shallow-water assemblages have naturally coloured protoplasm so it is not necessary to stain the sample (Le Calvez and Cesana 1972). However, the same problem of recently dead tests being treated as living will arise.

In conclusion, it must be emphasised that most of the criticism of staining techniques is because they are not perfect, that is, they do not stain just living forms but also some that have recently died. Nevertheless, at present there is no practical alternative to staining techniques and as long as there is an awareness of their limitations they still offer the best hope of distinguishing 'living' from dead tests.

Rose Bengal technique (after Walton 1952)

Dissolve 1 g of rose Bengal in 1 l of distilled water.

1. Wash the sample on a sieve.
2. Place the residue in a bowl, add rose Bengal solution, stir thoroughly and leave for at least 30 min.
3. Wash on the sieve to remove the surplus stain.
4. Dry at 60–100 °C.

Heated saturated Sudan black B (after Walker et al. 1974)

Add Sudan black B to 70% ethanol to make a saturated solution (~10 g l^{-1} of 70% ethanol).

1. Fix specimens in a solution of formaldehyde plus calcium chloride for 30 min (50 ml of 40% formaldehyde and 2 g calcium chloride in 1 l of filtered seawater, buffered to pH 8.3 using a buffer salt mixture).
2. Wash with distilled water.
3. Add stain heated to 40 °C and place in a water bath at this temperature for 30 min.
4. Siphon or decant excess stain, and wash sample with 70% ethonal.
5. Dry.

Heated acetylated Sudan black B (after Walker et al. 1974).

Dissolve 1 g of Sudan black B in 100 ml of diethyl ether, and filter solution through Whatman No. 1 paper. Heat the solution to boiling, add 0.5 ml acetic anhydride in 20 ml ether. Reflux the solution for 20 min. Cool, then filter. Transfer the mixture to a separation funnel, and extract repeatedly with cold water until the aqueous layer is no longer coloured, and not appreciably acidic when tested with universal indicator paper. Pour the final solution into an evaporating dish and allow the ether to evaporate. Scrape the metallic black crystals of acetylated Sudan black B from the dish, and prepare a 1% solution in 70% ethanol (yields ~100 ml of working stain solution).

1. Immediately upon collection, fix the samples for 30 min as described under the previous method.
2. Wash sample in distilled water.
3. Add stain heated to 40 °C and place in a water bath heated to 40 °C for 30 min.
4. Siphon, or decant, stain solution, and wash sample repeatedly with 70% ethanol.
5. Dry.

Laboratory processing

Tip the sample into a clean 63 μm (240-mesh) sieve and gently wash with tap-water to remove the preservative and fine-grained sediment. Place the residue retained on the sieve into a container (I use a stainless steel saucepan), stain (see above), wash again on a 63 μm sieve, return the residue to the container, dry at 60–100 °C and allow to cool.

In some cases it is necessary to concentrate the foraminifera using a heavy liquid separation. Heavy liquids are harmful and should only be used in a fume chamber. Use either carbon tetrachloride or trichloroethylene. It is necessary to set up a filter funnel in a retort stand over an empty beaker. Place a filter paper (labelled with the sample number) in the filter funnel. Put the cooled dried residue into a separate beaker, add two to three times its volume of heavy liquid and stir with a rod. The foraminifera will float to the surface to form a scum. Gently decant this into the filter paper. Add more heavy liquid and repeat the process until no more material is floated off. Remove the filter paper from the filter funnel and allow it to dry in the fume chamber. Tilt the beaker containing the residue on to its side and allow it to dry in the fume chamber. The heavy liquid in the beaker beneath the filter funnel can be reused. When dry the two parts of the sample should be stored in separate containers.

The flotation technique works well on smaller foraminifera which have hollow chambers. It is less effective for thick-walled, infilled or heavy tests. The residue should always be checked for these.

Note: Some authors use 125, 149 or even 250 μm sieves, but the case for using a fine (63 μm) mesh has been clearly stated by Schröder *et al.* (1987).

Data collection

Assemblage counts

For quantitative studies where it is desired to know the number of individuals per unit area (standing crop), it is necessary to count all the sample or a known proportion of it (assuming the sample is of known area). The easiest way is to weigh the foraminiferal concentrate and then take a weighed portion, e.g. one-half or one-third.

It is a matter of observation that when 250 or more individuals are counted, the relative proportions of the component species are reasonably constant. There is no point in counting large numbers (e.g. <500 individuals) from a sample if there is no gain in accuracy. Therefore, when deciding on the size of the subsample to be counted, it is sensible to select a size which will give >250 living tests. (However, it should be noted that Pielou (1979), disputes that a sample of even 300 individuals is adequate.)

The sample should be thinly spread on the gridded picking tray and individuals should be picked systematically either up and down the columns or from side to side along the rows. I normally mount living individuals in the top rows of a slide and dead individuals in the bottom rows. To obtain the standing crop, all the living individuals must be picked from the sample or subsample. The first 250 dead forms encountered will suffice for the dead assemblage. If it is desired to know the ratio of living to dead the number of living picked by the time 250 dead forms have been picked must be recorded. Likewise, to record the planktonic : benthic ratio, the proportions of each should be determined.

If the observer is familiar with the names of all the species encountered, it may not be neccessary to mount the material on a slide. It may suffice to count the individuals using a counter. However, great care must be taken to ensure the greatest possible accuracy. Also, if size data are required these must be collected during counting.

Biomass

One method of measuring biomass is to calculate the volume of the test. It is easier to approximate the complex shape of foraminifera to simple geometrical shapes than to develop a mathematical formula. Many are close to prolate or oblate sphaeroids, spheres or cones. As size increases arithmetically, volume increases logarithmically (Figs A.1, A.2). The simplest method is to divide given shapes into size groups, to select the middle-sized specimen of each group, to determine its size and volume and to multiply by the number in the group. The biomass is the sum of the group volumes of a given sample.

Relative and absolute abundance

Relative abundance refers to the proportion of a species of the entire living or dead assemblage, e.g. percentage. Absolute abundance refers to the number of individuals in a unit area of sea floor or volume of sediment. Standing crop is the absolute abundance of living forms. The foraminiferal number of Schott (1935) is the absolute abundance of foraminifera in 1 g dry weight of sediment.

It is important to realise that relative and absolute abundance measurements give different information. Peaks of abundance recorded by one method may not correspond with those of the other method. If a series of stations along a transect have a progressive increase in absolute abundance from one end to the other, a given species having a uniform absolute abundance throughout will have a higher percentage where the absolute abundance is low than where it is high. Likewise, another species increasing in absolute abundance at the same rate as the entire assemblage will, nevertheless, appear to have a uniform percentage occurrence.

Despite these difficulties, both methods are useful. Most dead assemblages are considered only in terms of relative abundance. So too are living assemblages from samples of unknown area. Where the sample area is known both relative and absolute methods may be used.

Species diversity

In its simplest form, diversity is a measure of the number of species (richness) in a sample. Ecologists

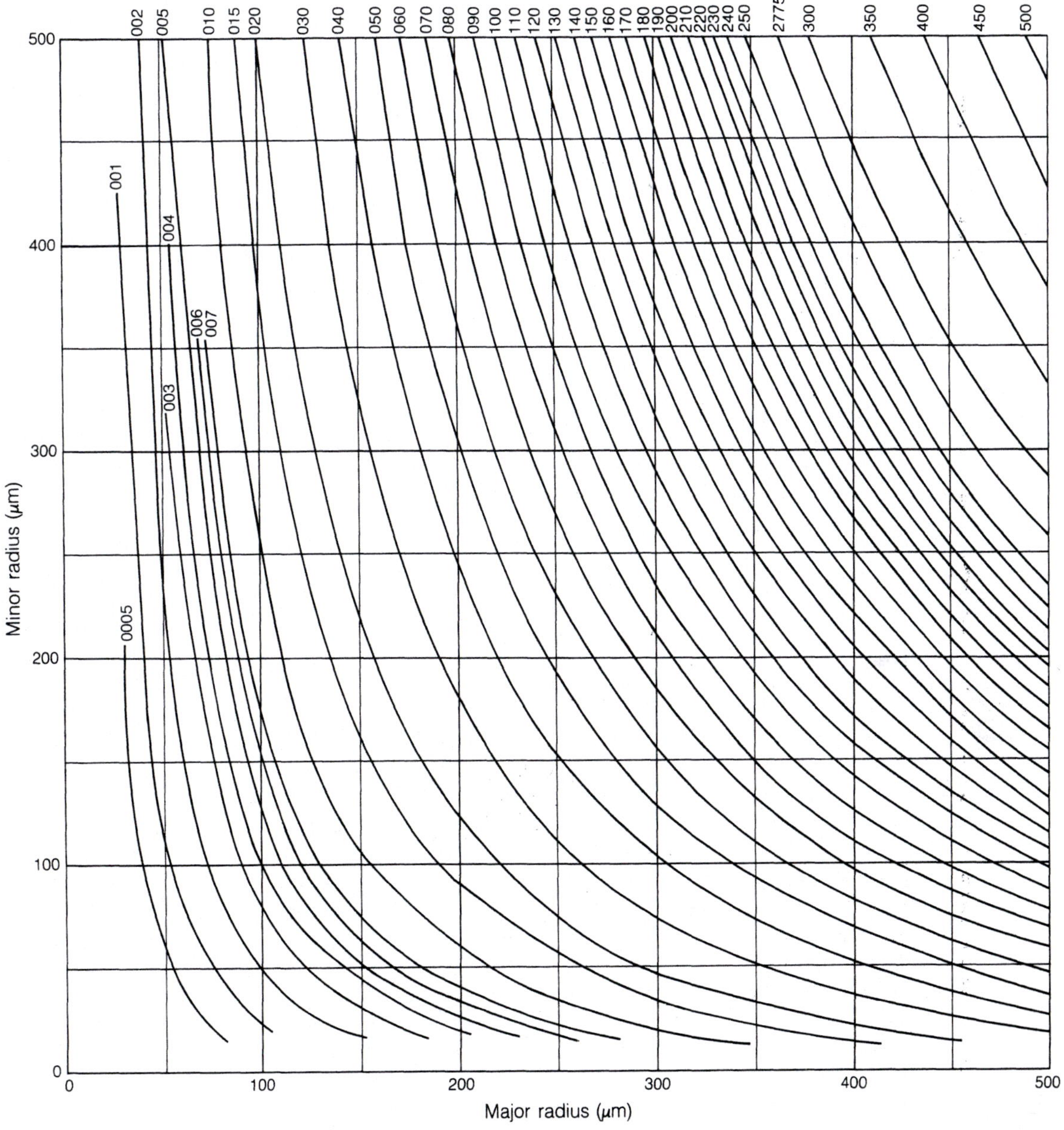

Figure A.1 Size–volume graph of sphere, oblate and prolate sphaeroids. Volume contours are in decimal parts of a cubic millimetre, e.g. 002 = 0.002 mm^3 (from Murray 1973)

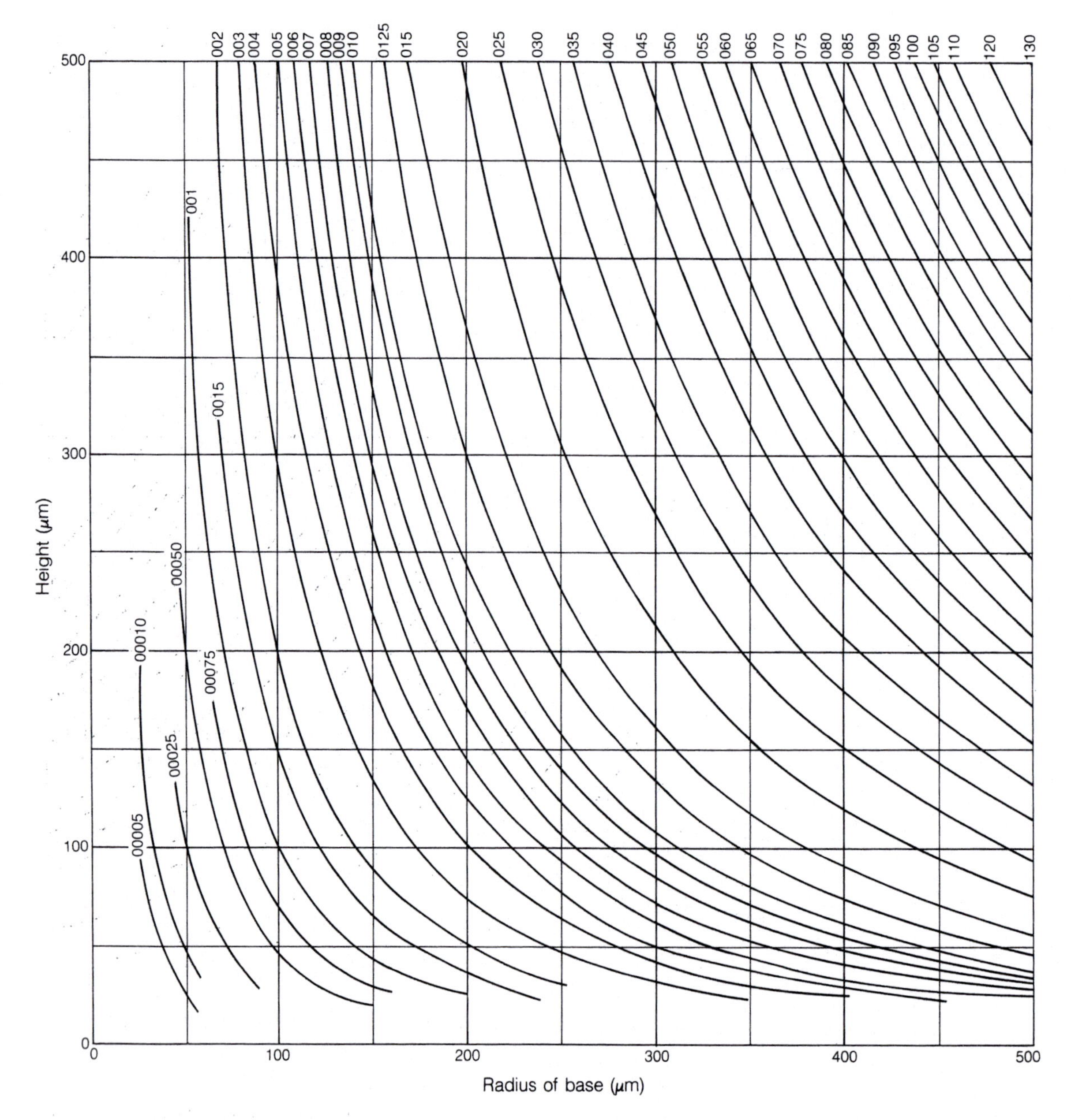

Figure A.2 Size–volume graph of cone. Volume contours as Fig. A.1 (from Murray 1973)

are also interested in heterogeneity or evenness. A useful review of available indices has been given by Peet (1974) but only those used in this book are considered here.

Species richness

The most frequently used measure is the α index first described by Fisher *et al.* (1943):

$$\alpha = \frac{n^1}{x},$$

where x ia a constant having a value <1 (this can be read from Fig. 125 of Williams 1964) and n_1 can be calculated from $N(1 - x)$, N being the size of the sample (number of individuals). This index assumes that the number of individuals of each species follow a logarithmic series. It takes the rarer species into account.

To test the α index, Murray (1968b) made successively large counts on three samples, and found that although the variation was not great there was a tendency for α to increase with sample size. Peet (1974) considers that this index is subject to the same limitations as other simple richness indices, because the shape of the dominance diversity curve is affected by the underlying pattern of niche division.

One advantage of the α index is that values can be read off a base graph by plotting the number of species against the number of individuals in a sample (Fig. A.3).

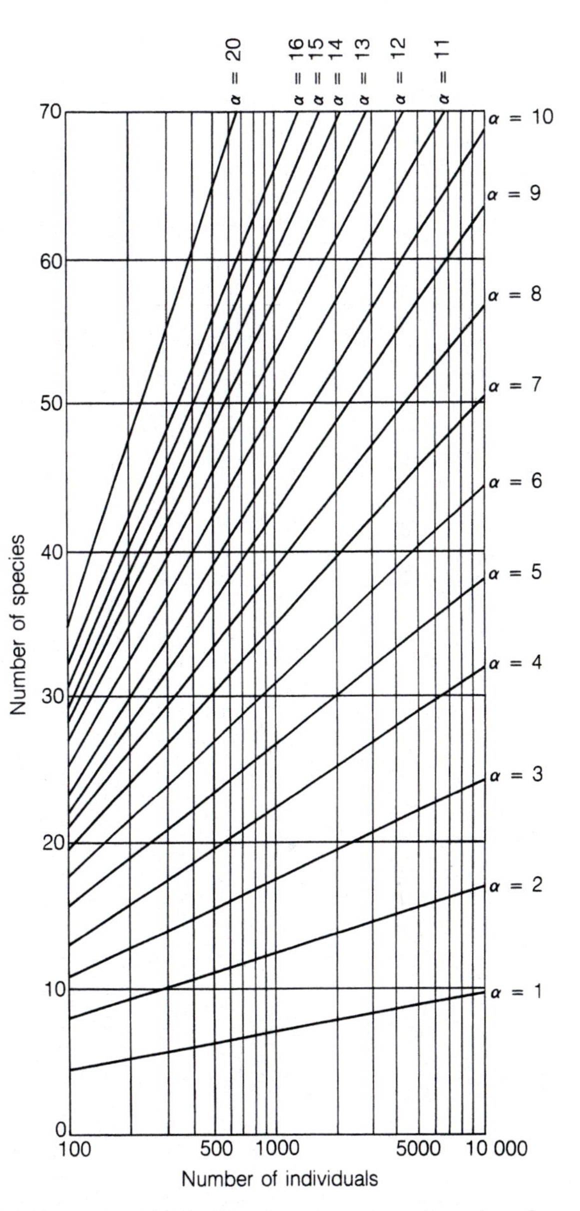

Figure A.3 Graph of number of species and number of individuals in an assemblage and lines of equal α diversity index (from Murray 1973)

Heterogeneity

Indices of heterogeneity take into account both the number of species and the distribution of individuals between species (equitability). The most commonly used measure is based on information theory (H) and uses the Shannon–Weaver formulation:

$$H(S) = -\sum_{i=1}^{s} p_i \ln p_i$$

where S is the number of species and p_i the proportion of the ith species (p = per cent divided by 100).

The maximum value of H for any given number of species is attained when all S species have equal abundances; then:

$$H(S)_{max} = \ln S.$$

The information function is a measure of uncertainty and therefore a reasonable measure of heterogeneity. However, it is valid only for an infinite sample. Pielou (1966) has argued that the Brillouin formula for information should be used for finite samples, but Peet (1974) does not consider this to be an acceptable index of heterogeneity.

There are difficulties associated with the use of the information function. First, the function depends strongly on the sample size and is almost invariably an underestimate for any given assemblage. This is partly due to the problem of inadequate sample size and the need to recognise every species present. Second, microenvironmental influences may obscure the picture of the macroenvironment so it may be necessary to take numerous samples and take the mean diversity as the indicator of the macroenvironment (Pielou 1979).

Notwithstanding these criticisms, the information function is commonly used as an index of diversity. Figure A.4 shows the change in values of $p_i \ln p_i$ with changes in percentage abundance. It can be seen that peak values are given between percentages of 17 and 61. Nevertheless, the value for 10% (–0.230) is <10 the value for 1% (–0.046) so the function is weighted in favour of the less abundant species. The main difficulty in interpreting the significance of any given value of $H(S)$ is that it is invariably smaller than

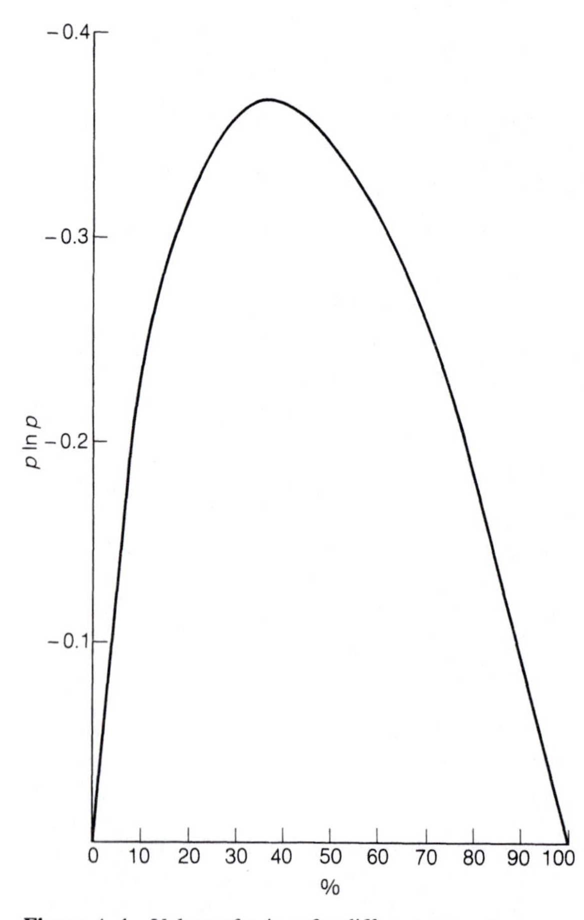

Figure A.4 Values of p_i in p_i for different percentages

Table A.1 $H(S)_{max}$ values for different numbers of species (S)

S	$H(S)_{max}$
1	0.00
5	1.60
10	2.30
20	2.99
30	3.40
40	3.69
50	3.91

$H(S)_{max}$. For example, an assemblage of 10 species may have an $H(S)$ of 1.60 but $H(S)_{max}$ is 2.30 (see Table A.1). This problem is considered under the next heading, equitability.

Equitability

Within the concept of heterogeneity is the role of evenness in the abundance of species, and this is termed 'equitability' (Peet 1974).

Equitability (E) can be measured by

$$E = e^{H(S)}$$

that is, e, the base of natural logarithms, raised to the $H(S)$ power. When all species are equally abundant $E = H(S)$. The ratio E/S is a measure of species equitability which relates the observed E to the maximum value (= 1) (Buzas and Gibson 1969). However, Peet (1974) cautions against drawing conclusions from the equitability of a sample unless the total number of species in the community is known (and this can never be the case).

Pielou's 'quick method' or 'abbreviated composition'

Pielou (1979) described a measure of diversity which is quick and easier to determine than either S, the number of species in a sample, or $H(S)$, the information function. Furthermore, it is claimed to be less subject to sample error.

It is a matter of observation that the proportions of the few dominant species in a sample are higher for those with a small number of species than for those with many species. Pielou suggested that the three dominant species would therefore give a measure of diversity if their values were recalculated to sum to 1:

$$\sum_{i=1}^{3} p_i = 1,$$

where p_i is the relative proportion of the ith species calculated from

$$p_i = \frac{n_i}{n_1 + n_2 + n_3},$$

with $i = 1, 2, 3$, and where n_i is the original abundance of the ith species.

For example, an assemblage has three dominant species with abundances $n_1 = 200$, $n_2 = 100$, $n_3 = 50$, therefore $p_1 = 0.57$, $p_2 = 0.29$ and $p_3 = 0.14$.

Because p is composed of three elements, it can be plotted on a triangular graph, but if $p_1 \geq p_2 \geq p_3$ then all the points plot either on the perimeter or within a small right-angled triangle (Fig. A.5).

Pielou made comparisons between polar shallow-water and warm-latitude deep-water assemblages, and these were plotted in different parts of the triangle although with considerable overlap. She also pointed out that along a traverse there is a wide scatter of values. In reality, this is a very crude diversity index which has little interpretational value.

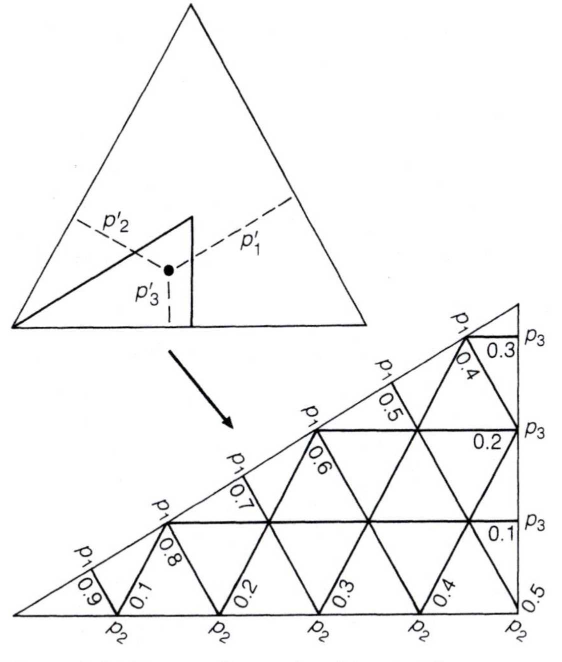

Figure A.5 Ternary diagram for p' (see text for discussion)

Wall structure

In the 1964 classification of foraminifera (Loeblich and Tappan 1964) all modern forms with hard tests fell into the three suborders – Textulariina, Miliolina and Rotaliina – which correspond with agglutinated, porcellaneous and hyaline wall structures respectively. The revised classification (Loeblich and Tappan 1984) divides the Rotaliina into four suborders (Fig. A.6). Extensive use of triangular plots was made

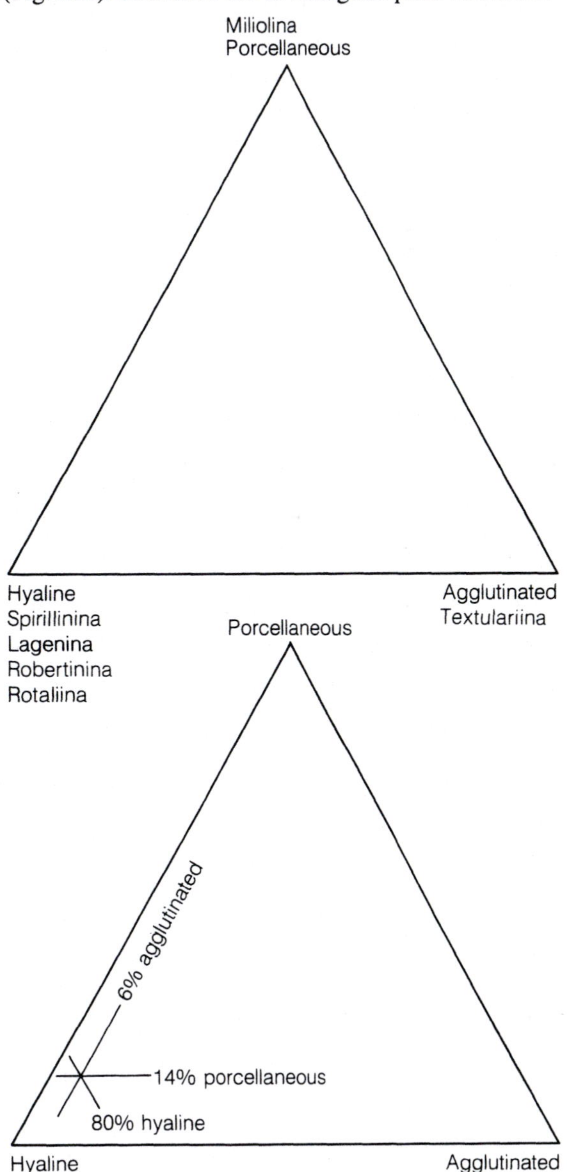

Figure A.6 (A) Ternary diagram of wall structure and suborders; (B) plotting a point

by Murray (1973) and they prove to be particularly useful for differentiating shallow-water environments.

Generic predominance

The idea of generic dominance, that is, the genus which forms the greatest percentage of the assemblage, was introduced in a study of the Gulf of Mexico by Walton (1964a). This approach was extended by Poag (1981) who preferred the term 'predominant'. Sometimes two or three genera are equally predominant. Fossil generic predominance patterns can often be matched with those of modern genera so this is a potentially useful palaeoecological technique (e.g. Sliter and Baker 1972).

Comparing samples

Comparison may be made to identify differences or similarities between two or more samples. All the methods available provide measures of similarity even though the results may be interpreted by the observer as difference. Some methods are based on the presence/absence of taxa while others are based on the relative proportions of the taxa. Raup and Crick (1979) have discussed the relative merits of different presence/absence methods.

Table A.2 The value of the index of affinity in model samples (Rogers 1976)

Number of species	Index of affinity (‰) Lower	Peak	Upper
5 : 5	88	94	97
10 : 10	83	89	94
20 : 20	80	84	88
30 : 30	74	81	86
30 : 20	58	63	68
30 : 10	27	33	39
20 : 10	44	50	56
10 : 5	44	50	56
20 : 5	20	26	29
30 : 5	12	17	21

The similarity index (Sanders 1960) or index of affinity (Rogers 1976) takes into account the relative abundance of taxa. To compare two samples, the lowest percentage occurrences of each species present are summed to give the percentage similarity. Rogers tested the validity of the method on 100 samples taken from an artificial assemblage, each containing 250 random numbers. This gave a range of values centred about a peak for each type of comparison made. Where two samples have an identical number of species, the index of affinity is >80% (peak values, Table A.2, upper part). Where two samples have very different numbers of species, the index of affinity is much lower (Table A.2 lower part). As a general rule of thumb, the minimum percentage value for the index of affinity of species combinations other than those given in Table A.2 can be calculated from

$$\frac{S_{min} \times 100}{S_{max}} - 6,$$

where S_{min} is the lower species number and S_{max} the higher species number. Thus for 24 : 18 species the value is 69%.

If the value of the index of affinity for a given sample pair is higher than the expected minimum value, then the two samples may be considered very similar.

Multivariate analysis

Large data bases can be subjected to a variety of multivariate techniques (principal component analysis, varimax factor analysis, correspondence analysis, etc.) using a computer. Useful references are Williamson (1972) and Pielou (1984). Such techniques are very useful and are generally considered to be objective. However, as pointed out by Gray (1981), this is not completely true. Normally only part of the data set is used because the computer cannot handle large numbers of species and especially rare ones, for zeros and ones skew the patterns and make interpretation difficult. Therefore, data sets with 100+ species are normally reduced to 30–40 taxa by the omission or grouping of rare species.

APPENDIX B

Ecological data for selected genera

The data are given in the following order: mode of life; substrate; mode of feeding; salinity; temperature; depth; environment. Definitions: hard substrates = rocks, shells, macrofauna, plants, etc.; salinity – brackish = 0–32‰, marine = 32–37‰, hypersaline = >37‰; temperature (of the bottom water) cold, temperate, warm (even in tropical areas the water deeper than ~100 m is temperate or cold); environment – shelf 0–180 m, bathyal 180 to ~4000 m, abyssal greater than ~4000 m.

The data relate to the main occurrence of each genus and excludes rare occurrences in other environments.

Acervulina epifaunal, attached; bryozoa, shells, etc.; passive suspension feeder, marine; temperate–warm; 0–60 m; inner shelf.

Adercotryma epifaunal, free; fine sand; detritivore; marine; cold; shelf–abyssal.

Alveolinella epifaunal, clinging; algal-covered carbonate gravel; herbivore, symbionts; marine; 18–26 °C; 3–70 m; inner shelf, lagoon.

Alveolophragmium epifaunal, free; sand; detritivore; marine; 10 °C; 20–700 m; shelf–upper bathyal.

Ammoastuta ? infaunal, free; sediment; ? detritivore; brackish; temperate; marshes.

Ammobaculites infaunal free; muddy sediment; detritivore; brackish-marine; temperate–tropical; brackish marshes and lagoons, inner shelf–upper bathyal.

Ammoglobigerina ? infaunal, free; sediment; ? detritivore; marine; cold; shelf–upper bathyal.

Ammonia infaunal, free; muddy sand; ? herbivore; brackish, marine, hypersaline; warm temperate–tropical; 0–30 °C; 0–50 m; brackish and hypersaline lagoons, inner shelf.

Ammoscalaria ? infaunal, free; sediment; detritivore; brackish–marine; cold–temperate; inner shelf.

Ammotium infaunal, free; muddy sediment; detritivore; brackish, 0–30 °C; 0–10 m; tidal marshes, brackish lagoons and estuaries, enclosed brackish shelf seas.

Amphisorus epifaunal, clinging; phytal, carbonate sediment; herbivore, symbionts; marine; 18–26 °C; 0–50 m; lagoon, nearshore.

Amphistegina epifaunal, free; phytal, coarse carbonate; herbivore, symbionts; marine; winter min. 14 °C, generally >20 °C; 0–130 m, coral reefs, lagoons.

Archaias epifaunal, clinging; phytal; herbivore, symbionts; marine; >22 °C; 0–20 m; inner shelf.

Arenoparrella epifaunal, free or clinging; muddy sediment; herbivore; brackish–hypersaline; 0–30 °C; mainly intertidal on marshes, but some on shelf–upper bathyal.

Articulina epifaunal, free; phytal or sediment; marine –hypersaline; >20 °C; 0–3500 m; inner shelf–bathyal.

Asterigerina epifaunal, free; sediment; ? herbivore; marine; subtropical–tropical; inner shelf.

Asterigerinata epifaunal, free; sediment; ? herbivore; marine; temperate; 0–100 m, inner shelf.

Asterorotalia epifaunal, free; sediment; ? herbivore; marine; subtropical–tropical, inner shelf.

Astrononion epifaunal–infaunal, free–clinging; muddy sediment; ? detritivore; marine; cold; shelf–bathyal.

Astrorhiza epifaunal, sessile; sediment; passive suspension feeder/carnivore; marine; cold–temperate; shelf.

Baculogypsina epifaunal, clinging; phytal; herbivore, symbionts; marine; >25 °C; 0–15 m, coral reefs, lagoons.

Bathysiphon infaunal, ? free; sediment; marine; cold; upper bathyal.
Bigenerina ? epifaunal, free; sediment; ? omnivore; marine; temperate–warm; shelf.
Bolivina infaunal–epifaunal, free; muddy sediment; ? detritivore; marine; cold to warm; inner shelf–bathyal.
Bolivinella ? infaunal, free; muddy sediment; marine; temperate–cold; shelf–upper bathyal.
Borelis epifaunal, free; algal-coated substrates; herbivore; symbionts; marine; 18–26 °C; 5–65 m; lagoon–reef.
Brizalina infaunal, free; muddy sediment; detritivore; marine; temperate–cold; marginal marine–bathyal.
Buccella ? infaunal, free; muddy sediment; ? detritivore: marine; cold–temperate; inner shelf–bathyal. 0–100 m; lagoons, inner shelf.
Bulimina ? infaunal, free; mud to fine sand; ? detritivore: marine; cold–temperate; inner shelf– bathyal.
Buliminella infaunal, free; muddy sediment; ? detritivore; marine; temperate; mainly shelf but also lagoons and upper bathyal.

Calcarina epifaunal, free; sediment; herbivore, symbionts; marine; 18–26 °C; 0–30 m; lagoons, reefs, inner shelf.
Cancris epifaunal, free; sediment; ? detritivore; marine; temperate–subtropical; 50–150 m; shelf.
Cassidella infaunal, free; mud; detritivore; marine; temperate; shelf.
Cassidulina infaunal, free; mud, sand; detritivore; marine; cold–temperate; shelf–bathyal.
Chilostomella infaunal, free; mud; detritivore; marine; outer shelf and bathyal.
Cibicides epifaunal, attached; hard substrates; ? passive suspension feeder; marine; cold–warm; 0 to > 2000 m; lagoons, shelf–bathyal.
Cibicidoides epifaunal, clinging; hard substrates; ? passive suspension feeder; marine; cold; shelf–bathyal.
Cribrostomoides epifaunal, free or clinging; sand; pebbles; detritivore; 30–35‰; <15 °C; shelf–bathyal.
Cuneata infaunal free; mud, sand; detritivore; 30–35‰; cold; shelf.
Cyclammina ? epifaunal, free; mud, sand; detritivore; marine; sigma-t 27.7; >100 m, outer shelf–abyssal.
Cycloclypeus epifaunal, free; muddy carbonate sediment; warm; 60–100 m; shelf.

Discorbis epifaunal, clinging or attached; firm substrates, coarse sand; herbivore; marine; temperate–warm; 0–50 m; inner shelf.
Discorinopsis infaunal, free; muddy sediment; detritivore; hypersaline; warm; marsh.

Eggerella infaunal, free; silt, fine sand; detritivore; 20–37‰; cold–temperate; lagoons–inner shelf, bathyal.
Eggerelloides infaunal, free; sand; detritivore; marine; temperate; shelf.
Ehrenbergina ? epifaunal, free; mud; ? detritivore; marine; cold; outer shelf–bathyal.
Eilohedra ? infaunal, free; mud; ? detritivore; marine; bathyal.
Elphidiella ? infaunal, free; sand; ? detritivore; 30–35‰; cold; inner shelf.
Elphidium keeled: epifaunal, free; sand, vegetation; herbivore; marine; temperate–warm; 35–70‰; 0–50 m; inner shelf; non-keeled: infaunal, free; mud, sand; herbivore–detritivore; 0–70‰; brackish –hypersaline marshes and lagoons, inner shelf (upper bathyal – *E. clavatum* only).
Epistominella epifaunal, semi-infaunal, free; mud; detritivore; marine; temperate–cold; shelf–bathyal.
Eponides epifaunal, free or clinging; sediment or hard substrates; ? detritivore; marine; cold to temperate; shelf–abyssal.

Fontbotia epifaunal, clinging; hard substrates; passive suspension feeder; marine; cold; bathyal.
Fursenkoina infaunal, free; mud; detritivore; 30–35‰, 0–1200 m; lagoons, shelf, upper bathyal.

Gaudryina epifaunal, attached; hard substrates; ? passive suspension feeder; marine; warm temperate; 50–500 m; shelf and upper bathyal.
Gavelinopsis epifaunal, clinging; hard substrates; ? passive suspension feeder; marine; temperate–cold; shelf–bathyal.
Glabratella epifaunal, clinging or attached; hard substrates; ? herbivore; hypersaline–marine; temperate –warm; 0–50 m; hypersaline tidal marshes and lagoons, marine inner shelf.
Globobulimina infaunal, free; mud; detritivore; marine; temperate–cold; shelf–bathyal.
Globocassidulina infaunal, free; mud; ? detritivore; marine; temperate–cold; shelf–bathyal.
Goësella epifaunal, free; sediment; detritivore; marine; temperate–warm; shelf–upper bathyal.
Gyroidina epifaunal, free; mud; ? detritivore; marine; cold; shelf–bathyal.

Hanzawaia epifaunal, clinging; hard substrates; ? passive suspension feeder; marine; temperate–warm; inner shelf.

Haplophragmoides infaunal, free; mud–sand; ? detritivore; marine; temperate–cold; marshes–bathyal.

Haynesina infaunal, free; mud, silt; ? herbivore; brackish, 0–30‰; cold–warm; marsh, lagoon, inner shelf.

Hemisphaerammina epifaunal, attached; hard substrates; passive herbivore; marine; temperate–cold; inner shelf.

Heterolepa epifaunal, ? clinging; hard substrates; ? passive suspension feeder; marine; temperate–cold; shelf, bathyal.

Heterostegina epifaunal, free; phytal, muddy carbonate sediments; herbiviore, symbionts; marine; 18–26 °C, 0–130 m; inner shelf, lagoon.

Hoeglundina infaunal, free; mud; ? detritivore; marine; cold; outer shelf–bathyal.

Homotrema epifaunal attached; hard substrates; passive suspension feeder; marine; temperate–warm; inner shelf.

Hormosina ? infaunal, free; mud; ? detritivore; marine; cold; bathyal, abyssal.

Hyalinea epifaunal, free; mud, silt; ? detritivore; marine; cold–temperate; 10–1000 m; shelf–upper bathyal.

Islandiella infaunal, free; mud, silt; ? detritivore; marine; <10 °C; >20 m; shelf–upper bathyal.

Jadammina epifaunal, free; mud; silt; ? herbivore or detritivore; 0–50‰; 0–30 °C; intertidal marshes.

Karreriella epifaunal, free; mud; silt; ? detritivore; marine; <10 °C; outer shelf–bathyal.

Lagenammina ? infaunal, free; sand; ? detritivore; 30–35‰; temperate–cold; shelf–bathyal.

Lenticulina epifaunal, free; mud; ? detritivore; marine; cold; outer shelf and bathyal.

Lepidodeuterammina epifaunal–infaunal, clinging; hard substrates, coarse sand; herbivore; marine; cold; shelf.

Marginopora epifaunal, attached; phytal, carbonate sand and gravel; herbivore, symbionts; marine; 18–26 °C; 0–45 m; inner shelf, lagoon, coral reefs.

Martinottiella ? epifaunal, free; mud; ? detritivore; marine; cold; >120 m; outer shelf–bathyal.

Massilina epiphytic, clinging; plants; herbivore; marine; temperate–warm; inner shelf.

Melonis infaunal, free; mud, silt; ? detritivore; marine; <10 °C; shelf–bathyal.

Miliammina infaunal, free; mud; silt; detritivore; 0–50‰; 0–30 °C; brackish–hypersaline marshes–upper bathyal.

Miliolinella epifaunal, clinging; plants and hard substrates; herbivore; 32–50‰, 10–30 °C; 0–100 m; hypersaline lagoons, normal marine lagoons and marshes, inner shelf.

Neoconorbina epifaunal, clinging; hard substrates; ? herbivore; marine; temperate; inner shelf.

Nonion infaunal, free; mud, silt; herbivore; 30–35‰; cold–warm; 0–180 m, shelf.

Nonionella infaunal, free; mud; ? detritivore; marine; temperate–warm; 10–1000 m; shelf–upper bathyal.

Nonionellina infaunal, free; mud–sand; detritivore; marine; cold; shelf.

Nonionoides infaunal, free; muddy sediment; ? detritivore; marine; temperate–warm; shelf.

Nummulites epifaunal, free; muddy carbonate sediment; herbivore; symbionts; marine; warm; 0–130 m; lagoon to inner shelf.

Nuttallides epifaunal, free or clinging; mud; ? detritivore; marine; <4 °C; bathyal–abyssal.

Operculina epifaunal, free; muddy carbonate sediment; herbivore, symbionts; marine–slightly hypersaline; warm 0–130 m; inner shelf, lagoons.

Oridorsalis epifaunal, free; mud; detritivore; marine; 4 °C; bathyal.

Pararotalia epifaunal, free; sand; herbivore; marine; warm; inner shelf.

Patellina epifaunal, clinging; hard substrates; ? herbivore; marine; cold–warm; 0–100 m, inner shelf.

Peneroplis epifaunal, clinging; plants and hard substrates; 35–53‰; 18–27 °C; 0–70 m; lagoons and innermost shelf.

Planorbulina epifaunal, attached; hard substrates; ? passive suspension feeder; marine; temperate–warm: 0–50 m; inner shelf.

Planulina epifaunal, clinging; hard substrates; ? passive suspension feeder; marine; cold–warm; shelf, bathyal.

Poroeponides epifaunal, clinging; hard substrates; ? herbivore/passive suspension feeder; marine; temperate–warm; inner shelf.

Portatrochammina epifaunal, free; sediment; detritivore; marine; temperate–cold; shelf–bathyal.

Pullenia infaunal, free; mud; detritivore; marine; cold; outer shelf to bathyal.

Pyrgo either epifaunal, free or clinging; plants or sediment; herbivore; marine; temperate–warm; inner shelf; *or* epifaunal, free; sediment; ? detritivore; marine; cold; shelf–bathyal.

Quinqueloculina epifaunal, free or clinging; plants or sediment; herbivore; marine–hypersaline; 32–65‰; cold–warm; hypersaline lagoons, marine shelf, rarely bathyal.

Recurvoides epifaunal, free; sediment; detritivore; marine; cold; shelf–bathyal.

Reophax infaunal, free; mud, sand; detritivore; marine; temperate–cold; lagoons, shelf–bathyal.

Rosalina epifaunal, clinging or attached; hard substrates; ? herbivore, omnivore; marine; temperate – warm; 0–100 m; lagoons, inner shelf.

Saccammina infaunal, free; sand; detritivore; marine; temperate–cold; 0–100 m; inner shelf.

Sorites epiphytic, clinging; plants, especially seagrass; herbivore, symbionts; 37–45‰; 18–26 °C; 0–70 m; lagoons and nearshore.

Spirillina epifaunal, clinging; hard substrates; marine; temperate–cold; 0–100 m; inner shelf.

Spirolina epiphytic, clinging; plants; herbivore; 37–50‰; 18–26 °C; 0–60 m; lagoons and nearshore.

Spiroloculina epifaunal, free or clinging; sediment or plants; herbivore; marine–hypersaline; temperate–warm; 0–40 m; lagoons, inner shelf.

Textularia epifaunal, ? clinging; hard substrates and sand; ? detritivore, marine cold–warm; 0–500 m; lagoons, shelf–bathyal.

Tiphotrocha epifaunal, free; sediment; herbivore or detritivore; brackish; temperate; intertidal; marshes.

Trifarina infaunal, free; mud, sand; ? detritivore; marine; temperate–cold; 0–400 m; shelf and upper bathyal.

Triloculina epifaunal, free or clinging; mud, sand, plants; herbivore, detritivore; marine – hypersaline, 32–?55‰; temperate–warm; mainly hypersaline lagoons or marine inner shelf, some bathyal species (cold).

Trochammina epifaunal or infaunal, free; sediment; herbivore or detritivore; 0–60‰; cold–warm, 0–30 °C; 0 to >6000 m; intertidal, shelf, bathyal, abyssal.

Trochulina epifaunal–infaunal, free; sediment; ? detritivore; marine; temperate warm; lagoons, shelf.

Uvigerina mainly infaunal, some epifaunal, free; muddy sediment; ? detritivore; marine; cold; 100 to >4500 m; shelf–bathyal.

APPENDIX C

Faunal reference list – recent taxa

The species names used in this book are given together with the original name where this differs. Generic determinations are based on Loeblich and Tappan (1988) with the following exceptions: *Angulogerina* had not been distinguished from *Trifarina* because the two cannot realistically be separated.

Acervulina inhaerens Schultze, 1854
Adercotryma glomerata (Brady) = *Lituola glomerata* Brady, 1878.
Affinetrina planciana (d'Orbigny) = *Triloculina planciana* d'Orbigny, 1839.
Alabaminella weddellensis (Earland) = *Eponides weddellensis* Earland, 1936.
Alabaminoides antarcticus Gudina and Saidova, 1967
Alliatina primitiva (Cushman and McCulloch) = *Cushmanella primitiva* Cushman and McCulloch, 1940.
Allogromia laticollaris Arnold, 1948.
Aveolinella quoyi d'Orbigny, 1826.
Alveolophragmium columbiense (Cushman) = *Haplophragmoides columbiensis* Cushman, 1925.
Alveolophragmium nitidum (Goës) = *Haplophragmium nitidum* Goës, 1896.
Ammoastuta inepta (Cushman and McCulloch) = *Ammobaculites ineptus* Cushman and McCulloch, 1939.
Ammoastuta salsa Cushman and Brönnimann, 1948.
Ammobaculites agglutinans (d'Orbigny) = *Spirolina agglutinas* d'Orbigny, 1846.
Ammobaculites americanus Cushman, 1910.
Ammobaculites barwonensis Collins, 1974.
Ammobaculites crassus Warren, 1957.
Ammobaculites dilatatus Cushman and Brönnimann, 1948.
Ammobaculites exiguus Cushman and Brönnimann, 1948.
Ammobaculites exilis Cushman and Brönnimann, 1948.
Ammobaculites filaformis Heron-Allen and Earland = *Ammobaculites agglutinans filaformis* Heron-Allen and Earland, 1934.
Ammobaculites persicus Lutze, 1974.
Ammodiscus minutissimus Cushman and McCulloch, 1939.
Ammodiscus siliceus (Terquem) = *Involutina silicea* Terquem, 1862.
Ammodiscus gullmarensis Höglund, 1948.
Ammodiscus tenuis Brady, 1884.
Ammoflintina argentea Echols, 1971.
Ammoglobigerina globigeriniformis (Parker and Jones) = *Lituola nautiloidea* Lamarck var. *globigeriniformis* Parker and Jones, 1865.
Ammoglobigerina globigeriniformis var. *pygmaea* (Höglund) = *Trochammina globigeriniformis* Parker and Jones var. *pygmaea* Höglund, 1947.
Ammoglobigerina globulosa (Cushman) = *Trochammina globulosa* Cushman, 1920.
Ammolagena clavata (Jones and Parker) = *Trochammina irregularis* (d'Orbigny) var. *clavata* Jones and Parker, 1860.
Ammomarginulina fluvialis (Parker) = *Ammoscalaria fluvialis* Parker, 1952.
Ammonia beccarii (Linné) = *Nautilus beccarii* Linné, 1758. Variants are included in this species.
Ammonia compressiuscula (Brady) = *Rotalia papillosa* Brady var. *compressiuscula* Brady, 1984.
Ammonia convexa (Collins) = *Streblus convexus* Collins, 1958.
Ammonia convexidorsa Zheng, 1978.
Ammonia ketienziensis (Ishizaki) = *Streblus ketienziensis* Ishizaki, 1943.
Ammonia papillosus (Brady) = *Rotalia papillosus* Brady, 1884.
Ammonia parkinsoniana (d'Orbigny) = *Rosalina*

parkinsoniana d'Orbigny, 1839.
Ammonia pauciloculata (Phleger and Parker) = *'Rotalia' pauciloculata* Phleger and Parker, 1951.
Ammonia perlucida (Heron-Allen and Earland) = *Rotalia perlucida* Heron-Allen and Earland, 1913.
Ammonia rolshauseni (Cushman and Bermudez) = *Rotalia rolshauseni* Cushman and Bermudez, 1946.
Ammonia sarmientoi (Redmond) = *Rotalia sarmientoi* Redmond, 1953.
Ammonia tepida (Cushman) = *Rotalia beccarii* (Linné) var. *tepida* Cushman, 1926. See *Ammonia beccarii* (Linné).
Ammoscalaria fluvialis Parker, 1952.
Ammoscalaria pseudospiralis (Williamson) = *Proteonina pseudospiralis* Williamson, 1858.
Ammoscalaria runiana (Heron-Allen and Earland) = *Haplophragmium runianum* Heron-Allen and Earland, 1916.
Ammotium cassis (Parker) = *Lituola cassis* Parker, in Dawson, 1870.
Ammotium planissimum (Cushman) = *Haplophragmoides planissimum* Cushman, 1927.
Ammotium pseudocassis (Cushman and Brönnimann) = *Ammobaculites pseudocassis* Cushman and Brönnimann, 1948.
Ammotium salsum (Cushman and Brönnimann) = *Ammobaculites salsum* Cushman and Brönnimann, 1948.
Ammotrochoides bignoti Janin, 1984.
Amphicoryna scalaris (Batsch) = *Nautilus scalaris* Batsch, 1791.
Amphisorus hemprichii Ehrenberg, 1839.
Amphistegina bicirculata Larsen, 1976.
Amphistegina gibbosa d'Orbigny, 1839.
Amphistegina lessonii d'Orbigny, 1826.
Amphistegina lobifera Larsen, 1976.
Amphistegina madagascariensis d'Orbigny, 1903.
Amphistegina papillosa Said, 1949.
Amphistegina quoyi d'Orbigny, 1926.
Amphistegina radiata (Fichtel and Moll) = *Nautilus radiatus* Fichtel and Moll, 1798.
Anomalinoides minimus (Forster) = *Truncatulina minima* Forster, 1892.
Archaias angulatus (Fichtel and Moll) = *Nautilus angulatus* Fichtel and Moll, 1798.
Archimerismus subnodosus (Brady) = *Hyperammina subnosa* Brady, 1884.
Arenoparrella mexicana (Kornfeld) = *Trochammina inflata* (Montagu) var. *mexicana* Kornfeld, 1931.
Arenoparrella mexicana (Kornfeld) var. *asiatica* Polski, 1959.
Arenoparrella oceanica Uchio, 1960.
Articulina lineata Brady, 1884.
Articulina mexicana Cushman, 1922.
Articulina tubulosa (Seguenza) = *Quinqueloculina tubulosa* Seguenza, 1862.
Asterigerina carinata d'Orbigny, 1839.
Asterigerinata mamilla (Williamson) = *Rotalia mamilla* Williamson, 1858.
Asterorotalia dentata (Parker and Jones) = *Rotalia beccarii* (Linné) var. *dentata* Parker and Jones, 1865.
Asterorotalia inflata (Millett) = *Rotalia schroeteriana* Parker and Jones var. *inflata* Millett, 1904.
Astrononion gallowayi Loeblich and Tappan, 1953.
Astrononion italicum Cushman and Edwards, 1937.
Astrononion sidebottomi Cushman and Edwards, 1937.
Astrononion stelligerum (d'Orbigny) = *Nonionina stelligera* d'Orbigny, 1839.
Astrorhiza arenaria Norman, 1876.
Astrorhiza limicola Sandahl, 1858.
Atlantiella atlantica (Parker) = *Trochamminella atlantica* Parker, 1952.
Aubignyna hamanakoensis (Ishiwada) = *Anomalina hamanakoensis* Ishiwada, 1958.

Baculogypsina sphaerulata (Parker and Jones) = *Orbitolina concava* Lamarck var. *sphaerulata* Parker and Jones, 1860.
Baculogypsinoides spinosus Yabe and Hanzawa, 1930.
Bathysiphon crassatina (Brady) = *Astrorhiza crassatina* Brady, 1884.
Bathysiphon discreta (Brady) = *Rhabdammina discreta* Brady, 1884.
Bathysiphon filiformis M. Sars, 1872.
Bigenerina irregularis Phleger and Parker, 1951.
Bigenerina nodosaria d'Orbigny, 1826.
Bigenerina taiwanica Nakamura, 1937.
Bolivina acuminata Natland, 1946.
Bolivina alata (Seguenza) = *Vulvulina alata* Seguenza, 1862.
Bolivina argentea Cushman var. *monicana* Zalesny, 1959.
Bolivina cincta Heron-Allen and Earland, 1932.
Bolivina compacta Sidebottom, = *Bolivina robusta* var. *compacta* Sidebottom, 1905.
Bolivina costata d'Orbigny, 1839.
Bolivina decussata Brady, 1884.
Bolivina dilatata Reuss, 1850.
Bolivina dilatata Reuss var. *dilitatissima* Silvestri, 1893.
Bolivina doniezi Cushman and Wickenden, 1929.

Bolivina goësii Cushman, 1922.
Bolivina interjuncta Galloway and Wissler, 1927.
Bolivina pacifica Cushman and McCulloch, 1942.
Bolivina paula Cushman and Cahill, 1932.
Bolivina perrini Kleinpell, 1938.
Bolivina persiensis Lutze, 1974.
Bolivina plicata d'Orbigny, 1839.
Bolivina pseudoplicata Heron-Allen and Earland, 1930.
Bolivina punctata d'Orbigny, 1839.
Bolivina quadrata Cushman and McCulloch, 1942.
Bolivina rankini Kleinpell, 1938.
Bolivina robusta Brady, 1881.
Bolivina seminuda Cushman, 1911.
Bolivina spissa Cushman, 1926.
Bolivina subadvena Cushman, 1926.
Bolivina subargentea Uchio, 1960.
Bolivina subreticulata Parr, 1932.
Bolivina thalmanni Renz, 1948.
Bolivina tongi Cushman, 1929.
Bolivinellina pseudopunctata (Höglund) = *Bolivina pseudopunctata* Höglund, 1947.
Bolivinellina tessellata (Phleger and Parker) = *Virgulina tessellata* Phleger and Parker, 1951.
Bolivinita minuta (Natland) = *Bolivina minuta* Natland, 1938.
Borelis pulchra (d'Orbigny) = *Alveolina pulchra* d'Orbigny, 1840.
Borelis schlumbergeri (Reichel) = *Neoalveolina pygmaea* (Hanzawa) var. *schlumbergeri* Reichel, 1937.
Bottelina labyrinthica Brady, 1881.
Brizalina acutula (Bandy) = *Bolivina advena* Cushman var. *acutula* Bandy, 1953.
Brizalina alata (Seguenza) = *Vululina alata* Segeunza, 1862.
Brizalina albatrossi (Cushman) = *Bolivina albatrossi* Cushman, 1922.
Brizalina arctica (Herman) = *Bolivina arctica* Herman, 1974.
Brizalina argentea (Cushman) = *Bolivina argentea* Cushman, 1926.
Brizalina barbata (Phleger and Parker) = *Bolivina barbata* Phleger and Parker, 1951.
Brizalina catanensis (Seguenza) = *Bolivina catanensis* Seguenza, 1862.
Brizalina daggarius (Parker) = *Bolivina daggarius* Parker, 1955.
Brizalina difformis (Williamson) = *Textularia variabilis* var. *difformis* Williamson, 1858.
Brizalina fragilis Phleger and Parker, 1951.
Brizalina lowmani (Phleger and Parker) = *Bolivina lowmani* Phleger and Parker, 1951.
Brizalina minima (Phleger and Parker) = *Bolivina minima* Phleger and Parker, 1951.
Brizalina oceanica (Cushman) = *Bolivina oceanica* Cushman, 1933.
Brizalina ordinaria (Phleger and Parker) = *Bolivina ordinaria* Phleger and Parker, 1952 (new name for *Bolivina simplex* Phleger and Parker, 1951).
Brizalina skagerrakensis (Qvale and Nigam) = *Bolivina skagerrakensis* Qvale and Nigam, 1985.
Brizalina spathulata (Williamson) = *Textularia variabilis* Williamson var. *spathulata* Williamson, 1858.
Brizalina striatella (Cushman) = *Bolivina advena* Cushman var. *striatella* Cushman, 1925.
Brizalina striatula (Cushman) = *Bolivina striatula* Cushman, 1922.
Brizalina striatula (Cushman) var. *spinata* Cushman, 1936.
Brizalina subaenariensis (Cushman) = *Bolivina subaenariensis* Cushman, 1922.
Brizalina subaenariensis mexicana (Cushman) = *Bolivina subaenariensis mexicana* Cushman, 1922.
Brizalina subspinescens (Cushman) = *Bolivina subspinescens* Cushman, 1922.
Brizalina variabilis (Williamson) = *Textularia variabilis* Williamson, 1858.
Brizalina vaughani (Natland) = *Bolivina vaughani* Natland, 1938.
Buccella angulata Uchio, 1960.
Buccella frigida (Cushman) = *Pulvinulina frigida* Cushman, 1921.
Buccella granulata (di Napoli Alliata) = *Eponides frigidus* var. *granulatus* di Napoli Alliata, 1952.
Buccella hannai (Phleger and Parker) = *Eponides hannai* Phleger and Parker, 1951.
Buccella hannai arctica Voloshinova, 1960.
Buccella parkerae Anderson, 1953.
Buccella peruviana (d'Orbigny) = *Rotalina peruviana* d'Orbigny, 1839.
Buccella tenerrima (Bandy) = *Rotalia tenerrima* Bandy, 1950.
Bulimina aculeata d'Orbigny, 1826.
Bulimina alazanensis Cushman, 1927.
Bulimina clava Cushman and Parker, 1947.
Bulimina costata d'Orbigny, 1852.
Bulimina denudata Cushman and Parker, 1938.
Bulimina elegans d'Orbigny, 1826.
Bulimina elongata d'Orbigny, 1826.
Bulimina exilis Brady = *Bulimina elegans* d'Orbigny var. *exilis* Brady, 1884.
Bulimina gibba Fornasini, 1902.
Bulimina marginata d'Orbigny, 1826.

Bulimina marginata d'Orbigny var. *biserialis* Millett, 1900.
Bulimina patagonica d'Orbigny, 1839.
Bulimina spicata Phleger and Parker, 1951.
Bulimina striata d'Orbigny, 1826.
Bulimina striata d'Orbigny var. *mexicana* Cushman, 1922.
Bulimina submarginata Parr, 1950.
Bulimina translucens Parker, 1953.
Buliminella elegantissima (d'Orbigny) = *Bulimina elegantissima* d'Orbigny, 1839.
Buliminella morgani Anderson, 1961.
Buliminella silviae Bermudez and Seiglie, 1963.
Buliminella tenuata Cushman = *Buliminella subfusiformis* Cushman var. *tenuata* Cushman, 1927.

Calcarina calcar d'Orbigny, 1826.
Calcarina calcarinoides (Cheng and Zheng) = *Rotalia calcarinoides* Cheng and Zheng, 1978.
Calcarina defrancii d'Orbigny, 1826.
Calcarina delicata Todd and Post, 1954.
Calcarina hispida Brady, 1876.
Calcarina spengleri (Gmelin) = *Nautilus spengleri* Gmelin, 1788.
Cancris auricula (Fichtel and Moll) = *Nautilus auricula* Fichtel and Moll, 1798.
Cancris congolensis Margerel and Kouyoumontzakis, 1978.
Cancris hainanensis Li, 1985.
Cancris oblonga (Williamson) = *Rotalina oblonga* Williamson, 1858.
Cancris panamensis Natland, 1938.
Cancris sagra (d'Orbigny) = *Rosalina sagra* d'Orbigny, 1839.
Caribeanella elatensis Perelis and Reiss, 1975.
Carpenteria monticularis Carter, 1877.
Carpenteria proteiformis Goës = *Carpenteria balaniformis* Gray var. *proteiformis* Goës, 1882.
Carpenteria utricularis (Carter) = *Polytrema atriculare* Carter, 1876.
Carterina spiculotesta (Carter) = *Rotalia spiculotesta* Carter, 1877.
Cassidella complanata (Egger) = *Virgulina schreibersiana* Czjzek var. *complanata* Egger, 1895.
Cassidella panikkari Seibold and Seibold, 1973.
Cassidulina algida Cushman, 1944.
Cassidulina bradshawi Uchio, 1960.
Cassidulina braziliensis Cushman, 1922.
Cassidulina californica Cushman and Hughes, 1925.
Cassidulina carinata Cushman, 1922 = *Cassidulina laevigata* d'Orbigny var. *carinata* Cushman, 1922.
Cassidulina complanata Ujiié and Kusukawa, 1969.
Cassidulina crassa d'Orbigny, 1839.
Cassidulina curvata Phleger and Parker, 1951.
Cassidulina cushmani Stewart and Stewart, 1930.
Cassidulina delicata Cushman and Hughes, 1925.
Cassidulina depressa Asano and Nakamura = *Cassidulina subglobosa depressa* Asano and Nakamura, 1937.
Cassidulina grandis (Troitskaya) = *Cassidulina grandis* Troitskaya, 1970.
Cassidulina islandica Nørvang, 1945 – see *Islandiella islandica*.
Cassidulina laevigata d'Orbigny, 1826.
Cassidulina limbata Cushman and Hughes, 1925.
Cassidulina lomitensis Galloway and Wissler, 1927.
Cassidulina minuta Cushman, 1933.
Cassidulina neocarinata Thalmann, 1950.
Cassidulina obtusa Williamson, 1858.
Cassidulina reniforme Nørvang = *Cassidulina crassa* d'Orbigny var. *reinforme* Nørvang, 1945.
Cassidulina subacuta (Gudina) = *Cassilamellina subacuta* Gudina, 1966.
Cassidulina subcarinata Uchio, 1960.
Cassidulina teretis Tappan, 1951.
Cassidulina tortuosa Cushman and Hughes, 1925.
Cassidulina yabei Asano and Nakamura, 1937.
Cassidulinoides bradyi (Norman) = *Cassidulina bradyi* Norman, 1880.
Cassidulinoides cornuta (Cushman) = *Virgulina cornuta* Cushman, 1913.
Cassidulinoides crassa (Earland) = *Ehrenbergina crassa* Earland, 1929.
Cassidulinoides parkerianus (Brady) = *Cassidulina parkeriana* Brady, 1881.
Cassidulinoides porrectus Parr, 1950.
Chilostomella cushmani Chapman, 1941.
Chilostomella oolina Schwager, 1878.
Chilostomella ovoidea Reuss, 1850.
Chrysalidinella dimorpha (Brady) = *Chrysalidina dimorpha* Brady, 1881.
Cibicides acknerianus (d'Orbigny) = *Rotalia ackneriana* d'Orbigny, 1846.
Cibicides antilleanus Drooger = *Cibicides pseudoungerianus* (Cushman) var. *antilleana Drooger*, 1953.
Cibicides boueanum (d'Orbigny) = *Truncatulina boueana* d'Orbigny, 1846.
Cibicides corpulentus Phleger and Parker, 1952.
Cibicides deprimus Phleger and Parker, 1951.
Cibicides dispars (d'Orbigny) = *Truncatulina dispars* d'Orbigny, 1839.
Cibicides dorsopustulosus LeRoy, 1939.
Cibicides fletcheri Galloway and Wissler, 1927.

Cibicides floridanus (Cushman) = *Truncatulina floridana* Cushman, 1918.
Cibicides gallowayi Cushman and Valentine, 1930.
Cibicides haidingeri (d'Orbigny) var. *pacificus* Cushman, 1924.
Cibicides lobatulus (Walker and Jacob) = *Nautilus lobatulus* Walker and Jacob, 1798.
Cibicides mabahethi Said, 1949.
Cibicides marlboroughensis Vella, 1957.
Cibicides mckannai Galloway and Wissler var. *suppressus* Martin, 1952.
Cibicides mollis Phleger and Parker, 1951.
Cibicides praecinctus (Karrer) = *Truncatulina praecinctus* Karrer, 1868.
Cibicides protruberans Parker, 1954.
Cibicides refulgens Montfort, 1808.
Cibicides umbonatus Phleger and Parker, 1951.
Cibicides ungerianus (d'Orbigny) = *Rotalina ungeriana* d'Orbigny, 1846.
Cibicidoides bradyi (Trauth) = *Truncatulina bradyi* Trauth, 1918.
Cibicidoides 'floridanus' (Cushman) see Poag (1981).
Cibicidoides kullenbergi (Parker) = *Cibicides kullenbergi* Parker, 1953.
Cibicidoides pseudoungerianus (Cushman) = *Truncatulina pseudoungeriana* Cushman, 1922.
Conotrochammina bullata (Höglund) = *Trochamminella bullata* Höglund, 1947.
Conotrochammina kennetti Echols, 1971.
Cornuspira involvens (Reuss) = *Operculina involvens* Reuss, 1850.
Cribrostomoides arenacea (Heron-Allen and Earland) = *Truncatulina lobatula* (Walker and Jacob) var. *arenacea* Heron-Allen and Earland, 1922.
Cribrostomoides crassimargo (Norman) = *Haplophragmium crassimargo* Norman, 1892.
Cribrostomoides jeffreysii (Williamson) = *Nonionina jeffreysii* Williamson, 1858.
Cribrostomoides kosterensis (Höglund) = *Labrospira kosterensis* Höglund, 1947.
Cribrostomoides scitulus (Brady) = *Haplophragmium scitulum* Brady, 1881.
Cribrostomoides sphaeriloculus (Cushman) = *Haplophragmoides sphaeriloculum* Cushman, 1910.
Cribrostomoides subglobosum (Sars) = *Lituola subglobosa* M. Sars, 1868.
Cuneata arctica (Brady) = *Reophax arctica* Brady, 1881.
Cyclammina cancellata Brady, 1879.
Cyclammina orbicularis Brady, 1881.
Cyclammina pusilla Brady, 1884.
Cyclocibicides vermiculatus (d'Orbigny) = *Planorbulina vermiculata* d'Orbigny, 1826.
Cycloclypeus carpenteri Brady, 1881.
Cyclorbiculina compressa (d'Orbigny) = *Orbiculina compressa* d'Orbigny, 1839.
Cymbaloporella tabellaeformis (Brady) = *Cymbalopora tabellaeformis* Brady, 1884.
Cymbaloporetta bradyi (Cushman) = *Cymbalopora poeyi* (d'Orbigny) var. *bradyi* Cushman, 1915.
Cymbaloporetta squamosa (d'Orbigny) = *Rosalina squamosa* d'Orbigny, 1839.
Cystammina pauciloculata Brady, 1884.

Dendronina arborescens Heron-Allen and Earland, 1922.
Deuterammina glabra (Heron-Allen and Earland) = *Trochammina glabra* Heron-Allen and Earland, 1932.
Discanomalina semipunctata (Bailey) = *Rotalina semipunctata* Bailey, 1851.
Discanomalina vermiculata (d'Orbigny) = *Truncatulina vermiculata* d'Orbigny, 1839.
Discorbis candeianus (d'Orbigny) = *Rosalina candeianus* d'Orbigny, 1839.
Discorbis chinensis Huang, 1970.
Discorbis floridensis Cushman = *Discorbis bertheloti* (d'Orbigny) var. *floridensis* Cushman, 1931.
Discorbis mira Cushman, 1922. See *Neoeponides auberi* (d'Orbigny).
Discorbis monicana Zalesny, 1959.
Discorbis villardeboanus (d'Orbigny) = *Rosalina vilardeboana* d'Orbigny, 1839.
Discorinopsis aguayoi (Bermudez) = *Discorbis aguayoi* Bermudez, 1935.
Dyocibicides biserialis Cushman and Valentine, 1930.

Eggerella advena (Cushman) = *Verneuilina advena* Cushman, 1921.
Eggerella bradyi (Cushman) = *Verneuilina bradyi* Cushman, 1911.
Eggerella parkerae (Uchio) = *Karreriella parkerae* Uchio, 1960.
Eggerelloides medius (Höglund) = *Verneulina media* Höglund, 1947.
Eggerelloides scabrus (Williamson) = *Bulimina scabra* Williamson, 1858.
Ehrenbergina compressa Cushman, 1927.
Ehrenbergina glabra Heron-Allen and Earland = *Ehrenbergina hystrix* var. *glabra* Heron-Allen and Earland, 1922.
Ehrenbergina pacifica Cushman, 1927.
Ehrenbergina pupa (d'Orbigny) = *Cassidulina pupa* d'Orbigny, 1839.

Ehrenbergina spinea Cushman, 1935.
Ehrenbergina trigona Goës, 1896.
Eilohedra levicula (Resig) = *Epistominella levicula* Resig, 1958.
Elphidiella arctica (Parker and Jones) = *Polystomella arctica* Parker and Jones, 1864.
Elphidiella hannai (Cushman and Grant) = *Elphidium hannai* Cushman and Grant, 1927.
Elphidiella nitida Cushman, 1936.
Elphidiella recens (Stschedrina) = *Elphidium oregonense recens* Stschedrina, 1955.
Elphidium advenum (Cushman) = *Polystomella advena* Cushman, 1922.
Elphidium albiumbilicatulum (Weiss) = *Nonion pauciloculum* Cushman subsp. *albiumbilicatulum* Weiss, 1954.
Elphidium articulatum (d'Orbigny) = *Polystomella articulata* d'Orbigny, 1839.
Elphidium asterineus (Troitskaya) = *Cribroelphidium asterineus* Troitskaya, 1973.
Elphidium bartletti Cushman, 1933.
Elphidium batialis Saidova, 1961.
Elphidium cariacoense Bermudez and Seiglie, 1963.
Elphidium charlottensis Vella, 1957.
Elphidium clavatum Cushman = *Elphidium incertum* (Williamson) var. *clavatum* Cushman, 1930. = *Elphidium excavatum* (Terquem) forma *clavatum.*
Elphidium complanatum (d'Orbigny) = *Polystomella complanata* d'Orbigny, 1839.
Elphidium craticulatus (Fichtel and Moll) = *Nautilus craticulatus* Fichtel and Moll, 1798.
Elphidium crispum (Linné) = *Nautilus crispus* Linné, 1758.
Elphidium cuvillieri Levy, 1966.
Elphidium decipiens Cushman, 1939.
Elphidium delicatulum Bermudez, 1949.
Elphidium discoidale (d'Orbigny) = *Polystomella discoidalis* d'Orbigny, 1839.
Elphidium etigoensis Husezima and Maruhasi, 1944.
Elphidium excavatum (Terquem) = *Polystomella excavata* Terquem, 1875.
Elphidium excavatum (Terquem) forma *alba* Feyling-Hanssen, 1972.
Elphidium fax Nicol, 1944.
Elphidium frigidum Cushman, 1933.
Elphidium galvestonense Kornfeld = *Elphidium gunteri* Cole var. *galvestonensis* Kornfeld, 1931 (part).
Elphidium gerthi Van Voorthuysen, 1957.
Elphidium granosum (d'Orbigny) = *Nonionina granosa* d'Orbigny, 1826.
Elphidium gunteri Cole, 1931.
Elphidium hughesi Cushman and Grant, 1927.
Elphidium imperatrix (Brady) = *Polystomella imperatrix* Brady, 1881.
Elphidium incertum (Williamson) = *Polystomella umbilicatula* var. *incerta* Williamson, 1858.
Elphidium iranicum (Yassini and Ghahreman) = *Cribroelphidium iranicum* Yassini and Ghahreman, 1977.
Elphidium jenseni Cushman = *Polystomella macella* (Fichtel and Moll) var. *jenseni* Cushman, 1905.
Elphidium koeboeense LeRoy, 1939.
Elphidium kugleri (Cushman and Brönnimann) = *Cribroelphidium kugleri* Cushman and Brönnimann, 1948.
Elphidium lene Cushman and McCulloch = *Elphidium incertum* (Williamson) var. *lene* Cushman and McCulloch, 1940.
Elphidium lessonii (d'Orbigny) = *Polystomella lessonii* d'Orbigny, 1839.
Elphidium lidoense Cushman, 1936.
Elphidium macellum (Fichtel and Moll) = *Nautilus macellus* Fichtel and Moll, 1798.
Elphidium magellanicum Heron-Allen and Earland, 1932.
Elphidium margaritaceum Cushman = *Elphidium advenum* (Cushman) var. *margaritaceum* Cushman, 1930.
Elphidium matagordanum (Kornfeld) = *Nonion depressula* (Walker and Jacob) var. *matagordana* Kornfeld, 1931.
Elphidium microgranulosum (Thalmann) = *Themeon granulosus* Galloway and Wissler, 1927.
Elphidium miyakoensis (Ujiié and Kusukawa) = *Cribrononion miyakoense* Ujiié and Kusukawa, 1969.
Elphidium morenoi Bermudez, 1935.
Elphidium novozealandicum Cushman, 1936.
Elphidium oceanensis (d'Orbigny) = *Polystomella oceanensis* d'Orbigny, 1826.
Elphidium oceanicum Cushman, 1933.
Elphidium oregonense Cushman and Grant, 1927.
Elphidium pauciloculum (Cushman) = *Nonion pauciloculum* Cushman, 1944. (also = *E. subarcticum* Cushman, 1944 according to Buzas, 1969).
Elphidium poeyanum (d'Orbigny) = *Polystomella poeyana* d'Orbigny, 1839.
Elphidium porisuturalis (Zheng) = *Cribrononion porisuturalis* Zheng, 1965.
Elphidium reticulosum Cushman, 1933.
Elphidium rugulosum Cushman and Wickenden, 1929.
Elphidium sandiegoensis Lankford, 1962.
Elphidium selseyense (Heron-Allen and Earland)

synonym of *E. excavatum* (Terquem).
Elphidium simplex Cushman, 1933.
Elphidium somaense Takayanagi, 1955.
Elphidium subarcticum Cushman, 1944.
Elphidium subgranulosum Asano, 1938.
Elphidium subincertum Asano, 1950.
Elphidium tokyoensis Aoki, 1967.
Elphidium translucens Natland, 1938.
Elphidium tsudai Chiji and Nakaseko, 1950.
Elphidium tumidum Natland, 1938.
Elphidium varium Buzas, 1965.
Elphidium williamsoni Haynes, 1973.
Eoeponidella pulchella (Parker) = *Pninaella pulchella* Parker, 1952.
Epistomaroides mexicana (Parker) = *Anomalinoides mexicana* Parker, 1954.
Epistominella arctica Green, 1960.
Epistominella bradyana (Cushman) = *Pulvinulinella bradyana* Cushman, 1927.
Epistominella exigua (Brady) = *Pulvinalina exigua* Brady, 1844.
Epistominella naraensis (Kuwano) = *Pseudoparrella naraensis* Kuwano, 1950.
Epistominella obesa Bandy and Arnal, 1957.
Epistominella pacifica (Cushman) = *Pulvinulinella pacifica* Cushman, 1927.
Epistominella sandiegoensis Uchio, 1960.
Epistominella smithi (Stewart and Stewart) = *Pulvinulinella smithi* Stewart and Stewart, 1930.
Epistominella takayanagi Iwasa, 1955.
Epistominella tamana (Kuwano) = *Pseudoparrella tamana* Kuwano, 1950.
Epistominella vitrea Parker, 1953.
Eponides babsae Brenner, 1962.
Eponides pusillus Parr, 1950.
Eponides regularis Phleger and Parker, 1951.
Eponides repandus (Fichtel and Moll) = *Nautilus repandus* Fichtel and Moll, 1798.
Eponides subtener (Galloway and Wissler) = *Rotalia subtenera* Galloway and Wissler, 1927.
Eponides tumidulus (Brady) = *Truncatulina tumidula* Brady, 1884.
Eponides tumidulus (Brady) var. *horvathi* Green, 1960.
Eponides turgidus Phleger and Parker, 1951.
Eratidus foliaceum (Brady) = *Haplophragmium foliaceum* Brady, 1881.
Evolvocassidulina orientalis (Cushman) = *Cassidulina orientalis* Cushman, 1922.

Fijiella simplex (Cushman) = *Trimosina simplex* Cushman, 1929.
Fissurina lucida (Williamson) = *Entosolenia marginata* (Montagu) var. *lucida* Williamson, 1848.
Fissurina marginata (Montagu) = *Vermiculum marginatum* Montagu, 1803.
Fissurina orbignyana Seguenza, 1862.
Fissurina semimarginata (Boomgart) = *Entosolenia semimarginata* Boomgart, 1949.
Flintina bradyana Cushman, 1921.
Flintinoides labiosa (d'Orbigny) = *Triloculina labiosa* d'Orbigny, 1839.
Fontbotia wuellerstorfi (Schwager) = *Anomalina wuellerstorfi* Schwager, 1866.
Fursenkoina apertura (Uchio) = *Virgulina apertura* Uchio, 1960.
Fursenkoina bramletti (Galloway and Morrey) = *Virgulina bramletti* Galloway and Morrey, 1929.
Fursenkoina complanata (Egger) = *Virgulina shreibersiana* Czjzek var. *complanata* Egger, 1893.
Fursenkoina davisi (Chapman and Parr) = *Virgulina davisi* Chapman and Parr, 1931.
Fursenkoina earlandi (Parr) = *Bolivina earlandi* Parr, 1950.
Fursenkoina fusiformis (Williamson) = *Bulimina pupoides* d'Orbigny var. *fusiformis* Williamson, 1858.
Fursenkoina loeblichi (Feyling-Hanssen) = *Virgulina loeblichi* Feyling-Hanssen, 1954.
Fursenkoina pontoni (Cushman) = *Virgulina pontoni* Cushman, 1932.
Fursenkoina rotundata (Parr) = *Virgulina rotundata* Parr, 1950.
Fursenkoina sandiegoensis (Uchio) = *Virgulina sandiegoensis* Uchio, 1960.
Fursenkoina schreibersiana (Czjzek) = *Virgulina schreibersiana* Czjzek, 1848.
Fursenkoina spinicostata Phleger and Parker, 1951.
Fursenkoina spinosa (Heron-Allen and Earland) = *Virgulina schreibersiana* Czjzek var. *spinosa* Heron-Allen and Earland, 1932.

Gaudryina aequa Cushman, 1947.
Gaudryina arenaria Galloway and Wissler, 1927.
Gaudrying atlantica (Bailey) var. *pacifica* Cushman and McCulloch, 1939.
Gaudryina exilis Cushman and Brönnimann, 1948.
Gaudryina rudis Wright, 1900.
Gaudryina wrightiana Millett, 1900.
Gavelinopsis companulata (Galloway and Wissler) = *Globoratalia campanulata* Galloway and Wissler, 1927.
Gavelinopsis praegeri (Heron-Allen and Earland) = *Discorbina praegeri* Heron-Allen and Earland, 1913.

Glabratella ornatissima (Cushman) = *Discorbis ornatissima* Cushman, 1925.
Glabratella patelliformis (Brady) = *Discorbina patelliformis* Brady, 1884.
Glabratella pyramidalis (Heron-Allen and Earland) = *Discorbina pyramidalis* Heron-Allen and Earland, 1924.
Glabratella subopercularis (Asano) = *Discorbis subopercularis* Asano, 1951.
Glabratella wrightii (Brady) = *Discorbina wrightii* Brady, 1881.
Glandulina nipponica Asano, 1951.
Glaphyrammina americana (Cushman) = *Ammobaculites americanus* Cushman, 1910.
Globobulimina affinis (d'Orbigny) = *Bulimina affinis* d'Orbigny, 1839.
Globobulimina auriculata (Bailey) = *Bulimina auriculata* Bailey, 1851.
Globobulimina hanzawai Asano, 1958.
Globobulimina pacifica Cushman, 1927.
Globobulimina turgida (Bailey) = *Bulimina turgida* Bailey, 1851.
Globocassidulina biora (Crespin) = *Cassidulina biora* Crespin, 1960.
Globocassidulina canalisuturata Eade, 1967.
Globocassidulina oblonga (Reuss) = *Cassidulina oblonga* Reuss, 1850.
Globocassidulina subglobsa (Brady) = *Cassidulina subglobosa* Brady, 1881.
Glomospira gordialis (Jones and Parker) = *Trochammina squamata*, var. *gordialis* Jones and Parker, 1860.
Göesella flintii Cushman, 1936.
Göesella mississippiensis Parker, 1954.
Gypsina plana (Carter) = *Polytrema planum* Carter, 1876.
Gypsina vesicularis (Parker and Jones) = *Orbitolina concava* var. *vesicularis* Parker and Jones, 1860.
Gyroidina altiformis R E and K C Stewart, 1930.
Gyroidina cushmani Boomgaart, 1949.
Gyroidina io Resig, 1958.
Gyroidina neosoldanii Brotzen, 1936.
Gyroidina orbicularis d'Orbigny, 1826.

Hansenisca soldanii d'Orbigny, 1826.
Hanzawaia concentrica (Cushman) = *Cibicides concentricus* Cushman, 1918.
Hanzawaia nipponica Asano, 1944.
Hanzawaia nitidula (Bandy) = *Cibicides basiloba* (Cushman) var. *nitidula* Bandy, 1953.
Haplophragmoides bonplandi Todd and Brönnimann, 1957.
Haplophragmoides bradyi (Robertson) = *Trochammina bradyi* Robertson, 1891.
Haplophragmoides canariensis (d'Orbigny) = *Nonionina canariensis* d'Orbigny, 1839.
Haplophragmoides hancocki Cushman and McCulloch, 1939.
Haplophragmoides labukensis Dhillon, 1968.
Haplophragmoides parkerae (Uchio) = *Recurvoidella parkerae* Uchio, 1960.
Haplophragmoides quadratus Earland, 1934.
Haplophragmoides rotulatum (Brady) = *Haplophragmium rotulatum* Brady, 1881.
Haplophragmoides subinvolutum Cushman and McCulloch, 1839.
Haplophragmoides wilberti Anderson, 1953.
Hauerina pacifica Cushman, 1917.
Haynesina germanica (Ehrenberg) = *Nonion germanica* Ehrenberg, 1840.
Haynesina orbiculare (Brady) = *Nonionina orbiculare* Brady, 1881.
Haynesina schmitti (Cushman and Wickenden) = *Elphidium schmitti* Cushman and Wickenden, 1929.
Haynesina tuberculata (d'Orbigny) = *Nonionina tuberculata* d'Orbigny, 1846.
Helenina andersoni (Warren) = *Pseudoeponides andersoni* Warren, 1957.
Hemisphaerammina bradyi Loeblich and Tappan, 1957.
Heterillina cribrostoma (Heron-Allen and Earland) = *Miliolinella circularis* (Bornemann) var. *cribrostoma* Heron-Allen and Earland, 1915.
Heterocyclina tuberculata (Möbius) = *Heterostegina tuberculata* Möbius, 1880.
Heterolepa dutemplei (d'Orbigny) = *Rotalina dutemplei* d'Orbigny, 1846.
Heterolepa mediocris (Finlay) = *Cibicides mediocris* Finlay, 1940.
Heterostegina antillarum d'Orbigny, 1839.
Heterostegina apogama Röttger, 1987.
Heterostegina depressa d'Orbigny, 1826.
Heterostegina operculinoides Hofker, 1927.
Heterostegina suborbicularis d'Orbigny, 1826.
Hippocrepinella hirudinea Heron-Allen and Earland, 1932.
Hippocrepina pusilla Heron-Allen and Earland, 1930.
Hoeglundina elegans (d'Orbigny) = *Rotalina elegans* d'Orbigny, 1826.
Homotrema rubra (Lamarck) = *Millepora rubra* Lamarck, 1816.
Hopkinsina pacifica Cushman, 1933.
Hormosinella distans (Brady) = *Reophax distans*

Brady, 1881.
Hormosina globulifera Brady, 1879.
Hormosina normani Brady, 1879.
Hyalinea balthica (Schröter) = *Nautilus balthicus* Schröter, 1783.
Hyalinea florenceae McCulloch, 1977.
Hyperammina elongata Brady, 1878.
Hyperammina friabilis Brady, 1844.

Islandiella helenae Feyling-Hanssen and Buzas, 1976.
Islandiella islandica (Nørvang) = *Cassidulina islandica* Nørvang, 1945.
Islandiella japonica (Asano and Nakamura) = *Cassidulina japonica* Asano and Nakamura, 1937.
Islandiella kasiwazakiensis (Husezima and Maruhasi) = *Cassidulina kasiwazakiensis* Husezima and Maruhasi, 1966.
Islandiella norcrossi (Cushman) = *Cassidulina norcrossi* Cushman.
Islandiella norcrossi australis Phleger and Parker, 1951.
Islandiella quadrata (Cushman and Hughes) = *Cassidulina subglobosa* Brady var. *quadrata* Cushman and Hughes, 1925.
Islandiella smechovi (Volashinova) = *Cassidulina smechovi* Voloshinova, 1952.
Islandiella umbonata Voloshinova, 1970.

Jadammina macrescens (Brady) = *Trochammina inflata* (Montagu) var. *macrescens* Brady, 1870.
Jullienella foetida Schlumberger, 1890.

Karreriella parkerae Uchio, 1960.
Karrerulina apicularis (Cushman) = *Gaudryina apicularis* Cushman, 1911.

Laevipeneroplis proteus (d'Orbigny) = *Peneroplis proteus* d'Orbigny, 1839.
Lagena apiopleura Loeblich and Tappan, 1953.
Lagena striata (d'Orbigny) = *Oolina striata* d'Orbigny, 1839.
Lagenammina difflugiformis (Brady) = *Reophax difflugiformis* Brady, 1879.
Laticarinina pauperata (Parker and Jones) = *Pulvinulina repanda* var. *pauperata* Parker and Jones, 1865.
Lenticulina cushmani (Galloway and Wissler) = *Robulus cushmani* Galloway and Wissler, 1927.
Lepidodeuterammina ochracea (Williamson) = *Rotalina ochracea* Williamson, 1858.
Loxostomum pseudobeyrichi (Cushman) = *Bolivina pseudobeyrichi* Cushman, 1926.

Marginopora vertebralis Quoy and Gaimard, in de Blainville, 1830.
Marsupulina schultzei Rhumbler, 1904.
Martinottiella antarctica (Parr) = *Schenckiella antarctica* Parr, 1950.
Martinottiella communis (d'Orbigny) = *Clavulina communis* d'Orbigny, 1826.
Martinottiella gymnesica (Colom) = *Listerella gymnesica* Colom, 1964.
Martinottiella minuta (Hofker) = *Listerella minuta* Hofker, 1951.
Massilina decorata Cushman, 1922.
Massilina protea Parker, 1953.
Massilina secans (d'Orbigny) = *Quinqueloculina secans* d'Orbigny, 1826.
Melonis barleeanum (Williamson) = *Nonionina barleeana* Williamson, 1858.
Melonis pompilioides (Fichtel and Moll) = *Nautilus pompilioides* Fichtel and Moll, 1798.
Melonis zaandamae (Van Voorthuysen) = *Anomalinoides barleeanum* (Williamson) *zaandamae* Van Voorthuysen, 1952.
Miliammina arenacea (Chapman) = *Miliolina oblonga* (Montagu) var. *arenacea* Chapman, 1916.
Miliammina lata Heron-Allen and Earland, 1930.
Miliammina fusca (Brady) = *Quinqueloculina fusca* Brady, 1870.
Miliammina herzensteini (Schlumberger) = *Sigmoilina herzensteini* Schlumberger, 1894.
Miliammina lata Heron-Allen and Earland, 1929.
Miliammina petila Saunders, 1958.
Miliolinella circularis (Bornemann) = *Triloculina circularis* Bornemann, 1855.
Miliolinella elongata Kruit = *Miliolinella circularis* (Bornemann) var. *elongata* Kruit, 1955.
Miliolinella microstoma Warren, 1957.
Miliolinella subrotunda (Montagu) = *Vermiculum subrotundum* Montagu, 1803.
Miniacina miniacea Pallas, 1766.
Murrayinella erinacea Heron-Allen and Earland, 1915. New name for *Discorbina imperiator* var. *globosa* Millett, 1903.
Murrayinella minuta (Takayanagi) = *Rotalia* ? *minuta* Takayanagi, 1955.
Murrayinella murrayi (Heron-Allen and Earland) = *Rotalia murrayi* Heron-Allen and Earland, 1915.

Neoconorbina pacifica Hofker, 1951.
Neoconorbina terquemi (Rzehak) = *Discorbina terquemi* Rzehak, 1888.
Neoeponides auberi (d'Orbigny) = *Rosalina auberi* d'Orbigny, 1839.

Nodellum membranaceum (Brady) = *Reophax membranaceum* Brady, 1913.
Nodobaculariella atlantica Cushman and Hanzawa, 1937.
Nodosaria raphanus (Linné) = *Nautilus rephanus* Linné, 1758.
Nodulina dentaliniformis (Brady) = *Reophax dentaliniformis* Brady, 1884.
Nonion akitaensis Asano, 1950.
'Nonion asterizans' (Fichtel and Moll) (see Loeblich and Tappan 1988: 640).
Nonion commune (d'Orbigny) = *Nonionina communis* d'Orbigny, 1846.
Nonion depressulus (Walker and Jacob) = *Nautilus depressulus* Walker and Jacob, 1798.
Nonion nicobarensis Cushman, 1936.
Nonion pacificum (Cushman) = *Nonionina umbilicatula* (Montagu) var. *pacifica* Cushman, 1924.
Nonionella atlantica Cushman, 1947.
Nonionella basiloba Cushman and McCulloch, 1940.
Nonionella basispinata (Cushman and Moyer) = *Nonion pizarrense* var. *basispinatum* Cushman and Moyer, 1930.
Nonionella bradii (Chapman) = *Nonionina scapha* var. *bradii* Chapman, 1916.
Nonionella fragilis Uchio, 1960.
Nonionella globosa Ishiwada, 1950.
Nonionella iridea Heron-Allen and Earland, 1932.
Nonionella monicana Zalesny, 1959.
Nonionella opima Cushman, 1947.
Nonionella pulchella Hada, 1931.
Nonionella stella Cushman and Moyer = *Nonionella miocenica* Cushman var. *stella* Cushman and Moyer, 1930.
Nonionella turgida (Williamson) = *Rotalina turgida* Williamson, 1858.
Nonionellina flemingi (Vella) = *Nonion flemingi* Vella, 1957.
Nonionellina labradorica (Dawson) = *Nonionina labradorica* Dawson, 1860.
Nonionoides boueanum (d'Orbigny) = *Nonionina boueanum* d'Orbigny, 1846.
Nonionoides elongatum (d'Orbigny) = *Nonionina elongata* d'Orbigny, 1852.
Nonionoides grateloupi (d'Orbigny) = *Nonionina grateloupi* d'Orbigny, 1826.
Nonionoides scaphum (Fichtel and Moll) = *Nautilus scapha* Fichtel and Moll, 1798.
Nonionoides sloanii d'Orbigny = *Nonionina sloanii* d'Orbigny, 1839.
Notodendrodes antarctikos Delaca, Lipps and Hessler, 1980.
Nouria polymorphinoides Heron-Allen and Earland, 1914.
Nubecularia lucifuga Defrance, 1825.
Nummulites cumingii (Carpenter) = *Amphistegina cumingii* Carpenter, 1860.
Nuttallides decoratus (Phleger and Parker) = *Pseudoparrella*? *decorata* Phleger and Parker, 1951.
Nuttallides rugosus (Phleger and Parker) = *Pseudoparrella*? *rugosa* Phleger and Parker, 1951.
Nuttallides umboniferus (Cushman) = *Pulvinulina umbonifera* Cushman, 1933.

Operculina ammonoides (Gronovius) = *Nautilus ammonoides* Gronovius, 1781.
Operculina complanata (Defrance) = *Lenticulites complanatus* Defrance, 1822.
Operculina gaimardii d'Orbigny, 1826.
Oridorsalis tener (Brady) = *Truncatulina tenera* Brady, 1884.
Oridorsalis umbonatus (Reuss) = *Rotalina umbonatus* Reuss, 1851.
Osangularia culter (Parker and Jones) = *Planorbulina farcta* (Fichtel and Moll) var. *ungeriana* (d'Orbigny) subvar. *culter* Parker and Jones, 1865.

Pacinonion novozealandicum (Cushman and Edwards) = *Astrononion novozealandicum* Cushman and Edwards, 1937.
Palmerinella gardenislandensis (Akers) = *Eponidella gardenislandensis* Akers, 1952.
Palmerinella palmerae Bermudez, 1934.
Pararotalia nipponica (Asano) = *Rotalia nipponica* Arano, 1937.
Parasorites 'orbitolinoides' (Hofker) = *Praesorites orbitolinoides* Hofker,
Paratrochammina simplissima (Cushman and McCulloch) = *Trochammina simplissima* Cushman and McCulloch, 1948.
Patellina corrugata Williamson, 1858.
Pateoris hauerinoides (Rhumbler) = *Quinqueloculina subrotunda* (Montagu) forma *hauerinoides* Rhumbler, 1936.
Peneroplis carinatus d'Orbigny, 1839.
Peneroplis pertusus (Forskål) = *Nautilus pertusus* Forskål, 1775.
Peneroplis planatus (Fichtel and Moll) = *Nautilus planatus* Fichtel and Moll, 1798.
Placopsilina bradyi Cushman and McCulloch, 1939.
Placopsilina confusa Cushman, 1920.
Placopsilinella aurantiaca Earland, 1934.
Planoglabratella opercularis (d'Orbigny) = *Rosalina opercularis* d'Orbigny, 1826.

Planogypsina squamiformis (Chapman) = *Gypsina vesicularis* (Parker and Jones) var. *squamiformis* Chapman, 1900.
Planorbulina acervalis Brady, 1884.
Planorbulina mediterranensis d'Orbigny, 1826.
Planorbulina variabilis (d'Orbigny) = *Truncatulina variabilis* d'Orbigny, 1826.
Planorbulinopsis parasita Banner, 1971.
Planulina ariminensis d'Orbigny, 1826.
Planulina exorna Phleger and Parker, 1951.
Planulina foveolata (Brady) = *Anomalina foveolata* Brady, 1884.
Planulina ornata (d'Orbigny) = *Truncatulina ornata* d'Orbigny, 1839.
Polysaccammina ipohalina Scott, 1976.
Polystomammina nitida (Brady) = *Trochammina nitida* Brady, 1884.
Poroeponides lateralis (Terquem) = *Rosalina lateralis* Terquem, 1878.
Portatrochammina antarctica (Parr) = *Trochammina antarctica* Parr, 1950.
Portatrochammina eltaninae Echols, 1971.
Portatrochammina lobata (Cushman) = *Trochammina lobata* Cushman, 1944.
Portatrochammina malovensis (Heron-Allen and Earland) = *Trochammina malovensis* Heron-Allen and Earland, 1929.
Portatrochammina simplissima (Cushman and McCulloch) = *Trochammina pacifica* Cushman var. *simplissima* Cushman and McCulloch, 1948.
Portatrochammina wiesneri (Parr) = *Trochammina wiesneri* Parr, 1950.
Praeglobobulimina spinescens (Brady) = *Bulimina pyrula* d'Orbigny var. *spinescens* Brady 1884.
Prolixoplecta exilis (Cushman) = *Dorothia exilis* Cushman, 1936.
Protoschista findens (Parker) = *Lituola findens* Parker, 1870.
Psammophax consociata Rhumbler, 1931.
Psammosphaera bowmani Heron-Allen and Earland, 1912.
Psammosphaera fusca Schultze, 1875.
Pseudobolivina antarctica Wiesner, 1931.
Pseudobulimina chapmani Heron-Allen and Earland, 1922.
Pseudoclavulina gracilis Zheng, 1978.
Pseudoclavulina robusta Zheng, 1978.
Pseodonodosinella nodulosa (Brady) = *Reophax nodulosa* Brady, 1879.
Pseudoparrella subperuviana Cushman and ten Dam, 1948.
Pseudopolymorphina novangliae (Cushman) = *Polymorphina lactea* (Walker and Jacob) var. *novangliae* Cushman, 1923.
Pseudothurammina limnetis (Scott and Medioli) = *Thurammina*? *limnetis* Scott and Medioli, 1980.
Pullenia bulloides (d'Orbigny) = *Nonionina bulloides* d'Orbigny, 1846.
Pullenia quinqueloba (Reuss) = *Nonionina quinqueloba* Reuss, 1851.
Pullenia simplex Rhumbler, 1931.
Pullenia subcarinata (d'Orbigny) = *Nonionina subcarinata* d'Orbigny, 1839.
Pullenia subsphaerica Parr, 1950.
Pyrgo denticulata (Brady) = *Biloculina ringens* Lamarck var. *denticulata* Brady, 1884.
Pyrgo elongata (d'Orbigny) = *Biloculina elongata* d'Orbigny, 1826.
Pyrgo murrhina (Schwager) = *Biloculina murrhina* Schwager, 1866.
Pyrgo subphaerica (d'Orbigny) = *Biloculina subsphaerica* d'Orbigny, 1840.
Pyrgo williamsoni (Silvestri) = *Biloculina williamsoni* Silvestri, 1923.

Quinqueloculina ackneriana d'Orbigny, 1846.
Quinqueloculina arctica Cushman, 1933.
Quinqueloculina aspera d'Orbigny, 1826.
Quinqueloculina auberiana d'Orbigny, 1826.
Quinqueloculina compta Cushman, 1947.
Quinqueloculina costata d'Orbigny, 1826.
Quinqueloculina crassicarinata Collins, 1958.
Quinqueloculina curta Cushman = *Quinqueloculina disparalis* var. *curta* Cushman, 1917.
Quinqueloculina dimidiata Terquem, 1876.
Quinqueloculina dunkerquiana (Heron-Allen and Earland) = *Miliolina dunkerquiana* Heron-Allen and Earland, 1930.
Quinqueloculina dutemplei d'Orbigny, 1846.
Quinqueloculina impressa Reuss, 1851.
Quinqueloculina intrica Terquem, 1878.
Quinqueloculina laevigata d'Orbigny, 1839.
Quiqueloculina lamarckiana d'Orbigny, 1839.
Quinqueloculina lata Terquem, 1876.
Quinqueloculina milletti (Weisner) = *Miliolina milletti* Wiesner, 1912.
Quinqueloculina oblonga (Montagu) = *Vermiculum oblongum* Montagu, 1893.
Quinqueloculina poeyana d'Orbigny, 1839.
Quinqueloculina rugosa d'Orbigny, 1839.
Quinqueloculina schlumbergeri (Wiesner) = *Miliolina schlumbergeri* Wiesner, 1923.
Quinqueloculina seminula (Linné) = *Serpula seminlum* Linné, 1758.

Quinqueloculina subcuneata Cushman = *Quinqueloculina crassa* d'Orbigny var. *subcuneata* Cushman, 1921.
Quinqueloculina triangularis d'Orbigny, 1846.
Quinqueloculina undulata d'Orbigny, 1852.
Quinqueloculina ungeriana d'Orbigny, 1846.
Quinqueloculina viennensis Le Calvez and Le Calvez, 1958.
Quinqueloculina vulgaris d'Orbigny, 1826.

Rectobolivina raphana (Parker and Jones) = *Uvigerina (Sagrina) raphanus* Parker and Jones, 1865.
Rectuvigerina arquatensis Papp, 1963.
Rectuvigerina nicoli Mathews, 1945.
Rectuvigerina phlegeri Le Calvez, 1958.
Recurvoides contortus Earland, 1934.
Recurvoides trochamminiforme Höglund, 1947.
Recurvoides turbinatus (Brady) = *Haplophragmium turbinatum* Brady, 1881.
Remaneica plicata (Terquem) = *Patellina plicata* Terquem, 1876.
Reophanus ovicula (Brady) = *Hormosina ovicula* Brady, 1879.
Reophax barwonensis Collins, 1974.
Reophax bilocularis Flint, 1899.
Reophax calcareus (Cushman) = *Proteonina difflugiformis* (Brady) var. *calcarea* Cushman, 1947.
Reophax caribensis Seiglie and Bermudez, 1969.
Reophax comprima (Phleger and Parker) = *Proteonina comprima* Phleger and Parker, 1951.
Reophax curtus Cushman, 1920.
Reophax cylindrica Brady, 1884.
Reophax excentricus Cushman, 1910.
Reophax fusiformis (Williamson) = *Proteonina fusiformis* Williamson, 1858.
Reophax gracilis (Kiaer) = *Nodulina gracilis* Kiaer, 1900.
Reophax guttifer (Brady) = *Lituola (Reophax) guttifera* Brady, 1881.
Reophax hispidulus Cushman, 1920.
Reophax micaceous Earland, 1934.
Reophax moniliformis Siddall, 1886.
Reophax nana Rhumbler, 1913.
Reophax pilulifer Brady, 1884.
Reophax scorpiurus Montfort, 1808.
Reophax scottii Chaster, 1892.
Reophax subdentaliniformis Parr, 1950.
Reophax subfusiformis Earland, 1933.
Repmanina charoides (Jones and Parker) = *Trochammina squamata* Jones and Parker var. *charoides* Jones and Parker, 1860.
Reussella atlantica Cushman = *Reussella spinulosa* (Reuss) var. *atlantica* Cushman, 1947.
Reussella pacifica Cushman and McCulloch, 1948.
Reussella spinulosa (Reuss) = *Verneuilina spinulosa* Reuss, 1850.
Rhabdammina abyssorum Sars, 1869.
Rhabdammina linearis Brady, 1879.
Rhabdamminella cylindrica (Brady) = *Marsipella cylindrica* Brady, 1882.
Robertina translucens Cushman and Parker, 1936.
Robertinoides normani (Göes) = *Bulimina normani* Göes, 1894.
Rosalina adhaerens Murray, 1965.
Rosalina anomala Terquem, 1875.
Rosalina bradyi (Cushman) = *Discorbis globularis* var. *bradyi* Cushman, 1915.
Rosalina campanulata (Galloway and Wissler) = *Discorbis campanulata* Galloway and Wissler, 1927.
Rosalina carnivora Todd, 1965.
Rosalina columbiensis (Cushman) = *Discorbis columbiensis* Cushman, 1925.
Rosalina floridana (Cushman) = *Discorbis floridana* Cushman, 1922.
Rosalina globularis d'Orbigny, 1826.
Rosalina leei Hedley and Wakefield, 1967.
Rosalina obtusa d'Orbigny, 1846.
Rosalina posidonicola (Colom) = *Conorboides posidonicola* Colom, 1942.
Rosalina villardeboana d'Orbigny, 1839.
Rosalina williamsoni (Chapman and Parr) = *Discorbis williamsoni* Chapman and Parr, 1932.
Rotaliammina mayori Cushman, 1924.
Rotaliammina squamiformis (Cushman and McCulloch) = *Trochammina squamiformis* Cushman and McCulloch, 1939.
Rotalidium annectens (Parker and Jones) = *Rotalia annectens* Parker and Jones, 1865.
Rotalidium japonicum (Hada) = *Rotalia japonica* Hada, 1931.
Rotalinoides gaimardi (d'Orbigny) = *Rotalia (Turbinularia) gaimardi* d'Orbigny, 1826.
Rupertina stabilis (Wallich) = *Rupertia stabilis* Wallich, 1877.

Saccammina alba Hedley, 1962.
Saccammina atlantica (Cushman) = *Proteonina atlantica* Cushman, 1944.
Saccammina spherica G. O. Sars, 1871.
Saccammina tabulata Rhumbler, 1931.
Saccorhiza ramosa (Brady) = *Hyperammina ramosa* Brady, 1879.
Sagrina pulchella d'Orbigny, 1839.

Saidovina karreriana (Brady) = *Bolivina karreriana* Brady, 1881.
Sigmoihauerina bradyi (Cushman) = *Hauerina bradyi* Cushman, 1917.
Sigmoilinita tenuis (Czjzek) = *Sigmoilina tenuis* Czjzek, 1848.
Sigmoilina umbonata Heron-Allen and Earland, 1922.
Sigmoilopsis schlumbergeri (Silvestri) = *Sigmoilina schlumbergi* Silvestri, 1904.
Siphonaperta horrida (Cushman) = *Quinqueloculina horrida* Cushman, 1947.
Siphonaperta minuta (Collins) = *Massilina minuta* Collins, 1958.
Siphotrochammina elegans Zaninetti, Brönnimann, Beurlen and Moura, 1977.
Siphotrochammina lobata Saunders, 1957.
Siphouvigerina interrupta (Brady) = *Uvigerina interrupta* Brady, 1879.
Sorites marginalis (Lamarck) = *Orbulites marginalis* Lamarck, 1816.
Sorites orbiculus (Forskål) = *Nautilus orbiculus* Forskål, 1775.
Sorites variabilis Lacroix, 1941.
Sphaerogypsina globula (Reuss) = *Ceriopora globulus* Reuss, 1848.
Sphaeroidina bulloides d'Orbigny, 1826.
Spirillina vivipara Ehrenberg, 1843.
Spirolina acicularis (Batsch) = *Nautilus (Lituus) acicularis* Batsch, 1791.
Spirolina arietina (Batsch) = *Nautilus (Lituus) arietinus* Batsch, 1791.
Spiroloculina angulosa Terquem, 1878.
Spiroloculina scita Cushman and Todd, 1944.
Spiroloculina soldanii Fornasini, 1886.
Spiroloculina turcomanica Brodskij, 1928.
Spirophthalmidium acutimargo Brady, 1884. = *Textularia agglutinans* d'Orbigny *biformis* Parker and Jones, 1865.
Spiroplectinella wrightii (Silvestri) = *Spiroplecta wrightii* Silvestri, 1903.
Stetsonia horvathi Green, 1960.
Stetsonia minuta Parker, 1954.
Streblus batavus Hofker – see *Ammonia beccarii* (Linné).
Subreophax aduncus (Brady) = *Reophax aduncus* Brady, 1882.
Suggrunda eckisi Natland, 1950.

Textularia agglutinans d'Orbigny, 1839.
Textularia bocki Höglund, 1947.
Textularia calva Lalicker, 1940.
Textularia candeina d'Orbigny, 1840.
Textularia conica d'Orbigny, 1839.
Textularia earlandi Parker, 1952.
Textularia foliacea Heron-Allen and Earland, 1915.
Textularia gramen d'Orbigny, 1846.
Textularia lateralis Lalicker, 1935.
Textularia mayori Cushman, 1922.
Textularia mexicana Cushman, 1922.
Textularia millettii Cushman, 1911.
Textularia panamensis Cushman, 1918.
Textularia parvula Cushman, 1922.
Textularia pseudogramen Chapman and Parr, 1937.
Textularia pseudorugosa Lacroix, 1932.
Textularia pseudoturris Cushman, 1922.
Textularia sagittula Defrance, 1824.
Textularia sandiegoensis Uchio, 1960.
Textularia schencki Cushman and Valentine, 1930.
Textularia secasensis Lalicker and McCulloch, 1940.
Textularia torquata Parker, 1952.
Textularia truncata Höglund, 1947.
Textularia vola Lalicker and McCulloch, 1940.
Textularia wiesneri Earland, 1933.
Tholosina bulla (Brady) = *Placopsilina bulla* Brady, 1881.
Thurammina faerleensis Höglund, 1948.
Thurammina papillata Brady, 1879.
Thurammina sphaerica Höglund, 1947.
Tiphotrocha comprimata (Cushman and Brönnimann) = *Trochammina comprimata* Cushman and Brönnimann, 1948.
Tolypammina vagans (Brady) = *Hyperammina vagans* Brady, 1879.
Tretomphaloides concinnus (Brady) = *Discorbina concinna* Brady, 1884.
Tretomphalus bulloides (d'Orbigny) = *Rosalina bulloides* d'Orbigny, 1839.
Trichohyalus tropicus (Collins) = *Discorinopsis tropica* Collins, 1958.
Trifarina angulosa (Williamson) = *Uvigerina angulosa* Williamson, 1858.
Trifarina bella (Phleger and Parker) = *Angulogerina bella* Phleger and Parker, 1951.
Trifarina bradyi Cushman, 1923.
Trifarina carinata (Cushman) = *Angulogerina carinata* Cushman, 1927.
Trifarina earlandi (Parr) = *Angulolerina earlandi* Parr, 1950.
Trifarina fluens Todd, 1948.
Trifarina fornasinii (Selli) = *Angulogerina fornasinii* Selli, 1948.
Trifarina jamaicensis Cushman and Todd, 1945.
Trifarina kokozuraensis (Asano) = *Angulogerina kokozuraensis* Asano, 1950.

Trifarina occidentalis (Cushman) = *Uvigerina occidentalis* Cushman, 1923.
Trilocularena patensis Closs, 1963.
Triloculina brevidentata Cushman, 1944.
Triloculina fichteliana d'Orbigny, 1826.
Triloculina frigida Lagoe, 1977.
Triloculina inflata d'Orbigny, 1826.
Triloculina insignis (Brady) = *Miliolina insignis* Brady, 1844.
Triloculina longirostra (d'Orbigny) = *Quinqueloculina longirostra* d'Orbigny, 1826.
Triloculina oblonga (Montagu) = *Vermiculum oblongum* Montagu, 1803.
Triloculina rotunda d'Orbigny, 1839.
Triloculina sidebottomi (Martinotti) = *Miliolinella subrotunda* Sidebottom, 1904.
Triloculina trigonula (Lamarck) = *Miliolites trigonula* Lamarck, 1804.
Triloculina trihedra Loeblich and Tappan, 1953.
Triloculinella obliquinoda Riccio, 1950.
Triloculinella procera (Goës) = *Miliolina procera* Goës, 1896.
Tritaxis inhaerens (Wiesner) = *Haplophragmoides canariensis* (d'Orbigny) var. *inhaerens* Wiesner, 1931.
Trochammina advena Cushman, 1922.
Trochammina charlottensis Cushman, 1925.
Trochammina discorbis Earland, 1934.
Trochammina gabboensis Parr, 1950.
Trochammina hadai Uchio, 1962.
Trochammina inconspicua Earland, 1934.
Trochammina inflata (Montagu) = *Nautilus inflatus* Montagu, 1808.
Trochammina japonica Ishiwada, 1950.
Trochammina kellettae Thalmann, 1932.
Trochammina malovensis Heron-Allen and Earland, 1929.
Trochammina multiloculata Höglund, 1947.
Trochammina nana (Brady) = *Haplophragmium nana* Brady, 1881.
Trochammina pacifica Cushman, 1925.
Trochammina pygmaea Höglund, 1947.
Trochammina rhumbleri Uchio, 1960.
Trochammina rossensis Warthin, 1934.
Trochammina rotaliformis Heron-Allen and Earland, 1911.
Trochammina squamata Jones and Parker, 1860.
Trochammina tasmanica Parr, 1950.
Trochammina voluta Saidova, 1975.
Trochamminella bullata Höglund, 1947.
Trochamminopsis conica (Earland) = *Trochammina conica* Earland, 1934.
Trochamminopsis quadriloba (Höglund) = *Trochammina quadriloba* Höglund, 1948.
Trochamminita irregularis Cushman and Brönnimann, 1948.
Trochulina avalonensis (Natland) = *Rotalia avalonensis* Natland, 1938.
Trochulina dimidiata (Parker and Jones) = *Discorbina dimidiata* Parker and Jones, 1951.
Trochulina lomaensis (Bandy) = *Rotalia lomaensis* Bandy, 1953.
Trochulina rosea (d'Orbigny) = *Rotalia rosea* d'Orbigny, 1826.
Trochulina translucens (Phleger and Parker) = *'Rotalia' translucens* Phleger and Parker, 1951.
Trochulina turbinata (Cushman and Valentine) = *Rotalia turbinata* Cushman and Valentine, 1930.
Trochulina versiformis (Bandy) = *Rotalia versiformis* Bandy, 1953.
Turritellella shoneana (Siddall) = *Trochammina shoneana* Siddall, 1878.

Uvigerina akitaensis Asano, 1950.
Uvigerina asperula Czjzek, 1848.
Uvigerina auberiana d'Orbigny, 1839.
Uvigerina bassensis Parr, 1950.
Uvigerina bononiensis Fornasini, 1888.
Uvigerina curticosta Cushman = *Uvigerina pigmea* d'Orbigny var. *curticosta* Cushman, 1927.
Uvigerina cushmani Todd, 1948.
Uvigerina excellens Todd, 1948.
Uvigerina finisterrensis Colom, 1952.
Uvigerina hispido-costata Cushman and Todd, 1945.
Uvigerina hollicki Thalmann, 1950.
Uvigerina incilis Todd, 1948.
Uvigerina juncea Cushman and Todd, 1941.
Uvigerina laevis Goës = *Uvigerina auberiana* d'Orbigny forma *laevis* Goës, 1896.
Uvigerina mediterranea Hofker, 1932.
Uvigerina peregrina Cushman, 1923.
Uvigerina peregrina Cushman var. *bradyana* Cushman, 1923.
Uvigerina peregrina Cushman var. *dirupta* Todd, 1948.
Uvigerina peregrina Cushman var. *parvula* Cushman, 1923.
Uvigerina proboscidea Schwager, 1866.
Uvigerina pygmaea d'Orbigny, 1826.
Uvigerina senticosa Cushman, 1927.
Uvigerinella glabra (Millett) = *Amphicoryne glabra* Millett, 1903.

Valvulina oviedoiana d'Orbigny, 1839.

Valvulineria araucana (d'Orbigny) = *Rosalina araucana* d'Orbigny, 1839.
Valvulineria arctica Green, 1960.
Valvulineria bradyi Brotzen, 1936.
Valvulineria complanata (d'Orbigny) = *Rosalina complanata* d'Orbigny, 1846.
Valvulineria hamanakoensis (Ishiwada) = *Anomalina hamanakoensis*, Ishiwada, 1958.
Valvulineria inflata (d'Orbigny) = *Valvulina inflata* d'Orbigny, 1839.
Valvulineria sadonica Asano, 1951.
Vertebralina striata d'Orbigny, 1826.
Vulvulina pennatula (Batsch) = *Nautilus* (*Orthoceras*) *pennatula* Batsch, 1791.

Warrenita palustris (Warren) = *Sulcophax palustris* Warren, 1957.
Zeaflorilus parri (Cushman) = *Nonionella parri* Cushman, 1936.

APPENDIX D

Glossary

Assemblage. The populations of all the species present in a sample.
Brackish. Salinity <32‰ (hyposaline).
Calcite compensation depth (CCD). Depth at which the supply of biogenic calcite is matched by the rate of dissolution.
Dead assemblage. That part of the foraminiferal fauna lacking cytoplasmic test contents and which therefore do not take up biological stain.
Density. The number of foraminifera living on or beneath a unit area of sea floor at any one time (= standing crop).
Dissolution. The loss of calcium carbonate from calcareous tests through the corrosive action of undersaturated water.
Diversity index. A measure of the relationship between the number of individuals and the number of species in an assemblage.
Epifaunal. Living on the sediment surface or other substrates.
Epiphytic. Living on plant substrates.
Eutrophic. Rich in nutrients/food.
Hyposaline. Salinity <32‰ (brackish).
Hypersaline. Salinity >37‰.
Infaunal. Living within the sediment.
Living assemblage. That part of the foraminiferal fauna with cytoplasmic test contents which stain with an appropriate reagent such as rose Bengal or Sudan black B.
Macrotidal. Tidal range >4 m.
Marginal marine. General term for coastal, lagoon, estuary, marsh environments.
Mesotidal. Tidal range 2–4 m.
Microtidal. Tidal range <2 m.
Niche. Traditionally the habitat with ecological parameters within the range for the successful existence of a species. The realised niche is that part of the total potential niche occupied by a given species.
Normal marine. Salinity 32–37‰.
Oligotrophic. Low in nutrients/food.
Population. The individuals of a single species.
Production. The number of new individuals resulting from reproduction during a unit of time. For benthic foraminifera it is conveniently defined as the number of test contributed to the sea floor during a period of time (such as 1 year).
Standing crop. The number of foraminifera living on or beneath a unit area of sea floor at any one time (= density).
Total assemblage. Stained and unstained tests from a sample, i.e. a mixture of living and dead.

References

Adams C G, 1983 Speciation, phylogenesis, tectonism, climate and eustacy: factors in the evolution of Cenozoic larger foraminiferal bioprovinces. In Sims R W, Price J H and Whalley P E S (eds) Evolution, time and space: the emergence of the biosphere. *Systematics Association Special Volume* **23**: 255–89

Adegoke O S, Omatsola N E, Salami N B 1976 Benthic foraminiferal biofacies off the Niger Delta. In Schafer C T and Pelletier B R (eds) *First International Symposium on Benthonic Foraminifera of Continental Margins. Maritime Sediments Special Publications* **1**: 279–92

Akers W H 1971 Estuarine foraminiferal associations of the Beaufort area, North Carolina. *Tulane Studies in Geology and Paleontology* **8**: 147–65

Akpati B N 1975 Foraminiferal distribution and environmental variables in eastern Long Island Sound, New York. *Journal of Foraminiferal Research* **5**: 127–44

Aksu A E, Vilks G 1988 Stable isotopes in planktonic and benthic foraminifera from Arctic Ocean surface sediments. *Canadian Journal of Earth Science* **25**: 701–9

Albani A D 1968 Recent Foraminiferida from Port Hacking, New South Wales. *Contributions from the Cushman Foundation for Foraminiferal Research* **19**: 85–119

Albani A D 1970 A foraminiferal fauna from the eastern continental shelf of Australia. *Contributions from the Cushman Foundation for Foraminiferal Research* **21**: 71–7

Albani A D 1974 New benthonic foraminifera from Australian waters. *Journal of Foraminiferal Research* **4**: 33–9

Albani A D 1978 Recent foraminifera of an estuarine environment in Broken Bay, New South Wales. *Australian Journal of Marine and Freshwater Research* **29**: 355–98

Albani A D, Favero V, Barbero R S 1984 Benthonic foraminifera as indicators of intertidal environments. *Geo-Marine Letters* **4**: 43–7

Albani A D, Johnson K R 1976 Resolution of foraminiferal biotopes in Broken Bay, New South Wales. *Journal of the Geological Society of Australia* **22**: 435–46

Albani A D, Serandrei Barbero R 1982 A foraminiferal fauna from the lagoon of Venice, Italy. *Journal of Foraminiferal Research* **12**: 234–41

Alexander S P, Banner F T 1984 The functional relationship between skeleton and cytoplasm in *Haynesina germanica* (Ehrenberg). *Journal of Foraminiferal Research* **14**: 159–70

Alexander S P, Delaca T E 1987 Feeding adaptations of the foraminifera *Cibicides refulgens* living epizoically and parasitically on the Antarctic scallop *Adamussium colbecki*. *Biological Bulletin* **173**: 136–59

Alexandersson E T 1978 Destructive diagenesis of carbonate sediments in the eastern Skagerrak, North Sea. *Geology* **6**: 324–7

Allen R A, Roda R S 1977 Benthonic foraminifera from LaHave estuary. *Maritime Sediments* **13**: 67–72

Aller J Y, Aller R C 1986 Evidence for localised enhancement of biological activity associated with tube and burrow structures in deep-sea sediments at the HEBBLE site, western North Atlantic. *Deep-Sea Research* **33**: 755–90

Almeida F, Setty M G A P 1972 Agglutinated foraminifera from the shelf of east coast of India. *Proceedings of the 11th Indian Colloquium on Micropalaeontology and Stratigraphy* pp 93–102

Altenbach A V 1987 The measurement of organic carbon in foraminifera. *Journal of Foraminiferal Research* **17**: 106–110

Altenbach A V 1988 Deep sea benthic foraminifera and flux rate of organic carbon. *Revue de Paléobiologie vol spéc* **2**: 719–20

Alve E, Nagy J 1986 Estuarine foraminiferal distribution in Sandebukta, a branch of the Oslo Fjord. *Journal of Foraminiferal Research* **16**: 261–84

Anan H S 1984 Littoral recent foraminifera from the Qosseir–Marsa Alam stretch of the Red Sea coast, Egypt. *Revue de Paléobiologie* **3**: 235–42

Anderson G J 1963 Distribution patterns of Recent foraminifera of the Bering Sea. *Micropalaeontology* **9**: 305–17

Anderson J B 1975a Ecology and distribution of foraminifera in the Weddell Sea of Antarctica. *Micropaleontology* **21**: 69–96

Anderson J B 1975b Factors controlling $CaCO_3$ dissolution in the Weddell Sea from foraminiferal distribution patterns. *Marine Geology* **19**: 315–32

Anderson O R, Bé A W H 1978 Recent advances in foraminiferal fine structure research. In Hedley R H, Adams C G (eds) *Foraminifera* **3**: 121–202

Andrew N L, Mapstone B D 1987 Sampling and the description of spatial pattern in marine ecology. In Barnes M (ed) *Oceanography and Marine Biology Annual Review* **25**: 39–90

Antony A 1968 Studies on the shelf water foraminifera of the Kerala coast. *Bulletin of the Department of Marine Biology and Oceanography, University of Kerala* **4**: 11–154

Antony A 1975 Foraminifera from the Kayamkulam Lake. *Bulletin of the Department of Marine Science, University of Cochin* **7**: 257–62

Antony A 1980a Interstitial foraminifera of the sandy beaches of the southwest coast of India. *Bulletin of the Department of Marine Science, University of Cochin* **11**: 103–32

Antony A 1980b Foraminifera of the Vembanad estuary. *Bulletin of the Department of Marine Science, University of Cochin* **11**: 25–63

Antony A, Kurian C V 1975 Seasonal occurrence of *Rotalia beccarii* (Linné) (Foraminifera) in the Vembanad estuary. *Bulletin of the Department of Marine Science, University of Cochin* **7**: 235–41

Apthorpe M 1980 Foraminiferal distribution in the estuarine Gippsland Lakes system, Victoria. *Proceedings of the Royal Society of Victoria* **91**: 207–32

Arnal R E, Quintero P J, Conomos T J, Gram R 1980 Trends in the distribution of Recent foraminifera in San Francisco Bay. *Special Publication, Cushman Foundation for Foraminiferal Research* **19**: 17–39

Arnold A J 1983 Foraminiferal thanatacoenoses on the continental slope off Georgia and South Carolina. *Journal of Foraminiferal Research* **13**: 79–90

Arnold A J, d'Escrivan F, Parker W C 1985 Predation and avoidance responses in the foraminifera of the Galapagos hydrothermal mounds. *Journal of Foraminiferal Research* **15**: 38–42

Arnold A J, Sen Gupta D K 1981 Diversity changes in the foraminiferal thanotocoenoses off the Georgia–South Carolina continental slope. *Journal of Foraminiferal Research* **11**: 268–76

Arnold Z M 1955 An unusual feature of miliolid reproduction. *Contributions from the Cushman Foundation for Foraminiferal Research* **6**: 94–6

Atkinson K 1969 The association of living Foraminifera with algae from the littoral zone, south Cardigan Bay, Wales. *Journal of Natural History* **3**: 517–42

Atkinson K 1971 The relationship of Recent foraminifera to the sedimentary facies in the turbulent zone, Cardigan Bay. *Journal of Natural History* **5**: 385–439

Ausseil-Badie J 1983 Distribution écologique des Foraminifères de l'estuaire et de la mangrove du fleuve Sénégal. *Archives des Sciences, Genève* **36**: 437–50

Ayala-Castanares A 1963 Sistematica y distribucion de los foraminiferos recientes de la Laguna de Terminos, Campeche, Mexico. *Universidad Nacional Autonoma de Mexico, Instituto de Geologia* **67**(3): 1–130

Ayala-Castanares A, Segura L R 1968 Ecologia y distribucion de los foraminiferos recientes de la Laguna Madre, Tamaulipas, Mexico. *Buletin del Instituto Geologico Mexico* **87**: 1–89

Ayala-Castanares A, Segura L R 1981 Foraminiferos recientes de la Laguna de Tamiahua, Veracruz, Mexico. *Anales Instituto de Ciencas del Mar y Limnologia* **8**: 103–57

Baccaert J 1976 Scientific report of the Belgian expedition to the Australian Great Barrier Reef, 1967. Foraminifera: 1. Soritidae of the Lizard Island reef complex, a preliminary report. *Annales de la Société Géologique de Belgique* **99**: 237–62

Bahafzallah A B K 1979 Recent benthic foraminifera from Jiddah Bay, Red Sea (Saudi Arabia). *Neues Jahrbuch für Geologie und Paläontologie 1979* pp 385–98

Bailey J W 1851 Microscopical examination of soundings made by the US Coast Survey off the Atlantic coast of the US. *Smithsonian Contributions to Knowledge* **2**: 161–82

Baines P G 1982 On internal tide generation models. *Deep-Sea Research* **29**: 307–38

Bandy O L 1953a Ecology and paleoecology of some California foraminifera. Part 1. The frequency distribution of recent Foraminifera off California. *Journal of Paleontology* **27**: 161–99

Bandy O L 1953b Ecology and paleoecology of some California foraminifera. Part II. Foraminiferal evidence of subsidence rates in the Ventura Basin. *Journal of Paleontology* **27**: 200–3

Bandy O L 1954 Distribution of some shallow-water foraminifera in the Gulf of Mexico. *Professional Paper of the US Geological Survey* **254-F**: 125–41

Bandy O L 1956 Ecology of foraminifera in northeastern Gulf of Mexico. *Professional Paper of the US Geological Survey* **274-G**: 179–204

Bandy O L 1961 Distribution of foraminifera, radiolaria and diatoms in sediments of the Gulf of California. *Micropaleontology* **7**: 1–26

Bandy O L 1963a Dominant paralic foraminifera of southern California and the Gulf of California. *Contributions from the Cushman Foundation for Foraminiferal Research* **14**: 127–34

Bandy O L 1963b Larger living foraminifera of the continental borderland of southern California.

Contributions from the Cushman Foundation for Foraminiferal Research **14**: 121–6
Bandy O L 1964a General correlation of foraminiferal structure with environment. In Imbrie J, Newell D (eds) *Approaches to Paleoecology* John Wiley, New York pp 75–90
Bandy O L 1964b Foraminiferal biofacies in sediments of Gulf of Batabano, Cuba, and their geologic significance. *Bulletin of the American Association of Petroleum Geologists* **48**: 1666–79
Bandy O L 1964c Foraminiferal trends associated with deep-water sands, San Pedro and Santa Monica basins, California. *Journal of Paleontology* **38**: 138–48
Bandy O L Arnal R E 1957 Distribution of recent foraminifera off west coast of Central America. *Bulletin of the American Association of Petroleum Geologists* **41**: 2037–53
Bandy O L, Echols J 1964 Antarctic foraminiferal zonation. *Antarctic Research Series* **1**: 73–91
Bandy O L, Ingle J C, Resig J M 1964a Foraminifera, Los Angeles County outfall area, California. *Limnology and Oceanography* **9**: 124–37
Bandy O L, Ingle J C, Resig J M 1964b Foraminiferal trends, Laguna Beach outfall area, California. *Limnology and Oceanography* **9**: 112–23
Bandy O L, Ingle J C, Resig J M 1964c Facies trends, San Pedro Bay, California. *Bulletin of the Geological Society of America* **75**: 403–24
Bandy O L, Ingle J C, Resig J M 1965a Foraminiferal trends, Hyperion outfall, California. *Limnology and Oceanography* **10**: 314–32
Bandy O L, Ingle J C, Resig J M, 1965b Modification of foraminiferal distribution by the Orange County outfall, California. *Ocean Science and Ocean Engineering* pp 55–76
Bandy O L, Lindenberg H G, Vincent E 1972 History of research, Indian Ocean foraminifera. *Journal of the Marine Biological Association of India* **13**: 86–105
Bandy O L, Rodolfo K S 1964 Distribution of foraminifera and sediments, Peru–Chile trench area. *Deep-Sea Research* **11**: 817–37
Banner F T 1971 A new genus of Planorbulinidae. An endoparasite of another foraminifer. *Revista Española de Micropaleontologia* **3**: 113–28
Banner F T 1978 Form and function in coiled benthic foraminifera. *The British Micropalaeontologist* **8**: 11–12
Banner F T, Culver S J 1978 Quaternary *Haynesina* n. gen. and Paleogene *Protelphidium* Haynes; their morphology, affinities and distribution. *Journal of Foraminiferal Research* **8**: 177–207
Banner F T, Pereira C P G 1981 A temporal and spatial analysis of foraminiferal diversity from the fringing reefs off Mombasa, East Africa. In Neale J W, Brasier M D (eds) *Microfossils from Recent and Fossil Shelf Seas* Ellis Horwood, Chichester pp 350–66
Banner F T, Sheeham R, Williams E 1973 The organic skeletons of rotaline foraminifera: a review. *Journal of Foraminiferal Research* **3**: 30–42
Banner F T, Williams E 1973 Test structure, organic skeleton and extrathalamous cytoplasm of *Ammonia* Brünnich. *Journal of Foraminiferal Research* **3**: 49–69
Barnes P W, Lien R 1988 Icebergs rework shelf sediment to 500 m off Antarctica. *Geology* **16**: 1130–3
Bartlett G A 1965 Preliminary investigation of benthonic foraminiferal ecology in Tracadie Bay, Prince Edward Island. Unpublished manuscript, Bedford Institute of Oceanography Report 65-3
Bartlett G A 1966 Distribution and abundance of foraminifera and thecamoebina in Miramichi River and Bay. Unpublished manuscript, Bedford Institute of Oceanography Report 66-2 pp 1–107
Basov I A 1975 Benthonic foraminifera of the areas of the S Sandwich Trench and of the Falkland Islands. *Academy of Sciences USSR, Oceanology* **103**: 94–100
Basov I A 1976 Benthic foraminiferans in shelf sediments of the Gulf of Guinea. *Biologiya Morya* **1**: 22–9
Basov I A 1981 Benthic foraminifera in the recent sediments of interior seas of the Malay Archipelago. *Oceanology* **21**: 66–70
Basov I A, Khusid T A 1983 Benthic foraminifers in Sea of Okhotsk sediments. *Marine Biology, Valdivostok* **6**: 31–43
Basson P W, Burchard J E, Hardy J T, Price A R G 1977 *Biotopes of the Western Arabian Gulf.* Arabian American Oil Company, Saudi Arabia p 284
Bates J M, Spencer R S 1979 Modification of foraminiferal trends by the Chesapeake–Elizabeth sewage outfall, Virginia Beach, Virginia. *Journal of Foraminiferal Research* **9**: 125–40
Bathurst R G C 1971 Carbonate sediments and their diagenesis. *Developments in Sedimentology* Elsevier, Amsterdam vol **12**, 620 pp.
Battistini R, Gayet J, Jouannic C, Labracherie M, Peypouquet J P, Pujol C, Pujos-Lamy A, Turon J L 1976 Étude des sédiments de la microfaune des îles Glorieuses (Canal de Mozambique). *Cahiers Office de la Recherche Scientifique et Technique Outre-Mer*, Sér. Geol. **8**: 147–71
Belanger P E, Curry W B, Matthews R K 1981 Core-top evaluation of benthic foraminiferal isotopic ratios for palaeoceanographic interpretation. *Palaeogeography, Palaeoclimatology, Palaeoecology* **33**: 205–20
Belanger P E, Streeter S S 1980 Distribution and ecology of benthic foraminifera in the Norwegian–Greenland Sea. *Marine Micropaleontology* **5**: 401–28
Benda W K, Puri H S 1962 The distribution of foraminifera and Ostracoda off the Gulf coast of the Cape Romano area, Florida. *Transactions of the Gulf Coast Association of Geological Societies* **12**: 303–41
Bergen F W, O'Neil P 1979 Distribution of Holocene foraminifera in the Gulf of Alaska. *Journal of Paleontology* **53**: 1267–92
Berger W H 1981 Oxygen and carbon isotopes in

foraminifera: an introduction. *Palaeogeography, Palaeoclimatology, Palaeoecology* **33**: 3–7

Berger W H, Soutar A 1970 Preservation of plankton shells in an anaerobic basin off California. *Bulletin of the Geological Society of America* **81**: 275–82

Berger W H, Vincent E 1986 Deep-sea carbonates: reading the carbon-isotope signal. *Geologische Rundschau* **75**: 249–69

Berger W H, Winterer E L 1974 Plate stratigraphy and the fluctuating carbonate line. In Hsü K J, Jenkyns H C (eds) *Pelagic Sediments: on Land and under the Sea*. Blackwell, Oxford pp 11–98

Bernhard J M 1986 Characteristic assemblages and morphologies of benthic foraminifera from anoxic, organic-rich deposits: Jurassic through Holocene. *Journal of Foraminiferal Research* **16**: 207–15

Bernhard J M 1987 Foraminiferal biotopes in Explorers Cove, McMurdo Sound, Antarctica. *Journal of Foraminiferal Research* **17**: 286–97

Bernhard J M 1988 Postmortem vital staining in benthic foraminifera: duration and importance in population and distributional studies. *Journal of Foraminiferal Research* **18**: 143–6

Bernstein B B, Hessler R R 1978 Spatial dispersion of benthic foraminifera in the abyssal central North Pacific. *Limnology and Oceanography* **23**: 401–16

Bernstein B B, Meador J P 1979 Temporal persistence of biological patch structure in an abyssal benthic community. *Marine Biology* **51**: 179–83

Berthois L, Crosnier A, Le Calvez Y 1968 Contribution a l'étude sédimentologique de plateau continental dans la Baie de Biafra *Cahiers Orstom séries Océanographique* **6**: 55–86

Berthois L, Le Calvez Y 1966 Étude sédimentologique des dépôts à *Jullienella foetida* de la région d'Abidjan (Côte d'Ivoire). *Bulletin Bureau de Recherches Géologiques et Minières* **1**: 45–55

Betjeman K J 1969 Recent foraminifera from western continental shelf of Western Australia. *Contributions from the Cushman Foundation for Foraminiferal Research* **20**: 119–38

Bhalla S N 1967 Recent foraminifera from Vishakapatnam beach sands and its relation to the known foramgeographical provinces in the Indian Ocean. *Bulletin of the National Institute of Sciences of India* **38**: 376–92

Bhalla S N 1970 Foraminifera from marine beach sands, Madras, and faunal provinces of the Indian Ocean. *Contributions from the Cushman Foundation for Foraminiferal Research* **21**: 156–63

Bhalla S N, Nigam R 1979 A note on recent foraminifera from Calanguie Beach sand, Goa. *Bulletin of the Indian Geological Association* **12**: 239–40

Bhatia S B 1956 Recent foraminifera from shore sands of western India. *Contributions from the Cushman Foundation for Foraminiferal Research* **7**: 15–24

Bhatia S B, Kumar S 1976 Recent benthonic foraminifera from the inner shelf area around Anjidiv Island, off Binge, west coast of India. *Maritime Sediments Special Publication* **1**: 239–49

Bignot G, Lamboy M 1980 Les foraminifères épibiontes à test calcaire hyalin des encroûtements polymétalliques de la marge continentale au nord-ouest de la péninsule ibérique. *Revue de Micropaléontologie* **23**: 3–15

Bilyard G R 1974 The feeding habits and ecology of *Dentalium entale stimpsoni* Henderson (Mollusca: Scaphopoda). *The Veliger* **17**: 126–38

Bizon G, Bizon J J 1984a Distribution des foraminifères sur le plateau continental au large du Rhône. *Pétrole et Techniques* **301**: 84–94

Bizon G, Bizon J J 1984b Les foraminifères des sediments, profonds. *Pétrole et Techniques* **301**: 104–20

Bjerkli K, Östmo-Saeter J S 1973 Formation of glauconie in foraminiferal shells on the continental shelf off Norway. *Marine Geology* **14**: 169–78

Blanc-Vernet L 1969 Contribution à l'étude des foraminifères de Méditerranée. *Recueil des Travaux de la Station Marine d'Endoume* **64-48**: 5–281

Blanc-Vernet L 1974 Microfaune de quelques dragages et carottes effectués devant les côtes de Tunisie (golfe de Gabès) et de Libye (Tripolitaine). *Géologie Méditerranéenne* **1**: 9–26

Blanc-Vernet L, Clairefond P, Orsolini P 1979 La Mer Pélagienne. Les foraminifères. *Géologie Méditerranéenne* **6**(1): 171–209

Blanc-Vernet L, Pujos M, Rosset-Moulinier M 1983 Les biocénoses de foraminifères benthiques des plateaux continentaux français (Manche, sud-Gascogne, Ouest-Provence). In Oertli H (ed) *Benthos '83, Second International Symposium on Benthic Foraminifera* (Pau, April 1983) pp 71–9

Bock W D 1970 *Thalassia testudinarium*, a habitat and means of dispersal for shallow-water benthonic foraminifera. *Transactions of the Gulf Coast Association of Geological Societies* **19**: 337–40

Bock W D 1976 Distribution and significance of foraminifera in the MALFA area. *Maritime Sediments, Special Publication* **1**: 221–37

Bock W D 1982 Coexistence of deep- and shallow-water foraminiferal faunas off Panama City, Florida. *Bulletin of the Geological Society of America* **93**: 246–51

Bock W D, Moore D R 1968 A commensal relationship between a foraminifer and a bivalve mollusk. *Gulf Research Rept* **2**: 273–9

Boichard R, Burollet P F, Lambert B, Villain J M 1985 La platforme carbonatée du Pater Noster est de Kalimantan (Indonésie) étude sédimentologique et écologique. *Notes et Mémoires, Total, Compagnie Française des Pétroles Paris* **20**: 1–103

Boltovskoy E 1954 Foraminiferos de la Bahia San Blas. *Revista del Instituto Nacional de Investigacion de las Ciencas Naturales, Geologicas* **3**: 247–300

Boltovskoy E 1958 The foraminiferal fauna of the Rio de

la Plata and its relation to the Caribbean area. *Contributions from the Cushman Foundation for Foraminiferal Research* **9**: 17–21

Boltovskoy E 1959 Foraminiferos recientes del sur de Brasil y sus relaciones con los de Argentina e India del oeste. *Armada Argentina, Servicio de Hidrografia Naval* **H1005**: 1–120

Boltovskoy E 1961 Foraminiferos de la plataforma continental entre el Cabo Santo Tome y la desembocadura del Rio de la Plata. *Revista del Museo Argentino de Ciencas Naturales 'Bernardino Rivadavia', Zoologicas* **6**: 249–346

Boltovskoy E 1963 The littoral foraminiferal biocoenoses of Puerto Deseado (Patagonia, Argentina). *Contributions from the Cushman Foundation for Foraminiferal Research* **14**: 58–70

Boltovskoy E 1964 Seasonal occurrences of some living foraminifera in Puerto Deseado (Patagonia, Argentina). *Journal du Conseil International pour l'Exploration de la Mer* **39**: 136–45

Boltovskoy E 1965 *Los foraminiferos recientes*. Editorial Universitaria de Buenos Aires, Argentina 510pp

Boltovskoy E 1966 Depth at which foraminifera can survive in sediments. *Contributions from the Cushman Foundation for Foraminiferal Research* **17**: 43–5

Boltovskoy E 1970 Distribution of the marine littoral foraminifera in Argentina, Uruguay and southern Brazil. *Marine Biology* **6**: 335–44

Boltovskoy E 1971 Relationship between benthonic foraminiferal fauna and substrate in the littoral zone. *Journal of Marine Geology* **7**: 26–30

Boltovskoy E 1972 Nota sobre los valores mininos de oxigenacion que pueden soportar los foraminiferos bentonicos. *Boletin de la Sociedad de Biologia de Concepcion* **44**: 135–43

Boltovskoy E 1976 Distribution of recent foraminifera of the South American region. In Hedley R H, Adams C G (eds) *Foraminifera* Academic Press, London vol 2 pp 171–236

Boltovskoy E, Boltovskoy A 1968 Foraminiferos y tecamebas de la parte inferior del Rio Quequen Grande Provincia de Buenos Aires, Argentina. *Revista del Museo Argentino de Ciencas naturales 'Bernardino Rivadavia', Hidrobiologia* **2**: 127–64

Boltovskoy E, Giussani de Kahn G, Watanabe S 1983 Variaciones estacionales y standing crop de los foraminiferos bentonicos de Ushuaia, Tierra del Fuego. *Physis* **A41**: 113–27

Boltovskoy E, Giussani G, Watanabe S, Wright R 1980 *Atlas of Benthic Shelf Foraminifera of the Southwest Atlantic*. Junk, The Hague pp 147

Boltovskoy E, Gualancanay E 1975 Foraminiferos bentonicos actuales de Ecuador 1. Provincia Esmeraldas. *Armada del Ecuador Instituta de Oceanographico Publicacion* pp 1–56

Boltovskoy E, Hincape de Martinez S 1983 Foraminiferos del manglar de Tesca, Cartagena, Colombia. *Revista Espanola de Micropaleontologia* **15**: 205–20

Boltovskoy E, Lena H 1966 Unrecorded foraminifera from the littoral of Puerto Deseado. *Contributions from the Cushman Foundation for Foraminiferal Research* **17**: 144–9

Boltovskoy E, Lena H 1969a Microdistribution des foraminifères benthoniques vivants. *Revue de Micropaléontology* **12**: 177–85

Boltovskoy E, Lena H 1969b Seasonal occurrences, standing crop and production in benthic foraminifera of Puerto Deseado. *Contributions from the Cushman Foundation for Foraminiferal Research* **20**: 87–95

Boltovskoy E, Lena H 1969c Les epibiontes de 'Macrocystis' flotante como indicadores hidrologicos. *Neotropica* **15**: 135–7

Boltovskoy E, Lena H 1970a Additional note on unrecorded foraminifera from littoral of Puerto Deseado (Patagonia, Argentina). *Contributions from the Cushman Foundation for Foraminiferal Research* **21**: 148–55

Boltovskoy E, Lena H 1970b On the decomposition of the protoplasm and the sinking velocity of the planktonic foraminifers. *Internationale Revue der gesamten Hydrobiologie und Hydrographie* **55**: 797–804

Boltovskoy E, Lena H 1971 The Foraminifera (except family Allogromiidae) which dwell in fresh water. *Journal of Foraminiferal Research* **1**: 71–6

Boltovskoy E, Lena H 1974 Foraminiferos del Rio de la Plata. *Armada Argentina Servicio de Hidrografia Naval* **H661**: 1–22

Boltovskoy E, Theyer F 1970 Foraminiferos recientes de Chile Central. *Revista del Museo Argentino de Ciencas Naturales 'Bernardino Rivadavia' Hidrobiologia* **2**: 279–380

Boltovskoy E, Totah V 1985 Diversity, similarity and dominance in benthic foraminifera along one transect of the Argentina shelf. *Revue de Micropaléontologie* **28**: 22–31

Boltovskoy E, Totah V I 1987 Relacion entre masas de agua y foraminiferos bentonicos en el Pacifico sudoriental. *Physis* (Buenos Aires) **A45**: 37–46

Boltovskoy E, Vidarte L M 1977 Foraminiferos de la zona de manglar de Guayaquil (Ecuador). *Revista del Museo Argentino de Ciencas Naturales 'Bernardino Rivadavia', Hidrobiologia* **5**: 31–40

Boltovskoy E, Wright R 1976 *Recent foraminifera*. Junk, The Hague p 515

Boltovskoy E, Zapata A 1980 Foraminiferos bentonicos como alimento de otros organismos. *Revista Española de Micropaleontologia* **12**: 191–8

Bornhold B D, Giresse P 1985 Glauconitic sediments on the continental shelf off Vancouver Island, British Columbia, Canada. *Journal of Sedimentary Petrology* **55**: 653–64

Bowser S S 1985 Invasive activity of *Allogromia* pseudopodial networks: skyllocytosis of a gelatin/agar gel. *Journal of Protozoology* **32**: 9–12

Bowser S S, Delaca T E 1985 'Skeletal' elements involved in prey capture by the Antarctic foraminiferan *Astrammina rara*. In Bailey G W (ed) *Proceedings of the 43rd Annual Meeting Electron Microscopy Society of America* pp 484–5

Bowser S S, Israel H A, McGee-Russell S M, Rieder C L 1984 Surface transport properties of pseudopodia: do intracellular and extracellular motility share a common mechanism? *Cell Biology International Reports* **8**: 1051–63

Boyle E A 1984 Sampling statistic limitations on benthic foraminifera: chemical and isotopic data. *Marine Geology* **58**: 213–24

Bradshaw J S 1957 Laboratory studies on the rate of growth of the foraminifer '*Streblus beccarii* (Linné) var. *tepida* (Cushman)'. *Journal of Paleontology* **31**: 1138–47

Bradshaw J S 1961 Laboratory experiments on the ecology of foraminifera. *Contributions from the Cushman Foundation for Foraminiferal Research* **12**: 87–106

Bradshaw J S 1968 Environmental parameters and marsh foraminifera. *Limnology and Oceanography* **13**: 26–38

Brady H B 1884 Report on the foraminifera dredged by HMS Challenger during the years 1873–1876. Reports of the Scientific Results of the Voyage of HMS *Challenger* during the years 1873–1876. *Reports of the Scientific Results of the Voyage of H M S Challenger* **9** (zoology): 1–814

Brady H B, Parker W K, Jones T R 1888 On some foraminifera from the Abrohlos Bank. *Transactions of the Zoological Society of London* **12**: 211–40

Braga J M 1942 Protozoários foraminiferos. In Nobre A, Braga J M Notas sôbre a fauna das ilhas Berlengas e Farilhões. *Memorias e estudos do Museu Zoologico* **138**: 53–66

Braga J M, Galhano M H 1965 Foraminiferos do Arquipélago da Maderia. *Publicações do Instituto de Zoologia 'Dr Augusto Nobre'* **94**: 1–134

Brand T E, Lipps J H 1982 Foraminifera in the trophic structure of shallow-water Antarctic marine communities. *Journal of Foraminiferal Research* **12**: 96–104

Brasier M D 1975a Ecology of sediment-dwelling and phytal foraminifera from the lagoons of Barbuda, West Indies. *Journal of Foraminiferal Research* **5**: 42–62

Brasier M B 1975b The ecology and distribution of recent foraminifera from the reefs and shoals around Barbuda, West Indies. *Journal of Foraminiferal Research* **5**: 193–210

Brasier M B 1981 Microfossil transport in the tidal Humber Basin. In Neale J W, Brasier M D (eds) *Microfossils from Recent and Fossil Shelf Seas* Ellis Horwood, Chichester pp 314–22

Brasier M D 1982 Architecture and evolution of the foraminiferid test – a theoretical approach. In Banner F T, Lord A R (eds) *Aspects of Micropalaeontology* Allen and Unwin, London pp 1–41

Brasier M 1986 Why do lower plants and animals biomineralize? *Paleobiology* **12**: 241–50

Bremer M L, Lohmann G P 1982 Evidence for primary control on the distribution of certain Atlantic Ocean benthonic foraminifera by degree of carbonate saturation. *Deep-Sea Research* **29**: 987–98

Brenner G J 1962 Results of the Puritan–American Museum of Natural History Expedition to western Mexico 14. A zoogeographic analysis of some shallow-water foraminifera in the Gulf of California. *Bulletin of the American Museum of Natural History* **123**: 253–97

Brodniewicz I 1965 Recent and some Holocene foraminifera of the southern Baltic Sea. *Acta Palaeontologica Polonica* **10**: 131–248

Bromley R G 1970 Predation and symbiosis in some Upper Cretaceous clionid sponges. *Meddelelser fra Dansk Geolgisk Forening* **19**: 398–405

Bromley R G, Nordmann E 1971 Maastrichtian adherent foraminifera encircling clionid pores. *Bulletin of the Geological Society of Denmark* **20**: 362–68

Brönnimann P 1978 Recent benthonic foraminifera from Brazil. Morphology and ecology part III. *Note du Laboratoire de Paléontologie de l'Université de Genève* **1**: 1–8

Brönnimann P 1979 Recent benthonic foraminifera from Brazil. Morphology and ecology Part IV. Trochamminids from the Campos shelf with a description of *Paratrochammina* n. gen. *Paläontologische Zeitschrift* **53**: 5–25

Brönnimann P, Buerlen G 1977a Recent benthonic foraminifera from Brazil. Morphology and ecology Part I. *Archives des Sciences, Genève* **30**: 77–90

Brönnimann P, Buerlen G 1977b Recent benthonic foraminifera from Brazil. Morphology and ecology Part II. *Archives des Sciences, Genève* **30**: 243–62

Brönnimann P, Dias-Brito D, Moura J A 1981b. Foraminiferos da facies mangue da planicie de maré de Guaratiba, Rio de Janeiro, Brazil. *Anas 11 Congresso Latino–Americano Paleontologia, Porto Alegre, Abril 1981* pp 877–91

Brönnimann P, Keij A J 1986 Agglutinated foraminifera (Lituolacea and Trochamminacea) from brackish waters of the State of Brunei and of Sabah, Malaysia, northwest Borneo. *Revue de Paleobiologie* **5**: 11–31

Brönnimann P, Keij A J, Zaninetti L 1983 *Bruneica clypea* n. gen., n. sp., a recent remaneicid (Foraminiferida: Trochamminacea) from brackish waters of Brunei, northwest Borneo. *Revue de Paléobiologie* **2**: 35–41

Brönnimann P, Moura J A, Dias–Brito D 1981a Estudos ecologicos na Baia de Sepetiba, Rio de Janeiro, Brazil: Foraminiferos. *Anais 11 Congresso Latino–Americano Paleontologia, Porto Alegre, Abril 1981* pp 861–75

Brönnimann P, Zaninetti L 1984 Agglutinated Foraminifera mainly Trochamminacea from the Baia de

Sepetiba, near Rio de Janeiro, Brazil. *Revue de Palébiologie* **3**: 62–115

Brönnimann P, Zaninetti L, Moura J A 1979 New recent 'allogromiine' and primitive textulariine foraminifera from brackish waters of Brazil. *Notes du Laboratoire de Paléontologie de l'Université de Genève* **4**: 27–36

Brooks A L 1967 Standing crop, vertical distribution and morphometrics of *Ammonia beccarii* (Linné). *Limnology and Oceanography* **12**: 667–84

Brooks W W 1973 Distribution of recent foraminifera from the southern coast of Puerto Rico. *Micropaleontology* **19**: 385–416

Brouwer J 1973 Foraminiferal faunas from deep-sea sediments in the Gulf of Guinea. *Verhandelingen van het Konink nederlandsch geologisch-mijnbouwkundig genootschap* (Geol. Ser) **30**: 19–55

Brunner C A, Ledbetter M T 1987 Sedimentological and micropaleontological detection of turbidite muds in hemipalegic sequences: an example from the Late Pleistocene level of Monterey Fan, central California continental margin. *Marine Micropaleontology* **12**: 233–39

Buchanan J B 1960 On *Jullienella* and *Schizammina*, two genera of arenaceous foraminifera from the tropical Atlantic, with a description of a new species. *Journal of the Linnaean Society, Zoology* **44**: 270–7

Buchanan J B, Hedley R H 1960 A contribution to the biology of *Astrorhiza limicola* (Foraminifera). *Journal of the Marine Biological Association of the United Kingdom* **39**: 549–60

Buchart B, Hansen H J 1977 Oxygen isotope fractionation and alga symbiosis in benthic foraminifera from the Gulf of Elat, Israel. *Bulletin of the Geological Society of Denmark* **26**: 185–94

Buckley D E, Owens E H, Schafer C T, Vilks G, Cranston R E, Rashid M A, Wagner E J E, Walker D A 1974 Canso Strait and Chedabucto Bay: a multidisciplinary study of the impact of man on the marine environment. *Geological Survey of Canada Paper 74–30*, **1**: 133–60

Burke S C 1981 Recent benthic foraminifera of the Ontong Java Plateau. *Journal of Foraminiferal Research* **11**: 1–19

Burmistrova I I 1969 Quantitative distribution of benthonic foraminifera in recent sediments of north region of the Indian Ocean. *Akademia NAUK, SSSR* pp 176–86

Burmistrova I L 1976 Benthonic foraminifera in the deep-sea sediments of the Arabian Sea. *Okeanologiya* **lb**: 685–9 (in Russian)

Buzas M A 1965 The distribution and abundance of foraminifera in Long Island Sound. *Smithsonian Miscellaneous Collections* **149**: 1–89

Buzas M A 1967 An application of canonical analysis as a method of comparing faunal areas. *Journal of Animal Ecology* **36**: 563–77

Buzas M A 1968a On the spatial distribution of foraminifera. *Contributions from the Cushman Foundation for Foraminiferal Research* **19**: 1–11

Buzas M A 1968b Foraminifera from the Hadley Harbor complex, Massachusetts. *Smithsonian Miscellaneous Collections* **152** : 1–26

Buzas M A 1969 Foraminiferal species densities and environmental variables in an estuary. *Limnology and Oceanography* **14**: 411–22

Buzas M A 1970 Spatial homogeneity: statistical analyses of unispecies and multispecies populations of foraminifera. *Ecology* **51**: 874–9

Buzas M A 1972 Patterns of species diversity and their explanation. *Taxon* **21**: 275–86

Buzas M A 1974 Vertical distribution of *Ammobaculites* in the Rhode River, Maryland. *Journal of Foraminiferal Research* **4**: 144–7

Buzas M A 1977 Vertical distribution of foraminifera in the Indian River, Florida. *Journal of Foraminiferal Research* **7**: 234–7

Buzas M A 1978 Foraminifera as prey for benthic deposit feeders: results of predator exclusion experiments. *Journal of Marine Research* **36**: 617–71

Buzas M A 1982 Regulation of foraminiferal densities by predation in the Indian River, Florida. *Journal of Foraminiferal Research* **12**: 66–71

Buzas M A, Carle K J 1979 Predators of foraminifera in the Indian River, Florida. *Journal of Foraminiferal Research* **9**: 336–40

Buzas M A, Collins L S, Richardson S L, Severin K 1989 Experiments on predation, substrate preference and colonisation of benthic foraminifera at the shelf break of Ft Pierce Inlet, Florida. *Journal of Foraminiferal Research* **19**: 146–52

Buzas M A, Culver S J 1980 Foraminifera: distribution of provinces in western North Atlantic. *Science* **209**: 687–9

Buzas M A, Culver S J, Isham L B 1985. A comparison of fourteen elphidiid (Foraminiferida) taxa. *Journal of Paleontology* **59**: 1075–90

Buzas M A, Gibson T G 1969 Species diversity: benthonic foraminifera in western North Atlantic. *Science* **163**: 72–5

Buzas M A, Severin K P 1982 Distribution and systematics of foraminifera in the Indian River, Florida. *Smithsonian Contributions to the Marine Sciences* **16**: 1–52

Buzas M A, Smith R K, Beem K A 1977 Ecology and systematics of foraminifera in two *Thalassia* habitats, Jamaica, West Indies. *Smithsonian Contributions to Paleobiology* **31**: 1–139

Cann J H, DeDeckker P 1981 Fossil Quaternary and living foraminifera from athalassic (non-marine) saline lakes, southern Australia. *Journal of Paleontology* **55**: 660–70

Caralp M, Lamy A, Pujos M 1970 Contribution à la connaissance de la distribution bathymétrique des foraminifères dans le Golfe de Gascogne. *Revista Española de Micropaleontologia* **2**: 55–84

Carbonel P, Pujos M 1982 Les variations architecturales des microfaunas de lac de Tunis; relations avec l'environnement. *Oceanologica Acta: Proceedings International Symposium on Coastal Lagoons* pp 79–85

Carter L, Schafer C T, Rashid M A 1979. Observations on depositional environments and benthos of the continental slope and rise, east of Newfoundland. *Canadian Journal of Earth Sciences* **16**: 831–46

Cato I, Olsson I, Rosenberg R 1980 Recovery and decontamination of estuaries. In Olausson E, Cato I (eds) *Chemistry and Biochemistry of Estuaries* John Wiley, Chichester pp 403–40

Cavelier-Smith T, Lee J J 1985 Protozoa as hosts for endosymbioses and the conversion of symbionts into organelles. *Journal of Protozoology* **32**: 376–9

Cearreta A 1988a Population dynamics of benthic foraminifera in the Santoña estuary, Spain. *Revue de Paléobiologie, vol spéc* **2**: 721–4

Cearreta A 1988b Distribution and ecology of benthic foraminifera in the Santoña estuary, Spain. *Revista Española de Paleontolgia* **3**: 23–38

Cebulski D E 1969 Foraminiferal populations and faunas in barrier-reef tract and lagoon, British Honduras. *Memoir of the American Association of Petroleum Geologists* **11**: 311–28

Cedhagen T 1988 Position in the sediment and feeding of *Astrorhiza limicola* Sandahl, 1857 (Foraminiferida). *Sarsia* **73**: 43–7

Chang S K 1983 Benthic foraminifera of the subtidal zone of Asan Bay, Korea. *Journal of the Oceanological Society of Korea* **18**: 125–41 (in Korean)

Chang S K 1984a Recent benthic foraminifera from Gwangyang Bay, Korea. In Oertli H J (ed) *Benthos '83* pp 141–6

Chang S K 1984b Recent benthic foraminifera as a sedimentary tool. In Oertli H J (ed) *Benthos '83* pp 147–51

Chang S K 1986 Implications of the recent benthic foraminifera in Gwangyang Bay, Korea. *Journal of the Oceanological Society of Korea* **21**: 1–12 (in Korean)

Chang S K 1987 Sample treatment and basic analysis for the study of benthic foraminifera. *Journal of the Oceanological Society of Korea* **22**: 153–67

Chang S K, Lee K S 1982 Recent benthic foraminifera from the intertidal flats of the vicinity of Inchon, Korea. *Bulletin of Korea Ocean Research and Development Institute* **4**: 63–72 (in Korean)

Chang S K, Lee K S 1983 Recent benthic foraminifera and its implications in the intertidal flat of Gyunggi Bay, Korea. *Journal of the Geological Society of Korea* **19**: 169–89 (in Korean)

Chang S K, Lee K S 1984 A study on the recent benthic foraminifera of the intertidal flats of Asan Bay, Korea. *Journal of the Geological Society of Korea* **20**: 171–88 (in Korean)

Chang Y M, Kaesler R L 1974 Morphological variation of the foraminifer *Ammonia beccarii* (Linné) from the Atlantic coast of the United States. *Kansas University Paleontological Contributions Paper* **69**: 1–23

Chapman F 1941 Report on foraminiferal soundings and dredgings of the FIS *Endeavour* along the continental shelf of the southeast coast of Australia. *Transactions of the Royal Society of South Australia* **65**: 145–211

Chasens S A 1981 Foraminifera of the Kenya coastline. *Journal of Foraminiferal Research* **11**: 191–202

Cheng T, Zheng S 1978 The recent foraminifera of the Xisha Islands, Guangdong Province, China, 1. *Studia Marina Sinica* **12**: 149–266 (in Chinese with English summary)

Cherif O H, El-Sheik H, Labib F 1988 The foraminifera of the Pleistocene coastal ridges of the area to the west of Alexandria, Egypt. *Revue de Paléobiologie, vol spéc* No 2, *Benthos '86* pp 735–40

Chierici M A, Busi M T, Cita M B 1962 Contribution à une étude écologique des foraminifères dans la mer Adriatique. *Revue de Micropáleontologie* **5**: 123–40

Chiji M, Lopez S M 1968 Regional foraminiferal assemblages in Tanabe Bay, Kii Peninsula, Central Japan. *Publications of the Seto Marine Biological Laboratory* **16**: 85–125

Christiansen B 1958 The foraminifer fauna in the Dröbak Sound in the Oslo Fjord (Norway). *Nytt Magasin for Zoologi* **6**: 5–91

Christiansen B 1964 *Spiculosiphon radiata*, a new foraminifera from northern Norway. *Astarte* **25**: 1–8

Christiansen O 1971 Notes on the biology of foraminifera. *Vie et Milieu* suppl **22**: 465–78

Cimmerman F, Drobne K, Ogorelec B 1988 L'association de foraminifères benthiques des vases de la baie de Veliko Jezero sur l'île de Mijlet et de la falaise Lenga, ouverte vers la mer (Adriatique moyenne). *Revue de Paléobiologie, vol spéc* No 2, *Benthos '86* pp 741–53

Cita M B, Zocchi M 1975 Faunal density and faunal diversity in benthic foraminifera on the floor of the Mediterranean Sea. *Rapport Commission International de la Mer Mediterrannée* **23** (4a): 157–60

Cita M B, Zocchi M 1978 Distribution patterns of benthic foraminifera on the floor of the Mediterranean Sea. *Oceanologica Acta* **1**: 445–62

Closs D 1963 Foraminiferos e Tecamebas de Lagoa dos Patos (RGS). *Boletin Escola de Geologia Porto Allegre* **11**: 1–130

Closs D, Barberena M C 1960 Foraminiferos recentes da Praia de Barra (Salvador, Bahia). *Escola de Geologia de Porto Allegre* **6**: 1–50

Closs D, Barberena M C 1962 Faunal studies of recent foraminifera from the shore sands of the state of Rio Grande do Sul in southern Brazil. *Contributions from the Cushman Foundation for Foraminiferal Research* **8**: 74–8

Closs D, Madiera M 1962 Tecamebas e foraminiferos do Arroio Chui (Santa Vitoria do Palmar, Rio Grande do Sul, Brazil). *Iheringia, Zoologia* **19**: 3–43

Closs D, Madeira M 1966 Foraminifera from the Paranagua Bay, State of Parana, Brazil. *Boletim da Universidade Federal do Paraña Conselho de Pesquisas, Zoologia* **2**: 139–62

Closs D, Madeira M 1967 Foraminiferos e Tecamebas aglutinantes da Lagoa de Tremandai, No Rio Grande do Sul. *Iheringia, Zoologia* **35**: 7–31

Closs D, Madeira M L 1968 Seasonal variations of brackish foraminifera in the Patos lagoon, southern Brazil. *Escola de Geologia Porto Allegre Publicaciones Especial* **15**: 1–151

Closs D, Madeiros V M F 1965 New observations on the ecological subdivisions of the Patos lagoon in southern Brazil. *Boletim Instituto de Ciencias Naturais, Rio Grande do Sul* **24**: 7–33

Closs D, Madeiros V M F 1967 Thecamoebina and foraminifera from the Mirim lagoon, southern Brazil. *Iheringia, Zoologia* **35**: 75–88

Cockbain, A E 1963 Distribution of foraminifera in Juan de Fuca and Georgia Straits, British Columbia, Canada. *Contributions from the Cushman Foundation for Foraminiferal Research* **14**: 37–57

Coleman A R 1980 Test structure and function of the agglutinated foraminifera *Clavulina*. *Journal of Foraminiferal Research* **10**: 143–52

Collins A C 1958 Foraminifera. *Great Barrier Reef Expedition 1928–29 Scientific Reports* **6**: 335–437

Collins A C 1974 Port Philip Survey 1957–63. Foraminiferida. *Memoirs of the National Museum of Victoria, Australia* **35**: 1–62

Collison P 1980 Vertical distribution of foraminifera off the coast of Northumberland, England. *Journal of Foraminiferal Research* **10**: 75–8

Colom D G 1964 Estudios sobre la sedimentación costera Balear (Mallorca y Menorca). *Memorias de la Real Academia de Ciencas y Artes de Barcelona* **34**: 495–550

Colom G 1970 Estudio de los foraminiferos de muestras de fondo de la costa de Barcelona. *Investigacion Pesquera* **34**: 355–84

Colom G 1974 Foraminiferos Ibericos. *Investigacion Pesquera* **38**(1): 1–245

Conover R J 1981 Nutritional strategies for feeding on small suspended particles. In Longhurst A R (ed) *Analysis of Marine Ecosystems* Academic Press, London pp 363–95

Cooper J A G, McMillan I K 1987 Foraminifera of the Mgeni estuary, Durban, and their sedimentological significance. *South African Journal of Geology* **90**: 489–98

Cooper S S 1964 Benthonic foraminifera of the Chukchi Sea. *Contributions from the Cushman Foundation for Foraminiferal Research* **15**: 79–104

Cooper W C 1961 Intertidal foraminifera of the California and Oregon coast. *Contributions from the Cushman Foundation for Foraminiferal Research* **12**: 47–63

Corliss B H 1976 Recent deep-sea benthonic foraminiferal distributions in the southeast Indian Ocean. *Antarctic Journal* **11**: 165–67

Corliss B H 1978 Studies of deep-sea benthonic foraminifera in the southeast Indian Ocean. *Antarctic Journal* **13**: 116–18

Corliss B H 1979a Recent deep-sea benthonic foraminiferal distributions in the southeast Indian Ocean: inferred bottom-water routes and ecological implications. *Marine Geology* **31**: 115–38

Corliss B H 1979b Taxonomy of recent deep-sea benthonic foraminifera from the southeast Indian Ocean. *Micropaleontology* **25**: 1–19

Corliss B H 1979c Size variation in the deep-sea benthonic foraminifer *Globocassidulina subglobosa* (Brady) in the southeast Indian Ocean. *Journal of Foraminiferal Research* **9**: 50–60

Corliss B H 1983 Distribution of Holocene deep-sea benthonic foraminifera in the southwest Indian Ocean. *Deep-Sea Research* **30**: 95–117

Corliss B H 1985 Microhabitats of benthic foraminifera within deep-sea sediments. *Nature* **314**: 435–38

Corliss B H, Chen C 1988 Morphotype patterns of Norwegian Sea deep-sea benthic foraminifera and ecological implications. *Geology* **16**: 716–79

Corliss B H, Honjo S 1981 Dissolution of deep-sea benthonic foraminifera. *Micropaleontology* **27**: 356–78

Cotley T L, Hallock P 1988 Test surface degradation in *Archaias angulatus*. *Journal of Foraminifera Research* **18**: 187–202

Coulbourn W T, Lutze G F 1988 Benthic foraminifera and their relation to the environment offshore of northwest Africa: a multivariate statistical analysis. *Revue de Paléobiologie, vol spéc* **2**: 755–64

Coulbourn W T, Resig J M 1975 On the use of benthic foraminifera as sediment tracers in a Hawaiian bay. *Pacific Science* **29**: 99–115

Crouch R W, Poag C W 1987 Benthic foraminifera of the Panamanian Province: distribution and origins. *Journal of Foraminiferal Research* **17**: 153–76

Culver S J 1980 Bibliography of North American recent benthic foraminifera. *Journal of Foraminiferal Research* **10**: 286–302

Culver S J 1988 New foraminiferal depth zonation of the northwestern Gulf of Mexico. *Palaios* **3**: 69–85

Culver S J, Banner F T 1978 Foraminiferal assemblages as Flandrian palaeoenvironmental indicators. *Palaeogeography, Palaeoclimatology, Palaeoecology* **24**: 53–72

Culver S J, Buzas M A 1980 Distribution of recent benthic foraminifera off the North American Atlantic coast. *Smithsonian Contributions to the Marine Sciences* **6**: 1–512

Culver S J, Buzas M A 1981a Foraminifera distribution of provinces in the Gulf of Mexico. *Nature* **290**: 328–9

Culver S J, Buzas M A 1981b Distribution of recent

benthic foraminifera in the Gulf of Mexico. *Smithsonian Contributions to the Marine Sciences* **8**(1): 1–411, (2): 412–898

Culver S J, Buzas M A 1981c Recent benthic foraminiferal provinces on the Atlantic continental margin of North America. *Journal of Foraminiferal Research* **11**: 217–40

Culver S J, Buzas M A 1982a Recent benthic foraminiferal provinces between Newfoundland and Yucatan. *Bulletin of the Geological Society of America* **93**: 269–77

Culver S J, Buzas M A 1982b Distribution of recent benthic foraminifera in the Caribbean region. *Smithsonian Contributions to the Marine Sciences* **14**: 1–382

Culver S J, Buzas M A 1983a Recent benthic foraminiferal provinces in the Gulf of Mexico. *Journal of Foraminiferal Research* **13**: 21–31

Culver S J, Buzas M A 1983b Benthic foraminifera at the shelf break: North American Atlantic and Gulf margins. *Society of Economic Paleontologists and Mineralogists Special Publication* **33**: 359–71

Culver S J, Buzas M A 1985 Distribution of recent benthic foraminifera off the North American Pacific coast from Oregon to Alaska. *Smithsonian Contributions to the Marine Sciences* **26**: 1–234

Culver S J, Buzas M A 1986 Distribution of recent benthic foraminifera off the North American Pacific coast from California to Baja. *Smithsonian Contributions to the Marine Sciences* **28**: 1–634

Culver S J, Buzas M A 1987 Distribution of recent benthic foraminifera off the Pacific coast of Mexico and Central America. *Smithsonian Contributions to the Marine Sciences* **30**: 1–184

Cushman J A 1932–1942 The foraminifera of the tropical Pacific collections of the *Albatross*, 1899–1900. *United States National Museum Bulletin* **161**(1): 1–85, (2): 1–79, (3): 1–67

Cushman J A, Todd R, Post R J 1954 Recent foraminifera of the Marshall Islands. *United States Geological Survey Professional Paper* **260-H**: 319–84

Daniels C H 1970 Quantitative ökologische Analyse der zeitlichen und räumlichen Verteilung rezenter Foraminiferen im Limksi Kanal bei Rovinj (nördliche Adria). *Göttinger Arbeiten zur Geologie und Paläontologie* **8**: 1–109

Daniels C H 1971 Jahreszeitliche ökologische Beobachtungen an Foraminiferen im Limski Kanal bei Rovinj, Jugoslavien (nördliche Adria). *Geologische Rundschau* **60**: 192–204

Davis R A 1964 Foraminiferal assemblages of Alacran Reef, Campeche Bank, Mexico. *Journal of Palaeontology* **38**: 417–21

Dayton P K, Hessler R R 1972 Role of biological disturbance in maintaining diversity in the deep sea. *Deep-Sea Research* **19**: 199–208

Deacon G E R 1963 The Southern Ocean. In Hill M N (ed) *The Sea* vol 2, Interscience Publishers, New York pp 281–96

Debenay J P 1985a Le lagon sud-ouest et la marge insulaire sud de Nouvelle-Calédonie: importance et répartition des foraminifères de grande taille. *Cahiers d'Océanographie Tropicale* **20**: 171–92

Debenay J P 1985b Le genre *Amphistegina* dans le lagon de Nouvelle-Calédonie (SW Pacifique). *Revue de Micropaléontologie* **28**: 167–80

Debenay J P 1986 Un modéle de lagon actuel transposable à des paléonenvironnements récifaux: le lagon sud-ouest de Nouvelle-Calédonie. *Compte Rendu Hebdomadaire des Séances de l'Académie des Sciences* **303**: 63–6

Debenay J P 1987a Répartition des sédiments carbonatés et relation avec l'hydrodynamisme dans un environnement récifal complexe: le lagon sud-ouest de Nouvelle-Calédonie. *Bulletin de la Société Géologique de France* **3**(8): 769–76

Debenay J P 1987b Sedimentology in the southwestern lagoon of New Caledonia, SW Pacific. *Journal of Coastal Research* **3**: 77–91

Debenay J P 1988a Foraminifera larger than 0.5 mm in the southwestern lagoon of New Caledonia: distribution related to abiotic properties. *Journal of Foraminiferal Research* **18**: 158–75

Debenay J P 1988b Dynamique sédimentaire au debouche de la Baie du Prony (Nouvelle-Calédonie): dispersion des lutites et des tests d'un foraminifère: *Operculina bartschi* Cushman. *Revue de Paléobiology, vol spéc* **2**: 765–70

Debenay J P 1988c Recent foraminifera tracers of oceanic water movements in the southwestern lagoon of New Caledonia. *Palaeogeography, Palaeoclimatology, Palaeoecology* **65**: 59–72

Debenay J P, Konate S 1987 Les foraminifères actuels des îles de Los (Guinée), premier inventaire, comparaison avec les microfaunes voisines. *Revue de Paléobiologie* **6**: 213–27

Debenay J P, Maryline B A, Ababacar L Y, Isabelle S Y 1987 Les écosystèmes paraliques au Sénégal. Description, repartition des peuplements de foraminifères benthiques. *Revue de Paléobiologie* **6**: 229–55

Debenay J P, Pagès J 1987 Foraminifères et thécamoebiens de l'estuaire hyperhalin du fleuve Casamance (Sénégal). *Revue de Hydrobiologie Tropicale* **20**: 233–56

Delaca T E 1982 Use of dissolved amino acids by the foraminifer *Notodendrodes antarctikos*. *American Zoologist* **22**: 683–90

Delaca T E, Karl D M, Lipps J H 1981 Direct use of dissolved organic carbon by agglutinated benthic foraminifera. *Nature* **289**: 287–9

Delaca T E, Lipps J H 1972 The mechanism and adaptive significance of attachment and substrate pitting in the foraminiferan *Rosalina globularis* d'Orbigny. *Journal of Foraminiferal Research* **2**: 68–72

Delaca T E, Lipps J H 1976 Shallow-water marine associations, Antarctic Peninsula. *Antarctic Journal of the United States* **11**: 12–20

Delaca T E, Lipps J H, Hessler R R 1980 The morphology and ecology of a new large agglutinated Antarctic foraminifer (Textulariina: Notodendrodidae *nov.*) *Zoological Journal of the Linnaean Society* **69**: 205–24

Dennison J M, Hay W H 1967 Estimating the needed sampling area for subaquatic ecologic studies. *Journal of Paleontology* **41**: 706–8

Deonarine B 1979 Foraminiferal distribution in two Nova Scotia marshes. *Maritime Sediments* **15**: 35–46

Detling M R 1958 Some littoral foraminifera from Sunset Bay, Coos County, Oregon. *Contributions from the Cushman Foundation for Foraminiferal Research* **9**: 25–31

Dhillon D S 1968 Notes on the foraminiferal sediments from the Lupar and Labuk estuaries, east Malaysia. *Borneo Region, Malaysia Geological Survey Bulletin* **9**: 56–73

Dieckmann G, Hemleben C, Spindler M 1987 Biogenic and mineral inclusions in a green iceberg from the Weddell Sea, Antarctica. *Polar Biology* **7**: 31–3

Dobson M, Haynes J R 1973 Association of foraminifera with hydroids on the deep shelf. *Micropaleontology* **19**: 78–90

d'Onofrio S, Marabini F, Vivaldi P 1976 Foraminiferi di alcune lagune del delta del Po. *Giornale di Geologia* ser 2a, **15**: 267–76

Douglas R G 1976 Benthic foraminiferal ecology and paleoecology: a review of concepts and methods. In Lipps L H, Berger W H, Buzas M A, Douglas R G, Ross C A (eds) *Foraminiferal Ecology and Paleoecology* Society of Economic Paleontologists and Mineralogists Short Course notes vol 6 pp 21–53

Douglas R G 1981 Paleoecology of continental margin basins: a modern case history from the borderland of southern California. In *Depositional Systems of Active Continental Margin Basins* Society of Economic Paleontologists and Mineralogists, Pacific Section, Short Course notes pp 121–56

Douglas R G, Heitman H L 1979 Slope and basin benthic foraminifera of the California borderland. *Society of Economic Paleontologists and Mineralogists Special Publication* **27**: 231–46

Douglas R G, Liestman J, Walch C, Blake C, Cotton M L 1980 The transition from live to sediment assemblage in benthic foraminifera from the southern California borderland. In Field M, Bouma A, Colburn I, Douglas R C, Ingle J (eds) *Pacific Coast Paleogeography Symposium* Pacific Section **4**: 256–80

Drobne K, Cimmerman F 1984 Die vertekale Verbreitung der Lituolaceen und Miliolaceen (Foraminifera) an einem Unterwasserkliff in der Adria (Jugoslawien). *Facies* **11**: 157–72

Drooger G W, Kaasschieter J P H 1958 Foraminifera of the Orinoco–Trinidad–Paria shelf. *Verhandelingen der Koninklijke Nederlandse Akademie van Wetenschappen, afd. Natuurkunde*, Eerste Reeks **22**: 1–108

Duguay L E, Taylor D L 1978 Primary production and calcification by the soritid foraminifer *Archaias angulatus* (Fichtel and Moll). *Journal of Protozoology* **25**: 356–61

Dunbar R B, Leventer A R, Stockton W L 1989 Biogenic sedimentation in McMurdo Sound, Antarctica. *Marine Geology* **85**: 155–79

Dunbar R B, Wefer G 1984 Stable isotope fractionation in benthic foraminifera from the Peruvian continental margin. *Marine Geology* **59**: 215–25

Dupeuble P A, Mathieu R, Momeni I, Poignant A, Rosset-Moulinier M, Rouvillois A, Ubaldo M 1971 Recherches sur les foraminifères actuels des côtes françaises de la Manche et la Mer du Nord. *Revue de Micropaléontologie* **5**: 83–95

Eade J V 1967 New Zealand recent foraminifera of the families Islandiellidae and Cassidulinidae. *New Zealand Journal of Marine and Freshwater Research* **1**: 421–54

Earland A 1934a Foraminifera, Part II. South Georgia. *Discovery Reports* **7**: 27–138

Earland A 1934b Foraminifera, Part III. The Falklands sector of the Antarctic (excluding South Georgia). *Discovery Reports* **10**: 1–208

Earland A 1936 Foraminifera, Part IV. Additional records from the Weddell Sea section, from material obtained by the SY *Scotia*. *Discovery Reports* **13**: 1–76

Echols R J 1971 Distribution of foraminifera in sediments of the Scotia Sea area, Antarctic waters. *American Geophysical Union, Antarctic Research Series* **15**: 93–168

Echols R J, Armentrout J M 1980 Holocene foraminiferal distribution patterns on the shelf and slope. Yakataga–Yakutat area, northern Gulf of Alaska. In Field M E, Bouma A H *et al.* (eds), *Quaternary Depositional Environments of the Pacific Coast, Pacific Coast Paleogeography Symposium* Society of Economic Paleontologists and Mineralogists pp 281–303

Edwards P G 1982 Ecology and distribution of selected foraminiferal species in the North Minch Channel, northwest Scotland. In Banner F T, Lord A R (eds) *Aspects of Micropalaeontology* Allen and Unwin London pp 111–41

Ehrenberg C G 1843 Verbreitung und Einfluss des mikroscopischen Lebens in Süd-und Nord-Amerika. *Abhandlungen der Koenigliche Akademie der Wissenschaften Berlin* (1841) pp 291–446

Ellison R L 1972 *Ammobaculites*, foraminiferal proprietor of Chesapeake Bay estuaries. *Memoir of the Geological Society of America* **133**: 247–62

Ellison R L 1984 Foraminifera and meiofauna on an intertidal mudflat, Cornwall, England: populations; respiration and secondary production; and energy budget. *Hydrobiologica* **109**: 131–48

Ellison R L, Murray J W 1987 Geographical variation in the distribution of certain agglutinated foraminifera along the North Atlantic margins. *Journal of Foraminiferal Research* **17**: 123–31

Ellison R L, Nichols M M 1970 Estuarine foraminifera from the Rappahannock River, Virginia. *Contributions from the Cushman Foundation for Foraminiferal Research* **21**: 1–17

Ellison R L, Peck G E 1983 Foraminiferal recolonisation on the continental shelf. *Journal of Foraminiferal Research* **13**: 231–41

Emery K O, Uchupi E 1972 Western North Atlantic Ocean: topography, rocks, structure, water, life and sediments. *Memoir of the American Association of Petroleum Geologists* **17**: 1–532

Epstein S, Buchsbaum R, Lowenstam H A, Urey H C 1953 Revised carbonate-water isotopic temperature scale. *Bulletin of the Geological Society of America* **64**: 1315–26

Erez J 1978 Vital effect on stable-isotope composition seen in Foraminifera and coral skeletons. *Nature* **273**: 199–202

Erskian M G, Lipps J H 1977 Distributions of foraminifera in the Russian River estuary, northern California. *Micropaleontology* **23**: 453–69

Erskian M G, Lipps J H 1987 Population dynamics of the foraminiferan *Glabratella ornatissima* (Cushman) in northern California. *Journal of Foraminiferal Research* **17**: 240–56

Fairbanks R G, Matthews R K 1978 The marine oxygen isotope record in Pleistocene coral, Barbados, West Indies. *Quaternary Research* **10**: 181–96

Fairbridge R W 1966 *The Encyclopedia of Oceanography* Dowden, Hutchinson and Ross, Stroudsburg Pa p 1021

Fermont W J J, Kreulen R, Van Der Zwaan G J 1983 Morphology and stable isotopes as indicators of productivity and feeding patterns in recent *Operculina ammonoides* (Gronovius). *Journal of Foraminiferal Research* **13**: 122–8

Feyling-Hanssen R W 1972 The foraminifer *Elphidium excavatum* (Terquem) and its variant forms. *Micropaleontology* **18**: 337–54

Field M E, Edwards B D 1980 Slopes of southern California continental borderland: a regime of mass transport. In Field M E, Bouma A H *et al.* (eds) *Quaternary Depositional Environments of the Pacific coast, Pacific Coast Palaeogeography Symposium* Society of Economic Paleontologists and Mineralogists pp 169–84

Fierro G 1961 Foraminiferi di sedimenti del Mar Ligure. *Rapports et Procés Verbaux des Réunions – Commission Internationale pour l'Exploration Scientifique de la Mer Mediterranée* **16**: 737–44

Finger K L 1981 Faunal reference list for Gulf of Mexico deep-water foraminifers recorded by Pflum and Frerichs in 1976. *Journal of Foraminiferal Research* **11**: 241–51

Finger K L, Lipps J H 1981 Foraminiferal decimation and repopulation in an active volcanic caldera, Deception Island, Antarctica. *Micropaleontology* **27**: 111–39

Fisher R A, Corbet A S, Williams C B 1943 The relationship between the number of species and the number of individuals in a random sample of an animal population. *Journal of Animal Ecology* **12**: 42–58

Flemming B, Wefer G 1973 Tauchbeobachtungen an Wellenrippeln und Abrasionserscheinungen in der Westlichen Ostsee südöstlich Bokniseck. *Meyniana* **23**: 9–18

Foldvik A, Gammelsrod T 1988 Notes on Southern Ocean hydrography, sea-ice and bottom water formation. *Palaeogeography, Palaeoclimatology, Palaeoecology* **67**: 3–17

Forti I R S, Roettger E 1967 Further observations on the seasonal variations of mixohaline foraminifera from the Patos Lagoon, southern Brazil. *Archivio di Oceanografia e Limnologia* **15**: 55–61

Frankel L 1974 Observations and speculations on the habitat of *Trochammina ochracea* (Williamson) in subsurface sediments. *Journal of Paleontology* **48**: 143–8

Frankel L 1975 Pseudopodia of surface and subsurface dwelling *Miliammina fusca* (Brady). *Journal of Foraminiferal Research* **5**: 211–17

Frerichs W E 1970 Distribution and ecology of benthonic foraminifera in the sediments of the Andaman Sea. *Contributions from the Cushman Foundation for Foraminiferal Research* **21**: 123–47

Fursenko A V, Fursenko K B 1973 Foraminifera and their associations of Busse Lagoon. *Academy of Sciences of the USSR, Siberian Branch, Transactions of the Institute of Geology and Geophysics* **62**: 49–118 (in Russian)

Fursenko A V, Trotskaya T S, Yebchek Y K, Nesmeroea O N, Poloboda T P, Fursenko K B 1979 Foraminifera of the Far Eastern seas of Russia. *Academia Nauk CCCP, Siberian Division, Institute of Geology and Geophysics* **387**: 398 (in Russian)

Gabel B 1971 Die Foraminiferen der Nordsee. *Helgoländer wissenschaftiche Meeresuntersuchungen* **22**: 1–65

Gabrie C, Montaggioni L 1982 Sedimentary facies from the modern coral reefs, Jordan Gulf of Aqaba, Red Sea. *Coral Reefs* **1**: 115–24

Galhano M H 1963 Foraminiferos da costa de Portugal (Algarve). *Publicaçoes do Instituto de Zoologia 'Dr Augusto Nobre'* **89**: 1–110

Galloway J J, Wissler S G 1927 Pleistocene foraminifera from the Lomita Quarry, Palos Verdes Hills, California. *Journal or Paleontology* **1**: 35–87

Gamito S L, Berge J A, Gray J S 1988 The spatial distribution patterns of the foraminiferan *Pelosina* cf. *arborescens* Pearcey in a mesocosm. *Sarsia* **73**: 33–8

Ganapati P N, Sarojini D 1959 Ecology of foraminifera off Visakhapatnam coast. *Procedings All-India Congress of the Zoological Institute* **2**: 311–15

Ganapati P N, Satyavati P 1958 Report on foraminifera in

bottom sediments in the Bay of Bengal off the east coast of India. *Andhra University Memoirs in Oceanography* **2**: 100–27

Ganssen G 1981 Isotopic analysis of foraminifera shells: interference from chemical treatment. *Palaeogeography, Palaeoclimatology, Palaeoecology* **33**: 271–6

Ganssen G 1983 Dokumentation von küstennahmen Auftrieb anhand stabiler Isotope in rezenten foraminiferen vor Nordwest-Afrika. *'Meteor' Forschungs Ergebnisse* **C37**: 1–46

Gardner W D, Richardson M J, Hinga K R, Biscaye P E 1984 Resuspension measured with sediment traps in a high-energy environment. *Earth and Planetary Science Letters* **66**: 262–78

Gary A C, Healy-Williams N 1988 Analysis of ecophenotypes in benthic foraminifera via quatitative image analysis. *Revue de Paléobiologie vol spéc* **2**: 771–5

Gebelein C D 1977 *Dynamics of Recent Carbonate Sedimenation and Ecology, Cape Sable, Florida* Brill, Leiden 120 pp

George M, Murray J W 1977 Glauconite in Celtic Sea sediments. *Proceedings of the Ussher Society* **4**: 94–101

Gevirtz J L, Park R A, Friedman G M 1971 Paraecology of benthonic foraminifera and associated micro-organisms of the continental shelf off Long Island, New York. *Journal of Paleontology* **45**: 153–77

Gibson T G 1988 Assemblage characteristics of modern benthic foraminifera and application to environmental interpretation of Cenozoic deposits of eastern North America. *Revue de Paléobiologie vol. spéc* **2**: 777–87

Gibson T G, Buzas M A 1973 Species diversity: patterns in modern and Miocene foraminifera of the eastern margin of North America. *Bulletin of the Geological Society of America* **84**: 217–38

Giussani de Kahn G, Watanabe S 1980 Foraminiferos bentonicos como indicadores de la corriente de Malvinas. *Revista Española de Micropaleontologia* **12**: 169–77

Glenn C R, Arthur M A 1988 Petrology and major element geochemistry of Peru margin phosphorites and associated diagenetic minerals: authigenesis in modern organic-rich sediments. *Marine Geology* **80**: 231–67

Glenn C, McManus J W, Alino P M, Talaue L L, Banzon V F 1981 Distributions of live foraminifers on a portion of Apo Reef, Mindoro, Philippines. *Proceedings of the Fourth International Coral Reef Symposium, Manila* **2**: 775–80

Goldstein S T, Frey R W 1986 Salt marsh foraminifera, Sapelo Island, Georgia. *Senckenbergiana maritima* **18**: 97–121

Golik A, Phleger F B 1977 Benthonic foraminifera from the Gulf of Panama. *Journal of Foraminiferal Research* **7**: 83–99

Gooday A J 1986a Soft-shelled foraminifera in meiofaunal samples from the bathyal northeast Atlantic. *Sarsia* **71**: 275–87

Gooday A J 1986b Meiofaunal foraminifera from the bathyal Porcupine Seabight (northeast Atlantic): size structure, standing crop, taxonomic composition, species diversity and vertical distribution in the sediment. *Deep-Sea Research* **33**: 1345–73

Gooday A J 1988 A response by benthic foraminifera to the deposition of phytodetritus in the deep sea. *Nature* **332**: 70–3

Gooday A J, Cook P L 1984 An association between komakiacean foraminifers (Protozoa) and paludicelline ctenostomes (Bryozoa) from the abyssal northeast Atlantic. *Journal of Natural History* **18**: 765–84

Gostin V A, Haib J R, Belperio A P 1984 The sedimentary framework of northern Spencer Gulf, South Australia. *Marine Geology* **61**: 111–38

Goudie A S, Sperling C H B 1977 Long distance transport of foraminiferal tests by wind in the Thar desert, northwest India. *Journal of Sedimentary Petrology* **47**: 630–3

Gould H R, Stewart R H 1955 Continental terrace sediments in northeastern Gulf of Mexico. In Hough J L (ed) *Finding Ancient Shorelines* Society of Economic Paleontologists and Mineralogists Special Publication **3**: 2–20

Grabert B 1971 Zur Eignung von Foraminiferen als Indikatoren für Sandwanderung. *Sonderdruck aus der Deutschen Hydrographischen Zeitschrift* **24**: 1–14

Graham D W, Corliss B H, Bender M L, Keigwin L D 1981 Carbon and oxygen isotopic disequilibria of recent deep-sea benthic foraminifera. *Marine Micropaleontology* **6**: 483–97

Graham J J, Militante P J 1959 Recent foraminifera from the Puerto Galera area, northern Mindoro, Philippines. *Stanford University Publications, Geological Sciences* **6**(2): 1–170

Grant K, Hoare T B, Ferrall K W, Steinker D C 1973 Some habitats of foraminifera, Coupon Bight, Florida. *Compass* **59**: 11–6

Gray J S 1981 *The Ecology of Marine Sediments. An Introduction to the Structure and Function of Benthic Communities* Cambridge University Press, Cambridge 185 pp

Green K E 1960 Ecology of some Arctic foraminifera. *Micropaleontology* **6**: 57–78

Gregory M R 1973 Benthonic foraminifera from a mangrove swamp, Whangaparapara, Great Barrier Island. *Tane (University of Auckland Field Club) Anniversary Issue* pp 193–204

Greiner G O G 1969 Recent benthonic foraminifera: environmental factors controlling their distribution. *Nature* **223**: 168–70

Greiner G O G 1974 Environmental factors controlling the distribution of benthonic foraminifera. *Breviora* **420**: 1–35

Grell K G 1979 Cytogenetic systems and evolution in

foraminifera. *Journal of Foraminiferal Research* **9**: 1–13
Grobe F, Fütterer D 1981 Zur Fragmentierung benthischer Foraminiferen in der Kieler Bucht (Westliche Ostsee). *Meyniana* **33**: 85–96
Grossman E L 1984a Stable isotope fractionation in live benthic foraminifera from the southern California borderland. *Palaeogeography, Palaeoclimatology, Palaeoecology* **47**: 301–27
Grossman E L 1984b Carbon isotope fractionation in live benthic foraminifera – comparison with inorganic precipitate studies. *Geochimica Cosmochimica Acta* **48**: 1505–12
Grossman E L 1987 Stable isotopes in modern benthic foraminifera: a study a vital effects. *Journal of Foraminiferal Research* **17**: 48–61
Grossman S, Benson R H 1967 Ecology of Rhizopoda and Ostracoda of southern Pamlico Sound region, North Carolina. *University of Kansas Paleontological Contributions* **44**: 1–82
Grünig A, Herb R 1980 Paleoecology of Late Eocene benthonic foraminifera from Possagno (Treviso – northern Italy). *Cushman Foundation Special Publication* **18**: 65–85
Guzman H M, Obando V L, Cortes J 1987 Meiofauna associated with a Pacific coral reef in Costa Rica. *Coral Reefs* **6**: 107–12
Haake F W 1962 Untersuchungen an der Foraminiferen-fauna im Wattgebiet zwishen Langeoog und dem Festland. *Meyniana* **12**: 25–64
Haake F W 1967 Zum Jahresgang von Populationen einer Foraminiferen-Art in der westlichen Ostee. *Meyniana* **17**: 13–27
Haake F W 1970 Zur Tiefenverteilung von Miliolinen (Foram) im Persischen Golf. *Paläontologische Zeitschrift* **44**: 196–200
Haake F W 1975 Miliolinen (Foram) in Oberfläschensedimenten des Persischen Golfes. *'Meteor' Forschungs Ergebnisse* **C21**: 15–51
Haake F W 1977 Living benthic foraminifera in the Adriatic Sea: influence of water depth and sediment. *Journal of Foraminiferal Research* **7**: 62–75
Haake F W 1980a Benthische Foraminiferen in Oberfläschen-Sedimenten und Kernen des Ostatlantiks vor Senegal/Gambia (Westafrika). *'Meteor' Forschungs Ergebnisse* **C32**: 1–29
Haake F W 1980b Sedimentologische und faunistische Untersuchungen an Watten in Taiwan. *Senckenbergiana Maritima* **12**: 247–55
Haake F W 1982 Occurrences of living and dead salt marsh foraminifera in the interior of northern Germany. *Senckenbergiana Maritima* **14**: 217–25
Haake F W, Coulbourn W T, Berger W H 1982 Benthic foraminifera: depth distribution and redeposition. In Rad V, Hinz K, Sarnthein M, Seibold E (eds) *Geology of the Northwest African Continental Margin* Springer-Verlag, Berlin pp 632–57
Haig D W 1988 Miliolid foraminifera from inner neritic sand and mud facies of the Papuan lagoon, New Guinea. *Journal of Foraminiferal Research* **18**: 203–36
Haig D W, Burgin S 1982 Brackish-water foraminiferids from the Purari River delta, Papua New Guinea. *Revista Española de Micropaleontologia* **14**: 359–66
Hald M, Vorren T O 1984 Modern and Holocene foraminifera and sediments on the continental shelf off Troms, North Norway. *Boreas* **13**: 133–54
Halicz E, Noy N, Reiss Z 1984 Foraminifera from Shura Arwashie Mangrove. In Por F D, Dor I (eds) *Aquatic Biology of the Mangal Ecosystem.* Developments in Hydrobiology Junk, Den Haag pp 145–50
Halicz E, Reiss Z 1979 Recent Textulariidae from the Gulf of Elat (Aqaba), Red Sea. *Revista Española de Micropaleontologia* **40**: 295–320
Hallock P 1979 Trends in test shape with depth in large symbiont-bearing foraminifera. *Journal of Foraminiferal Research* **9**: 61–9
Hallock P 1981a Light dependence in *Amphistegina. Journal of Foraminiferal Research* **11**: 40–6
Hallock P 1981b Production of carbonate sediments by selected large benthic foraminifera on two Pacific coral reefs. *Journal of Sedimentary Petrology* **51**: 467–74
Hallock P 1984 Distribution of selected species of living algal symbiont-bearing foraminifera on two Pacific coral reefs. *Journal of Foraminiferal Research* **14**: 250–61
Hallock P 1985 Why are larger foraminifera large? *Paleobiology* **11**: 195–208
Hallock P 1988a The role of nutrient availability in bioerosion: consequences to carbonate buildups. *Palaeogeography, Palaeoclimatology, Palaeoecology* **63**: 275–91
Hallock P 1988b Diversification in algal symbiont-bearing foraminifera: a response to oligotrophy? *Revue de Paléobiology vol. spéc.* **2**: 789–97
Hallock P, Cottey T L, Forward L B, Halas J 1986 Population biology and sediment production of *Archaias angulatus* (Foraminiferida) in Largo Sound, Florida. *Journal of Foraminiferal Research* **16**: 1–8
Hallock P, Forward L B, Hansen H J 1986 Influence of environment on the test shape of *Amphistegina. Journal of Foraminiferal Research* **16**: 224–31
Hallock P, Glenn E C 1986 Larger foraminifera: a tool for paleoenvironmental analysis of Cenozoic carbonate depositional facies. *Palaios* **1**: 55–64
Hallock P, Hansen H J 1979 Depth adaptation in *Amphistgina*: change in lamellar thickness. *Bulletin of the Geological Society of Denmark* **27**: 99–104
Hallock P, Schlager W 1986 Nutrient excess and the demise of coral reefs and carbonate platforms. *Palaios* **1**: 389–98
Hallock Muller P 1978 Carbon fixation and loss in a foraminiferal–algal symbiont system. *Journal of Foraminiferal Research* **8**: 35–41

Haman D 1969 Seasonal occurrence of *Elphidium excavatum* (Terquem) in Llandanwg Lagoon (North Wales, UK). *Contributions from the Cushman Foundation for Foraminiferal Research* **20**: 139–42
Haman D 1971 Foraminiferal assemblages in Tremadoc Bay, North Wales, UK. *Journal of Foraminiferal Research* **1**: 126–43
Haman D 1973 Récents Elphidiidae, Elphidiinae provenant des stations littorales de l'Alaska de l'Ouest situées sous une haute latitude. *Revue de Micropaléontologie* **16**: 176–83
Haman D 1983 Modern Textulariina (Foraminiferida) from the Balize Delta, Louisiana. *Proceedings of the First Workshop on Arenaceous Foraminifera* 7–9 September 1981, Continental Shelf Institute, Norway, Publication No 108 pp 59–87
Hamsa K M S A 1972 Foraminifera of the Palk Bay and Gulf of Mannar. *Journal of the Marine Biological Association of India* **14**: 418–23
Hansen H J 1965 On the sedimentology and the quantitative distribution of living foraminifera in the northern part of the Øresund. *Ophelia* **2**: 323–31
Hansen H J, Buchardt B 1977 Depth distribution of *Amphistegina* in the Gulf of Elat, Israel. *Utrecht Micropalaeontological Bulletin* **15**: 205–24
Hansen H J, Dalberg P 1979 Symbiotic algae in milioline foraminifera: CO_2 uptake and shell adaptation. *Bulletin of the Geological Society of Denmark* **28**: 47–55
Hansen H J, Lykke-Anderson A L 1976 Wall structure and classification of fossil and recent elphidiid and nonionid foraminifera. *Fossils and Strata* **10**: 1–37
Harman R A 1964 Distribution of foraminifera in the Santa Barbara Basin, California. *Micropaleontology* **10**: 81–96
Harmelin J G, Vacelet J, Vasseur P 1985 Les grottes sous-marines obscures: un milieu extrême et un remarquable biotope refuge. *Téthys* **11**: 214–29
Hasegawa S 1979 Foraminifera of the Himi Group, Hokuriku Province, central Japan. *Science Reports of the Tohoku University, Sendai, Second Series (Geology)* **49**: 89–163
Hasewaga S 1988 Distribution of recent foraminiferal fauna in Toyama Bay, central Japan. *Revue de Paléobiologie, vol spéc* **2**: 803–13
Hasegawa S, Takayanagi Y 1981 Notes on homotrematid foraminifera from Toyana Bay, central Japan. *Science Reports of the Tohoku University Sendai, Second Series (Geology)* **51**: 67–86
Haward J B, Haynes J R 1976 *Chlamys opercularis* (Linnaeus) as a mobile substrate for foraminifera. *Journal of Foraminiferal Research,* **6**: 30–8
Haynes J 1965 Symbiosis, wall structure and habitat in foraminifera. *Contributions from the Cushman Foundation for Foraminiferal Research* **16**: 40–3
Haynes J R 1973 Cardigan Bay recent Foraminifera. *Bulletin of the British Museum (Natural History) Zoology*, Supplement **4**: 1–245
Haynes J R 1981 *Foraminifera* Macmillan, London 433pp
Haynes J, Dobson M 1969 Physiography, foraminifera and sedimentation in the Dovey Estuary (Wales). *Geological Journal* **6**: 217–56
Hayward B W 1979 An interitdal *Zostera* pool community at Kawerua, Northland, and its foraminiferal microfauna. *Tane* **25**: 173–86
Hayward B W 1981 Foraminifera in nearshore sediments of the eastern Bay of Islands, northern New Zealand. *Tane* **27**: 123–34
Hayward B W 1982 Associations of benthic foraminifera (Protozoa : Sarcodina) of inner shelf sediments around the Cavalli Islands, northeast New Zealand. *New Zealand Journal of Marine and Freshwater Research* **16**: 27–56
Hayward B W, Brazier R C 1980 Taxonomy and distribution of present-day *Bolivinella. Journal of Foraminiferal Research* **10**: 102–16
Hayward B W, Grace R V, Bull V H 1984 Soft bottom macrofauna, foraminifera and sediments off the Chicken Islands, northern New Zealand. *Tane* **30**: 141–64
Hedley R H, Hurdle C M, Burdett I D J 1965 A foraminiferal fauna from the western continental shelf, North Island, New Zealand. *New Zealand Oceanographic Institute Memoir* **25**: 1–48
Hedley R H, Hurdle C M, Burdett I D J 1967 The marine fauna of New Zealand: intertidal foraminifera of the *Corallina officinalis* zone. *New Zealand Oceanographic Institute Memoir* **38**: 1–86
Henbest L G 1963 Biology, mineralogy, and diagenesis of some typical late Paleozoic sedentary foraminifera and algal-foraminiferal colonies. *Special Publication from the Cushman Foundation for Foraminiferal Research* **6**: 1–44
Herb R 1971 Distribution of recent benthonic foraminifera in the Drake Passage. *American Geophysical Union, Antarctic Research Series* **17**: 251–300
Hermelin J O 1983 Biogeographic patterns of modern *Reophax dentaliniformis* Brady (arenaceous benthic Foraminifera) from the Baltic Sea. *Journal of Foraminiferal Research* **13**: 155–62
Hermelin J O R, Scott D B 1985 Recent benthic foraminifera from the central North Atlantic. *Micropaleontology* **31**: 199–220
Heron-Allen E, Earland A 1913 Clare Island foraminifera. *Proceedings of the Royal Irish Academy* **31**(64): 1–188
Heron-Allen E, Earland A 1914, 1915 The Foraminifera of the Kerimba Archipelago (Portuguese East Africa). *Transactions of the Zoological Society of London* **20**: 363–90, 543–790
Heron-Allen E, Earland A 1932 Foraminifera Part 1. The ice-free area of the Falkland Islands and adjacent seas. *Discovery Reports* **4**: 293–460
Hickman C S, Lipps J H 1983 Foraminiferivory: selective ingestion of foraminifera and test alterations produced by the neogastropod *Olivella. Journal of Foraminiferal Research* **13**: 108–14
Hiltermann H 1973 Zur Sociologie von agglutinierenden

Foraminiferen polar, skandinavischer und äquatorialer Meer. *Geologische Jahrbuch* **A6**: 101–20
Hiltermann H 1982 Meereskundliche und paläokologische Biozönotik. *Paläontologische Zeitschrift* **56**: 153–64
Hiltermann H 1984 Bathymetrische und synokölogische Verteiling von Benthos-Foraminiferen. *Erdöl und Kohle* **37**: 11–4
Hiltermann H 1987 Syn–ecological study of benthic foraminifera of the Gulf of Mexico. *Erdöl und Kohle* **40**: 9–14
Hiltermann H, Brönnimann P, Zaninetti L 1981 Neue Biozönosen in der Sedimenten der Mangrove bei Acupe, Bahia, Brasilien. *Notes de Laboratoire de Paléontologie de l'Université de Genève* **8**: 1–6
Hiltermann J, Haman D 1985 Sociology and synecology of brackish-water foraminifera and thecamboebinids of the Balize Delta, Louisiana. *Facies* **13**: 287–94
Hiltermann H, Tüxen J 1978 Die Biozönosen der Phlegerschen Benthos-Faunen vom Mississippi–Sund und-Delta. *Paläontologische Zeitschrift* **52**: 271–9
Hofker J 1968 Foraminifera from the Bay of Jakarta, Java. *Bijdragen Tot de Dierkunde* **37**: 11–59
Hofker J 1977 The foraminifera of Dutch tidal flats and salt marshes. *Netherlands Journal of Sea Research* **11**: 223–96
Hofker J 1978 Biological results of the Snellius Expedition XXX. The foraminifera collected in 1929 and 1930 in the eastern part of the Indonesian Archipelgao. *Zoologische Verhandelingen Rijksmuseum van Natuurlijke Histoire te Leiden* **161**: 1–69
Hofker J 1983 Zoological exploration of the continental shelf of Surinam: the foraminifera of the shelf of Surinam and the Guyanas. *Zoologische Verhandelingen* **201**: 1–75
Höglund H 1947 Foraminifera in the gullmar Fjord and the Skagerrak. *Zoologiska Bidrag från Uppsala* **26**: 1–328
Holme N A, McIntyre A D 1984 *Methods for the Study of Marine Benthos*. IBP Handbook 16 2nd edn Blackwell Scientific Publications, Oxford
Hooper K 1969 A re-evaluation of eastern Mediterranean foraminifera using factor-vector analysis. *Contributions from the Cushman Foundation for Foraminiferal Research* **20**: 147–51
Hooper K 1975 Foraminiferal ecology and associated sediments of the lower St Lawrence estuary. *Journal of Foraminiferal Research* **5**: 218–38
Hoskin I 1974 A constant-volume sampling device, the Hosbac minisampler. *Micropaleontology* **20**: 110–11
Hottinger L 1977 Distribution of larger Peneroplidae, *Borelis* and Nummulitidae in the Gulf of Elat, Red Sea. *Utrecht Micropalaeontological Bulletin* **15**: 35–109
Hottinger L 1980 Repartition comparée des grands foraminifères de la Mer Rouge at de l'Océan Indien. *Annali dell'Universita di Ferrara* (NS Sez IV) Supplement **6**: 1–13
Hottinger L 1982 Larger foraminifera, giant cells with a historical background. *Naturwissenschaften* **69**: 361–71
Hottinger L 1983 Processes determining the distribution of larger foraminifera in space and time. *Utrecht Micropaleontological Bulletin* **30**: 239–53
Howarth R J, Murray J W 1969 The Foraminiferida of Christchurch Harbour, England: a reappraisal using multivariate techniques. *Journal of Paleontology* **43**: 660–75
Huang T 1971 Foraminiferal trends in the surface sediments of Taiwan Strait. *Technical Bulletin, ECAFFE* **4**: 23–62
Huang T 1983 Foraminiferal biofacies of the Taiwan Strait, ROC. *Bollettino della Società Paleontologica Italiana* **22**: 151–77
Hueni C M, Anepohl J, Gevirtz J, Casey R 1978 Distribution of living benthonic foraminifera as indicators of oceanographic processes of the South Texas outer continental shelf. *Transaction of the Gulf Coast Association of Geological Societies* **28**: 193–200
Hughes G W 1977 Recent foraminifera from the Honiara Bay area, Solomon Islands. *Journal of Foraminiferal Research* **7**: 45–57
Hughes G W 1985 Recent foraminifera and selected biometrics of *Heterostegina* from Ontong Java Atoll, Solomon Islands, southwest Pacific. *Journal of Foraminiferal Research* **15**: 13–17
Hughes G W 1988 Modern bathyal agglutinating foraminifera from the Vella Gulf and Blanche Channel, New Georgia, Solomon Islands, southwest Pacific. *Journal of Foraminiferal Research* **18**: 304–10
Hughes-Clark M W, Keij A J 1973 Organisms as producers of carbonate sediment and indicators of environment in the southern Persian Gulf. In Purser B H (ed) *The Persian Gulf* Springer-Verlag, Berlin pp 33–56
Iaccarino S 1967 Ricerche sui foraminiferi contenuti in sei carote prelevate nel Mar Ligure (La Spezia). *Bollettino della Societa Geologica Italiana* **86**: 59–88
Iaccarino S 1969 I foraminiferi di campioni di fondo prelevati nel Golfo di Taranto (M Junio). *Ateno Parmense, Acta Naturalia* **5**: 73–99
Ikeya N 1970 Population ecology of benthonic foraminifera in Ishikari Bay, Hokkaido, Japan. *Records of Oceanographic Works in Japan* **10**: 173–91
Ikeya N 1971a Species diversity of recent benthonic foraminifera off the Pacific coast of North Japan. *Reports of Faculty of Science, Shizuoka University* **6**: 179–201
Ikeya N 1971b Species diversity of benthonic foraminifera, off the Shimokita Peninsula, Pacific coast of North Japan. *Records of Oceanographic Works in Japan* **11**: 27–37
Imbrie J, Van Donk J, Kipp N G 1973 Paleoclimatic investigation of a late Pleistocene Caribbean deep-sea core: comparison of isotopic and faunal methods. *Quaternary Research* **3**: 10–38
Ingle J C 1967 Foraminiferal biofacies variation and the Miocene–Pliocene boundary in southern California. *Bulletin of American Paleontology* **52**(236): 217–394

Ingle J C, Keller G 1980 Benthic foraminiferal biofacies of the eastern Pacific margin between 40° S and 32° N. In Field M E, Bouma A H *et al.* (eds) *Quaternary Depositional Environments of the Pacific Coast, Pacific Coast Paleogeography Symposium* Society of Economic Paleontologists and Mineralogists pp 341–55

Ingle J C, Keller G, Kolpack R L 1980 Benthic foraminiferal biofacies, sediments and water masses of the southern Peru–Chile Trench area, southeastern Pacific Ocean. *Micropaleontology* **26**: 113–50

Ishiwada Y 1958 Recent foraminifera from the brackish lake Hamana-ko. Studies on the brackish water III. *Geological Survey of Japan Report* **180**: 1–19

Jahn T L, Rinaldi R A 1959 Protoplasmic movement in the foraminiferan, *Allogromia laticollaris* and a theory of its mechanism. *Biological Bulletin* **117**: 100–18

Janin M C 1984 *Ammotrochoides bignoti* n. gen. n. sp., foraminifère des croûtes feromanganésifères des marges de l'océan Atlantique nord-est. In Oertli H J (ed) *Benthos '83* pp 327–37

Jarke J 1958 Sedimente und Mikrofauna im Bereich der Grenzschwelle zweier ozeanische Raüme, dargestellt an einem Schnitt über den Island Faröer-Rücken (Nordatlantischer Ozean-Rosengarten-Europäisches Nord-Mer). *Geologische Rundschau* **47**: 234–49

Jarke J 1960 Beitrag zur kenntnis der Foraminiferen fauna der mittleren und westlichen Barents Sea. *Internationale Revue der gesamten Hydrobiolgie und Hydrographie* **45**: 581–654

Jarke J 1961a Beobachtungen über kalkauflosung an Schalen von Mikrofossilien in Sedimenten der Westlichen Ostsee. *Deutsche Hydrographische Zeitschrift* **14**: 6–11

Jarke J 1961b Die Beziehungen zwischen hydrographischen Verhältnissen, Faziesentwicklung und Foraminiferen-verbreitung in der heutigen Nordsee als vorbild für die Verhältnisse während der Miocän-Zeit. *Meyniana* **10**: 21–36

Jell J S, Maxwell W H G, McKellar R G 1965 The significance of the larger foraminifera in the Heron Island reef sediments. *Journal of Paleontology* **39**: 273–9

Jepps M W 1942 Studies on *Polystomella* Lamarck (Foraminifera). *Journal of the Marine Biological Association of the United Kingdom* **25**: 607–66

Jindrich V 1983 Structure and diagenesis of recent algal–foraminiferal reefs, Fernando de Noronha, Brazil. *Journal of Sedimentary Petrology* **53** : 449–59

John A W G 1987 The regular occurrence of *Reophax scottii* Chaster, a benthic foraminiferan, in plankton samples from the North Sea. *Journal of Micropalaeontology* **6** : 61–3

Johnson J H 1950 A Permian algal-foraminiferal consortium from west Texas. *Journal of Paleontology* **24**: 6–12

Johnson K R, Albani A D 1973 Biotopes of recent benthonic foraninifera in Pitt Water, Broken Bay, NSW (Australia). *Palaeoegeography, Palaeoclimatology, Palaeoecology* **14**: 265–76

Jones G D, Ross C A 1979 Seasonal distribution of foraminifera in Samish Bay, Washington. *Journal of Paleonotology* **53**: 245–57

Jones J R, Cameron B 1987 Surface distribution of foraminifera in a New England salt marsh: Plum Island, Massachusetts. *Maritime Sediments and Atlantic Geology* **23**: 131–40

Jones R S, Charnock M A 1985 'Morphogroups' of agglutinating foraminifera. Their life positions and feeding habits and potential applicability in (paleo) ecological studies. *Revue de Paléobiologie* **4**: 311–20

Jones R W 1986 Distribution of morphogroups of recent agglutinating foraminifera in the Rockall Trough – a synopsis. *Proceedings of the Royal Society of Edinburgh* **88B**: 55–8

Jorissen F 1987 The distribution of benthic foraminifera in the Adriatic Sea. *Marine Micropaleontology* **12**: 21–48

Jorissen J F 1988 Benthic foraminifera from the Adriatic Sea: principles of phenotypic variation. *Utrecht Micropaleontological Bulletin* **37**: 1–174

Kaesler R L 1966 Quantitative re-evaluation of ecology and distribution of recent foraminifera and Ostracoda of Todos Santos Bay, Baja California, Mexico. *Paleontological Contributions, University of Kansas* **10**: 1–50

Kaminski M A 1985 Evidence for control of abyssal agglutinated foraminiferal community structure by substrate disturbance: results from the Hebble area. *Marine Geology* **66**: 113–31

Kaminski M A, Grassle J F, Whitlash R B 1988 Life history and recolonisation among agglutinated foraminifera in the Panama Basin. *Abhandlungen der Geologischen Bundesanstalt* **41**: 229–43

Kane H E 1967 Recent microfaunal biofacies in Sabine Lake and environs, Texas and Louisiana. *Journal of Paleontology* **41**: 947–64

Keller G H 1982 Organic matter and the geotechnical properties of submarine sediments. *Geo-Marine Letters* **2** : 191–8

Kellogg D E, Kellogg T B 1987 Microfossil distributions in modern Amundsen Sea sediments. *Marine Micropaleontology* **12**: 203–22

Kellogg T B, Kellogg D E 1988 Antarctic cryogenic sediments: biotic and inorganic facies of ice shelf and marine-based ice sheet environments. *Palaeogeography, Palaeoecology, Palaeoclimatology* **67**: 51–74

Kennett J P 1967 New foraminifera from the Ross Sea, Antarctica. *Contributions from the Cushman Foundation for Foraminiferal Research* **18** : 133–5

Kennett J P 1968 The fauna of the Ross Sea. Part 6 Ecology and distribution of foraminifera. *New Zealand Department of Scientific and Industrial Research Bulletin* **186**: 1–46

Kennett J P 1982 *Marine Geology* Prentice-Hall,

Englewood Cliffs pp 1–813
Kim B K, Han J H 1972 A foraminiferal study of the bottom sediments off the southeastern coast of Korea. *United Nations ECAPE, CCOP Technical Bulletin* **6**: 13–29
Kitazato H 1981 Observations of behaviour and mode of life of benthic foraminifers in laboratory. *Geoscience Reports of Shizuoka University* **6**: 61–71
Kitazato H 1984 Microhabitats of benthic foraminifera and their application to fossil assemblages. In Oertli H J (ed) *Benthos '83* pp 339–44
Kitazato H 1988a Locomotion of some benthic foraminifera in and on sediments. *Journal of Foraminiferal Research* **18**: 344–9
Kitazato H 1988b Ecology of benthic foraminifera in the tidal zone of a rocky shore. *Revue de Paléobiologie, vol spéc* **2**: 815–25
Kloos D 1980 Studies on the foraminifer *Sorites orbiculus. Geologie en Mijnbouw* **59**: 375–83
Kloos D 1982 Destruction of tests of the foraminifer *Sorites orbiculus* by endolithic micro-organisms in a lagoon on Curaçao (Netherlands Antilles). *Geologie en Mijnbouw* **61**: 201–5
Knight R, Mantoura R F C 1985 Chlorophyll and carotenoid pigments in foraminifera and their symbotic algae: analysis by high performance liquid chromatography. *Marine Ecology Progress Series* **23**: 241–9
Knudsen K L 1988 Marine interglacial deposits in the Cuxhaven area, NW Germany: a comparison of Holsteinian, Eemian and Holocene foraminiferal faunas. *Eiszeitalter und Gegenwart* **38**: 69–77
Kohl B 1985 Early Pliocene benthic foraminifers from the Salina Basin, southeastern Mexico. *Bulletins of American Paleontology* **88** (322): 1–173
Kontrovitz M, Snyder S W, Brown R J 1978 A flume study of the movement of foraminifera tests. *Palaeogeography, Palaeoclimatology, Palaeoecology* **23**: 141–50
Kouyoumontzakis G 1981 Les associations de foraminifères benthiques du plateau continental Congolais : une radiale au large de la Lagune Conkouati. *Téthys* **10**: 121–8
Kouyoumontzakis G 1982a Le biotope à *Cancris congolensis* aux abords de l'estuaire de fleuve Congo. *Océanographie Tropicale* **17**: 139–44
Kouyoumontzakis G 1982b Les associations de foraminifères du plateau congolais: foraminifères benthiques. *Cahiers de Micropaléontologie* **2**: 155–62
Kouyoumontzakis G 1984 Les Amphisteginidae (Foraminifera) du plateau continental Congolais dans le cadre de la marge ouest Africaine. *Revue de Micropaléontologie* **27**: 196–208
Kremer B P, Schmaljohann R, Röttger R 1980 Features and nutritional significance of photosynthates produced by unicellular algae symbiotic with larger foraminifera. *Marine Ecology Progress Series* **2**: 225–8
Kroopnick P M 1985 The distribution of ^{13}C of ΣCO_2 in the world oceans. *Deep-Sea Research* **32**: 57–84
Kruit C 1955 Sediments of the Rhône Delta. 1 Grain size and microfauna. *Verhandelingen van het Koninklijk Nederlandsch Geologisch-Mijnbouwkundig Genootschap Geologische Serie* **15**: 357–514
Kuile B T, Erez J 1984 *In situ* growth rate experiments on the symbiont-bearing foraminifera *Amphistegina lobifera* and *Amphisorus hemprichii. Journal of Foraminiferal Research* **14**: 262–76
Kuile B H, Erez J, Lee J J 1987 The role of feeding in the metabolism of larger symbiont bearing foraminifera. *Symbiosis* **4**: 335–50
Kustanowich S 1964 Foraminifera of Milford Sound. *New Zealand Oceanographic Institute Memoir* **17**: 49–63
Lagoe M 1976 Species diversity of deep-sea benthic foraminifera from the central Arctic Ocean. *Bulletin of the Geological Society of America* **87**: 1678–83
Lagoe M B 1977 Recent benthic foraminifera from Prudhoe Bay, Alaska. *Journal of Paleontology* **53**: 258–29
Lagoe M B 1979a Modern benthic foraminifera from Prudhoe Bay, Alaska. *Journal of Paleontology* **53**: 258–62
Lagoe M B 1979b Recent benthonic foraminiferal biofacies in the Arctic Ocean. *Micropaleontology* **25**: 214–24
Lagoe M B 1980 Recent Arctic foraminifera: an overview. In Field M D (ed) *Quaternary Depositional Environments of the Pacific Coast, Pacific Coast Paleogeography Symposium* **4**: 33–42. Society of Economic Paleontologists and Mineralogists
Langer M 1988 Recent epiphytic foraminifera from Vulcano (Mediterranean Sea). *Revue de Paléobiologie, vol spéc* No 2 *Benthos '86* pp 827–32
Langus B G, Medioli F, Watkins C 1972 Sedimentation of foraminiferal tests. *Rivista Italiana di Paleontologie e Stratigrafia* **78**: 681–92
Lankford R R 1959 Distribution and ecology of foraminifera from East Mississippi Delta margin. *Bulletin of the American Association of Petroleum Geologists* **43**: 2088–99
Lankford R R 1959 Distribution and ecology of foraminifera from East Mississippi Delta margin. *Bulletin of the American Association of Petroleum Geologists* **43**: 2088–99
Lankford R R, Phleger F B 1973 Foraminifera from the nearshore turbulent zone, western North America. *Journal of Foraminiferal Research* **3**: 102–32
Larsen A R 1976 Studies of recent *Amphistegina.* Taxonomy and some ecological aspects. *Israel Journal of Earth Sciences* **25**: 1–26
Larsen A R 1982 Foraminifera from off the west coast of Africa. *Atlantide Report* **13**: 49–149
Le Calvez J 1947 *Entosolenia marginata* (Walker and Boys) foraminifère apogamique extoparasite d'une autre foraminifère: *Discorbis villardeboanus* (d'Orbigny).

Compte Rendu Hebdomadaire des Séances de l'Academie des Sciences, Paris **224**: 1448–50

Le Calvez J, Le Calvez Y 1951 Contribution à l'étude des foraminifères des eaux saumâtres: I. Étangs de Canet et de Salses. *Vie and Milieu* **2**: 237–45

Le Calvez J, Le Calvez Y 1958 Répartition des foraminifères dans la Baie de Villefranche 1 – Miliolidae. *Annales de l'Institut Océanographique Monaco*, **35**: 159–234

Le Calvez Y 1963 Contribution á l'ètude des foraminifères de la region d'Abidjan (Côte d'Ivoire) *Revue de Micropaléontologie* **6**: 41–50

Le Calvez Y 1965 Les foraminifères. In Guilcher A, Berthois L, Le Calvez Y, Battistini R, Crosnier A. Les recifs coralliens et le lagon de l'île Mayotte (Archipel des Comores, Océan Indien). *Mémoire Office de la Recherche Scientifique et Téchnique Outre-Mer* pp 181–201

Le Calvez Y, Boillot G 1967 Étude des foraminifères contenus dans le sédiments actuels de la Manche occidentale. *Revue de Géographie Physique et de Géologie Dynamique* **9**: 391–408

Le Calvez Y, Cesana D 1972 Détection de l'état de vie chez les foraminifères. *Annales de Paléontologie, Invertebrès* **58**: 129–34

Le Campion J 1970 Contribution à l'étude des foraminifères du Bassin d'Arcachon et du proche océan. *Bulletin de l'Institute Géologique du Bassin Aquitaine* **8**: 3–98

Lee J J 1974 Towards understanding the niche of foraminifera. In Hedley R H, Adams C G (eds) *Foraminifera* **1**: 207–60

Lee J J 1980 Nutrition and physiology of the foraminifera. *Biochemistry and Physiology of Protozoa* **3**: 43–66 (2nd edn)

Lee J J, Bock W D 1976 The importance of feeding in two species of soritid foraminifera with algal symbionts. *Bulletin of Marine Science* **26**: 530–7

Lee J J, Crockett L J, Hagen J, Stone, R J 1974 The taxonomic identity and physiological ecology of *Chlamydomonas hedleyi* sp. nov., algal flagellate symbiont from the foraminifer *Archaias angulatus*. *British Phycological Journal* **9**: 407–22

Lee J J, McEnery M E, Garrison J R 1980 Experimental studies of larger foraminifera and their symbionts from the Gulf of Elat on the Red Sea. *Journal of Foraminiferal Research* **10**: 31–47

Lee J J, McEnery M, Pierce S, Muller W A 1966 Prey and predator relationships in the nutrition of certain littoral foraminifera. *Journal of Protozoology Supplement* **13**: 23 (Abstract 86)

Lee J J, Muller W A, Stone R J, McEnery M E, Zucker W 1969 Standing crop of foraminifera in sublittoral epiphytic communities of a Long Island salt marsh. *Marine Biology* **4**: 44–61

Lee J J, Pierce S, Tentchoff M, McLaughlin J J 1961 Growth and physiology of foraminifera in the laboratory: Part 1 collection and maintenance. *Micropaleontology* **7**: 461–6

Lee J J, Tietjen J H, Mastropaolo C, Rubin H 1977 Food quality and the heterogeneous spatial distribution of meiofauna. *Helgoländer Wissenschaftliche Meeresuntersuchungen* **30**: 272–82

Lee J J, Tietjen J H, Stone R J, Muller W A, Rullman J, McEnery M 1970 The cultivation and physiological ecology of members of salt marsh epiphytic communities. *Helgoländer Wissenschaftliche Meeresuntersuchungen* **20**: 136–56

Le Furgey A, St Jean J 1976 Foraminifera in brackish-water ponds designed for waste control and aquaculture studies in North Carolina. *Journal of Foraminiferal Research* **6**: 274–94

Lena H 1966 Foraminiferos recientes de Ushuaia (Tierra del Fuego, Argentina). *Ameghiniana* **4**: 311–36

Lena E 1975 Foraminiferos bentonicos del area de Isla Elefante (Antartida). *Physis* **A34**: 505–31

Lena H 1976 Distribucion de los foraminiferos bentonicos en el area oceanica adyacente al Rio de la Plata. *Physis* **A34**: 135–44

Lena H, L'Hoste S G 1974 Foraminiferos de Aguas salobres (Mar Chiquita, Argentina). *Revista Española de Micropaeontologia* **7**: 539–48

Lessard R H 1980 Distribution patterns of intertidal and shallow-water foraminifera of the tropical Pacific Ocean. *Cushman Foundation Special Publication* **19**: 40–58

Lesslar P 1987 Computer–assisted interpretation of depositional palaeoenvironments based on foraminifera. *Geological Society of Malaysia Bulletin* **21**: 103–19

Leutenegger S 1977 Ultrastructure and motility of dinophyceans symbiotic with larger, imperforated foraminifera. *Marine Biology* **4**: 157–64

Leutenegger S 1984 Symbiosis in benthic foraminifera: specificity and host adaptions. *Journal of Foraminiferal Research* **14**: 16–35

Leutenegger S, Hansen H J 1979 Ultrastructural and radiotracer studies of pore function in foraminifera. *Marine Biology* **54**: 11–16

Levy A 1971 Eaux saumâtres et milieux margino–littoraux. *Revue de Géographie Physique et du Géologie Dynamique* **13**: 269–78

Levy A, Mathieu R, Poignant A, Rosset-Moulinier M, Rouvillois A 1975 Sur quelques foraminifères actuels des plages de Dunkerque et des environs: néotypes et espèces nouvelles. *Revue de Micropaléontologie* **17**: 171–81

Levy A, Mathieu R, Poignant A, Rosset-Moulinier M, Rouvillois A 1980 Distribution des foraminifères de la marge continentale algérienne, Baie de Bou-Ismail. *Annales des Mines et de la Géologie Tunis* **3**(28): 443–67

Levy A, Mathieu R, Poignant A, Rosset-Moulinier M, Rouvillois A 1982a Foraminifères benthiques actuels de sédiments profonds du Golfe de Guinée. *Cahiers de*

Micropaléontologie **2**: 123–33
Levy A, Mathieu R, Poignant A, Rosset-Moulinier M, Rouvillois A 1982b Contribution à la connaissance des foraminifères du littoral Kenyan (Océan Indien). *Cahiers de Micropaléontologie* **2**: 135–48
Levy A, Mathieu R, Poignant A and Rosset-Moulinier M 1988 Les Soritidae et les Peneroplidae dans les biofacies de la plate-forme des Bahamas. *Revue de Paléobiologie, vol spéc* **2**: 833–41
Lewis K B, Jerkins C 1969 Geographical variation of *Noninellina flemingi. Micropalaeontology* **15**: 1–12
Li Y F 1986 Taphonomic foraminiferal communities in the reaches of Haihe River and Jiyunhe River near the sea and its palaeogeographic significance. *Acta Micropalaeontologica Sinica* **3**: 251–8 (in Chinese with English abstract)
Li Y F, Zhang Q S 1986 Recent foraminifers from Great Wall Bay, King George Island, Antarctica. *Acta Micropaleontologica Sinica* **3**: 335–46
Lidz L 1966 Planktonic foraminifera in the water column of the mainland shelf off Newport beach, California. *Limnology and Oceanography* **11**: 257–63
Lindenberg H G, Auras A 1984 Distribution of arenaceous foraminifera in depth profiles of the Southern Ocean (Kerguelen Plateau area). *Palaeogeography, Palaeoclimatology Palaeoecology* **48**: 61–106
Lipps J H 1982 Biology/paleobiology of foraminifera. In Broadhead T W (ed) *Foraminifera. Notes for a Short Course* Paleontological Society, New Orleans pp 1–21
Lipps J H 1983 Biotic interactions in benthic foraminifera. In Trevesz M J S, McCall P L (eds) *Biotic Interactions in Recent and Fossil Benthic Communities* Plenum Press, New York pp 331–76
Lipps J H, Krebs W N, Temnikov N K 1977 Microbiota under Antarctic ice shelves. *Nature* **265**: 232–3
Lipps J H, Ronan T E, DeLaca T E 1979 Life below the Ross Ice Shelf, Antarctica. *Science* **203**: 447–9
Lipps J H, Delaca T E 1980 Shallow-water foraminiferal ecology, Pacific Ocean. In Field M E, Bouma A H *et al.* (eds) *Quaternary Depositional Environments of the Pacific Coast, Pacific Coast Paleogeography Symposium* Society of Economic Paleontologists and Mineralogists pp 325–40
Lipps J H, Severin K P 1985 *Alveolinella quoyi*, a living fusiform foraminifera, at Motupore Island, Papua New Guinea. *Science in New Guinea* **11**: 126–37
Loeblich A R, Tappan H 1953 Studies of Arctic foraminifera. *Smithsonian Miscellaneous Collections* **12**: 1–150
Leoblich A R Jr, Tappan H 1964 Sarcodina, chiefly thecamoebians and Foraminiferida. In Moore R C (ed) *Treatise on Invertebrate Paleontology* Geological Society of America, New York Part C vols 1–2 900 pp
Leoblich A R Jr, Tappan H 1984 Suprageneric classification of the Foraminiferida (Protozoa). *Micropaleontology* **30**: 1–70
Leoblich A R Jr, Tappan H 1988 *Foraminiferal Genera and their Classification* van Nostrand Reinhold, New York
Lohmann G P 1978 Abyssal benthonic foraminifera as hydrographic indicators in the western South Atlantic Ocean. *Journal of Foraminiferal Research* **8**: 6–34
Loose T 1970 Turbulent transport of benthonic foraminifera. *Contributions from the Cushman Foundation for Foraminiferal Research* **21**: 161–6
Lopez E 1979 Algal chloroplasts in the protoplasm of three species of benthic foraminifera: taxonomic affinity, viability and persistence. *Marine Biology* **53**: 201–11
Love L G, Murray J W 1963 Biogenic pyrite in recent sediments of Christchurch Harbour, England. *American Journal of Science* **261**: 433–48
Ludwich J C, Walton W R 1957 Shelf–edge, calcareous prominences in northeastern Gulf of Mexico. *Bulletin of the American Association of Petroleum Geologists* **41**: 2054–2101
Lukashina N P 1987 Benthic foraminifera and their relationship to water masses on sills of the North Atlantic Ocean. *Oceanology* **27**: 201–5
Lutze G F 1964a Zum Färben rezenter Foraminiferen. *Meyniana* **14**: 43–7
Lutze G F 1964b Statistical investigations on the variability of *Bolivina argentea* Cushman. *Contributions from the Cushman Foundation for Foraminiferal Research* **15**: 105–16
Lutze G F 1965 Zur Foraminiferen–fauna der Ostsee. *Meyniana* **15**: 75–142
Lutze G F 1968a Jahresgang der Foraminiferen-Fauna in der Bottsand Lagune (Westliche Ostsee). *Meyniana* **18**: 13–30
Lutze G F 1968b Siedlungs–Strukturen rezenter Foraminiferen. *Meyniana* **18**: 31–4
Lutze G F 1974a Foraminiferen der Kieler Bucht (Westliche Ostsee): 1 'Hausgartengebiet' des Sonderforschungsbereiches 95 der Universität Kiel. *Meyniana* **26**: 9–22
Lutze G F 1974b Benthische Foraminiferen in Oberfläschen-Sedimenten des Persischen Golfes. Teil 1: Arten. *'Meteor' Forschungs Ergebnisse* **C17**: 1–66
Lutze G F 1980 Depth distribution of benthic foraminifera on the continental margin off NW Africa. *'Meteor' Forschungs-Ergebnisse* **C32**: 31–80
Lutze G F, Coulbourn W T 1984 Recent benthic foraminifera from the continental margin of northwest Africa: community structure and distribution. *Marine Micropaleontology* **8**: 361–401
Lutze G F, Grabert B, Seibold E 1971 Lebendbeobachtunges an Grossforminiferen (*Heterostegina*) aus dem Persischen Golf. *'Meteor' Forschungs–Ergebnisse* **C6**: 21–40
Lutze G F, Mackensen A, Wefer G 1983 Foraminiferen der Kieler Bucht: 2. Salinitätsansprüche von *Eggerella scabra* (Williamson). *Meyniana* **35**: 55–65
Lutze G F, Thiel H 1989 Epibenthic foraminifera from

elevated microhabitats: *Cibicides wuellerstorfi* and *Planulina ariminensis*. *Journal of Foraminiferal Research* **19**: 153–8

Lutze G F, Wefer G 1980 Habitat and sexual reproduction of *Cyclorbiculina compressa* (d'Orbigny), Soritidae. *Journal of Foraminiferal Research* **10**: 251–60

Luz B, Reiss Z, Zohary-Segev T 1983 Seasonal variation of stable isotopes in *Amphorisorus* from the Gulf of Aqaba (Elat), Red Sea. *Utrecht Micropalaeontological Bulletins* **30**: 229–37

Lynts G W 1962 Distribution of recent foraminifera in Upper Florida Bay and associated sounds. *Contributions from the Cushman Foundation for Foraminiferal Research* **13**: 127–44

Lynts G W 1966 Variation of foraminiferal standing crop over short lateral distances in Buttonwood Sound, Florida Bay. *Limnology and Oceanography* **11**: 562–6

Lynts G W 1971 Distribution and model studies on foraminifera living in Buttonwood Sound, Florida Bay. *Memoir of the Miami Geological Society* **1**: 73–115

Macarovici N, Cehan-Ionesi B 1962 Distribution des foraminifères sur la plate-forme continentale du nord-ouest de la Mer Rouge. *Travaux du Museum d'Historie Naturelle 'Gr. Antipa', Buccuresti* **3**: 45–60

McCave B 1988 Biological pumping upwards of the coarse fraction of deep-sea sediments. *Journal of Sedimentary Petrology* **58**: 148–58

McCulloch I 1977 *Qualitative Observations on Recent Foraminiferal Tests with Emphasis on the Eastern Pacific* University of Southern California, Los Angeles Parts 1–3 pp 1–1079

Macdonald A G 1975 *Physiological Aspects of Deep Sea Biology* Cambridge University Press, Cambridge 450pp

McGlasson R H 1959 Foraminiferal biofacies around Santa Catalina Island, California. *Micropaleontology* **5**: 217–40

McIntyre A D 1961 Quantitative differences in the fauna of boreal mud associations. *Journal of the Marine Biological Association of the United Kingdom* **41**: 599–616

McKee E D, Chronic J, Leopold E B 1959 Sedimentary belts in lagoon of Kapingamarangi Atoll. *Bulletin of the American Association of Petroleum Geologists* **43**: 501–62

Mackensen A 1987 Benthische Foraminiferen auf dem Island-Schottland Rücken: Umwelt-Anzeiger an der Grenze zweier ozeanischer-Raüme. *Paläontologische Zeitschrfit* **61**: 149–79

Mackensen A, Hald M 1988 *Cassidulina teretis* Tappan and *C. laevigata* d'Orbigny: their modern and late Quaternary distribution in northern seas. *Journal of Foraminiferal Research* **18**: 16–24

Mackensen A, Sejrup H P, Jansen E 1985 The distribution of living benthic foraminifera on the continental slope and rise southwest Norway. *Marine Micropaleontology* **9**: 275–306

McKenzie K G 1962 A record of foraminifera from Oyster Harbour, near Albany, Western Australia. *Royal Society of Western Australia* **45**: 117–32

McKnight W M 1962 The distribution of foraminifera on parts of the Antarctic coast. *Bulletin of American Paleontology* **44**: 65–154

Madiera M 1969 Foraminifera from São Francisco do Sul, State of Santa Catarina, Brazil. *Iheringia, Zoologia* **37**: 3–29

Mageau N C, Walker D A 1976 Effects of ingestion of foraminifera by larger invertebrates. *Maritime Sediments Special Publication* **1**: 89–105

Maiklem, W R 1967 Black and brown speckled foraminiferal sand from the southern part of the Great Barrier Reef. *Journal of Sedimentary Petrology* **37**: 1023–30

Maiklem W R 1968 Some hydraulic properties of bioclastic carbonate grains. *Sedimentology* **10**: 109–09

Malmgren B A 1984 Analysis of the environmental influence on the morphology of *Ammonia beccarii* (Linné) in southern European salinas. *Geobios* **17**: 737–46

Marle L J Van 1988 Bathymetric distribution of benthic foraminifera on the Australina–Irian Jaya continental margin, eastern Indonesia. *Marine Micropalaeontology* **13**: 97–152

Marshall P R 1976 Some relationships between living and total foraminiferal faunas on Pedro Bank, Jamaica. *Maritime Sediments Special Publication* **1**: 61–70

Marszalek D S, Wright R C, Hay W W 1969 Function of the test in foraminifera. *Transactions of the Gulf Coast Association of Geological Societies* **19**: 341–52

Martin R A 1981 Benthic foraminifera from the Orange-Lüderitz shelf, southern African continental margin. *Marine Geoscience Unit, University of Cape Town, Bulletin* **11**: 1–75

Martin R E 1986 Habitat and distribution of the foraminifer *Archaias angulatus* (Fichtel and Moll) (Miliolina, Soritidae), northern Florida Keys. *Journal of Foraminiferal Research* **16**: 201–6

Martin R E 1988 Benthic foraminiferal zonation in deep-water carbonate platform margin environments, northern Little Bahama Bank. *Journal of Paleontology* **62**: 1–8

Martin R E, Steinker D C 1973 Evaluation of techniques of recognition of living foraminifera. *Comprass* **50**: 26–30

Martin R E, Wright R C 1988 Information loss in the transition from life to death assemblages of foraminifera in back reef environments, Key Largo, Florida. *Journal of Paleontology* **62**: 399–410

Massiota R, Cita M B, Mancuso M 1976 Benthonic foraminifers from bathyal depths in the Eastern Mediterranean. *Maritime Sediments, Special Publication* **1**: 251–62

Matera N J, Lee J J 1972 Environmental factors affecting the standing crop of foraminifera in sublittoral and psammolittoral communities of a Long Island salt

marsh. *Marine Biology* **14**: 89–103
Mateu G 1970 Estudio sistemática y bioecológico de los foraminiferos vivientes de los litorales de Cataluña y Baleares. *Trabajos del Instituto Español de Oceanografia* **38**: 1–84
Mateu G 1971 Les foraminifères de la mer d'Alboran. Leur importance comme indicateur des différentes masses d'eau qui confluent dans cette mer et l'influence des courantes marins dans la distribution de leurs biocénoses planctoniques et bentoniques. *Rapports Commission Internationale pour l'Exploration de la Mer Mediterranée* **20**: 211–13
Mateu G 1974 Foraminíferos recientes de la isla de Menorca (Baleares) y su aplicacion como indicadores biológicos de contaminacion litoral. *Boletin de la Sociedad de Historia Natural de Baleares* **19**: 89–112
Mathieu P 1970 Distribution des foraminifères benthoniques sur le plateau continental Ivorien aux abords de l'embouchure du Bandama en relation avec la nature sédimentologique. *Quatrième Colloque Africain de Micropaléontologie* pp 268–87
Mathieu R 1971 Les associations des foraminifères de plateau continental atlantique de maroc au large de Casablanca. *Revue de Micropaléontologie* **14**: 55–61
Mathieu R 1988 Foraminifères actuels et résurgences côtièrs sur la marge continentale Atlantique de Maroc. *Revue de Paléobiologie, vol spéc* **2**: 845–50
Matoba Y 1970 Distribution of recent shallow water foraminifera of Matsushima Bay, Miyagi Prefecture, northeast Japan. *Science Reports of the Tôhoku University, Sendai, Second Series (Geology)* **42**: 1–85
Matoba Y 1976a Foraminifera from off Noshiro, Japan and postmortem destruction of tests in the Japan Sea. *Progress in Micropaleontology American Museum of Natural History* pp 168–89
Matoba Y 1976b Distribution of foraminifera around the Oga Peninsula. *Reports on the Influence of the Waste from Steel Industries on Environments, Akita Prefecture* **3**: 182–216
Motoba Y 1976c Recent foraminiferal assemblages off Sendai, northeast Japan. *Maritime Sediments, Special Publication* **1**: 205–20
Matoba Y 1984 Paleoenvironment of the Sea of Japan. In Oertli H J (ed) *Benthos '83* pp 409–14
Matoba Y, Honma N 1986 Depth distribution of recent benthic foraminifera off Nishitsugaru, eastern Sea of Japan. In Matoba Y, Kato M (eds) *Studies on Cenozoic Benthic Foraminifera in Japan, Akita* pp 53–78
Matoba Y, Nakagawa H 1972 Recent foraminiferal assemblages from the continental shelf and slope off Akita, Japan Sea coast of northeast Japan *Prof Jun-Ichi Iwai Memorial Volume* pp 657–71
Matsumaru K, Matsuo Y 1976 Short note on the recent benthonic foraminiferids from the beach sediments of the subtropical and tropical island in the western Pacific region. *Journal of Saitama University, Faculty of Education., (Mathematics and Natural Science)* **25**: 15–26
Mead G A 1985 Recent benthic foraminifera in the Polar Front region of the southwest Atlantic. *Micropaleontology* **31**: 221–48
Mead G A, Kennett J P 1987 The distribution of recent benthic foraminifera in the Polar Front region, southwest Atlantic. *Marine Micropaleontology* **11**: 343–60
Mearns D L, Hine A C, Riggs S R 1988 Comparison of sonographs taken before and after Hurricane Diana, Onslow Bay, North Carolina. *Geology* **16**: 267–70
Medioli F S, Schafer C T, Scott D B 1986 Distribution of recent benthonic foraminifera near Sable Island, Nova Scotia. *Canadian Journal of Earth Sciences* **23**: 985–1000
Mello J F, Buzas M 1968 An application of cluster analysis as a method of determining biofacies. *Journal of Paleontology* **42**: 747–58
Michie M G 1987 Distribution of foraminifera in a macrotidal tropical estuary: Port Darwin, Northern Territory of Australia. *Australian Jounal of Marine and Freshwater Research* **38**: 249–59
Mikhalevich V I 1972 Bottom foraminifers from the continental shelf of west–equatorial Africa and some aspects of zoogeography. *Okealogija, Akademia Nauk SSSR* **12**: 520–6
Mikhalevich V I 1983 The bottom foraminifera from the shelves of the tropical Atlantic. *USSR Academy of Sciences, Zoological Institute* pp 1–245
Milam R W, Anderson J B 1981 Distribution and ecology of recent benthonic foraminifera of the Adélie–George V continental shelf and slope, Antarctica. *Marine Micropaleontology* **6**: 297–325
Miller A A L, Scott D B, Medioli F S 1982 *Elphidium excavatum* (Terquem): ecophenotypic versus subspecific variation. *Journal of Foraminiferal Research* **12**: 116–44
Miller D J, Ellison R L 1982 The relationship of foraminifera and submarine topography on the New Jersey–Delaware continental shelf. *Bulletin of the Geological Society of America* **93**: 239–45
Miller K G, Lohman G P 1982 Environmental distribution of recent benthic foraminifera on the northeast United States continental slope. *Bulletin of the Geological Society of America* **93**: 200–6
Millett F W 1898–1904 Report on the recent foraminifera of the Malay Archipelago Parts I–XVII. *Journal of the Royal Microscopical Society* 1898: I 258–69, II 459–513, III 607–14; 1899: IV 249–55, V 357–65, VI 557–64; 1900: VII 6–13, VIII 273–81, IX 539–49; 1901: X 1–11, XI 485–97, XII 619–28; 1902: XIII 509–28; 1903: XIV 253–75, XV 685–704; 1904: XVI 489–506, XVII 597–609. Whole reprinted as one volume by Antiquariaat Junk, Netherlands 1970.
Milliman J 1976 *Miniacina miniacea*: modern foraminiferal sands on the outer Moroccan shelf. *Sedimentology* **23**: 415–19
Miro M D de, Gualancañay E 1974 Foraminiferos

bentonicos de la plataforma continental de la Provincia de Esmeraldas, Ecuador. *Armada de Ecuador Instituto Oceanografica Publicacion* pp 1–12

Möbius K A 1880 Foraminifera von Mauritius. In Möbius K A, Richter R, von Martens R (eds) *Beitrag zur Meeresfauna der Insel Mauritius und der Seychellen* Gutman, Berlin pp 65–136

Moncharmont-Zei M 1964 Studio ecologico sui foraminiferi del Golfo di Puzzuoli (Napoli). *Pubblicazioni della Stazione Zoologica di Napoli* **34**: 160–84

Monier C 1973 Note Préliminaire sur les foraminifères benthiques du platier interne du grand récif de Tuléar (Madagascar). *Téthys* **5**: 241–50

Montaggioni L F 1981 Les associations de foraminifères dans les sédiments récifaux de l'archipel des Mascareignes (Océan Indien). *Annales de l'Institute Océanographique, Paris* 57: 41–62

Moore P G 1985 *Cibicides lobatulus* (Protozoa: Foraminifera) epizoic on *Astacilla longicornis* (Crustacea: Isopoda) in the North Sea. *Journal of Natural History* **19**: 129–33

Morariu A, Hottinger L 1988 Amphisteginid specific identifications, dimorphism, coiling direction and provincialism. *Revue de Paléobiologie, vol spéc.* **2**: 695–8

Morishima M, Chiji M 1952 The foraminiferal thanatocoenoses of Akkeshi Bay and its vicinity. *Memoirs of the College of Science, University of Kyoto* **B20**: 113–17

Morris P H 1982 Distribution and palaeoecology of Middle Jurassic Foraminifera from the Lower Inferior Oolite of the Cotswolds. *Palaeogeography, Palaeoclimatology, Palaeoecology* **37**: 319–47

Moulinier M 1967 Répartition des foraminifères benthiques dans les sédiments de la Baie de Seine entre le Cotentin et le meridien de Ouistreham. *Cahiers Océanographiques* **19**: 477–94

Moura A R 1965 Foraminiferos da ilha da Inhaca. *Revista dos Estudos Gerais Universitarios de Moçambique* ser 2, **2**: 3–74

Muller P H 1974 Sediment production and population biology of the benthic foraminifer *Amphistegina madagascariensis. Limnology and Oceanography* **19**: 802–9

Muller P H 1976 Sediment production by shallow-water, benthic foraminifera at selected sites on Oahu, Hawaii. *Maritime Sediments, Special Publication* **1**: 263–5

Muller W A 1975 Competition for food and other niche-related studies of three species of salt-marsh foraminifera. *Marine Biology* **31**: 339–51

Muller W A, Lee J J 1969 Apparent indispensability of bacteria in foraminiferan nutrition. *Journal of Protozoology* **16**: 471–8

Mullineaux L S 1987 Organisms living on manganese nodules and crusts: distribution and abundance at three North Pacific sites. *Deep-Sea Research* **34**: 165–84

Mullineaux L S 1988 Taxonomic notes on large agglutinated foraminifers encrusting manganese nodules, including the description of a new genus, *Chrondrodapis* (Komakiacea). *Journal of Foraminiferal Research* **18**: 46–53

Mullineaux L S, DeLaca T E 1984 Distribution of Antarctic benthic foraminifers settling on the pecten *Adamussium colbecki. Polar Biology* **3**: 185–9

Mullins H T, Thompson J B, McDougall K, Vercoutere T L 1985 Oxygen-minimum zone edge effects: evidence from the central California coastal upwelling system. *Geology* **13**: 491–4

Murray J W 1963 Ecological experiments on Foraminiferida. *Journal of the Marine Biological Association of the United Kingdom* **43**: 621–42

Murray J W 1965a The Foraminiferida of the Persian Gulf 2. The Abu Dhabi region *Palaeogeography, Palaeoclimatology, Palaeoecology* **1**: 307–32

Murray J W 1965b Two species of British recent Foraminiferida. *Contributions from the Cushman Foundation for Foraminiferal Research* **16**: 148–50

Murray J W 1965c On the foraminifera of the Plymouth Region. *Journal of the Marine Biological Association of the United Kingdom* **45**: 481–505

Murray J W 1966d Significance of benthic foraminiferids in plankton samples. *Journal of Paleontology* **39**: 156–7

Murray J W 1966a The Foraminiferida of the Persian Gulf 3. The Halat al Bahrani region. *Palaeogeography, Palaeoclimatology, Palaeoecology* **2**: 59–68

Murray J W 1966b The Foraminiferida of the Persian Gulf 4. Khor al Bazam. *Palaeogeography, Palaeoclimatology, Palaeoecology* **2**: 153–69

Murray J W 1966c The Foraminiferida of the Persian Gulf 5. The shelf off the Trucial coast. *Palaeogeography, Palaeoclimatology, Palaeoecology* **2**: 267–78

Murray J W 1967 Production in benthic foraminiferids. *Journal of Natural History* **1**: 61–8

Murray J W 1968a The living Foraminiferida of Christchurch Harbour, England. *Micropaleontology* **14**: 83–96

Murray J W 1968b Living foraminifers of lagoons and estuaries. *Micropaleontology* **14**: 435–55

Murray J W 1969 Recent foraminifers from the Atlantic continental margin of the United States. *Micropaleontology* **15**: 401–9

Murray J W 1970a The Foraminiferida of the Persian Gulf 6. Living forms in the Abu Dhabi region. *Journal of Natural History* **4**: 55–67

Murray J W 1970b The Foraminiferida of the hypersaline Abu Dhabi lagoon, Persian Gulf. *Lethaia* **3**: 51–68

Murray J W 1970c Foraminifers of the Western Approaches to the English Channel. *Micropaleontology* **16**: 471–85

Murray J W 1971a *An Atlas of British Recent Foraminiferids* Heinemann, London 245 pp

Murray J W 1973 *Distribution and Ecology of Living Benthic Foraminiferids* Heinemann, London 288 pp

Murray J W 1976a Comparative studies of living and dead benthic foraminiferal distributions. In Hedley R H, Adams C G (eds) *Foraminifera* **2**: 45–109

Murray J W 1976b A method of determining proximity of marginal seas to an ocean. *Marine Geology* **22**: 103–19

Murray J W 1979a Recent benthic foraminiferids of the Celtic Sea. *Journal of Foraminiferal Research* **9**: 193–209

Murray J W 1979b British nearshore foraminiferids. In Kermack D M and Barnes R S K(eds) *Synopses of the British Fauna* (new series) No 16 Academic Press, London 62 pp

Murray J W 1980 The foraminifera of the Exe Estuary. Essays on the Exe Estuary, *Devonshire Association Special Volume* **2**: 89–115

Murray J W 1982 Benthic foraminifera: the validity of living, dead or total assemblages for the interpretation of palaeoecology. *Journal of Micropalaeontology* **1**: 137–40

Murray J W 1983 Population dynamics of benthic foraminifera: results from the Exe Estuary, England. *Journal of Foraminiferal Research* **13**: 1–12

Murray J W 1984a Benthic foraminifera: some relationships between ecological observations and palaeoecological interpretations. In Oertli H (ed) *Benthos '83 Second International Symposium Benthic Foraminifera* (Pau April 1983) pp 465–69

Murray J W 1984b Paleogene and Neogene benthic foraminifers from Rockall Plateau. In Roberts D, Schnitker *et al. Initial Reports of the Deep Sea Drilling Project* **81**: 504–34

Murray J W 1985 Recent foraminifera from the North Sea (Forties and Ekofisk areas) and the continental shelf west of Scotland. *Journal of Micropalaeontology* **4**: 117-25

Murray J W 1986 Living and dead Holocene foraminifera of Lyme Bay, southern England. *Journal of Foraminiferal Research* **16**: 347–52

Murray J W 1987a Biogenic indicators of suspended sediment transport in marginal marine environments: quantitative examples from SW Britain. *Journal of the Geological Society, London* **144**: 127–33

Murray J W 1987b Benthic foraminiferal assemblages: criteria for the distinction of temperate and subtropical carbonate environments. In Hart M B (ed) *Micropalaeontology of Carbonate Sediments* Ellis Horwood, Chichester pp 9–20

Murray J W 1988 Neogene bottom water-masses and benthic foraminifera in the NE Atlantic Ocean. *Journal of the Geological Society, London* **145**: 125–32

Murray J W 1989 Syndepositional dissolution of calcareous foraminifera in modern shallow water sediments. *Marine Micropaleontology* **15**: 117–21

Murray J W, Hawkins A B 1976 Sediment transport in the Severn Estuary during the past 8000–9000 years. *Journal of the Geological Society London* **132**: 385–98

Murray J W, Sturrock S, Weston J F 1982 Suspended load transport of foraminiferal tests in the tide- and wave-swept sea. *Journal of Foraminiferal Research* **12**: 51–65

Murray J W, Taplin C M 1984 Larger agglutinated foraminifera from the Faeroe Channel and Rockall Trough collected by W B Carpenter. *Journal of Micropalaeontology* **3**: 59–62

Murray J W, Weston J F, Haddon C A, Powell A D J 1986 Miocene to recent bottom water masses of the north-east Atlantic: an analysis of benthic foraminifera. In Summerhayes C P, Shackleton N J (eds) *North Atlantic Palaeoceanography*, Geological Society Special Publication **21**: 219–30

Murray J W, Wright C A 1970 Surface textures of calcareous foraminiferids. *Palaeontology* **13**: 184–7

Murray J W, Wright C A 1974 Palaeogene Foraminiferida and palaeoecology, Hampshire and Paris basins and the English Channel. *Special Papers in Palaeontology* **14**: 1–171

Murray W G, Murray J W 1987 A device for obtaining representative samples from the sediment–water interface. *Marine Geology* **76**: 313–17

Myers E H 1936 The life cycle of *Spirillina vivipara* Ehrenberg, with notes on morphogenesis, systematics and distribution of the foraminifera. *Journal of the Royal Microscopical Society* **56**: 120–46

Myers E H 1942 Biological evidence as the rate at which tests of foraminifera are contributed to the sediments. *Journal of Paleontology* **16**: 397–8

Myers E H 1943a Life activities of foraminifera in relation to marine ecology. *Proceedings of the American Philosophical Society* **86**: 439–58

Nagy J 1965 Foraminifera in some bottom samples from shallow water in Vestspitsbergen. *Årbok norsk Polarinstitut* 1963 pp 109–25

Nagy J, Alve E 1987 Temporal changes in foraminiferal faunas and impact of pollution in Sandebukta, Oslo Fjord. *Marine Micropaleontology* **12**: 109–28

Nagy J, Dypvik H, Bjaerke T 1984 Sedimentological and paleontological analyses of Jurassic North Sea deposits from deltaic environments. *Journal of Petroleum Geology* **7**: 169–88

Naidu T Y, Rao D C, Rao M S 1985 Foraminifera as pollution indicators in the Visakhapatnam Harbour complex, east coast of India. *Bulletin of the Geological, Mining and Metallurgical Society Of India* **52**: 88–96

Naidu R Y, Rao M S 1988 Foraminiferal ecology of Bendi Lagoon, east coast of India. *Revue de Paléobiologie vol. spéc.* 2 pp 851–8

Narappa K V, Rao M S, Rao M P 1981 Living Foraminiferida from the estuarine complex of the Gautami and Nilarevu distributaries of river Godavari – Part 1. Living populations in relation to ecological factors. *Proceedings of the IX Indian Colloquium on Micropalaeontology and Stratigraphy* pp 49–68

Neagu T 1982 Foraminifères récents de la zone du récif

coralligène de l'île Mbudya (Côte orientale de la Tansanie). *Revista Española de Micropaleontologia* **14**: 99–136

Nichols M N, Ellison R L 1967 Sedimentary patterns of microfauna in a coastal plain estuary. In Lauff G H (ed) *Estuaries* American Association for the Advancement of Science, Washington DC, pp 23–88

Nichols M M, Norton W 1969 Foraminiferal populations in a coastal plain estuary. *Palaeogeography, Palaeoclimatology, Palaeoecology* **6**: 197–213

Nienstedt J C, Arnold A J 1988 The distribution of benthic foraminifera on seamounts near the East Pacific Rise. *Journal of Foraminiferal Research* **18**: 237–49

Nigam R 1984 Living benthonic foraminifera in a tidal environment: Gulf of Khambhat (India). *Marine Geology* **58**: 415–25

Nigam R 1986a Dimorphic forms of recent foraminifera: an additional tool in paleoclimatic studies. *Palaeogeography, Palaeoclimatology, Palaeoecology* **53**: 239–44

Nigam R 1986b Foraminiferal assemblages and their use as indicators of sediment movement: a study in the shelf region off Navapur, India. *Continental Shelf Research* **5**: 421–30

Nigam R, Rao A S 1987 Proloculus size variation in recent benthic foraminifera: implications for paleoclimatic studies. *Estuarine, Coastal and Shelf Science*, **24**: 649–55

Nigam R, Sarupria J S 1981 Cluster analysis and ecology of living benthonic foraminiferids from inner shelf off Ratnagiri, west coast, India. *Journal of the Geological Society of India* **22**: 175–80

Nigam R, Setty M G A P, Ambre N V 1979 A checklist of benthic foraminiferids from the inner shelf off Dabhol–Vengurla region, Arabian Sea. *Journal of the Geological Society of India* **20**: 244–47

Nigam R, Thiede J 1983 Recent foraminifers from the inner shelf of the central west coast, India: a reappraisal using factor analysis. *Proceedings of the Indian Academy of Science (Earth and Planetary Science)* **92**: 121–8

Nikoljuk V F 1968 Was bergen die Erdchichten der Wüste Kara-Kum? *Pedobiologia* **7**: 335–52

Nomura R 1981–82 List and bibliography of the recent benthonic Foraminifera of Japan, 1925–1981. *Memoirs of the Faculty of Education Shimane University* **15**: 31–60, **16**: 21–54

Nomura R 1984 Cassidulinidae (Foraminiferida) from the eastern part of Lützow–Holm Bay, Antarctica. *Transactions and Proceedings of the Paleontological Society of Japan,* NS **136**: 492–501

Nomura R 1988 Ecological significance of wall microstructure of benthic foraminifera in the southwestern sea of Japan. *Revue de Paléobiologie, vol spéc* **2**: 859–71

Nørvang A 1945 The zoology of Iceland. *Foraminifera* **2**(2): 1–79

Nowlin W D 1971 Water masses and general circulation of the Gulf of Mexico *Oceanology International* (1971 Feb) pp 28–33

Nyholm K G 1961 Morphogensis and biology of the foraminifer *Cibicides lobatulus. Zoologiska Bidrag från Uppsala* **33**: 157–96

Nyholm K G, Gertz I 1973 To the biology of the monothalamous foraminifer *Allogromia marina* n. sp. *Zoon* **1**: 89–93

Nyholm K G, Olsson I 1973 Seasonal fluctuations of the meiobenthos in an estuary on the Swedish west coast. *Zoon* **1**: 69–76

Nyholm K G, Olsson I, Andren L 1977 Quantitative investigations on the macro- and meiobenthic fauna in the Göta River estuary. *Zoon* **5**: 15–28

Nyong E E, Olsson R K 1984 A paleoslope model of Campanian to lower Maestrichtian foraminifera in the North American Basin and adjacent continental margin. *Marine Micropaleontology* **8**: 437–77

Oggioni E, Zandini L 1987 Response of benthic foraminifera to stagnant episodes – a quantitative study of core BAN 81-23, eastern Mediterranean *Marine Geology* **75**: 241–61

Olausson E 1965 Evidence of climatic changes in North Atlantic deep-sea cores. *Progress in Oceanography* **3**: 221–52

Olivieri R 1963 Rapporti quantitativi fra i foraminiferi recenti delle spiagge di Rimini, Porto Corsini e Lido de Venezia. *Bollettino della Societa Paleontologica Italiana* **2**: 94–108

Olsson I 1976 Distribution and ecology of the foraminiferan *Ammotium cassis* (Parker) in some Swedish estuaries. *Zoon* **4**: 137–47

Olsson I 1977 Wall structure of the foraminifer *Ammotium cassis* (Parker) and its ecological significance. *Zoon* **5**: 11–14

Olsson R K, Nyong E E 1984 A paleoslope model for Campanian–lower Maestrichtian foraminifera of New Jersey and Delaware. *Journal of Foraminiferal Research* **14**: 50–68

Østby K L, Nagy J 1981 Foraminiferal stratigraphy of Quaternary sediments in the western Barents Sea. In Neale J W, Brasier M B (eds) *Microfossils from Recent and Fossil Shelf Seas* Ellis Horwood, Chichester pp 261–73

Østby K L, Nagy J 1982 Foraminiferal distribution in the western Barents Sea, Recent and Quaternary. *Polar Reserch* **1**: 53–95

Osterman L E, Kellogg T B 1979 Recent benthic foraminiferal distributions from the Ross Sea, Antarctica: relation to ecologic and oceanographic conditions. *Journal of Foraminiferal Research* **9**: 250–69

Otvos E G 1978 Calcareous benthic foraminiferal faunas in a very low salinity setting, Lake Pontchartrain, Louisiana. *Journal of Foraminiferal Research* **8**: 262–9

Otvos E G, Bock W D 1976 Massive long-distance

transport and redeposition of Upper Cretaceous planktonic foraminifers in Quanternary sediments. *Journal of Sedimentary Petrology* **46**: 978–84

Painter P K, Spencer R S 1984 A statistical analysis of variants of *Elphidium excavatum* and their ecological control in southern Chesapeake Bay, Virginia. *Journal of Foraminiferal Research* **14**: 120–8

Palmieri V 1976a Recent and subrecent foraminifera from the Wynnum 1 : 2500 sheet area, Moreton Bay, *Queensland. Queensland Government Mining Journal* **77**: 365–84

Palmieri V 1976b Modern and relect foraminifera from the central Queensland continental shelf. *Queensland Government Mining Journal* **77**: 407–23

Parker F L 1948 Foraminifera of the continental shelf from the Gulf of Maine to Maryland. *Bulletin of the Museum of Comparative Zoology Harvard* **100**: 213–41

Parker F L 1952 Foraminiferal distribution in the Long Island Sound–Buzzards Bay area. *Bulletin of the Museum of Comparative Zoology Harvard* **106**: 427–73

Parker F L 1954 Distribution of the foraminifera in the north-eastern Gulf of Mexico. *Bulletin of the Museum of Comparative Zoology Harvard* **111**: 425–588

Parker F L 1958 Eastern Mediterranean foraminifera. *Swedish Deep-Sea Expedition* **8**: 219–83

Parker F L, Athearn W D 1959 Ecology of Marsh foraminifera in Poponesset Bay. *Journal of Paleontology* **33**: 333–43

Parker F L, Phleger F B, Peirson J F 1953 Ecology of foraminifera from San Antonion Bay and environs, southwest Texas. *Cushman Foundation for Foraminiferal Research Special Paper* **2**: 1–75

Parr W J 1945 Recent foraminifera from Barwon Heads, Victoria. *Proceedings of the Royal Society of Victoria* **56**: 189–227

Parr W J 1950 Foraminifera. *BANZARE* 1929–31, *Reports.* **B5**: 232–392

Pawlowski J, Lapierre L 1988 Foraminifères benthiques actuelles de l'Atlantique nord-est. *Revue de Paléobiologie, vol spéc* **2**: 873–8

Peet R K 1974 The measurement of species diversity. *Annual Review of Ecology and Systematics* **5**: 285–307

Pereira C A F D 1969 Recent foraminifera of southern Brazil collected by hydrographic vessel *Baependi. Iheringia, Zoologia* **37**: 37–95

Perelis L, Reiss Z 1975 Cibicididae in recent sediments from the Gulf of Elat. *Israel Journal of Earth Sciences* **24**: 73–96

Peryt T M, Peryt D 1975 Association of sessile tubular foraminifera and cyanophytic algae. *Geological Magazine* **112**: 612–14

Peterson L C 1984 Recent abyssal benthic foraminiferal biofacies of the eastern equatorial Indian Ocean. *Marine Micropaleonotology* **8**: 479–519

Pflum C E 1966 The distribution of foraminifera in the eastern Ross Sea, Amundsen Sea and Bellingshausen Sea, Antarctica. *Bulletin of American Paleontology* **50**: 146–209

Pflum C E, Frerichs W E 1976 Gulf of Mexico deep-water foraminifers. *Cushman Foundation for Foraminiferal Research Special Publication* **14**: 1–124

Phleger F P 1951 Ecology of foraminifera, northwest Gulf of Mexico. Part 1 foraminifera distribution. *Memoir of the Geological Society of America* **46**: 1–88

Phleger F B 1952a Foraminifera ecology off Portsmouth, New Hampshire. *Bulletin of the Museum of Comparative Zoology Harvard* **106**: 316–90

Phleger F P 1952b Foraminifera distribution in some sediment-samples from the Canadian and Greenland Arctic. *Contributions from the Cushman Foundation for Foraminiferal Research* **3**: 80–9

Phleger F B 1954 Ecology of foraminifera and associated micro-organisms from Mississippi Sound and environs. *Bulletin of the American Association of Petroleum Geologists* **38**: 584–647

Phleger F B 1955 Ecology of foraminifera in southeastern Mississippi Delta area. *Bulletin of the American Association of Petroleum Geologists* **39**: 712–52

Phleger F B 1956 Significance of living foraminiferal populations along the central Texas coast. *Contributions from the Cushman Foundation for Foraminiferal Research* **7**: 106–51

Phleger F B 1960a *Ecology and Distribution of Recent Foraminifera* Johns Hopkins Press, Baltimore 297 pp

Phleger F B 1960b Sedimentary patterns of microfaunas in northern Gulf of Mexico. In Shepard F P, Phleger F B, van Andel Tj H (eds) *Recent Sediments, Northwest Gulf of Mexico* American Association of Petroleum Geologists pp 267–301

Phleger F B 1960c Foraminiferal population in Laguna Madre, Texas. *Science Reports of the Tohoku University, Sendai,* Second Series (Geology), Special Volume **4**: 83–91

Phleger F B 1964a Patterns of living benthonic foraminifera, Gulf of California. *Memoir of the American Association of Petroleum Geologists* **3**: 377–94

Phleger F B 1964b Foraminiferal ecology and marine geology. *Marine Geology* **1**: 16–43

Phleger F B 1965a Patterns of marsh foraminifera, Galveston Bay, Texas. *Limnology and Oceanography,* Supplement **R10**: 169–84

Phleger F B 1965b Sedimentology of Guerrero Negro Lagoon, Baja California, Mexico. In Whittard W F, Bradshaw R(eds) *Submarine Geology and Geophysics* Colston papers pp 205–35

Phleger F B 1965c Depth patterns of benthonic foraminifera in the eastern Pacific. *Progress in Oceanography* **3**: 273–87

Phleger F B 1966a Patterns of living marsh foraminifera in south Texas coastal lagoons. *Boletin de la Sociedad Geologica Mexicana* **28**: 1–44

Phleger F B 1966b Living foraminifera from coastal marsh, southwestern Florida. *Boletin de la Sociedad Geologica Mexicana* **28**: 45–60

Phleger F B 1967 Marsh foraminiferal patterns, Pacific coast of North America. *Ciencia del Mar y Limnologia Mexico* pp 11–38
Phleger F B 1970 Foraminiferal populations and marine marsh processes. *Limnology and Oceanography* **15**: 522–34
Phleger F B 1976 Foraminifera and ecological processes in St Lucia lagoon, Zululand. *Maritime Sediments Special Publication* **1**: 195–204
Phleger F B, Bradshaw J S 1966 Sedimentary environment in a marine marsh. *Science* **154**: 1551–3
Phleger F B, Ewing G C 1962 Sedimentology and oceanography of coastal lagoons in Baja California, Mexico. *Bulletin of the Geological Society of America* **73**: 145–82
Phleger F B, Lankford R R 1957 Seasonal occurrences of living benthonic foraminifera in some Texas bays. *Contributions from the Cushman Foundation for Foraminiferal Research* **8**: 93–105
Phleger F B, Lankford R R 1978 Foraminifera and ecological processes in the Alvarado Lagoon area, Mexico. *Journal of Foraminiferal Research* **8**: 127–31
Phleger F B, Parker F L 1951 Ecology of foraminifera, northwest Gulf of Mexico. Part II Foraminifera species. *Geological Society of America Memoir* **46**: 1–64
Phleger F B, Parker F L, Peirson J F 1953 North Atlantic foraminifera. *Reports of the Swedish Deep-Sea Expedition* **7**: 1–122
Phleger F B, Soutar A 1973 Production of benthic foraminifera in three east Pacific oxygen minima. *Micropaleontology* **19**: 110–15
Phleger F P, Walton W R 1950 Ecology of marsh and bay foraminifera, Barnstable, Massachusetts. *American Journal of Science* **248**: 274–94
Pielou E C 1966 The measurements of diversity in different types of biological collections. *Journal of Theoretical Biology* **13**: 131–44
Pielou E C 1979 A quick method of determining the diversity of foraminiferal assemblages. *Journal of Paleontology* **53**: 1237–42
Pielou E C 1984 *The Interpretation of Ecological Data. A Primer on Classification and Ordination* John Wiley, New York 263 pp
Poag C W 1969 Dissolution of molluscan calcite by the attached foraminifer *Vasiglobulina*, new genus (Vasiglobuliminae, new subfamily). *Tulane Studies in Geology and Paleontology* **7**: 45–70
Poag C W 1972 Shelf-edge submarine banks in the Gulf of Mexico: paleoecology and biostratigraphy. *Transactions of the Gulf Coast Association of Geological societies* **22**: 267–87
Poag C W 1976 The foraminiferal community of San Antonio Bay. In Bouma A H (ed) *Shell Dredging and its Influence on Gulf Coast Environments* Houston Publishing pp 304–36
Poag C W 1978 Paired foraminiferal ecophenotypes in Gulf Coast estuaries: ecological and paleoecological implications. *Transactions of the Gulf Coast Association of Geological Societies* **28**: 395–421
Poag C W 1981 *Ecologic Atlas of Benthic Foraminifera of the Gulf of Mexico.* Hitchinson and Ross, Stroudsburg, Pa viii: 174 pp
Poag C W 1982 Environmental implications of test-to-substrate attachment among some modern sublittoral foraminifera. *Bulletin of the Geological Society of America* **93**: 252–68
Poag C W 1984 Distribution and ecology of deep-water benthic foraminifera in the Gulf of Mexico. *Palaeogeography, Palaeoclimatology, Palaeoecology* **48**: 25–37
Poag C W, Knebel H J, Todd R 1980 Distribution of modern benthic foraminifers on the New Jersey outer continental shelf. *Marine Micropaleontology* **5**: 43–69
Poag C W, Sweet W E 1972 Claypile Bank, Texas continental shelf. In Rezak R, Henry V J (eds) Contributions on the geological and geophysical oceanography of the Gulf of Mexico. *Texas A & M University Oceanography Series* **3**: 223–61
Poag C W, Tresslar R C 1979 Habitat associations among living foraminifers of West Flower Garden Bank, Texas continental shelf. *Transactions of the Gulf Coast Association of Geological Societies* **29**: 347–51
Poag C W, Tresslar R C 1981 Living foraminifers of West Flower Garden Bank, northernmost coral reef in the Gulf of Mexico. *Micropaleontology* **27**: 31–70
Polski W 1959 Foraminiferal biofacies off the North Asiatic coast. *Journal of Paleontology* **33**: 569–87
Price M V 1980 On the significance of test form in benthic salt-marsh foraminifera. *Journal of Foraminiferal Research* **10**: 129–35
Pujos M 1972 Répartition des biocoenoses de foraminifères benthiques sur le plateau continental du Golfe de Gascogne à l'ouest de l'embouchure de la Gironde. *Revista Española de Micropaleontologia* **4**: 141–56
Pujos M 1973 (for 1971) Mise en évidence des biocoenoses, faunes déplacées et paléothanatocoenoses de foraminifères benthiques sur un plateau continental: applications à la zone ouest-Girdonde (golfe de Gascogne). *Bulletin de la Société Géologique de France* **13**(7): 251–6
Pujos M 1976 Ecologie des foraminifères benthiques et des thécamoebiens de la Gironde et au plateau continental Sud-Gascogne. Application à la connaissance du Quaternaire terminal de la région Ouest-Gironde. *Memoires de l'Institut de Géologie du Bassin d'Aquitaine* **8**: 1–274
Pujos M 1983 *Jadammina polystoma*, temoin d'un environnement contraignant dans l'estuaire de la Gironde (France). In Oertli H (ed) *Benthos '83, Second International Symposium on Benthic Foraminifera* (Pau, April 1983) pp 511–17
Pujos-Lamy A 1973a Répartition bathymétrique des foraminifères benthiques profonds du Golfe de

Gascogne. Comparaision avec d'autres aires océaniques. *Revista Española de Micropaleontologia* **5**: 213–34

Pujos-Lamy A 1973b *Bolivina subaenariensis* Cushman, indicateur d'un milieu confiné dans le Gouf de Cap-Breton. *Comptes Rendus Académie Sciences (Paris)* **D277**: 2655–8

Pujos-Lamy A 1973c Le déplacement des faunes de foraminifères benthiques actuels sur la pente continentale et dans la plaine abyssale du Golfe de Gascogne. *Bulletin de la Société Géologique de France* **15** (7): 392–400

Purser B H 1973 *The Persian Gulf* Springer-Verlag, Berlin 471 pp

Quilty P G 1985 Distribution of foraminiferids in sediments of Prydz Bay, Antarctica. *Special Publication, South Australià Department of Mines and Energy* **5**: 329–40

Quinterno P, Carlson P R, Molnia B F 1980 Benthic foraminifers from the eastern Gulf of Alaska. In Field M E, Bouma A H *et al.* (eds) *Quaternary Depositional Environments of the Pacific Coast, Pacific Coast Paleogeography Symposium*, Society of Economic Paleontologists and Mineralogists pp 13–21

Quinterno P J, Gardner J V 1987 Benthic foraminifers on the continental shelf and upper slope, Russian River area, North California. *Journal of Foraminiferal Research* **17**: 132–52

Qvale G 1981 Distribution of foraminifers along the Norwegian continental margin – surface assemblages. In Neale J W, Brasier M D (eds) *Microfossils from Recent and Fossil Shelf Seas* Ellis Horwood, Chichester pp 323–35

Qvale G 1986a Distribution of benthic foraminifers in surface sediments along the Norwegian continental shelf between 62° and 72° N. *Norsk Geologisk Tidsskrift* **66**: 209–21

Qvale G 1986b Benthic foraminifers in the Norwegian Channel: a comparison of Upper Quaternary and recent zonations. *Norsk Geologisk Tidsskrift* **66**: 325–32

Qvale G, Markussen B, Thiede J 1984 Benthic foraminifers in fjords: response to water masses. *Norsk Geologisk Tidsskrift* **64**: 235–49

Qvale G, Nigam R 1985 *Bolivina skagerrakensis*, a new name for *Bolivina* cf. *B robusta*, with notes on its ecology and distribution. *Journal of Foraminiferal Research* **15**: 6–12

Qvale G, Van Weering T C E 1985 Relationships of surface sediments and benthic foraminiferal distribution patterns in the Norwegian Channel (northern North Sea). *Marine Micropaleontology* **9**: 496–88

Radford S 1976a Recent foraminifera from Tobago Island, West Indies. *Revista Española de Micropaleontologia* **8**: 193–218

Radford S 1976b Depth distribution of recent foraminifera in selected bays of Tobago Island, West Indies. *Revista Española de Micropaleontologia* **8**: 219–38

Ragothaman V, Kumar V 1985 Recent foraminifera from off the coast of Rameswaram, Palk Bay, Tamil Nadu. *Bulletin of the Geological, Mining and Metallurgical Society of India* **52**: 97–121

Ragothaman V, Manivannan V 1985 Recent foraminifera from off the coast of Mandapam, Tamil Nadu State. *Bulletin of the Geological, Mining and Metallurgical Society of India* **52**: 122–46

Ramanathan R 1970 Quantitative differences in the living benthic foraminifera of Vellar estuary, Tamil Nadu. *Journal of the Geological Society of India* **11**: 127–41

Rao K K 1969–71 Foraminifera of the Gulf of Cambay. *Journal of the Bombay Natural History Society* **66**: 584–96, **67**: 259–73, **68**: 9–19

Rao K K, Kutty M K, Panikkar B M 1985 Frequency distribution of foraminifera off Trivandrum, west coast of India. *Indian Journal of Marine Sciences* **14**: 74–8

Rao K K, Roa T S S 1979 Studies on pollution ecology of foraminifera of the Trivandrum coast. *Indian Journal of Sciences* **8**: 31–5

Rao M S, Vedantam D 1970 Recent foraminifera from off Pentakota, east coast of India. *Micropaleontology* **16**: 325–44

Rao T V, Rao M S 1974 Recent foraminifera of Suddagedda Estuary, east coast of India. *Micropaleontology* **20**: 398–419

Rasheed D A, Ragothaman V 1978 Ecology and distribution of recent foraminifera from the Bay of Bengal off Port-Novo, Tamil Nadu State, India. *Proceedings of the VII Indian Colloquium on Micropalaeontology and Stratigraphy* pp 263–98

Raup D M, Crick R E 1979 Measurement of faunal similarity in paleontology. *Journal of Paleontology* **53**: 1213–27

Reddy K R, Rao R J 1983 Diversity and dominance of living and total foraminiferal assemblages, Pennar Estuary, Andhra Pradesh. *Journal of the Geological Society of India* **24**: 594–603

Reddy K R, Rao R J 1984 Foraminifera – salinity relationship in the Pennar Estuary, India. *Journal of Foraminiferal Research* **14**: 115–19

Reddy K R, Rao R J, Naidu M G C 1975 Living and total foraminiferal fauna, Pennar Estuary, Andhra Pradesh, India. *Proceedings of the 4th Colloquium on Indian Microplaeontology and Stratigraphy* pp 16–20

Reinhöhl-Kompa S 1985 Zur Populationsökologie der Foraminiferen in nordöstlichen Ems-Dollart-Ästuar, südliche Nordsee. *Senckenbergiana Maritima* **17**: 147–61

Reiss Z 1977 Foraminiferal research in the Gulf of Elat – Aqaba. *Utrecht Micropaleontological Bulletin* **15**: 7–26

Reiss Z, Hottinger L 1984 The Gulf of Aqaba, ecological, micropaleontology. *Ecological Studies* **50**: 1–354 Springer-Verlag, Berlin

Reiter M 1959 Seasonal variations in intertidal foraminifera of Santa Monica Bay, California. *Journal of Paleontology* **33**: 606–30

Resig J M 1958 Ecology of foraminifera of the Santa Cruz Basin, California. *Micropaleontology* **4**: 287–308

Resig J M 1960 Foraminiferal ecology around ocean outfalls off southern California. In Pearson E A (ed) *Proceedings of the First International Conference on Waste Disposal in the Marine Environment* Pergamon Press, New York pp 104–21

Resig J M 1974 Recent foraminifera from a landlocked Hawaiian lake. *Journal of Foraminiferal Research* **4**: 69–76

Resig J 1981 Biogeography of benthic foraminifera of the northern Nazca plate and adjacent continental margin. *Memoir of the Geological Society of America* **154**: 619–66

Rex M A 1981 Community structure in the deep-sea benthos. *Annual Review of Ecology and Systematics* **12**: 331–53

Rey L (ed) 1982 *The Arctic Ocean* Macmillan, London 433 pp

Rhoads D C, Morse J W 1971 Evolutionary and ecologic significance of oxygen-deficient marine basins. *Lethaia* **4**: 413–38

Rhumbler L 1935 Rhizopoden der Kieler Bucht, gesammelt durch A Remane. I. Teil. *Schriften Naturwissenschaftlichen Vereins für Schleswig-Holstein* **21**: 143–94

Rhumbler L 1936 Rhizopoden der Kieler Bucht, gesammelt durch A. Remane. II. Teil. *Kieler Meeresforschungen* **1**: 179–242

Richter G 1964a Zur Ökologie der Foraminiferen I. Die Foraminiferen-Gesellchaften des Jadegebietes. *Natur und Museum, Frankfurt* **94**: 343–53

Richter G 1964b Zur Ökologie der Foraminiferen II. Lebensraum und Lebensweise von *Nonion depressulum, Elphidium excavatum und Elphidium selseyense. Natur und Museum, Frankfurt* **94**: 421–30

Richter G 1965 Zur Okologie der Foraminiferen III. Verdriftung und Transport in der Gezeitenzone. *Natur und Museum, Frankfurt* **95**: 51–62

Richter G 1967 Faziesbereiche rezenter und subrezenter Wattensedimente nach ihren Foraminiferen-Gemeinschaften. *Senckenbergiana Lethaea* **48**: 291–335

Rieman F 1983 Biological aspects of deep-sea manganese nodule formation. *Oceanologica Acta* **6**: 303–11

Risdal D 1964 Foraminiferfaunaens relasjon til dy-bdeforholdene i Olofjorden, med en diskusjon av de senkvartaere foraminifersoner. *Norges Geologiske Undersökelse* **226**: 1–142

Rocha A T, Torquato J R 1967 Contribuição para o conhecimentoda microfauna e da mineralogia das areias de praia das Ilhas da Taipa e de Coloane (Provincia Portugesa de Macau). *Boletin Instituto de Investigacao Cientifica de Angola* **4**: 89–104

Rocha A T, Ubaldo M L 1964 Contribution for the study of foraminifera from sands of Diu, Gogola and Simbor. *Garcia de Orta (Lisbon)* **12**: 407–20

Rodrigues C G, Hooper K 1982a Recent benthonic foraminiferal associations from offshore environments in the Gulf of St Lawrence. *Journal of Foraminiferal Research* **12**: 337–52

Rodrigues C G, Hooper K 1982b The ecological significance of *Elphidium clavatum* in the Gulf of St Lawrence, Canada. *Journal of Paleontology* **56**: 410–22

Rodrigues M A 1968 Foraminiferos recentes da Barra de Itabapoana, Estado do Rio de Janeiro. *Anais de Academia Brasileira de Ciencas* **40**: 555–69

Roettger E U 1970 Recent foraminifera from the continental shelf of Rio Grande do Sul collected by the hydrographic vessel *Canopus*. *Iheringia, Zoologia* **38**: 3–71

Rogers M J 1976 An evaluation of an index of affinity for comparing assemblages, in particular of foraminifera. *Palaeontology* **19**: 503–15

Ross C A 1972 Biology and ecology of *Marginopora vertebralis* (Foraminiferida), Great Barrier Reef. *Journal of Protozoology* **19**: 181–92

Ross C A 1974 Evolutionary and ecological significance of large calcareous Foraminiferida (Protozoa), Great Barrier Reef. *Proceedings of the Second International Coral Reef Symposium* **1**: 327–33

Ross C A, Ross J R P 1978 Adaptive evolution in the soritids *Marginopora* and *Amphisorus* (Foraminiferida). *Scanning Electron Microscopy* **2**: 53–60

Rosset-Moulinier M 1972 Études des foraminifères de côtes nord et ouest de Bretagne. *Travaux du Laboratoire de Géologie, École Normale Supérieure* **6**: 1–225

Rosset-Moulinier M 1981 Foraminiferal biocoenoses in an epicontinental sea of detrital sedimentation, the Channel. In Neale J W, Brasier M (eds) *Micropalaeontology of Shelf Seas* Ellis Horwood, Chichester pp 304–13

Rosset-Moulinier M 1986 Les populations de foraminifères benthiques de la Manche. *Cahiers de Biologie Marine* **27**: 387–440

Rosset-Moulinier M 1988 Comparaison de la répartition des foraminifères benthiques sur un plateau continental et dans une mer epicontinentale en climat tempère. *Revue de Paléobiologie vol spéc* **2**: 879–84

Rosset-Moulinier M, Roux P 1977 Application de quelques methodes d'analyse des donnees aux biocoenoses de foraminifères de la Baie de Saint-Brieuc (Côtes-du-Nord, France). *Revue de Micropaléontologie* **20**: 100–13

Rottgardt D 1952 Micropaläontologisch wichtige Bestandteile rezenter brackisher Sedimente an den Küsten Schleswig-Holstein. *Meyniana* **1**: 169–228

Röttger R 1972a Die Bedeutug der Symbiose von *Heterostegina depressa* (Foraminifer, Nummulitidae) für hohe Siedlungsdichte und Karbonatproduktion. *Verhandlunge der Deutschen Zoologischen Gessellschaft* **65**: 42–7

Röttger R 1972b Analyse von wachstumkurven von *Heterostegina depressa* (Foraminifera: Nummulitidae). *Marine Biology* **17**: 228–42

Röttger R 1974 Larger foraminifera: reproduction and early stages of development in *Heterostegina depressa. Marine Biology* **26**: 5–12

Röttger R 1976 Ecological observations of *Heterostegina depressa* (Foraminifera, Nummulitidae) in the laboratory and in its natural habitat. *Maritime Sediments Special Publication* **1**: 75–9

Röttger R 1978 Unusual multiple fission in the gamont of the larger foraminifera *Heterostegina depressa. Journal of Protozoology* **25**: 41–4

Röttger R 1981 Vielteiling bei Grossforaminiferen. *Mikrokosmos* **5**: 137–8

Röttger R, Fladung M, Schmaljohann R, Spindler M, Zacharies H 1986 A new hypothesis; the so-called megalospheric schizont of the larger foraminifer, *Heterostegina depressa* d'Orbigny, 1826, is a separate species. *Journal of Foraminiferal Research* **16**: 141–9

Röttger R, Irwan A, Schmaljohann R, Franzisket L 1980 Growth of the symbiont-bearing foraminifers *Amphistegina lessonii* d'Orbigny and *Heterostegina depressa* d'Orbigny (Protozoa). In Schwemmler W, Schenk H A (eds) *Endosymbiosis and Cell Biology* Water de Gruyter, Berlin vol 1 pp 125–32

Röttger R, Schmaljohann R 1976 Foraminifera: Gamogonie, Teil des Entwicklungsgangs der rezenten Nummulitidae *Heterostegina depressa. Die Naturwurissenschaften* **10**: 486

Röttger R, Spindler M, Schmaljohann R, Richwien M, Fladung M 1984 Functions of the canal system in the rotaliid foraminifer *Heterostegina depressa. Nature* **309**: 789–91

Rouvillois A 1966 Contribution à l'étude micropaléontologique de la Baie de Roi, au Spitzberg. *Revue de Micropaléontologie* **9**: 169–76

Rouvillois A 1967 Observations morphologiques, sédimentologiques et écologiques sur la plage del la ville Ger, dans l'estuaire de la Rance. *Cahiers Océanographiques* **19**: 375–89

Rouvillois A 1970 Biocoenose et taphocoenose de foraminifères sur le plateau continental Atlantique au large de l'île d'Yeu. *Revue de Micropaléontologie* **13**: 188–204

Rouvillois A 1972a Biocoenose de foraminifères en relation avec les conditions physico-chimiques du milieu dans les bassins et l'avant-port de Saint Malo (Ille-et-Vilaine). *Cahiers de Micropaléontologie* ser 3, **1**: 1–10

Rouvillois A 1972b Influence du barrage de l'usine marée–motrice sur la morphologie, l'écologie et la biocénose de la plage de la ville Ger dans l'estuaire de la Rance. *Quatrième Congrès International de la Mer* pp 115–23

Rouvillois A 1974 Sur quelques espèces rares de foraminifères dans l'estuaire de la Rance. *Cahiers de Micropaléontologie* **3**: 11–19

Ruiz J B, Sellier de Civrieux J M 1972 Distribucion y ecologia de las facies *Ammonia* y *Cribroelphidium* en la Laguna de las Maritas (Venezuela). *Boletin del Instituto Oceanografico de la Universidad Oriente* **11**: 83–96

Said R 1949 Foraminifera of the northern Red Sea. *Cushman Laboratory for Foraminiferal Research Special Publication* **26**: 1–44

Said R 1950a Additional foraminifera from the northern Red Sea. *Contributions from the Cushman Foundation for Foraminiferal Research* **1**: 4–9

Said R 1950b The distribution of foraminifera in the northern Red Sea. *Contributions from the Cushman Foundation for Foraminiferal Research* **1**: 9–29

Saidova H M 1961a Quantitative distribution of benthic foraminifera off Antarctica. *Oceanology* **139**: 769–9 (in Russian)

Saidova H M 1961b Ecology of the foraminifera and palaeogeography of the Far Eastern seas of the USSR and the northwestern part of the Pacific Ocean. *Akademia Nauk SSSR, Instituta Okeanologii* pp 1–232 (in Russian)

Saidova K M 1961c Quantitative distribution of bottom foraminifera in the northeastern part of the Pacific Ocean. *Trudy Instituta Okeanologii* **45**: 65–71 (in Russian)

Saidova K M 1962 Distribution of the principal species of calcareous foraminifera in the northwestern part of the Pacific Ocean. *Academia Nauk USSR, Voprosy Micropaleontology* **6**: 31–63 (in Russian)

Saidova K M 1963 Zonal quantitative distribution of bottom foraminifera in the Pacific Ocean. *Voprosy Mikropaleontologii* **7**: 196–208 (in Russian)

Saidova K M 1970 Benthonic foraminifera from the region of the Kurile–Kamchatka Trench (material from the 39th voyage of the *Vityaz). Trudy Instituta Okeanologii* **86**: 134–61 (in Russian)

Saidova K M 1975 *Benthonic foraminifera of the Pacific Ocean* Academy of Sciences of the USSR, P P Shirshov Institute of Oceanology pp 1–875 (in Russian)

Saidova K M 1976 Benthic foraminifera of the world ocean (zonal and quantitative distribution). *Akademia Nauk SSSR, Instituta Okeanologii* pp 1–161 (in Russian)

Saing C 1971 The distribution of recent foraminifera in the Bay of Bengal off the Arakan coast. *Union of Burma Journal of Life Sciences* **4**: 65–83

Saint-Marc P 1986 Qualitative and quantitative analysis of benthic foraminifers in Paleocene deep-sea sediments of the Sierra Leone Rise, central Atlantic. *Journal of Foraminiferal Research* **16**: 244–53

Sakai K, Nishihira M 1981 Population study of the benthic foraminifer, *Baculogypsina sphaerulata*, on an Okinawan reef flat and preliminary estimation of its annual production. *Proceedings of the Fourth International Coral Reef Symposium Manila* **2**: 763–6

Saks N M 1981 Growth, productivity and excretion of *Chlorella* spp. endosymbionts from the Red Sea:

implications for host foraminifera. *Botanica Marina* **24**: 445–9

Salami M B 1982 Bathyal benthonic foraminifera biofacies from the Nigeria sector of the Gulf of Guinea (West Africa). *Revista Española de Micropaleontologia* **14**: 455–61

Salvat B, Vénec–Peyré M T 1981 The living foraminifera in the Scilly Atoll lagoon (Society Islands). *Proceedings of the Fourth International Coral Reef Symposium Manila* **2**: 767–74

Sanchez-Ariza M C 1983 Specific associations of recent benthic foraminifera of the neritic zone in the Motrif–Nerja area, Spain, as a function of depth: diversity and constancy. *Journal of Foraminiferal Research* **13**: 13–20

Sanchez-Ariza M C 1984 Recent benthic foraminifera associations of the neritic zone, Motril–Nerja area, Spain: relationship with calcium carbonate content of surficial sediments. In Oertli H J (ed) *Benthos '83* pp 539–44

Sanders H L 1960 Benthic studies in Buzzards Bay III. The structure of the soft-bottom community. *Limnology and Oceanography* **5**: 138–53

Sanders H L 1986 Marine benthic diversity: a comparative study. *American Naturalist* **102**: 243–82

Schaaf A, Hoffert M, Karpoff A M, Wirrman D 1977 Association de structure stromatolithiques et de foraminifères sessiles dans un encroûtement ferromanganésifère à coeur granitique en provenance de l'Atlantique Nord. *Comptes Rendues Académie Sciences (Paris)* **D284**: 1705–8

Schafer C T 1967 Preliminary survey of the distribution of living benthonic foraminifera in Northumberland Strait. *Maritime Sediments* **3**: 105–8

Schafer C T 1970 Studies of benthonic foraminifera in Restigouche Estuary: 1. Faunal distribution patterns near pollution sources. *Maritime Sediments* **6**: 121–34

Schafer C T 1971 Sampling and spatial distribution of benthonic foraminifera. *Limnology and Oceanography* **16**: 944–51

Schafer C T 1982 Foraminiferal colonisation of an offshore dump site in Chaleur Bay, New Brunswick, Canada. *Journal of Foraminiferal Research* **12**: 317–26

Schafer C T, Cole F E 1974 Distributions of benthonic foraminifera: their use in delimiting local nearshore environments. *Geological Survey of Canada Paper* **74–30**: 103–8

Schafer C T, Cole F E 1976 Foraminiferal distribution patterns in the Restigouche Estuary. *Maritime Sediments Special Publication* **1**: 1–24

Schafer C T, Cole F E 1982 Living benthic foraminifera distribution on the continental slope and rise east of Newfoundland, Canada. *Bulletin of the Geological Society of America* **93**: 207–17

Schafer C T, Cole F E 1986 Reconnaissance survey of benthonic foraminifera from Baffin Island fjord environments. *Arctic* **39**: 232–9

Schafer C T, Cole F E 1988 Environmental associations of Baffin Island fjord agglutinated foraminifera. *Abhandlungen der Geologischen Bundesanstalt* **41**: 307–23

Schafer C T, Cole F E, Carter L 1981 Bathyal zone benthic foraminiferal genera off northeast Newfoundland. *Journal of Foraminiferal Research* **11**: 296–313

Schafer C T, Prakash A 1968 Current transport and deposition of foraminiferal tests, planktonic organisms and other lithogenic particles in Bedford Basin, Nova Scotia. *Maritime Sediments* **4**: 100–3

Schafer C T, Smith J N 1983 River discharge, sedimentation and benthic environment variations in Miramichi inner bay, New Brunswick. *Canadian Journal of Earth Sciences* **20**: 388–98

Schafer C T, Young J A 1977 Experiments on mobility and transportability of some nearshore benthonic foraminiferal species. *Geological Survey of Canada Paper* **77–1c**: 27–31

Schiffelbein P 1985 Extracting the benthic mixing impulse response function: a constrained deconvolution technique. *Marine Geology* **64**: 313–36

Schnitker D 1969 Distribution of foraminifera on a portion of the continental shelf of the Golfe de Gascogne (Gulf of Biscay). *Bulletin de Centre de Recherches de Pau – SNPA* **3**: 33–64

Schnitker D 1971 Distribution of foraminifera on the North Carolina continental shelf. *Tulane Studies in Geology and Paleontology* **8**: 169–215

Schnitker D 1974a Ecotypic variation in *Ammonia beccarii* (Linné). *Journal of Foraminiferal Research* **4**: 216–23

Schnitker D 1974b West Atlantic abyssal circulation during the past 120 000 years. *Nature* **248**: 385–87

Schnitker D 1979 The deep waters of the western North Atlantic during the past 24 000 years, and the re-initiation of the Western Boundary Undercurrent. *Marine Micropaleontology* **4**: 265–80

Schott W 1935 Die Foraminiferen in den Äquatorialen Teil des Atlantischen Ozeans. *Deutsche Atlantische Expedition* **6**: 411–616

Schrader H, Cheng G, Mahood R 1983 Preservation and dissolution of foraminiferal carbonate in an anoxic slope environment, southern Gulf of California. *Utrecht Micropaleontological Bulletin* **30**: 205–27

Schroder C J, Scott D B, Medioli F S 1987 Can smaller benthic foraminifera be ignored in paleoenvironmental analyses? *Journal of Foraminiferal Research* **17**: 101–5

Schroder C J, Scott D B, Medioli F S, Bernstein B B, Hessler R R 1988 Larger agglutinated foraminifera: comparison of assemblages from central North Pacific and western North Atlantic (Nares Abyssal Plain). *Journal of Foraminiferal Research* **18**: 25–41

Schwab D, Hofer H W 1979 Metabolism in the protozoan *Allogromia latticolaris* Arnold. *Zeitschrift für*

Mikroskopisch-anatomische Forschung, Leipzig **93**: 715–27

Schwemmler W 1980 Principles of endocytobiosis: structure, function, and information. In Schwemmler W, Schenk H E A (eds) *Endosymbiosis and Cell Biology* Walter de Gruyter, Berlin vol 1 pp 565–83

Scott D B 1974 Recent benthonic foraminifera from Samish and Padilla bays, Washington. *Northwest Science* **48**: 211–18

Scott D B 1976a Quantitative studies of marsh foraminiferal patterns in southern California and their application to Holocene stratigraphic problems. *Maritime Sediments Special Publication* **1**: 153–70

Scott D B 1976b Brackish-water foraminifera from southern California and description of *Polysaccammina ipohaline* n. gen., s. sp. *Journal of Foraminiferal Research* **6**: 312–21

Scott D, Gradstein F, Schafer C, Miller A, Williamson M 1983 The recent as a key to the past: does it apply to agglutinated foraminiferal assemblages? *Continental Shelf Institute, Norway, Publication* **108**: 147–57

Scott D B, Martini I P 1982 Marsh foraminifera zonations in western James and Hudson bays. *Naturaliste canadien* **109**: 399–414

Scott D B, Medioli F S 1980a Living vs. total foraminiferal populations: their relative usefulness in paleoecology. *Journal of Paleontology* **54**: 814–31

Scott D B, Medioli F S 1980b Quantitative studies of marsh foraminiferal distributions in Nova Scotia: implications for sea level studies. *Cushman Foundation for Foraminiferal Research Special Publication* **17**: 1–58

Scott D B, Medioli F S, Schafer C T 1977 Temporal changes in foraminiferal distributions in Miramichi River estuary, New Brunswick. *Canadian Journal of Earth Science* **14**: 1566–87

Scott D B, Mudie P J, Bradshaw J S 1976 Benthonic foraminifera of three southern Californian lagoons: ecology and recent stratigraphy. *Journal of Foraminiferal Research* **6**: 59–75

Scott D B, Piper D J W, Panagos A G 1979 Recent salt marsh and intertidal mudflat foraminifera from the western coast of Greece. *Revista Italiana de Paleontologia e Stratigrafia* **85**: 243–66

Scott D B, Schafer C T, Medioli F S 1980 Eastern Canadian estuarine foraminifera: a framework for comparison. *Journal of Foraminiferal Research* **10**: 205–234

Scott D B, Williamson M A, Duffett T E 1981 Marsh foraminifera of Prince Edward Island: their recent distribution and application to former sea level studies. *Maritime Sediments and Atlantic Geology* **17**: 98–129

Seibold I 1975 Benthonic foraminifera from the coast and lagoon of Cochin (south India). *Revista Española de Micropaleontologia* **7**: 175–213

Seibold I, Seibold E 1981 Offshore and lagoonal benthic foraminifera near Cochin (southwest India) distribution, transport, ecological aspects. *Neues Jahrbuch für Geologie und Paläontologie Abhandlungen* **162**: 1–56

Seiglie G A 1966 Distribution of foraminifers in the sediments of Araya-los-Testigos shelf and upper slope. *Caribbean Journal of Science* **6**: 93–117

Seiglie G A 1968a Foraminiferal assemblages as indicators of high organic carbon content in sediments and in polluted water. *Bulletin of the American Association of Petroleum Geologists* **52**: 2231–41

Seiglie G A 1968b Relationship between the distribution of *Amphistegina* and the submerged Pleistocene reefs off western Puerto Rico. *Tulane Studies in Geology* **6**: 138–47

Seiglie G A 1970 The distribution of the foraminifers in Yabucoa Bay, southeastern Puerto Rico and its paleoecological significance. *Revista Española de Micropaleontologia* **2**: 183–208

Seiglie G A 1971a Distribution of foraminifers in the Cabo Rojo Platform and their paleoecological significance. *Revista Española de Micropaleontologia* **3**: 5–33

Seiglie G A 1917b A preliminary note on the relationships between foraminifers and pollution in two Puerto Rican bays. *Caribbean Journal of Science* **11**: 93–8

Seiglie G A 1971c Foraminiferos de las Bahias de Mayagüez y Añasco, y sus alrededores, oeste de Puerto Rico. *Revista Española de Micropaleontologia* **3**: 255–76

Seiglie G A 1973 Pyritization in living foraminifers. *Journal of Foraminiferal Research* **3**: 1–6

Seiglie G A 1974 Foraminifers of Mayagüez and Añasco bays and its surroundings. *Caribbean Journal of Science* **14**: 1–68

Seiglie G A 1975 Foraminifers of Guayanilla Bay and their use as environmental indicators. *Revista Española de Micropaleontologia* **7**: 453–87

Seiglie G A, Bermudez P J 1963 Distribucion de los foraminiferos del Golfo de Cariaco. *Boletin del Instituto Oceanografico de la Universidad Oriente* **2**: 7–87

Seiler W G 1975 Tiefenverteilung benthischer Foraminiferen am portugiesischen Kontinentalhang. *'Meteor' Forschungs Ergebnisse* **C23**: 47–94

Sejrup H P, Fjaeran R T, Hald M, Beck L, Hagen J, Miljeteig I, Morvik I, Norvik J 1981 Benthonic foraminifera in surface samples from the Norwegian continental margin between 62° N and 65° N. *Journal of Foraminiferal Research* **11**: 277–95

Sejrup H P, Guibault J P 1980 *Cassidulina reniforme* and *C. obtusa* (Foraminifera), taxonomy, distribution and ecology. *Sarsia* **65**: 79–85

Sellier de Civrieux J M 1970 Biofacies bentonicas de foraminiferos en la plataforma continental de Cumana – Venezuela. *Boletin del Instituto Oceanografico de la Universidad Oriente* **9**: 21–70

Sellier de Civrieux J M 1977 Foraminiferos indicadores de communidades bentonicas recientes en Venezuela. Parte II: ecologia y distribucion de los foraminiferos mas frecuentes de la plataforma continental en el Parque Nacional Mochima. *Boletin del Instituto Oceanografico de la Universidad Oriente* **16**: 3–62
Sellier de Civrieux J M, Bermudez P J 1973 Ecologia y distribucion de foraminiferos bentonicos del Golfo de Santa Fe (Venezuela). *Revista Española de Micropaleontologia* **5**: 33–80
Sen Gupta B K 1971 The benthonic foraminifera of the tail of the Grand Banks. *Micropaleontology* **17**: 69–98
Sen Gupta B K, Hayes W B 1979 Recognition of Holocene benthic foraminiferal facies by recurrent group analysis. *Journal of Foraminiferal Research* **9**: 233–45
Sen Gupta B K, Kilbourne R T 1974 Diversity of benthic foraminifera on the Georgia continental shelf. *Bulletin of the Geological Society of America* **85**: 969–74
Sen Gupta B K, Kilbourne R T 1976 Depth distribution of benthic foraminifera on the Georgia continental shelf. *Maritime Sediments Special Publication* **1**: 25–38
Sen Gupta B K, Lee R F, May M S 1981 Upwelling and an unusual assemblage of benthic foraminifera on the northern Florida continental slope. *Journal of Paleontology* **55**: 853–7
Sen Gupta B K, McMullen R M 1969 Foraminiferal distribution and sedimentary facies on the Grand Banks of Newfoundland. *Canadian Journal of Earth Sciences* **6**: 475–87
Sen Gupta B K, Schafer C T 1973 Holocene benthonic foraminifera in leeward bays of St Lucia, West Indies. *Micropaleontology* **19**: 341–65
Sen Gupta B K, Shin I C, Wendler S T 1987 Relevance of specimen size in distribution studies of deep-sea benthic foraminifera. *Palaios* **2**: 332–8
Sen Gupta B K, Strickert D P 1982 Living benthic foraminifera of the Florida–Hatteras slope: distribution trends and anomalies. *Bulletin of the Geological Society of America* **93**: 218–24
Setty M G A P 1976 The relative sensitivity of benthonic foraminifera in the polluted marine environment of Cola Bay, Goa. *Proceedings VI Indian Colloquium on Micropalaeontology and Stratigraphy* pp 225–34
Setty M G A P 1978a Recent foraminiferal assemblages from the continental shelf sediments of Madras, Bay of Bengal. *Recent Researches in Geology* **7**: 53–7
Setty M G A P 1978b Shelf edge regime and foraminifera off Pondicherry, Bay of Bengal. *Indian Journal of Marine Sciences* **7**: 302–4
Setty M G A P 1982a Recent marine microfauna from the continental margin, west coast of India. *Journal of Scientific and Industrial Research India* **41**: 647–9
Setty M G A P 1982b Pollution effects monitoring with foraminifera as indices in the Thana Creek, Bombay area. *Journal of Environmental Studies* **18**: 205–9
Setty M G A P 1984a Benthic foraminiferal biocoenoses in the estuarine regimes of Goa. *Revista Italiana di Paleontologia e Stratigrafia* **89**: 437–45
Setty M G A P 1984b Larger foraminifera from a relict structure off Karwar, western Indian continental margin. *Proceedings of the Indian National Science Academy* **50A**: 127–38
Setty M G A P, Nigam R 1980 Microenvironments and anomalous benthic foraminiferal distribution within the neritic regime of Dabhol–Vengurla section (Arabian Sea). *Revista Italiana de Paleontologia e Stratigrafia* **86**: 417–28
Setty M G A P, Nigam R 1982 Foraminiferal assemblages and organic carbon relationship in benthic marine ecosystem of western Indian continental shelf. *Indian Journal of Marine Sciences* **11**: 225–32
Setty M G A P, Nigam R 1984 Benthic foraminifera as pollution indices in the marine environment of the west coast of India. *Revista Italiana de Paleontologia e Stratigrafi* **89**: 421–36
Setty M G A P, Nigam R 1986 Benthic foraminifers as indices of diversity and hyposalinity in a modern clastic shelf regime off Bombay, India. In Thompson M F, Sarogini R, Nagabhushanam R (ed) *Indian Ocean, Biology of Benthic Organisms* A A Balkema, Rotterdam pp 284–8
Setty M G A P, Nigam R, Ambre N V 1979 Graphic pattern of foraminiferal dominance in nearshore region of central west coast of India. *Mahasagar – Bulletin of the National Institute of Oceanography* **12**: 195–9
Severin K P 1983 Test morphology of benthic foraminifera as a discriminator of biofacies. *Marine Micropaleontology* **8**: 65–76
Severin K P 1987 Laboratory observations on the rate of subsurface movement of a small miliolid foraminifer. *Journal of Foraminiferal Research* **17**: 110–16
Severin K P 1987 Spatial and temporal variation in *Marginopora vertebralis* on seagrass in Papua New Guinea during a six–week period. *Micropaleontology* **33**: 368–77
Severin K P, Culver S J, Blanpied C 1982 Burrows and trails produced by *Quinqueloculina impressa* Reuss, a benthic foraminifer, in fine–grained sediment. *Sedimentology* **29**: 897–901
Severin K P, Erskian M G 1981 Laboratory experiments on the vertical movement of *Quinqueloculina impressa* Reuss through sand. *Journal of Foraminiferal Research* **11**: 133–6
Severin K P, Lipps J H 1989 The weight–volume relationship of the test of *Alveolinella quoyi*: implications for the taphonomy of large fusiform foraminifera. *Lethaia* **22**: 1–12
Sgarrella F, Barra D, Improta A 1985 The benthic foraminifers of the Gulf of Policastro (southern Tyrrhenian Sea, Italy). *Bollettino della Societa dei Naturalisti in Napoli* **92**: 67–144
Shackleton N J 1967 Oxygen isotope analyses and Pleistocene temperatures reassessed. *Nature* **215**: 15–17

Shackleton N J 1974 Attainment of isotopic equilibrium between ocean water and the benthonic foraminifera genus *Uvigerina*: isotopic changes in the ocean during the last glacial. *Colloques Internationaux du Centre National de Recherche Scientifique* **219**: 203–9

Shackleton N J 1986 The Plio-Pleistocene ocean: stable isotope history. *American Geophysical Union, Geodynamics Series* **15**: 141–53

Shackleton N J, Hall M A 1984 Oxygen and carbon isotope stratigraphy of Deep Sea Drilling Project Hole 552A: Plio-Pleistocene glacial history. In Roberts D G, Schnitker D *et al.* (eds) *Initial Reports of the Deep Sea Drilling Project,* **81**: 599–609

Shackleton N J, Kennett J P 1975 Paleotemperature history of the Cenozoic and initiation of Antarctic glaciation: oxygen and carbon isotope analyses in DSDP Sites 277, 279 and 281. In Kennett J P, Houtz R E, *et al.* (eds). *Initial Reports of the Deep Sea Drilling Project* **29**: 743–55

Shackleton N J, Opdyke N D 1973 Oxygen isotope and paleomagnetic stratigraphy of equatorial Pacific core V 28–238: oxygen isotope temperatures and ice volumes on a 10^5 year and 10^6 year scale. *Quaternary Research* **3**: 39–55

Shenton E H 1957 A study of the foraminifera and sediments of Matagorda Bay, Texas. *Transaction of the Gulf Coast Association of Geological Societies* **7**: 135–50

Shifflett E 1961 Living, dead and total foraminiferal faunas, Heald Bank Gulf of Mexico. *Micropaleontology* **7**: 45–54

Showers W J 1980 Biometry of the foraminifera *Rosalina globularis* (d'Orbigny) in Antarctic environments. *Journal of Foraminiferal Research* **10**: 61–74

Showers W J, Daniels R A, Laine D 1977 Marine biology at Palmer Station, 1975 austral winter. *Antarctic Journal of the United States* **12**: 22–5

Siddall J D 1878 The foraminifera of the River Dee. *Proceedings of the Chester Society for Natural Science* **2**: 42–56

Sliter W 1965 Laboratory experiments on the life cycle and ecological controls of *Rosalina globularis* d'Orbigny. *Journal of Protozoology* **12**: 210–15

Sliter W V 1970 *Bolivina doniezi* Cushman and Wickenden in clone culture. *Contributions from the Cushman Foundation for Foraminiferal Research* **21**: 87–99

Sliter W V 1971 Predation on benthic foraminifers. *Journal of Foraminiferal Research* **1**: 20–9

Sliter W V, Baker R A 1972 Cretaceous bathymetric distribution of benthic foraminifera. *Journal of Foraminiferal Research* **2**: 167–83

Smith D A, Scott D B, Medioli F S 1984 Marsh foraminifera in the Bay of Fundy: modern distribution and application to sea-level determinations. *Maritime Sediments and Atlantic Geology* **20**: 127–42

Smith D C 1980 Principles of colonisation of cells by symbionts as illustrated by symbiotic algae. In Schwemmler W, Schenk H E A (eds) *Endocytobiosis and Cell Biology.* Walter de Gruyter, Berlin vol 1 pp 317–32

Smith P B 1963 Recent foraminifera off Central America. Quantitative and qualitative analysis of the family Bolivinidae. *Professional Paper United States Geological Survey* **429A**: 1–39

Smith P B 1964 Recent foraminifera off Central America. Ecology of benthic species. *Professional Paper, United States Geological Survey* **429B**: 1–55

Smith P B 1973 Foraminifera of the North Pacific Ocean. *United States Geological Survey Professional Paper* **766**: 1–17

Smith P B, Emiliani C 1968 Oxygen-isotope analysis of recent tropical Pacific benthic foraminifera. *Science* **21**: 1335–6

Smith R K 1968 An intertidal *Marginopora* colony in Suva Harbour, Fiji. *Contributions from the Cushman Foundation for Foraminiferal Research* **19**: 12–17

Smith R K 1987 Fossilization potential in modern shallow-water benthic foraminiferal assemblages. *Journal of Foraminiferal Research* **17**: 117–22

Smyth M J 1988 The foraminifer *Cymbaloporella tabellaeformis* (Brady) bores into gastropod shells. *Journal of Foraminiferal Research* **18**: 277–84

Sneddea J W, Nummedal D, Amos A F 1988 Storm and fair-weather combined flow on the central Texas continental shelf. *Journal of Sedimentary Petrology* **58**: 580–95

Sournia A 1976 Primary production of sands in the lagoon of an atoll and the role of foraminiferan symbionts. *Marine Biology* **37**: 29–32

Spindler M 1980 The pelagic gulfweed *Sargassum natans* as a habitat for the benthic foraminifera *Planorbulina acervalis* and *Rosalina globularis*. *Neues Jahrbuch für Geologie und Paläontologie Monatheft* **9**: 569–80

Srivastava S S, Goel K K, Nath S S 1985 Recent foraminifera from Kavaratti Island (Lakshadweep), Arabian Sea, India. *Bulletin of the Geological, Mining and Metallurgical Society of India* **52**: 335–52

Staff G M, Stanton R J, Powell E N, Cummins G 1986 Time-averaging, taphonomy and their impact on paleocommunity reconstruction: death assemblages in Texas bays. *Bulletin of the Geological Society of America* **97**: 428–43

Stanley D J 1972 *The Mediterranean Sea: a Natural Sedimentation Laboratory* Dowden, Hutchinson and Ross, Stroudsburg, Pa

Stanley D J 1981 Unifites: structureless muds of gravity-flow origin in Mediterranean basins. *Geo-Marine Letters* **1**: 77–83

Stanley D J, Culver S J, Stubblefield W L 1986 Petrologic and foraminiferal evidence for active downslope transport in Wilmington Canyon. *Marine Geology* **69**: 207–18

Steinbeck P L, Bergstein J 1979 Foraminifera from

Hommocks salt-marsh, Larchmont Harbor, New York. *Journal of Foraminiferal Research* **9**: 147–58

Steinker D C 1976 Foraminifera of the rocky tidal zone, Moss Beach, California. *Maritime Sediments Special Publication* **1**: 181–93

Steinker D C 1977 Foraminiferal studies in tropical carbonate environments: South Florida and Bahamas. *Florida Scientist* **40**: 46–61

Steinker P J, Steinker D C 1976 Shallow-water foraminifera, Jewfish Cay, Bahamas. *Maritime Sediments Special Publication* **1**: 171–80

Stevenson J C, Ward L G, Kearney M S 1988 Sediment transport and trapping in marsh systems: implications of tidal flux studies. *Marine Geology* **80**: 37–59

Stephens A C 1923 Preliminary survey of the Scottish waters of the North Sea by Petersen Grab. *Scientific Investigations, Fishery Board of Scotland* **3**: 1–21 (for 1922)

Stockton W L 1973 Distribution of benthic foraminifera at Arthur Harbor, Anvers Island. *Antarctic Journal of the United States* **8**: 348–50

Streeter S S 1972 Living benthonic foraminifera of the Gulf of California: a factor analysis of Phleger's 1964 data. *Micropaleontology* **18**: 64–73

Streeter S S, Lavery S A 1982 Holocene and latest glacial benthic foraminifera from the slope and rise off eastern North America. *Bulletin of the Geological Society of America* **93**: 190–9

Strong D R, Simberloff D, Abele L G, Thistle A B (eds) 1984. *Ecological Communities; Conceptual Issues and Evidence*. Princeton University Press

Sturrock S, Murray J W 1981 Comparison of low energy and high energy marine middle shelf foraminiferal faunas, Celtic Sea and western English Channel. In Neale J W, Brasier M (eds) *Micropalaeontology of Shelf Seas* Ellis Horwood, Chichester pp 250–60

Sverdrup H V, Johnson M W, Fleming R H 1942 *The Oceans, their Physics, Chemistry and General Biology*. Prentice-Hall, New York 1087 pp

Swinbanks D D, Shirayama Y 1984 Burrow stratigraphy in relation to manganese diagenesis in modern deep-sea carbonates. *Deep-Sea Research* **31**: 1197–1223

Takayanagi Y, Hasegawa S 1987 *Checklist and Bibliography of Post-Paleozoic foraminifera Established by Japanese Workers, 1890–1986*. Institute of Geology and Paleontology pp 1–95

Tapley S 1969 Foraminiferal analysis of the Miramichi Estuary. *Maritime Sediments* **5**: 20–9

Temnikov N K 1976 US Antarctic Research Program, 1975–1976. *Antarctic Journal of the United States* **11**: 220–2

Tendal O S 1979 Aspects of the biology of Komokiacea and Xenophyophoria. *Sarsia* **64**: 13–17

Tendal O S, Thomsen E 1988 Observations on the life position and size of the large foraminifer *Astrorhiza arenaria* Norman, 1876 from the shelf off northern Norway. *Sarsia* **73**: 39–42

Theyer F 1971 Benthic foraminiferal trends, Pacific–Antarctic Basin. *Deep-Sea Research* **18**: 723–38

Thiede J, Qvale G, Skarboe O, Strand J E 1981. Benthonic foraminiferal distributions in a southern Norwegian fjord system: a re-evaluation of Oslo Fjord data. *Special Publications of the International Association of Sedimentologists* **5**: 469–95

Thomas F C, Schafer C T 1982 Distribution and transport of some common foraminiferal species in the Minas Basin, eastern Canada. *Journal of Foraminiferal Research* **12**: 24–38

Thompson L C 1978 Distribution of living benthic foraminifera, Isla de los Estados, Tierra del Fuego, Argentina. *Journal of Foraminiferal Research* **8**: 241–57

Tietjen J H 1971 Ecology and distribution of deep-sea meiobenthos off North Carolina. *Deep-Sea Research* **18**: 941–57

Tinoco I M 1958 Observações sôbre a Microfauna de Foraminiferos da Lagoa de Araruama, Estadio do Rio de Janeiro. *Anais de Academia Brasileira de Cencas* **30**: 575–84

Tinoco I M 1966a Contribuição a sedimentologia e microfauna da Baia de Sepetiba (Estado do Rio de Janeiro). *Trabalhos Oceanograficos Universidade Federal de Pernambuco* **7/8**: 123–36

Tinoco I M 1966b Foraminiferos do Atol das Rocas. *Trabalhos Oceanograficos Universidade Federal de Pernambuco* **7/8**: 91–114

Todd R 1960 Some observations on the distribution of *Calcarina* and *Baculogypsina* in the Pacific. *Tohoku University Science Reports, Second Series (Geology), Special Volume* **4**: 100–7

Todd R 1961 Foraminifera from Onotoa Atoll, Gilbert Islands. *United States Geological Survey Professional Paper* **354–H**: 171–91

Todd R 1965a A new *Rosalina* (Foraminifera) parasitic on a bivalve. *Deep-Sea Research* **12**: 831–7

Todd R 1976 Some observations about *Amphistegina* (Foraminifera). *Progress in Micropaleontology* pp 382–94

Todd R 1965 The foraminifera of the tropical Pacific collections of the *Albatross*, 1899–1900. *United States National Museum Bulletin* **161** (4): 1–139

Todd R, Low D 1966 Foraminifera from the Arctic Ocean off the eastern Siberian coast. *United States Geological Survey, Professional Paper* **550–C**: 79–85

Todd R, Low D 1967 Recent foraminifera from the Gulf of Alaska and southeastern Alaska. *United States Geological Survey, Professional Paper* **573–A**: 1–46

Todd R, Low D 1980 Foraminifera from the Kara and Greenland seas, and review of Arctic studies. *United States Geological Survey, Professional Paper* **1070**: 1–30

Todd R, Low D 1981 Marine flora and fauna of the northeastern United States. Protozoa: Sarcodina:

benthic foraminifera. *National Oceanic and Atmospheric Administration National Marine Fisheries Service Circular 439* pp 1–51

Travis J L, Bowser S S 1988 Optical approaches to the study of foraminiferan motility. *Cell Motility and the Cytoskeleton* **10**: 126–336

Troitskaya T S 1969 Environmental conditions and distribution of foraminifera in the Japan sea (families Elphidiidae, Cassidulinidae and Islandiellidae). *Academy of Sciences of the USSR Siberian Branch, Transaction of the Institute of Geology and Geophysics* **71**: 136–60 (in Russian)

Troitskaya T S 1973 Foraminifera of the Japan Sea west shelf and their environments. *Academy of the USSR Siberian Branch, Transactions of the Institute of Geology and Geophysics* **62**: 199–76 (in Russian)

Tufescu M 1968 *Ammonia tepida* (Cushman) (ord. Foraminifera). Some features of its variability in the Black Sea Basin. *Revue Roumaine de Biologie et Zoologie* **13**: 169–77

Tufescu M 1973 Les associations de foraminifères du Nord-ouest de la Mer Noire. *Revista Española de Micropaleontologia* **5**: 15–32

Turekian K K 1969 The oceans, streams and atmosphere. In Wedepohl K H (ed) *Handbook of Geochemistry* **1**: 297–323

Tyler P A 1988 Seasonality in the deep sea. *Oceanography and Marine Biology Annual Review* **26**: 227–58

Uchio T 1960a Biological results of the Japanese Antarctic Research Expedition. 12. Benthonic foraminifera of the Antarctic Ocean. *Special Publications from the Seto Marine Biological Laboratory* **12**: 1–20

Uchio T 1960b Ecology of living benthonic foraminifera from the San Diego California area. *Special Publications of the Cushman Foundation* **5**: 1–72

Uchio T 1962a Influence of the river Shinano on foraminifera and sediment grain size distribution. *Publications of the Seto Marine Biological Laboratory* **10**: 363–93

Uchio T 1962b Recent foraminifera thanatocoenoses of beach and nearshore sediments along the coast of Wakayama-Ken, Japan. *Publications of the Seto Marine Biological Laboratory* **10**: 133–44

Uchio T 1967 Foraminiferal assemblages in the vicinity of the Seto Marine Biological Laboratory, Shirahama-Cho, Wakayama-ken, Japan (Part 1). *Publications of the Seto Marine Biological Laboratory* **15**: 399–417

Ujiié H 1962 Introduction to statistical foraminiferal zonation. *Journal of the Geological Society of Japan* **68**: 431–50

Ujiié H 1988 Benthic foraminiferal changes related to techonic development of the Sea of Japan. *Revue de Paléobiologie, vol spéc* **2**: 895–901

Ujiié H, Kusukawa T 1969 Analysis of foraminiferal assemblages from Miyako and Yamada bays, northeastern Japan. *Bulletin of the National Science Museum, Tokyo* **12**: 735–72

Van Andel T H 1960 Source and dispersion of Holocene sediments, northern Gulf of Mexico. In Shepard F P, Phleger F B, Van Andel T H (eds) *Recent Sediments, Northwest Gulf of Mexico* American Association of Petroleum Geologists, Tulsa pp 34–55

Van der Zwaan G J 1983 Quantitative analyses and reconstruction of benthic foraminiferal communities. *Utrecht Micropaleontological Bulletins* **30**: 9–69

Van Morkhoven F P C M, Berggren W A, Edwards A S 1986 Cenozoic cosmopolitan deep-water benthic foraminifera. *Bulletin des Centres de Recherches Exploration-Production Elf-Aquitaine Memoir* **11**: 1–421

Van Voorthuysen J H 1960 Die Foraminiferen des Dollart-Ems-Estuarium. *Verhandelingen van het Konink Nederlandsch Geologisch–mijnbouwkundig Genootschap* (Geol ser) **19**: 237–69

Van Voorthuysen J H 1973 Foraminiferal ecology in the Ria de Arosa, Galicia, Spain. *Zoologische Verhandelingen* **123**: 1–68

Van Weering T C E, Qvale G 1983 Recent sediments and foraminiferal distribution in the Skagerrak, northeastern North Sea. *Marine Geology* **52**: 75–99

Vangerow E F 1965 Salzgehalt-und Temperaturmessungen im Lebensraum der Wattforaminiferen. *Natur und Museum* **95**: 63–6

Vangerow E F 1972 Ökologische beobachtungen an Foraminiferen des Brackwasserbereiches der Rhônemündung (Sudfrankreich). *Journal of Experimental Marine Biology and Ecology* **8**: 145–65

Vangerow E F 1974 Récentes observations écologiques des foraminifères dans la zone saumâtre de l'embouchure du Rhône. *Revue de Micropaléontologie* **17**: 95–196

Vella P 1957 Studies in New Zealand foraminifera. *New Zealand Geological Survey Palaeontological Bulletin* **28**: 1–64

Vénec-Peyré M T 1981 Les foraminiferes et la pollution: étude de la microfaune de la Cale du Dourduff (embrouchure de la rivière de Morlaix). *Cahiers de Biologie Marine* **22**: 25–33

Vénec–Peyrè M T 1983a Les foraminifères et le milieu: étude de trois ecosystemes. In Oertli H (ed) *Benthos '83* pp 573–81

Vénec-Peyré M T 1983b Étude de la croissance et de la variabilité chez un foraminifere benthique littoral, *Ammonia beccarii* (Linné), en Mediterranéen occidentale. *Cahiers de Micropaléontologie* **2**: 5–31

Vénec-Peyré M T 1984 Étude de la distribution des foraminifères vivant dans la Baie de Banyuls-sur-Mer. *Pétrole et Techniques* **301**: 22–43

Vénec-Peyré M T 1985a Étude de la distribution des foraminifères vivants dans le lagon de l'île haute volcanique de Moorea (Polynésie français). *Proceedings*

of the Fifth International Coral reef Congress, Tahiti **5**: 227–32
Vénec-Peyré M T 1985b Le rôle de certains foraminifères dans la bioerosion et la sédimentogenèse. *Comptes Rendus de l'Academie des Sciences, Paris* **300**: 83–8
Vénec-Peyré M T 1988a Two new species of bioeroding Trochamminidae (Foraminiferida) from French Polynesia. *Journal of Foraminiferal Research* **18**: 1–5
Vénec-Peyré M T 1988b Adaptation of foraminiferal populations to the water oligotrophy in French Polynesia. *Revue de Paléobiologie, vol spéc* **2**: 903–8
Vénec-Peyré M T, Bhaud M, Boucher G 1981 Observations sur les méthods d'echantillonnage des foraminifères benthiques vivants. *106e Congrès national des Sociétés savantes, Perpignan 1981, Sciences* **1**: 205–14
Vénec-Peyré M T, Le Calvez Y 1981 Étude des foraminifères de l'herbier à posidonies de Banyuls-sur-Mer. *106e Congrès national des Sociétés savantes, Perpignan 1981*, Science **1**: 191–203
Vénec-Peyré M T, Le Calvez Y 1986 Foraminifères benthiques et phénomenès de transfert: importance des études comparatives de la biocenose et de la thanatocènose. *Bulletin du Muséum d'Histoire Naturelle, Paris* ser 4 **8**: 171–84
Vénec–Peyré M T, Salvat B 1981 Les foraminifères de l'atoll de Scilly (Archipel de la Société): étude comparée de la biocénose et de la thanatocènose. *Anales de l'Institut Oceánographique* NS **57**: 79–110
Venkatachalapathy V, Shareef N A 1975 Morphology, distributional pattern and wall structure of *Ammonia beccarii* (Linné) from west coast of India. *Revista Española de Micropaleontologia* **7**: 51–62
Vilks G 1969 Recent foraminifera in the Canadian Arctic. *Micropaleontology* **15**: 35–60
Vilks G, Deonarine B, Wagner F J, Winters G V 1982 Foraminifera and Mollusca in surface sediments of the southeastern Labrador shelf and Lake Melville, Canada. *Bulletin of the Geological Society of America* **93**: 225–38
Vincent E, Killingley J S, Berger W H 1981 Stable isotope composition of benthic foraminifera from the equatorial Pacific. *Nature* **289**: 639–42
Vinot–Bertouille A C, Duplessy J C 1973 Individual isotopic fractionation of carbon and oxygen in benthic foraminifera. *Earth and Planetary Science Letters* **18**: 247–52
Voigt E 1970 Foraminiferen und (?) Phoronidea als Kommensalen auf den Hartgründen der Maasstrichter Tuffkreide. *Paläontologische Zeitschrift* **44**: 86–92
Voigt E, Bromley R G 1974 Foraminifera as commensals around clinoid sponge papillae: Cretaceous and Recent. *Senckenbergiana Maritima* **6**: 33–45
Walford L A 1946 A new graphic method of describing the growth of animals. *Biological Bulletin* **90**: 141–7
Walker D A 1976 An *in situ* investigation of life cycles on benthonic mid-littoral foraminifera. *Maritime Sediments Special Publication* **1**: 51–9
Walker D A, Linton A E, Schafer C T 1974 Sudan black B: a superior stain to rose Bengal for distinguishing living from non-living foraminifera. *Journal of Foraminiferal Research* **4**: 205–15
Waller H O 1960 Foraminiferal biofacies off the south China coast. *Journal of Paleontology* **34**: 1164–82
Walsh J J 1983 Death in the sea: enigmatic phytoplankton losses. *Progress in Oceanography* **12**: 1–86
Walsh J J 1988 *On the Nature of Continental Shelves* Academic Press, London 508pp
Walton W R 1952 Techniques for recognition of living foraminifera. *Contribution from the Cushman Foundation for Foraminiferal Research* **3**: 56–60
Walton W R 1955 Ecology of living benthonic foraminifera, Todos Santos Bay, Baja California. *Journal of Paleontology* **29**: 952–1018
Walton W R 1964a Recent foraminiferal ecology and paleoecology. In Imbrie J, Newell N D (eds) *Approaches to Paleoecology*, John Wiley, New York pp 151–237
Walton W R 1964b Ecology of benthonic foraminifera in the Tampa–Sarasota Bay area, Florida. In Miller R L (ed) *Papers* pp 429–54 *in Marine Geology* Macmillan, New York
Wang P (ed) 1980 *Papers on Marine Micropalaeontology* China Ocean Press pp 1–204 (in Chinese)
Wang P 1983 Verbreitung der Benthos-Foraminiferen im Elbe-Ästuar. *Meyniana* **35**: 67–83
Wang P (ed) 1985 *Marine Micropaleontology of China.* China Ocean Press, Beijing and Springer-Verlag, Berlin pp 1–370
Wang P, Lutze G F 1986 Inflated later chambers: ontogenetic changes of some recent hyaline benthic foraminifera. *Journal of Foraminiferal Research* **16**: 48–62
Wang P, Min Q, Bian Y 1978 *Distribution of foraminifera and Ostracoda in Bottom Sediments of the Northwestern Part of the Southern Yellow Sea and its Geological Significance.* Tongji University, Shanghai pp 1–115 (in Chinese)
Wang P, Min B Q, Bian Y H, Zhang J J 1979 A preliminary study of foraminiferal and ostracod assemblages distribution in bottom sediments of the East China Sea. *Journal of Tung-Chi University* **2**: 90–108 (in Chinese with English abstract)
Wang P, Min Q, Bian Y, Hua D 1980 Characteristics of foraminiferal and ostracod thanatocenoses from some river mouths of China and their geological significance. *Papers in Marine Micropalaeontology, Peking* pp 101–11 (in Chinese)
Wang P, Murray J W 1983 The use of foraminifera as indicators of tidal effects in estuarine deposits. *Marine Geology* **51**: 239–50
Wang P, Zhang J, Zhao Q, Min Q, Bian Y, Zheng L, Cheng X, Chen R 1988 *Foraminifera and Ostracoda in Bottom sediments off the East China Sea.* China

Ocean Press, Beijing pp 1–438 (in Chinese)
Ward B L, Barrett P J, Vella P 1987 Distribution and ecology of benthic foraminifera in McMurdo Sound, Antarctica. *Palaeogeography, Palaeoclimatology, Palaeoecology* **58**: 139–53
Warwick R M 1981 Survival strategies of meiofauna. In Jones N V, Wolff W J (eds) *Feeding and Survival Strategies of Estuarine Organisms* pp 39–52 Plenum Press
Wantland K F 1975 Distribution of Holocene benthonic foraminifera on the Belize shelf. In Wantland K F, Pusey W C (eds) *Belize Shelf-carbonate Sediments, Clastic Sediments and Ecology*. American Association of Petroleum Geologists pp 332–99
Watanabe S 1982 Foraminiferos bentonicos en muestras de plancton. *Physis* **A41**: 43–4
Watkins J G 1961 Foraminiferal ecology around the Orange County, California, ocean sewer outfall. *Micropaleontology* **7**: 199–206
Webb P N 1988 Upper Oligocene–Holocene benthic foraminifera of the Ross Sea region. *Revue de Paléobiologie, vol spéc* **2**: 589–603
Wefer G 1976a Umwelt, Produktion und Sedimentation benthischer Foraminiferen in der Westlichen Ostee. *Rept. Sondersforschungsbereich 95, Kiel* **14**: 1–103
Wefer G 1976b Environmental effects on growth rates of benthic foraminifera. *Maritime Sediments Special Publication* **1**: 39–50
Wefer G, Berger W H 1980 Stable isotopes in benthic foraminifera: seasonal variation in large tropical species. *Science* **209**: 803–5
Wefer G, Killingley J S, Lutze G F 1981 Stable isotopes in recent larger foraminifera. *Palaeogeography, Palaeoclimatology, Palaeoecology* **33**: 253–70
Wefer G, Lutze G F 1978 Carbonate production by benthic foraminifera and accumulation in the western Baltic. *Limnology and Oceanography* **23**: 992–96
Wefer G, Richter W 1976 Colonization of artificial substrates by foraminifera. *Keiler Meeresforschungen* **3**: 72–5
Weiss D 1976 Distribution of benthonic foraminifera in the Hudson River estuary. *Maritime Sediments, Special Publication* **1**: 119–29
Weiss D 1976 Distribution of benthonic foraminifera in diatom and bivalve distribution in recent sediments of the Hudson estuary. *Estuarine and Coastal Marine Science* **7**: 393–400
Werdelin L, Hermelin J O R 1983 Testing for ecophenotypic variation in a benthic foraminifer. *Lethaia* **16**: 303–7
Weston J F 1984 Wall structure of agglutinated foraminifera *Eggerella bradyi* (Cushman) and *Karreriella bradyi* (Cushman). *Journal of Micropalaeontology* **3**: 29–31
Weston J F 1985 Comparison between Recent benthic foraminiferal faunas of the Porcupine Seabight and Western Approaches continental slope. *Journal of Micropalaeontology* **4**(2): 165–83
Weston J F, Murray J W 1984 Benthic foraminifera as deep-sea water-mass indicators. In Oertli H J (ed) *Benthos '83* pp 605–10
Wetmore K L 1987 Correlations between test strength, morphology and habitat in some benthic foraminifera from the coast of Washington. *Journal of Foraminiferal Research* **17**: 1–13
Wetmore K L 1988 Burrowing and sediment movement by benthic foraminifera as shown by time-lapse cinematography. *Revue de Paléobiologie, vol spéc* **2**: 921–7
Wignall P B, Myers K J 1988 Interpreting benthic oxygen levels in mudrocks: a new approach. *Geology* **16**: 452–5
Wilcoxon J A 1964 Distribution of foraminifera off the southern Atlantic coast of the United States. *Contributions from the Cushman Foundation for Foraminiferal Research* **15**: 1–24
Williams C B 1964 *Patterns in the Balance of Nature* Academic Press, London 324 pp
Williams D F, Ehrlich R, Spero H D, Healy-Williams N, Gary A C 1988 Shape and isotopic differences between conspecific foraminiferal morphotypes and resolution of paleoceanographic events. *Palaeogeography, Palaeoclimatology, Palaeoecology* **64**: 153–62
Williams D F, Healy–Williams N 1984 Stable isotope gradients in modern benthic foraminifera of the Vema Channel, South Atlantic. *Marine Geology* **58**: 123–35
Williams D F, Röttger R, Schmaljohann R, Keigwin L 1981 Oxygen and carbon isotopic fractionation and algal symbiosis in the benthic foraminiferan *Heterostegina depressa*. *Palaeogeography, Palaeoclimatology, Palaeoecology* **33**: 231–51
Williamson M 1972 *The Analysis of Biological Populations*. Edward Arnold, London 180 pp
Williamson M A, Keen C E, Mudie P J 1984 Foraminiferal distribution on the continental margin off Nova Scotia. *Marine Micropalaeontology* **9**: 219–39
Wilson J B 1979 Biogenic carbonate sediments on the Scottish continental shelf and on Rockall Bank. *Marine Geology* **33**: M85–M93
Wilson J C 1982 Shelly faunas associated with temperate offshore tidal deposits. It Stride A H (ed) *Offshore Tidal Sands* Chapman and Hall London pp 126–71
Woodruff F, Savin S, Douglas R G 1980 Biological fractionation of oxygen and carbon isotopes by recent benthic foraminifera. *Marine Micropaleontology* **5**: 3–11
Wright R I 1968 Miliolidae (foraminiferos) recientes del estuario del Rio Quequen Grande (Provincia de Buenos Aires). *Revista del Museo Argentino de Ciencas Naturales 'Bernardino Rivadavia', Hidrobiologia* **2**: 225–56
Wright R 1978 Neogene benthic foraminifers from DSDP Leg 42A, Mediterranean Sea. *Initial Reports of the Deep Sea Drilling Project* **47**: 709–26

Wright R C, Hay W W 1971 The abundance and distribution of foraminifers in a back-reef environment, Molasses Reef, Florida. *Memoir of the Miami Geological Society* **1**: 121–74

Yamamoto S, Tokuyama H, Fujioka K, Takeuchi A, Ujiié H 1988 Carbonate turbidites deposited on the floor of the Palau trench. *Marine Geology* **82**: 217–33

Yassini I, Ghahreman A 1977 Récapitulation de la distribution des ostracodes et des foraminifères de lagoon de Pahlavi, province de Gilan, Iran de Nord. *Revue de Micropaléontologie* **19**: 172–90

Yoshida S 1954 The foraminifera of Lake Saroma. *Tokyo University of Education, Studies from the Geological and Mineralogical Institute* **3**: 149–58

Zahn R, Winn K, Sarnthein M 1986 Benthic foraminiferal $d^{13}C$ and accumulation rates of organic carbon: *Uvigerina peregrina* group and *Cibidoides wuellerstorfi*. *Paleoceanography* **1**: 27–42

Zalesny E R 1959 Foraminiferal ecology of Santa Monica Bay, California. *Micropaleontology* **5**: 101–26

Zaninetti L 1979 L'étude des foraminifères des mangroves actuelles: réflexion sur les objectifs et sur l'état des connaissances. *Archives des Sciences Genève* **32**: 151–61

Zaninetti L 1982 Les foraminifères des marais salants de Salin-de-Girand (Sud de la France): milieu de vie et transport dans le salin, comparaison avec les microfaunas marines. *Géologie Mediterranéenne* **9**: 447–70

Zaninetti L 1984 Les foraminifères du salin de Bras del Port (Santa Pola, Espagne), avec remarques sur la distribution des Ostracodes. *Revista d'Investigacions Geologiques* **38/39**: 123–38

Zaninetti L, Brönnimann P, Beurlen G, Moura J A 1976 La mangrove de Guaratiba et la Baie des Sepetiba, État de Rio de Janeiro, Brésil: foraminifères et écologie. Note préliminaire. *Comptes Rendus des Séances, Société de Physique et d'Histoire Naturelle, Genève*, NS **11**: 39–44

Zaninetti L, Brönnimann P, Beurlen G, Moura J A 1977 La mangrove de Guaratiba et la Baie de Sepetiba, État de Rio de Janeiro, Brésil: foraminifères et écologie. *Achives des Sciences Genève* **30**: 161–78

Zaninetti L, Brönnimann P, Dias-Brito D, Arai M, Casaletti P, Koutsoukos E, Silvira S 1979 Distribution écologique des foraminifères dans la Mangrove d'Acupe, État de Bahia, Brésil. *Notes de Laboratiore de Paléontologie de l'Université de Genève* **4**: 1–17

Zheng S 1979 The recent foraminifera of the Xisha Islands, Guangdong Province, China 11. *Studia Marina Sinica* **15**: 101–232 (in Chinese with English summary)

Zheng S 1980 The recent foraminifera of the Zhongsha Islands, Guandong Province, China. 1. *Studia Marina Sinica* **16**: 143–82 (in Chinese with English summary)

Zheng S, Fu Z 1988 The distribution of agglutinated foraminifera in the East China Sea. *Revue de Paléobiologie vol spéc* **2**: 929–49

Zieman J C 1975 Seasonal variation of turtle grass *Thalassia testudinum* König, with reference to temperature and salinity effects. *Aquatic Botany* **1**: 107–23

Zimmerman M A, Williams D F, Röttger R 1983 Symbiont-influenced isotopic disequilibrium in *Heterostegina depressa*. *Journal of Foraminiferal Research* **13**: 115–21

Zmiri A, Kahan D, Hochstein S, Reiss Z 1974 Phototaxis and thermotaxis in some species of *Amphitegina* (Foraminifera). *Journal of Protozoology* **21**: 133–8

Zobel B 1973 Biostratigraphische Untersuchungen an Sedimenten des indisch-pakistanischen Kontinentalrandes (Arabisches Meer). *Meteor Forschungs Ergebnisse* **C12**: 9–73

Zohary T, Reiss Z, Hottinger L 1980 Population dynamics of *Amphisorus hemprichii* (Foraminifera) in the Gulf of Elat (Aqaba), Red Sea. *Eclogae Geologicae Helvetiae* **73**: 1073–94

Zumwalt G S, Delaca T E 1980 Utilization of brachiopod feeding currents by epizoic foraminifera. *Journal of Paleontology* **54**: 477–84

Zweig-Strykowski M, Reiss Z 1975 Bolivinitidae from the Gulf of Elat. *Israel Journal of Earth Sciences* **24**: 97–111

General Index

Bold numbers indicate definitions

Geographic Index

Taxonomic index – suprageneric categories

In the text these are commonly used informally, e.g. miliolids rather than Miliolidae

Taxonomic index – genera and species

Figures in bold refer to named associations, illustrations or ecological details. For the entry in the faunal reference, list refer to pp. 327–40